QUANTUM ELECTRONICS FOR ATOMIC PHYSICS

# Quantum Electronics for Atomic Physics

Warren Nagourney
*Seattle, WA*

# OXFORD
### UNIVERSITY PRESS

Great Clarendon Street, Oxford OX2 6DP

Oxford University Press is a department of the University of Oxford.
It furthers the University's objective of excellence in research, scholarship,
and education by publishing worldwide in

Oxford New York

Auckland Cape Town Dar es Salaam Hong Kong Karachi
Kuala Lumpur Madrid Melbourne Mexico City Nairobi
New Delhi Shanghai Taipei Toronto

With offices in

Argentina Austria Brazil Chile Czech Republic France Greece
Guatemala Hungary Italy Japan Poland Portugal Singapore
South Korea Switzerland Thailand Turkey Ukraine Vietnam

Oxford is a registered trade mark of Oxford University Press
in the UK and in certain other countries

Published in the United States
by Oxford University Press Inc., New York



British Library Cataloguing in Publication Data

Data available

Library of Congress Cataloging in Publication Data

Data available

Typeset by SPI Publisher Services, Pondicherry, India
Printed in Great Britain
on acid-free paper by
CPI Antony Rowe, Chippenham, Wiltshire

ISBN 978–0–19–953262–9 (Hbk.)

1 3 5 7 9 10 8 6 4 2

*To my wife, Patricia.*

# Preface

This book is based upon a series of lectures which I gave for a graduate level class in quantum electronics in the University of Washington physics department. I assumed a working knowledge of intermediate electromagnetic theory, quantum mechanics and optics. I took a slightly different approach from most books and courses on the subject by deliberately slanting the material so that it would be more relevant to atomic physics experimentalists than, perhaps, to workers in the telecommunications industry.

As a result of my experimental atomic physics orientation, I have included topics not usually found in traditional texts on quantum electronics. An example of this is the application of nonlinear optics to the synthesis of coherent radiation in regions of the spectrum where lasers don't work very well or at all. Most books cover the theory of second harmonic generation, sum frequency mixing and parametric processes, but avoid discussions of details of a practical frequency synthesis system. We therefore discuss such matters as the optimum cavity geometry for "enhancement" of nonlinear processes, "impedance matching" into the cavity, assessment and correction of cavity astigmatism and practical techniques for mode-matching into the cavity. These issues receive scant attention in most books on quantum electronics.

Another area not discussed in most texts is the frequency stabilization of lasers to cavities and of cavities to lasers. The former situation is encountered when one constructs a frequency standard or seeks to observe an atomic transition over long periods of time. The latter is a necessary condition for the use of a build-up cavity in nonlinear frequency synthesis. Both approaches require some knowledge of control system theory, and a review of this subject is provided. In addition, two important techniques for generating a "discriminant" for frequency locking are discussed.

The remainder of the book covers topics in a manner which is similar to that of other textbooks on quantum electronics. Thus, we provide a fairly conventional discussion of Guassian beams, "standing-wave" cavities, continuous wave laser theory, electro-optical and acousto-optical modulation and some nonlinear optics theory. There is no formal treatment of optical waveguides or fibers even though they are used in several of the devices discussed in the text; there is a comprehensive and excellent literature on these subjects and I felt that an additional treatment would do little to enhance the reader's understanding of the devices discussed in this text.

## Acknowledgements

I would like to thank Dr. Eric Burt, Professor Steve Lamoreaux and Professor Ady Arie for reading portions of the manuscript and making helpful suggestions.

# Contents


**1 Gaussian beams** — 1
1.1 Introduction — 1
1.2 The paraxial wave equation — 1
1.3 Gaussian beam functions and the complex beam parameter, $q$ — 2
1.4 Some Gaussian beam properties — 3
1.5 The phase term: Gouy phase — 5
1.6 Simple transformation properties of the complex beam parameter — 6
1.7 Matrix formulation of paraxial ray optics: ABCD rule — 8
1.8 Further reading — 10
1.9 Problems — 11

**2 Optical resonators – geometrical properties** — 13
2.1 Introduction — 13
2.2 The two-mirror standing wave cavity — 13
2.3 Stability — 15
2.4 Solution for an arbitrary two-mirror stable cavity — 17
2.5 Higher-order modes — 19
2.6 Resonant frequencies — 21
2.7 The traveling wave (ring) cavity — 23
2.8 Astigmatism in a ring cavity — 26
2.9 Mode matching — 30
2.10 Beam quality characterization: the $M^2$ parameter — 32
2.11 Further reading — 34
2.12 Problems — 35

**3 Energy relations in optical cavities** — 36
3.1 Introduction — 36
3.2 Reflection and transmission at an interface — 36
3.3 Reflected fields from standing wave cavity — 37
3.4 Internal (circulating) field in a standing wave cavity — 38
3.5 Reflected and internal intensities — 39
3.6 The resonant character of the reflected and circulating intensities — 40
3.7 Impedance matching — 41
3.8 Fields and intensities in ring cavity — 44
3.9 A novel "reflective" coupling scheme using a tilted wedge — 45
3.10 Photon lifetime — 46
3.11 The quality factor, $Q$ — 47
3.12 Relation between $Q$ and finesse — 47
3.13 Alternative representation of cavity loss — 48
3.14 Experimental determination of cavity parameters — 48

| | | |
|---|---|---|
| 3.15 | Further reading | 50 |
| 3.16 | Problems | 51 |

**4 Optical cavity as frequency discriminator**    **53**

| | | |
|---|---|---|
| 4.1 | Introduction | 53 |
| 4.2 | A simple example | 53 |
| 4.3 | Side of resonance discriminant | 55 |
| 4.4 | The manipulation of polarized beams: the Jones calculus | 56 |
| 4.5 | The polarization technique | 58 |
| 4.6 | Frequency modulation | 61 |
| 4.7 | The Pound–Drever–Hall approach | 63 |
| 4.8 | Frequency response of a cavity-based discriminator | 67 |
| 4.9 | Further reading | 70 |
| 4.10 | Problems | 70 |

**5 Laser gain and some of its consequences**    **72**

| | | |
|---|---|---|
| 5.1 | Introduction | 72 |
| 5.2 | The wave equation | 72 |
| 5.3 | The interaction term | 73 |
| 5.4 | The rotating wave approximation | 74 |
| 5.5 | Density matrix of two-level system | 75 |
| 5.6 | The classical Bloch equation | 77 |
| 5.7 | Radiative and collision-induced damping | 79 |
| 5.8 | The atomic susceptibility and optical gain | 84 |
| 5.9 | The Einstein A and B coefficients | 88 |
| 5.10 | Doppler broadening: an example of inhomogeneous broadening | 92 |
| 5.11 | Comments on saturation | 94 |
| 5.12 | Further reading | 98 |
| 5.13 | Problems | 98 |

**6 Laser oscillation and pumping mechanisms**    **100**

| | | |
|---|---|---|
| 6.1 | Introduction | 100 |
| 6.2 | The condition for laser oscillation | 100 |
| 6.3 | The power output of a laser | 101 |
| 6.4 | Pumping in three-level and four-level laser systems | 103 |
| 6.5 | Laser oscillation frequencies and pulling | 106 |
| 6.6 | Inhomogeneous broadening and multimode behavior | 107 |
| 6.7 | Spatial hole burning | 109 |
| 6.8 | Some consequences of the photon model for laser radiation | 110 |
| 6.9 | The photon statistics of laser radiation | 112 |
| 6.10 | The ultimate linewidth of a laser | 117 |
| 6.11 | Further reading | 119 |
| 6.12 | Problems | 119 |

**7 Descriptions of specific CW laser systems**    **121**

| | | |
|---|---|---|
| 7.1 | Introduction | 121 |
| 7.2 | The He-Ne laser | 121 |

| | | |
|---|---|---|
| 7.3 | The argon ion laser | 123 |
| 7.4 | The continuous wave organic dye laser | 126 |
| 7.5 | The titanium-sapphire laser | 130 |
| 7.6 | The CW neodymium-yttrium-aluminum-garnet (Nd:YAG) laser | 132 |
| 7.7 | The YAG non-planar ring oscillator: a novel ring laser geometry | 134 |
| 7.8 | Diode-pumped solid-state (DPSS) YAG lasers | 135 |
| 7.9 | Further reading | 136 |
| **8** | **Laser gain in a semiconductor** | **137** |
| 8.1 | Introduction | 137 |
| 8.2 | Solid state physics background | 137 |
| 8.3 | Optical gain in a semiconductor | 148 |
| 8.4 | Further reading | 157 |
| 8.5 | Problems | 157 |
| **9** | **Semiconductor diode lasers** | **159** |
| 9.1 | Introduction | 159 |
| 9.2 | The homojunction semiconductor laser | 159 |
| 9.3 | The double heterostructure laser | 162 |
| 9.4 | Quantum well lasers | 167 |
| 9.5 | Distributed feedback lasers | 173 |
| 9.6 | The rate equations and relaxation oscillations | 179 |
| 9.7 | Diode laser frequency control and linewidth | 187 |
| 9.8 | External cavity diode lasers (ECDLs) | 192 |
| 9.9 | Semiconductor laser amplifiers and injection locking | 202 |
| 9.10 | Miscellaneous characteristics of semiconductor lasers | 208 |
| 9.11 | Further reading | 210 |
| 9.12 | Problems | 210 |
| **10** | **Mode-locked lasers and frequency metrology** | **212** |
| 10.1 | Introduction | 212 |
| 10.2 | Theory of mode locking | 212 |
| 10.3 | Mode locking techniques | 217 |
| 10.4 | Dispersion and its compensation | 221 |
| 10.5 | The mode-locked Ti-sapphire laser | 225 |
| 10.6 | Frequency metrology using a femtosecond laser | 228 |
| 10.7 | The carrier envelope offset | 230 |
| 10.8 | Further reading | 233 |
| 10.9 | Problems | 233 |
| **11** | **Laser frequency stabilization and control systems** | **235** |
| 11.1 | Introduction | 235 |
| 11.2 | Laser frequency stabilization – a first look | 235 |
| 11.3 | The effect of the loop filter | 237 |
| 11.4 | Elementary noise considerations | 238 |
| 11.5 | Some linear system theory | 241 |
| 11.6 | The stability of a linear system | 245 |

xii    Contents

| | |
|---|---:|
| 11.7  Negative feedback | 247 |
| 11.8  Some actual control systems | 256 |
| 11.9  Temperature stabilization | 262 |
| 11.10 Laser frequency stabilization | 266 |
| 11.11 Optical fiber phase noise and its cancellation | 275 |
| 11.12 Characterization of laser frequency stability | 277 |
| 11.13 Frequency locking to a noisy resonance | 283 |
| 11.14 Further reading | 285 |
| 11.15 Problems | 285 |
| **12  Atomic and molecular discriminants** | **287** |
| 12.1  Introduction | 287 |
| 12.2  Sub-Doppler saturation spectroscopy | 287 |
| 12.3  Sub-Doppler dichroic atomic vapour laser locking (sub-Doppler DAVLL) and polarization spectroscopy | 293 |
| 12.4  An example of a side-of-line atomic discriminant | 298 |
| 12.5  Further reading | 299 |
| 12.6  Problems | 299 |
| **13  Nonlinear optics** | **301** |
| 13.1  Introduction | 301 |
| 13.2  Anisotropic crystals | 301 |
| 13.3  Second harmonic generation | 309 |
| 13.4  Birefringent phase matching | 314 |
| 13.5  Quasi-phase-matching | 320 |
| 13.6  Second harmonic generation using a focused beam | 325 |
| 13.7  Second harmonic generation in a cavity | 332 |
| 13.8  Sum-frequency generation | 337 |
| 13.9  Parametric interactions | 338 |
| 13.10 Further reading | 351 |
| 13.11 Problems | 351 |
| **14  Frequency and amplitude modulation** | **352** |
| 14.1  Introduction | 352 |
| 14.2  The linear electro-optic effect | 352 |
| 14.3  Bulk electro-optic modulators | 354 |
| 14.4  Traveling wave electro-optic modulators | 359 |
| 14.5  Acousto-optic modulators | 360 |
| 14.6  Further reading | 372 |
| 14.7  Problems | 372 |
| **References** | 374 |
| **Index** | 378 |

# 1

# Gaussian beams

## 1.1 Introduction

Atomic physics calculations involving radiative interactions often model the radiation as a plane wave, which is appropriate when studying individual atoms but is a gross oversimplification when considering larger-scale interactions, such as laser gain and nonlinear optics. Studies of the latter sort require a more realistic description of the fields. The *Gaussian beam model* serves this purpose very well: it is simple, elegant and realistic, since it is an excellent approximation to the lowest transverse mode of a laser (and the lowest mode of an optical cavity).

A full description of Gaussian beams will be given, beginning with the use of the *complex beam parameter*, $q$, to completely describe such a beam and ending with the extremely useful $ABCD$ rule, which facilitates the study of Gaussian beams in the presence of optical elements such as lenses, spherical mirrors and crystals for laser gain or nonlinear optics. The approach will be fairly complete but with several slightly more "advanced" topics such as astigmatism and higher-order modes deferred until a later chapter.

## 1.2 The paraxial wave equation

In a region of space distant from currents and charges, it can be shown (using Maxwell's equations) that the six Cartesian components of the electric and magnetic fields satisfy the *scalar wave equation*

$$\nabla^2 u - \frac{1}{c^2}\frac{\partial^2 u}{\partial t^2} = 0, \tag{1.1}$$

where $u$ is any field component ($E_x, E_y, E_z, B_x$, etc.). If we assume that the wave is monochromatic with a time dependence given by $e^{i\omega t}$, then this equation becomes

$$\nabla^2 u + k^2 u = 0, \tag{1.2}$$

where $k$ is the magnitude of the *wave vector* of the wave. We assume at present that the propagation medium is homogeneous and isotropic and can be described by a *refractive index*, $n$, so that

$$k = \frac{\omega n}{c} = \frac{2\pi n}{\lambda}, \tag{1.3}$$

where $\lambda$ is the *vacuum* wavelength of the radiation and $c$ is the speed of light in a vacuum.

The *paraxial assumption* requires that the normals to the wavefronts make a small angle ($\ll 1$ radian) with some fixed axis, which we will take as the $z$-axis of a Cartesian coordinate system in all that follows. We will assume that the beam differs only slightly from a plane wave and can therefore be modeled as a plane wave which is *spatially modulated* by a slowly varying function of the coordinates, $\psi(x, y, z)$,

$$u(x, y, z) = \psi(x, y, z)e^{-ikz}. \tag{1.4}$$

In order for the paraxial condition to be satisfied, $\psi(x, y, z)$ must vary much more slowly along the $z$-axis than it does along the $x$- and $y$-axes. Thus, when substituting eqn 1.4 into the wave equation, $\partial^2 \psi / \partial z^2$ can be ignored compared to the other second derivatives and we obtain,

$$\text{Paraxial wave equation:} \quad \nabla_t^2 \psi - 2ik\frac{\partial \psi}{\partial z} = 0, \tag{1.5}$$

where $\nabla_t^2$ is the *transverse Laplacian* $(= \frac{\partial^2}{\partial x^2} + \frac{\partial^2}{\partial y^2})$.

## 1.3  Gaussian beam functions and the complex beam parameter, $q$

A particularly simple solution of the paraxial wave equation (eqn 1.5) can be obtained by substituting a *trial solution* of the form,

$$\psi(x, y, z) = \exp\left\{-i\left(P(z) + \frac{k}{2q(z)}r^2\right)\right\}, \quad r^2 = x^2 + y^2, \tag{1.6}$$

where the functions $P(z)$ and $q(z)$ are to be determined. We are looking for cylindrically symmetric solutions at present (hence, $\psi$ depends only upon $r$ and $z$); we will consider solutions with azimuthal dependence in the next chapter when we study higher-order modes. Comparing terms of equal power in $r$, one obtains:

$$\frac{dq(z)}{dz} = 1 \tag{1.7}$$

$$\text{and} \quad \frac{dP(z)}{dz} = -\frac{i}{q(z)}. \tag{1.8}$$

Equation 1.7 is trivially solved, yielding the very important $z$-dependence of $q$:

$$q(z_2) = q(z_1) + (z_2 - z_1). \tag{1.9}$$

where $z_1$ and $z_2$ are any two points on the $z$-axis. The *phase term*, $P(z)$, will be discussed later.

The *complex beam parameter*, $q$, can be written in terms of two real parameters, $R$ and $\omega$, using the following expression (which might appear somewhat arbitrary at first sight):

$$\frac{1}{q} = \frac{1}{R} - i\frac{\lambda}{n\pi\omega^2}. \tag{1.10}$$

The interpretation of the quantities $R$ and $\omega$ becomes clearer when eqn 1.10 is substituted into eqns 1.6 and 1.4,

$$u(x, y, z) = \exp\left\{-i\left(P(z) + kz + k\frac{r^2}{2R}\right) - \frac{r^2}{\omega^2}\right\}, \tag{1.11}$$

where we have used $k = 2\pi n/\lambda$. The field components, $u$, have a Gaussian radial dependence with $e^{-1}$ radius of $\omega$ and a *phase shift* given by the three bracketed terms in eqn 1.11. The interpretation of $R$ as the *wavefront radius* can be seen by considering the $r$-dependence of *surfaces of constant phase* (i.e., wavefronts):

$$z + \frac{r^2}{2R} = const, \tag{1.12}$$

where the term $P(z)$ has been left out since (as will be shown later) it varies with $z$ much more slowly than the $kz$ term. It should be readily evident that eqn 1.12 is an approximation to an equation for a sphere of radius $R$. The approximation is valid when $r \ll R$, which is equivalent to the paraxial assumption.

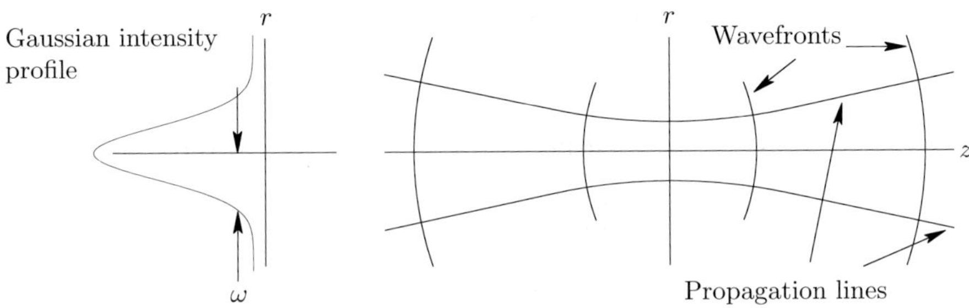


**Fig. 1.1** A Gaussian beam near a focus showing the $e^{-2r^2/\omega^2}$ radial intensity profile and the spherical wavefronts.

In summary, we have found one solution of the source-free wave equation in the paraxial approximation and shown that it has a Gaussian radial dependence (i.e., it is a *Gaussian beam*) and that all of the properties of the Gaussian beam are contained in a *complex beam parameter*, $q$, which depends upon the beam radius, $\omega$, and wavefront radius, $R$, and transforms very simply with $z$ as given by eqn 1.9.

## 1.4   Some Gaussian beam properties

Figure 1.1 illustrates the focusing property of a Gaussian beam: it does not come to a *focus* at a point but rather achieves a *minimum spot size* (called a *waist*, whose radius is $\omega_0$) when the wavefront becomes a plane (infinite radius). The corresponding value of $q$ at a waist is found by setting $R = \infty$ in eqn 1.10:

$$\text{At waist:}\quad q \equiv q_0 = i\frac{n\pi\omega_0^2}{\lambda}. \tag{1.13}$$

If $z$ is the distance from the waist, the other values of $q$ are (from eqn 1.9):

$$q(z) = q_0 + z = i\frac{n\pi\omega_0^2}{\lambda} + z. \tag{1.14}$$

From this, we can obtain the very useful results that, given an arbitrary $q(z)$,

$$\text{Distance to waist } = -Re\{q(z)\} \quad\text{and,} \tag{1.15}$$

$$\text{Radius of waist } = \sqrt{\frac{\lambda}{n\pi}Im\{q(z)\}}. \tag{1.16}$$

Using eqn 1.14 and eqn 1.10, we can obtain the dependence of $\omega$ and $R$ as a function of the distance ($z$) from a waist:

$$\omega(z) = \omega_0 \left[ 1 + \left( \frac{\lambda z}{n\pi\omega_0^2} \right)^2 \right]^{1/2} \tag{1.17}$$

$$R(z) = z \left[ 1 + \left( \frac{n\pi\omega_0^2}{\lambda z} \right)^2 \right]. \tag{1.18}$$

Equation 1.17 clearly shows the hyperbolic shape of the beam contour and allows one to define the *Rayleigh length* as the distance from the waist to the place where the spot size increases by a factor of $\sqrt{2}$ and the *confocal parameter* as twice the Rayleigh length:

$$\text{Rayleigh length} \equiv z_R = \frac{n\pi\omega_0^2}{\lambda} \tag{1.19}$$

$$\text{Confocal parameter} \equiv b = \frac{2n\pi\omega_0^2}{\lambda}. \tag{1.20}$$

These definitions allow us to rewrite the $z$-dependence of $\omega$ and $R$ in a much simpler form:

$$\omega(z) = \omega_0 \left[ 1 + \left( \frac{z}{z_R} \right)^2 \right]^{1/2} \tag{1.21}$$

$$R(z) = z + \frac{z_R^2}{z}. \tag{1.22}$$

The Rayleigh length and confocal parameter are illustrated in Fig. 1.2.

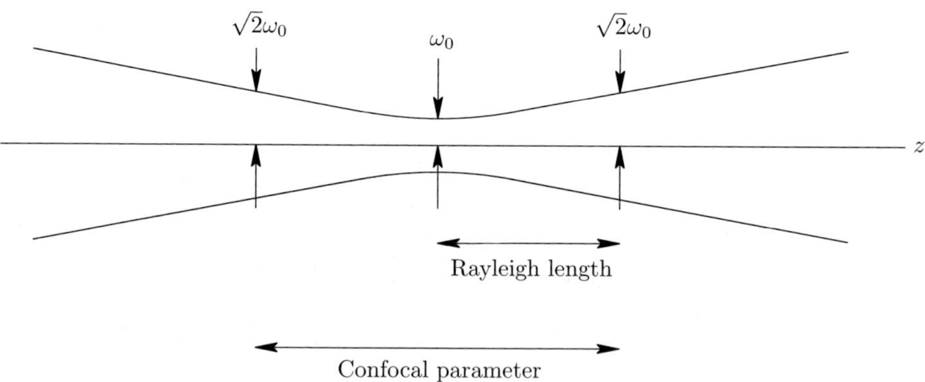


**Fig. 1.2**  An illustration of the definitions of the Rayleigh length, $z_R$, and confocal parameter, $b$.

It is often useful to work with $1/q$ which we will *define* as $q_1$. We can then easily show that the distance from the waist to the current position, the Rayleigh length and the waist size can be written as:

$$\text{Distance to waist:} \quad z = -\frac{Re\{q_1\}}{|q_1|^2} \qquad \left(q_1 \equiv \frac{1}{q}\right) \tag{1.23}$$

$$\text{Rayleigh length:} \quad z_R = -\frac{Im\{q_1\}}{|q_1|^2} \tag{1.24}$$

$$\text{Waist size:} \quad \omega_0^2 = \left(\frac{\lambda}{n\pi}\right) z_R = -\left(\frac{\lambda}{n\pi}\right)\frac{Im\{q_1\}}{|q_1|^2} \tag{1.25}$$

$$\text{Spot size:} \quad \omega^2 = -\frac{\lambda}{n\pi Im\{q_1\}}. \tag{1.26}$$

Equations 1.21 and 1.22 allow us to determine $R$ and $\omega$ at a distance $z$ from the waist. Occasionally we would like to know the size of the waist if we are given $1/q$ (actually, $R$ and $\omega$) at some place other than at the waist. Using eqn 1.25 and the definition of $1/q$, we obtain:

$$\omega_0 = \frac{\lambda R\omega}{\sqrt{(\pi n\omega^2)^2 + (\lambda R)^2}}. \tag{1.27}$$

## 1.5 The phase term: Gouy phase

The calculations up to this point have ignored the $P(z)$ term, justifying this by assuming that its $z$-dependence is much slower than that of the $kz$ term. Now that a solution for $q$ as a function of $z$ has been obtained, we can substitute it into eqn 1.8,

$$\frac{dP(z)}{dz} = -\frac{i}{q(z)} = -\frac{i}{z + i(n\pi\omega_0^2/\lambda)}. \tag{1.28}$$

Integrating this (from the waist, where $z = 0$) yields,

$$iP(z) = \ln[1 - i(\lambda z/n\pi\omega_0^2)] = \ln\sqrt{1 + \left(\frac{\lambda z}{n\pi\omega_0^2}\right)^2} - i\tan^{-1}\left(\frac{\lambda z}{n\pi\omega_0^2}\right). \tag{1.29}$$

The first term above gives the expected amplitude decrease with $z$ and the second term gives an additional *phase shift*, which is rapidly changing with $z$ when $z < z_R$. The complete expression for a field component of a Gaussian beam can now be written down (where an overall proportionality constant of 1 is assumed):

$$u = \frac{1}{\sqrt{1 + \left(\frac{z}{z_R}\right)^2}}\exp\left\{i\tan^{-1}\left(\frac{z}{z_R}\right) - ik\left(z + \frac{r^2}{2R}\right) - \frac{r^2}{\omega^2}\right\}. \tag{1.30}$$

The additional phase shift is often called the *Gouy* phase, after the French physicist who discovered in 1890 that a converging light beam underwent a 180° phase shift as one traversed the focus along the direction of propagation. The total phase evolution of the Gaussian beam (on axis) can be written as the sum of two components: the *normal* phase shift and the *Gouy phase*:

$$\text{Phase shift of Gaussian beam} = i\phi = \overbrace{ikz}^{\text{normal}} \; \overbrace{- \, i\tan^{-1}\frac{z}{z_R}}^{\text{Gouy}}. \tag{1.31}$$

The Gouy phase changes occur largely within one Rayleigh length of the waist, justifying the assumption that the normal phase shift is a much more rapid function of $z$ (since it varies significantly over the distance of one wavelength and $z_R \gg \lambda$ in the paraxial approximation). The Gouy phase will play a large role in determining the resonant frequencies of an optical cavity, where it causes the location of the resonances to be determined by the transverse mode number in addition to that of the axial mode. The $z$-dependence of the Gouy phase is shown graphically in Fig. 1.3.

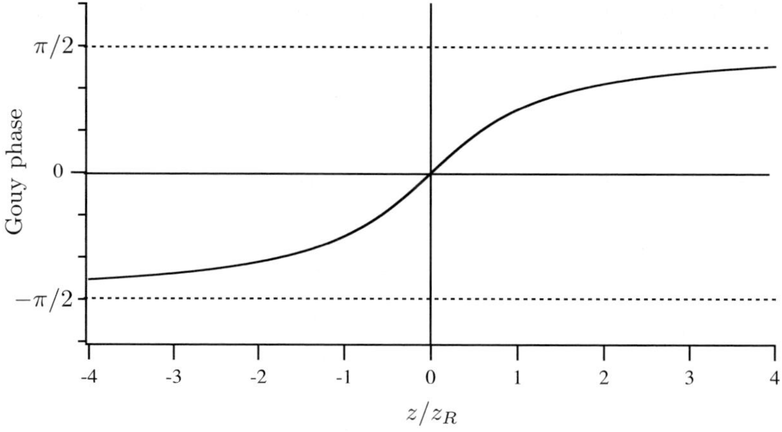

**Fig. 1.3** The $z$-dependence of the Gouy phase.

## 1.6   Simple transformation properties of the complex beam parameter

We have shown that the behavior of a Gaussian beam is determined solely by a single complex parameter, $q$, from which one can determine the beam's spot size and wavefront radius of curvature at all points along the beam's path. To complete our discussion of this beam model, we will determine how $q$ is transformed in the four most common situations encountered in the laboratory: free-space propagation, refraction by a *thin* lens, reflection by a *thin* spherical mirror, and transmission through a slab of some transparent material with refraction index, $n$. Using these transformation rules, one can completely determine the behavior of a Gaussian beam as it progresses through most pieces of apparatus found in the laboratory. This approach of directly calculating the transformations of $q$ is very useful but tedious; in the next section we will describe a much simpler technique for accomplishing the same task.

**Free-space:** Free space propagation is immediately obtained from eqn 1.9.

$$q(z_2) = q(z_1) + z_2 - z_1. \tag{1.32}$$

**Thin lens:** Assume that the spot size doesn't change. We are thus only interested in how the wavefront radius changes upon traversing a thin lens.

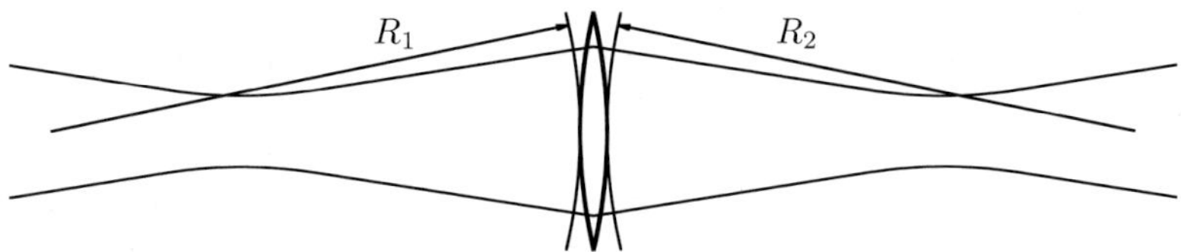

Since the wavefront radii obey the simple lens formula (where $i$ is the distance between the image and the lens, $o$ is the distance between the object and the lens and $f$ is the focal length),

$$\frac{1}{i} + \frac{1}{o} = \frac{1}{f} \longrightarrow \frac{1}{R_1} - \frac{1}{R_2} = \frac{1}{f} \tag{1.33}$$

(the sign change is due to our observance of the sign conventions for radii).

$$\text{Since } \omega_1 = \omega_2, \quad \frac{1}{q_1} - \frac{1}{q_2} = \frac{1}{f} \longrightarrow q_2 = \frac{q_1}{1 - q_1/f}. \tag{1.34}$$

**Thin spherical mirror:** A thin mirror is described by the same formula as the lens, with $f = R/2$, where $R$ is the radius of curvature of the mirror.

$$q_2 = \frac{q_1}{1 - 2q_2/R}. \tag{1.35}$$

**Slab of thickness, $d$, and index, $n$:** We assume that the beam is *normally incident* on the input and output faces of the slab (non-normal incidence introduces *astigmatism*, which will be treated in the next chapter). Three situations need to be considered: transition through the input face, translation by distance $d$ and transition through the output face. We assume that the slab is surrounded by a medium of unity refractive index. From the expression for $1/q$ (eqn 1.10), one can see that the imaginary term has a $1/n$ dependence. To demonstrate that the real part also has this dependence, one need only assume that the normals to the spherical wavefronts obey Snell's law. Thus, the transition $q_0 \to q_1$ across the input face becomes

$$q_1 = nq_0. \tag{1.36}$$

The internal traversal of the slab ($q_1 \to q_2$) satisfies eqn 1.9:

$$q_2 = q_1 + d. \tag{1.37}$$

Finally, the transition ($q_2 \to q_3$) across the output face satisfies

$$q_3 = \frac{1}{n} q_2. \tag{1.38}$$

Thus, the overall transition across the slab is described by

$$q_3 = q_0 + \frac{d}{n}. \tag{1.39}$$

## 1.7   Matrix formulation of paraxial ray optics: ABCD rule

We will make a slight digression to ray optics in order to introduce a very useful formalism for describing ray propagation using matrices. We will then discuss the remarkable utility of these same matrices in determining the propagation of Gaussian beams.

We will first consider a single paraxial ray in an axially symmetrical optical system. This ray is completely described by a two-element column vector containing the *height*, $y$, of the ray above the axis and the *slope*, $y'$, of the ray with respect to the $z$-axis:

$$\text{Ray vector:} \quad \begin{pmatrix} y \\ y' \end{pmatrix}. \tag{1.40}$$

When considering the four situations discussed earlier (free-space propagation, lenses, mirrors and slabs), the ray vector is subject to a *linear transformation* represented by a *ray matrix*, called the *ABCD matrix*:

$$\begin{pmatrix} y_2 \\ y_2' \end{pmatrix} = \begin{pmatrix} A & B \\ C & D \end{pmatrix} \begin{pmatrix} y_1 \\ y_1' \end{pmatrix}, \tag{1.41}$$

where the subscripts $_{1,2}$ are for the ray before and after the transformation, respectively. A description of our four transformations using ray matrices will now be obtained.

**Free-space propagation:** The ray matrix is simply obtained using the equation for a straight line (we assume the propagation distance is $d$):

$$y_2' = y_1', \quad y_2 = y_1 + y_1' d \quad \longrightarrow \quad \begin{pmatrix} A & B \\ C & D \end{pmatrix} = \begin{pmatrix} 1 & d \\ 0 & 1 \end{pmatrix}. \tag{1.42}$$

**Thin lens:** The calculation is similar to that for the $q$-transformation. Again, the ray height is unchanged ($y_2 = y_1$) and the slope change is obtained using the lens formula (eqn 1.33):

$$y_2 = y_1, \quad y_2' = -\frac{y_2}{i}, \quad y_1' = \frac{y_1}{o}. \tag{1.43}$$

Using the lens formula, we obtain:

$$y_2' = y_1' - y_1/f \quad \longrightarrow \quad \begin{pmatrix} A & B \\ C & D \end{pmatrix} = \begin{pmatrix} 1 & 0 \\ -1/f & 1 \end{pmatrix}. \tag{1.44}$$

**Thin spherical mirror:** The ray matrix is the same as for the thin lens, replacing $f$ with $R/2$:

$$\begin{pmatrix} A & B \\ C & D \end{pmatrix} = \begin{pmatrix} 1 & 0 \\ -2/R & 1 \end{pmatrix}. \tag{1.45}$$

(In order to use this approach with a mirror, one must redefine the *positive z-direction after* a reflection.)

**Slab of thickness, $d$:** From Snell's law in the paraxial approximation, the change in y can be easily determined:

$$y_2 = y_1 + y_3' d = y_1 + y_1' \frac{d}{n}, \tag{1.46}$$

where we assumed that sines of the angles are equal to the slopes and used Snell's law at the input face. The quantity $y_3'$ is the slope of the ray *inside the slab*. The slope doesn't change when traversing the crystal since the exit angle of the ray is the same is its entry angle (for parallel crystal faces). Thus,

$$\begin{pmatrix} A & B \\ C & D \end{pmatrix} = \begin{pmatrix} 1 & d/n \\ 0 & 1 \end{pmatrix}. \tag{1.47}$$

A useful property of all of the above ABCD matrices is the fact that they are *unimodular*:

$$\det \begin{pmatrix} A & B \\ C & D \end{pmatrix} = AD - BC = 1. \tag{1.48}$$

This property also holds for *composite* systems composed of any number of the above elements since the determinant of a product of matrices is the product of the determinants of the individual matrices.

A remarkable connection between the ray matrices and $q$-transformations can be seen in Table 1.1. From the slightly odd way in which the $q$-transformations are presented, it should be clear that for these four cases the following rule holds:

$$q_1 = \frac{A q_0 + B}{C q_0 + D} \iff \begin{pmatrix} A & B \\ C & D \end{pmatrix}. \tag{1.49}$$

Stated in words: *For a free-space propagation, thin lens, thin mirror or slab, knowledge of the ABCD matrix immediately allows one to write down the q-transformation using the expression on the left side of eqn 1.49.*

For this rule to be generally useful, we need to find its analogue (if it exists) for a composite system composed of varying numbers of all four of the elements discussed above. To do this, we will first consider a system composed of any two elements and investigate the composite $q$-transformation for such a system. We therefore have two transformations of the following form (labeling them by their *ABCD* matrices):

$$q_0 \xrightarrow{ABCD} q_1 \xrightarrow{A'B'C'D'} q_2. \tag{1.50}$$

The *ABCD* matrices for the two transformations are:

**Table 1.1** Relations between $q$-transformations and ray matrices.

| Case | $q$-transformation | Ray matrix |
|---|---|---|
| Free-space | $q_1 = q_0 + d = \dfrac{(1)q_0 + (d)}{(0)q_0 + (1)}$ | $\begin{pmatrix} 1 & d \\ 0 & 1 \end{pmatrix}$ |
| Thin lens | $q_1 = \dfrac{q_0}{1 - q_0/f} = \dfrac{(1)q_0 + (0)}{(-1/f)q_0 + (1)}$ | $\begin{pmatrix} 1 & 0 \\ -1/f & 1 \end{pmatrix}$ |
| Mirror | $q_1 = \dfrac{q_0}{1 - 2q_0/R} = \dfrac{(1)q_0 + (0)}{(-2/R)q_0 + (1)}$ | $\begin{pmatrix} 1 & 0 \\ -2/R & 1 \end{pmatrix}$ |
| Slab | $q_1 = q_0 + d/n = \dfrac{(1)q_0 + (d/n)}{(0)q_1 + (1)}$ | $\begin{pmatrix} 1 & d/n \\ 0 & 1 \end{pmatrix}$ |

$$q_0 \longrightarrow q_1 : \begin{pmatrix} A & B \\ C & D \end{pmatrix} \qquad q_1 \longrightarrow q_2 : \begin{pmatrix} A' & B' \\ C' & D' \end{pmatrix}. \tag{1.51}$$

Applying the rule twice:

$$q_2 = \frac{A'q_1 + B'}{C'q_1 + D'} = \frac{(A'A + B'C)q_0 + (A'B + B'D)}{(C'A + D'C)q_0 + (C'B + D'D)}. \tag{1.52}$$

The terms in brackets are just the components of *the matrix product of the two matrices*. Thus the rule is valid for composite systems provided that one uses the composite matrix formed by taking the matrix product of the individual matrices. It is obvious that an analogous rule holds for the ray optics case. We therefore obtain the *ABCD* rule:

**ABCD rule:** *The overall q-transformation for a complex system composed of thin lenses, thin mirrors, free spaces and slabs can be obtained by determining the ABCD matrices of the individual components, multiplying the matrices together and applying eqn 1.49 to the composite matrix.*

As a practical matter, it should be noted that the component *ABCD* matrices are written down in *the reverse order from left to right to the order in which the beam encounters the various components*. We will see several examples of this "rule" in the next few chapters.

## 1.8   Further reading

This chapter provided an essential introduction to Gaussian beams and their manipulation using the complex beam parameter and the *ABCD* rule. Much of the formalism is based upon the pioneering work on the subject conducted by the researchers at Bell Laboratories in the 1960s and is described in the classic paper by Kogelnik and Li (1966). Other books on the subject include the excellent textbook by Yariv (1989) the tome by Siegman (1986) and numerous other books on laser theory. Three excellent representative samples of the latter group are: Verdeyen (1995), Milonni and Erberly (1988), and Svelto (2004). The background material in classical optics can be obtained from the excellent textbooks by Hecht (2002) and Jenkins and White (1957) and from the well-known treatise by Born and Wolf (1980).

## 1.9  Problems

(1.1)  Calculate the ABCD matrix for a ray entering a spherical dielectric interface from a medium of refractive index $n_1$ to a medium of index $n_2$, with radius of curvature $R$ ($R > 0$ if center is to left of surface). Make the usual paraxial approximations.

(1.2)  You are given an optical system described by an ABCD matrix in which both the input plane and the output plane are at beam waists. If the input waist is $\omega_1$, determine the conditions on the matrix elements and the output waist, $\omega_2$ (in terms of ABCD and $\omega_1$).

(1.3)  Find the ABCD matrix of a "Gaussian" aperture whose field transmission $t(x, y)$ is given by

$$t(x, y) = \exp\left[-\frac{x^2 + y^2}{2\omega_a^2}\right].$$

(1.4)  It is often necessary to accurately measure the spot size of a Gaussian beam. Below are two possible methods for doing this in a laboratory:

   (a)  One way to measure the beam radius is to advance a knife edge into the beam with a micrometer and measure the power of the beam with a digital power meter downstream from the knife edge. Between what two percentages of power will the knife edge advance a distance equal to $2\omega$? (Make it symmetrical: i.e., $\Delta\%$ and $(100 - \Delta)\%$).
   (b)  A commercial instrument uses the scheme below to automate the spot measuring process. A cylindrical knife-edge rotates at angular speed $\Omega$. If the open aperture subtends an angle $\theta$ from the center and the radius of the tube is $R$, write down an expression for the "rise-time" of the signal at the detector as a function of the spot radius, where the rise-time is defined as being between the same two percentages as calculated above.

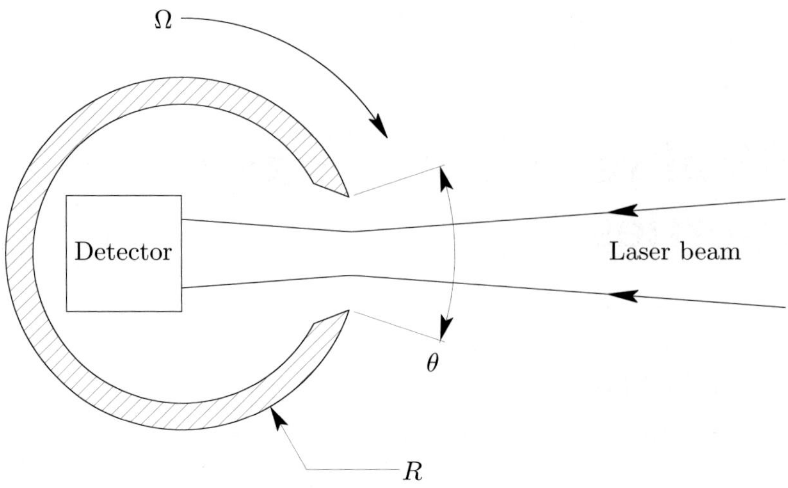
Ω
Detector
Laser beam
θ
R

# 2
# Optical resonators – geometrical properties

## 2.1 Introduction

Optical cavities (resonators) have numerous applications, perhaps most notably as the feedback mechanism in laser oscillators. In the atomic physics laboratory, they are also used for optical spectrum analysis, a stable frequency reference for laser frequency stabilization and for enhancement of the power of fundamental beams in nonlinear frequency synthesis.

We will consider two variants of the optical resonator: the standing wave cavity and the traveling wave cavity. Both types are used in lasers but traveling wave (or "ring") cavities are usually preferred for nonlinear frequency synthesis. Despite the apparent greater simplicity of the two-mirror standing-wave cavity compared to the four-mirror ring resonator, the analysis of the latter is actually no more difficult than that of the former. One problematic aspect of the ring resonator not shared by the standing wave cavity is the inevitable presence of astigmatism, since a practical realization of a ring cavity requires that mirrors (and often intracavity crystals as well) be used off-axis. We will analyze this in some detail.

This chapter will discuss resonator eigenmodes, which depend only upon the cavity geometry and the wavelength of the radiation. The next chapter will cover cavity energy relationships, which depend upon the reflectivity of the cavity mirrors and various loss mechanisms. The latter discussion will introduce the concepts of cavity finesse, $Q$, photon lifetime, etc. This division between *geometrical* properties and *energy* relationships is perhaps analogous to the division in classical mechanics between kinematics and dynamics.

## 2.2 The two-mirror standing wave cavity

The two-mirror resonator is essentially a classical Fabry–Perot interferometer with spherical mirrors. Calculation of the eigenmodes of this instrument is complicated by the open geometry. There are a number of ways of analyzing an open, two-mirror resonator – the most rigorous treatment is probably to use formal diffraction theory. The Fresnel–Kirchoff formulation of Huygens' principle leads to two integral equations which can be solved using some simplifying assumptions to make the problem more tractable. We will use a much simpler approach which is based upon the Gaussian beam theory worked out in the previous chapter together with a simple and intuitively obvious self-consistency approach. Our treatment of course implicitly acknowledges diffraction, since the latter is built into the theory of Gaussian beams.

The two-mirror resonator consists of two spherical mirrors with radii of curvature $R_1$ and $R_2$ (Fig. 2.1). We are currently only interested in the field configurations which form stable eigen-modes of the cavity. Thus, we assume that there is some means (e.g., partially transmitting coatings) by which radiation can be injected into the space between the mirrors but we are not interested at present in its details.

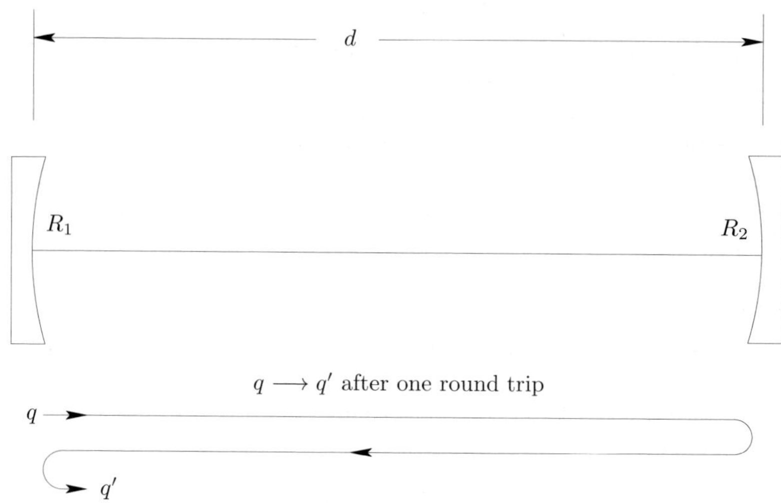


**Fig. 2.1** A two-mirror standing-wave cavity.

Our *self-consistent* approach requires that the complex beam parameter, $q$, replicates itself after one round trip in the cavity. If this happens, the field distribution also replicates itself after one round trip since $q$ contains everything there is to know about the Gaussian field (except for its *phase*, which will be discussed later when we determine the *resonant frequencies* of the cavity). In terms of the round trip ABCD matrix, self-consistency requires that

$$q = \frac{Aq + B}{Cq + D} \quad \text{or} \quad \frac{1}{q} = \frac{C + D\left(\frac{1}{q}\right)}{A + B\left(\frac{1}{q}\right)}. \tag{2.1}$$

Due to the way that $q$ is defined, it is more convenient to solve this equation for $q_1 = 1/q$. The resulting quadratic equation:

$$B\left(\frac{1}{q}\right)^2 + (A - D)\left(\frac{1}{q}\right) - C = 0, \tag{2.2}$$

has solutions:

$$q_1 = \frac{1}{q} = \frac{D - A}{2B} \pm \frac{1}{2B}\sqrt{(A - D)^2 + 4BC}. \tag{2.3}$$

Since the elements of the ABCD matrix are real, the first term must be associated with the real part of $1/q$ and the second term must be pure imaginary and be associated with the imaginary part of $1/q$. Thus, using eqn 1.10, one obtains:

$$R = \frac{2B}{D - A} \tag{2.4}$$

$$\omega = \sqrt{\frac{2\lambda|B|}{n\pi\sqrt{4 - (A + D)^2}}}, \tag{2.5}$$

where we used the fact that the determinant of an ABCD matrix is unity in deriving the second equation. The values so obtained are of course valid only at the *reference point* in the cavity (the beginning and end points of the round trip). The reference point is completely arbitrary and would normally be chosen for convenience (a judicious choice can reduce the number of matrices needed to be multiplied together). Values at other points in the cavity can be obtained using the transformation properties of $q$ or eqn 1.17 and eqn 1.18.

## 2.3  Stability

In order to obtain a meaningful value of $q$, we required that the expression under the square root in eqn 2.3 be negative. This leads to a physically reasonable (*real*) value for the radius in eqn 2.5. The radius will therefore be non-physical if the following inequality is violated:

$$\text{Stability criterion:} \quad |A + D| \le 2, \tag{2.6}$$

which is only valid in the paraxial approximation. This inequality is called the *stability criterion* for the given cavity geometry. If it is violated, it will be impossible for the field to replicate itself after one round trip.

Some additional insight can be obtained by using a *ray matrix* approach and by replacing the two mirror cavity by an *equivalent* periodic sequence of lenses having the same *powers* (reciprocals of the focal length) as the mirrors. In this way, the ray trajectory will be *unfolded* and a stable cavity can be shown to have a periodic ray path which remains close to the axis while an unstable cavity will have a ray path which deviates increasingly from the cavity axis.

We can approach this analytically by writing down the overall ray matrix for $n$ *units of periodicity*, each of which consists of a pair of lenses as shown in Fig. 2.2. We require the $n$th power of an ABCD matrix representing one unit of periodicity. Using *Sylvester's theorem* to evaluate the matrix corresponding to $n$ periods yields:

$$\begin{pmatrix} A & B \\ C & D \end{pmatrix}^n = \frac{1}{\sin\theta} \begin{pmatrix} A\sin n\theta - \sin(n-1)\theta & B\sin n\theta \\ C\sin n\theta & D\sin n\theta - \sin(n-1)\theta \end{pmatrix}, \tag{2.7}$$

where

$$\cos\theta = \frac{1}{2}(A + D) \implies \theta = \cos^{-1}\left\{\frac{1}{2}(A + D)\right\}. \tag{2.8}$$

In order for $\theta$ to be *real*, the expression inside the curly brackets must be between $+1$ and $-1$, which leads to eqn 2.6. Thus, for a stable system with a real value of $\theta$, the

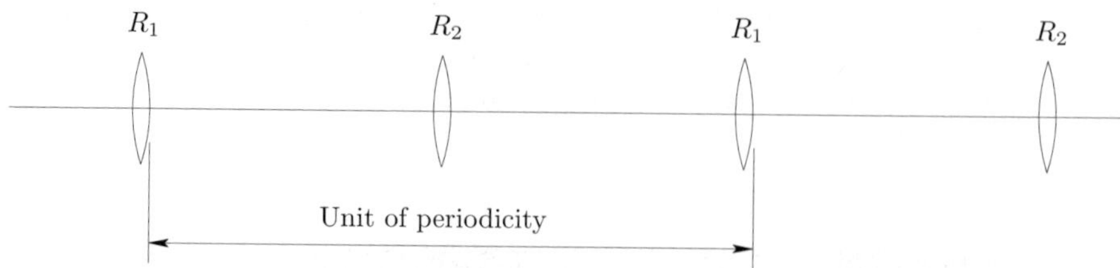

**Fig. 2.2** A lens sequence which is equivalent to a two-mirror cavity. Each *unit of periodicity* represents one round-trip path in the two mirror cavity.

matrix elements are simple sine and cosine functions of $n$ and a ray will have a periodic behavior. An unstable system will have an *imaginary* $\theta$ which yields *hyperbolic* sines and cosines and causes the ray to deviate exponentially from the axis (and would also violate the paraxial assumption).

We can determine the constraints on $R_1$ and $R_2$ required for stability by writing the round trip ABCD matrix as a product of four matrices. We take the (arbitrary) starting point just to the right of the left-hand mirror. Then we have a free-space propagation by a distance $d$, a reflection by the mirror whose radius of curvature is $R_2$, a second free-space propagation and a reflection by the mirror with radius $R_1$. We therefore obtain (writing the matrices in *reverse order*),

$$\begin{pmatrix} A & B \\ C & D \end{pmatrix} = \begin{pmatrix} 1 & 0 \\ -\frac{2}{R_1} & 1 \end{pmatrix} \begin{pmatrix} 1 & d \\ 0 & 1 \end{pmatrix} \begin{pmatrix} 1 & 0 \\ -\frac{2}{R_2} & 1 \end{pmatrix} \begin{pmatrix} 1 & d \\ 0 & 1 \end{pmatrix}. \tag{2.9}$$

In what follows, it is convenient to describe the cavity *geometry* in terms of two dimensionless parameters, $g_1$ and $g_2$:

$$g_1 \equiv 1 - \frac{d}{R_1} \tag{2.10}$$

$$g_2 \equiv 1 - \frac{d}{R_2}. \tag{2.11}$$

Then,

$$\begin{pmatrix} A & B \\ C & D \end{pmatrix} = \begin{pmatrix} 1 & 0 \\ \frac{2}{d}(g_1 - 1) & 1 \end{pmatrix} \begin{pmatrix} 1 & d \\ 0 & 1 \end{pmatrix} \begin{pmatrix} 1 & 0 \\ \frac{2}{d}(g_2 - 1) & 1 \end{pmatrix} \begin{pmatrix} 1 & d \\ 0 & 1 \end{pmatrix}$$

$$= \begin{pmatrix} 2g_2 - 1 & 2g_2 d \\ \frac{2}{d}(2g_1 g_2 - g_1 - g_2) & 4g_1 g_2 - 2g_2 - 1 \end{pmatrix}. \tag{2.12}$$

The *stability criterion* becomes:

$$|A + D| \leq 2 \implies |4g_1 g_2 - 2| \leq 2. \tag{2.13}$$

We therefore obtain:

$$\text{Stability criterion:} \quad 0 \leq g_1 g_2 \leq 1. \tag{2.14}$$

This expression can be plotted (Fig. 2.3) on a two-dimensional coordinate system whose axes are $g_1$ and $g_2$. The plot indicates *regions of stability* bounded by the lines $g_1 = 0$ and $g_2 = 0$ together with the hyperbolas in eqn 2.14. The geometrical interpretation of $g_{1,2}$ should be clear: these parameters vary from 1 (plane mirrors) through 0 (*confocal*, where the mirror spacing equals the mirror radii) to -1 (spherical, where the mirrors are part of a sphere).

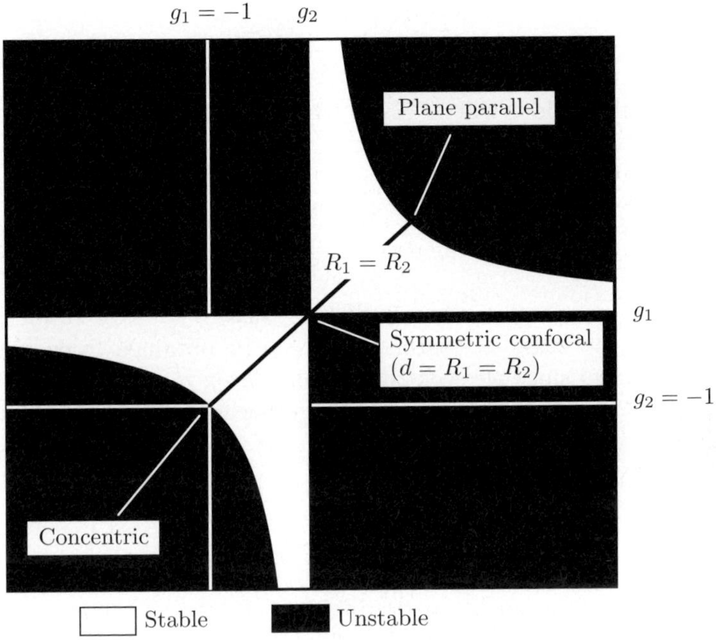


**Fig. 2.3** A *stability diagram* displaying regions of stability in terms of the parameters $g_{1,2}$ in a two-mirror cavity.

## 2.4 Solution for an arbitrary two-mirror stable cavity

The self-consistent solution for a two-mirror cavity will be obtained by first writing down the solution for $q_1 = 1/q$ using eqn 2.3 and the product matrix (eqn 2.12):

$$q_1 = \frac{1}{q} = \frac{g_1 - 1}{d} \pm \frac{1}{g_2 d} \sqrt{g_1 g_2 (g_1 g_2 - 1)}. \tag{2.15}$$

The spot size at the left-hand mirror (the reference point) can be obtained by using this result and eqn 1.10:

$$\omega^2 = \left( \frac{\lambda d}{n\pi} \right) \sqrt{\frac{g_2}{g_1 (1 - g_1 g_2)}}. \tag{2.16}$$

The wavefront radius at the left-hand mirror can be obtained by substituting the elements of the ABCD matrix into eqn 2.4. It should come as no surprise that the result is $-R$, which is just the mirror radius (with a sign which is consistent with the *sign convention*, which states that the radius is positive if the center of curvature is to the left of the wavefront and negative otherwise). It would be a mistake, however, to conclude that this is always the case: in a ring cavity, the mirror radius and wavefront radius are different. The waist size can be obtained by substituting the imaginary part of $q_1$ from eqn 2.15 into eqn 1.25. After a bit of algebra, we get:

$$\omega_0^2 = \left(\frac{\lambda d}{n\pi}\right)\frac{\sqrt{g_1 g_2(1 - g_1 g_2)}}{g_1 + g_2 - 2g_1 g_2}. \tag{2.17}$$

A few additional relations will be useful when we calculate the resonant frequencies of the cavity. These are the Rayleigh length, $z_R$, and the $z$-coordinates, $z_1$ and $z_2$, of mirrors 1 and 2 with respect to the waist (these all appear in the *Gouy phase* term for the resonant frequency). The Rayleigh length can be obtained from eqn 1.24 (or directly from eqn 2.17):

$$z_R = d\frac{\sqrt{g_1 g_2(1 - g_1 g_2)}}{g_1 + g_2 - 2g_1 g_2}. \tag{2.18}$$

The distance from the waist to mirror 1 ($z_1$) can be obtained from eqn 1.23 (where two minus signs cancel since we seek the *distance from the waist to the mirror*). The distance to mirror 2, $z_2$, is obtained from $z_1$ together with the relationship $z_2 - z_1 = d$. The results are:

$$z_1 = d\frac{g_2(g_1 - 1)}{g_1 + g_2 - 2g_1 g_2} \tag{2.19}$$

$$z_2 = d\frac{g_1(1 - g_2)}{g_1 + g_2 - 2g_1 g_2}. \tag{2.20}$$

It would be instructive at this point to analyze the two-mirror cavity for the very simple case of a *symmetric resonator*, where $R_1 = R_2 = R$. As we can see from Fig. 2.3, all symmetrical configurations between the plane mirror cavity and the spherical mirror cavity are stable. To find the spot sizes, we set $g_1 = g_2 = g = 1 - d/R$ and obtain:

$$\text{At mirror:} \quad \omega^2 = \left(\frac{\lambda d}{n\pi}\right)\sqrt{\frac{1}{1 - g^2}} = \left(\frac{\lambda R}{n\pi}\right)\sqrt{\frac{d}{2R - d}} \tag{2.21}$$

$$\text{Waist:} \quad \omega_0^2 = \left(\frac{\lambda d}{2n\pi}\right)\sqrt{\frac{1 + g}{1 - g}} = \left(\frac{\lambda}{n\pi}\right)\sqrt{\frac{dR}{2} - \frac{d^2}{4}}. \tag{2.22}$$

A very common configuration is the one with $d = R$. This is the *confocal* resonator whose position on the stability diagram is halfway between the plane mirror cavity and the spherical cavity (constrained to lie on the line tilted at 45° in Fig. 2.3). The values of the radius at the mirror and waist are:

$$\omega_0 = \left(\frac{\lambda d}{2n\pi}\right)^{1/2} \tag{2.23}$$

$$\omega_{mirror} = \left(\frac{\lambda d}{n\pi}\right)^{1/2} = \sqrt{2}\omega_0. \tag{2.24}$$

Note that $d$ in this case is just the *confocal parameter*, which is perhaps the origin of its name.

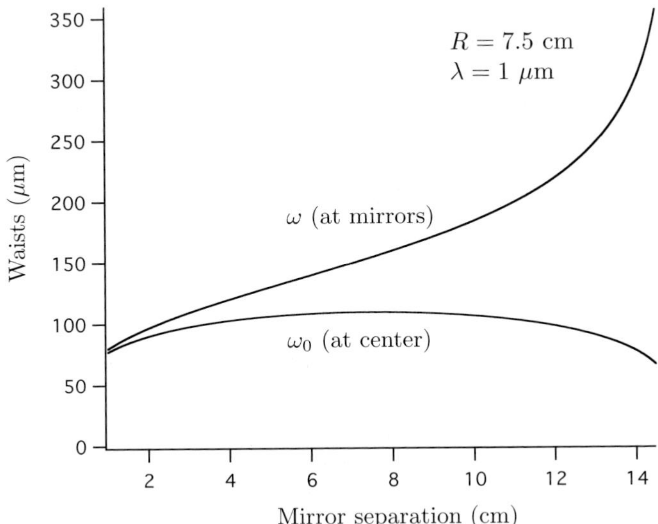


**Fig. 2.4** The spot sizes at the mirrors and waist sizes as a function of the mirror separation for a symmetrical two-mirror cavity.

The spot radii as a function of mirror separation are plotted in Fig. 2.4 for a cavity whose mirror radii are 7.5 cm. Notice that the cavity is unconditionally stable for small mirror separations (and $\omega$ approaches $\omega_0$ as the beam becomes collimated) but the spot size at the mirror increases very rapidly as one approaches the instability point (spherical cavity).

## 2.5   Higher-order modes

In deriving the solutions to the paraxial wave equation (eqn 1.5), we have considered cylindrically symmetric field configurations with a very simple Gaussian dependence on $r$. In order to describe an arbitrary field configuration, we need to find a complete set of functions, each one of which is a solution of the paraxial wave equation. We can then expand a general beam in a linear combination of these functions, which are called *modes*. The simple Gaussian beam studied in the previous chapter is only one (the lowest) of an infinite number of possible modes. We will now describe a complete set of solutions to the paraxial wave equation in *Cartesian coordinates*. One might expect a better coordinate system to be *cylindrical coordinates*, but the *eigenmodes* of the paraxial wave equation in the Cartesian system turn out to be more often observed in the laboratory than the eigenmodes in cylindrical coordinates.

After a great deal of algebra, one can show that the following functions are solutions of the paraxial wave equation:

$$u(x, y, z)_{nm} =$$

$$\frac{\omega_0}{\omega(z)} H_n(\sqrt{2}\frac{x}{\omega}) H_m(\sqrt{2}\frac{y}{\omega})$$

$$\times \exp\left\{-i(kz - \Phi(m, n; z)) - i\frac{k}{2q}(x^2 + y^2)\right\}, \qquad (2.25)$$

where

$$\frac{\omega_0}{\omega(z)} = \frac{1}{\sqrt{1 + (z/z_R)^2}} \qquad (2.26)$$

$$\Phi(n, m; z) = (n + m + 1)\tan^{-1}(z/z_R) \quad \text{(Gouy phase)}. \qquad (2.27)$$

The indices $n$ and $m$ are integers, and the functions $H_n()$ and $H_m()$ are *Hermite polynomials* of order $n$ and $m$. The origin of the coordinate system ($z = 0$) is at the waist. When $n = m = 0$, this expression reduces to that of the *lowest order* Gaussian beam, eqn 1.30.

We should note that, since all of the transverse coordinates in eqn 2.25 appear divided by $\omega(z)$, the *cross-sectional shape* of the beam is independent of the position along the beam and that the size of the profile is scaled by the spot radius, $\omega(z)$. These field distributions are called *Hermite–Gaussian modes*. Since the free-space electric and magnetic fields are transverse to the $k$-vector, the modes are called TEM$_{mnq}$ (transverse electromagnetic). The indices $m$ and $n$ are defined above; $q$ is an axial mode number which will be defined shortly (and should be distinguished from the complex beam parameter – the notation is occasionally unfortunate). Plots of three different modes are shown in Fig. 2.5.

The Hermite–Gaussian modes have the following properties:

- They are all characterized by the *same* complex beam parameter, $q$, defined by

$$\frac{1}{q(z)} = \frac{1}{R(z)} - i\frac{\lambda}{n\pi\omega^2(z)}. \qquad (2.28)$$

- They all satisfy the same ABCD rule as (lowest order) Gaussian beams. The *mode numbers*, $m$ and $n$ are preserved under all of the transformations discussed in this book.
- The mode *shape* is independent of $z$ and scales with $\omega(z)$.
- The mode of index $m$ has a half-width, $x_m$, (in one transverse coordinate), where

$$x_m \approx \sqrt{m} \times \omega. \qquad (2.29)$$

Thus, $\omega$ is no longer the beam size in higher-order modes. A focused spot has its size *degraded* by $\sqrt{m}$.

- The maximum mode number ($m_{max}$) at a waist which will "fit" into an aperture of radius $a$ is:

$$m_{max} \approx (a/\omega_0)^2. \qquad (2.30)$$

This *spatial filtering* behavior of small apertures allows one to filter out modes whose mode number (in either coordinate) is greater than $m_{max}$.

- The far-field half-angle spread, $\theta_m$ is also degraded compared to that $(\theta_0)$ of the lowest order mode:

$$\theta_m \approx \sqrt{m} \times \theta_0. \tag{2.31}$$

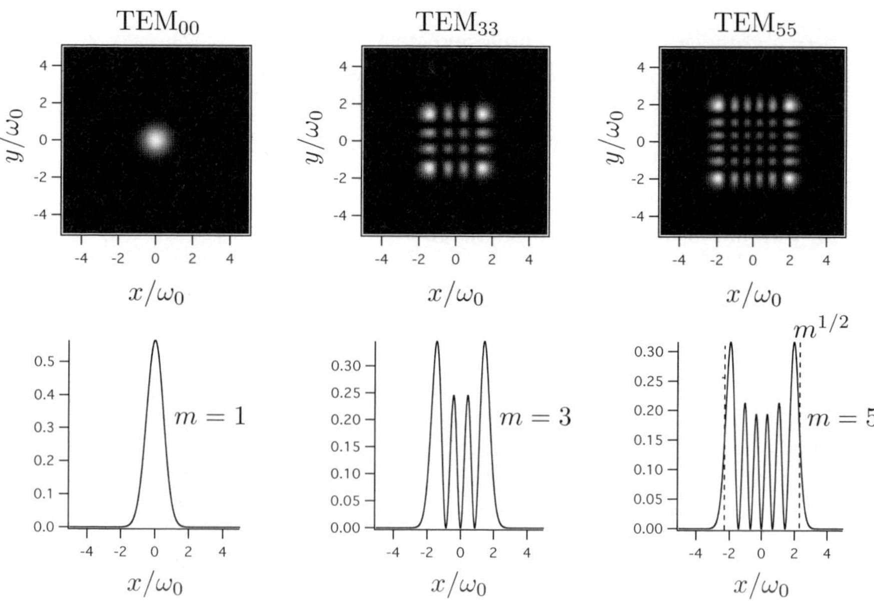

**Fig. 2.5** Three Hermite–Gaussian modes together with their intensity distributions in each dimension.

Since the definition of $q$ and the ABCD rule are not dependent upon the mode number and the mode numbers are not altered by reflection from spherical mirrors, the self-consistancy procedure for determining the possible higher order *cavity modes* is exactly the same as the procedure we have already outlined for the lowest-order Gaussian cavity mode. Thus, for every lowest-order Gaussian mode, there are an infinite number of higher-order modes with the same value of $q$. As we will shortly see, the resonant frequencies will in general be *different* for different modes.

## 2.6 Resonant frequencies

We derived the possible cavity modes by insisting that the field *replicate* itself after one round trip. So far, we have only considered the replication of the field amplitude distribution, determined by the complex beam parameter, $q$. It is also necessary for the *phase* to be the same (modulo integral multiples of $2\pi$) after a round trip in order for the field to constructively interfere with itself after multiple passes (this is actually the only requirement when analyzing the classical Fabry Perot interferometer). From this condition, we can determine the resonant frequencies of the resonator. (In the following, we assume the index of refraction, $n$, is unity.)

If the mirror separation is $d$, the round trip phase change, $\delta$, on axis can be determined from eqn 2.25:

$$\delta = 2kd - 2(n + m + 1)(\tan^{-1}(z_2/z_R) - \tan^{-1}(z_1/z_R)), \tag{2.32}$$

where $z_1$ and $z_2$ are the $z$-coordinates of the two mirrors measured from the waist. The two $\tan^{-1}$ functions can be combined using eqns 2.18, 2.19, and 2.20 together with the following trigonometric identity which holds for any $x_1$ and $x_2$:

$$\tan^{-1} x_1 - \tan^{-1} x_2 = \cos^{-1}\left\{ \frac{1 + x_1 x_2}{\sqrt{(1 + x_1^2)(1 + x_2^2)}} \right\}. \tag{2.33}$$

After a fair amount of algebra, the result is:

$$\tan^{-1}(z_2/z_R) - \tan^{-1}(z_1/z_R) = \cos^{-1} \pm\sqrt{g_1 g_2}, \tag{2.34}$$

where $g_{1,2}$ are defined in eqns 2.10 and 2.11, and the plus is taken when $g_1, g_2 > 0$ while the minus is taken when $g_1, g_2 < 0$. A cavity resonance will occur when the round trip phase shift is an integral multiple of $2\pi$:

$$\delta = q(2\pi) \implies \frac{\omega d}{c} - (n + m + 1)\cos^{-1} \pm\sqrt{g_1 g_2} = q\pi, \tag{2.35}$$

where $q$ is an integer. Solving for the resonant frequencies, $\nu_{nmq}$,

$$\nu_{nmq} = \left( q + (n + m + 1)\frac{\cos^{-1} \pm\sqrt{g_1 g_2}}{\pi} \right) \frac{c}{2d}. \tag{2.36}$$

For geometries which are close to planar, confocal or spherical, the Gouy phase terms are:

$$\frac{\cos^{-1} \pm\sqrt{g_1 g_2}}{\pi} \approx \begin{cases} 0 & : \quad g_1, g_2 \to 1 \quad \text{(near-planar)} \\ 1/2 & : \quad g_1, g_2 \to 0 \quad \text{(near-confocal)} \\ 1 & : \quad g_1, g_2 \to -1 \quad \text{(near-spherical)} \end{cases} \tag{2.37}$$

The spacing between axial modes ($q$ values) is called the *free spectral range* and is given by:

$$\text{Free spectral range: FSR} = \frac{c}{2d}. \tag{2.38}$$

It is interesting to note that the mode spacing (both axial *and* transverse) in a confocal resonator is one half the free spectral range since the Gouy phase term is $1/2$. This geometry exhibits a great deal of degeneracy between transverse and axial modes and has the advantage of providing a spectrum which is largely independent of which modes are excited by the incoming radiation. As a result of this, the confocal geometry is preferred in commercial optical spectrum analyzers since the alignment is relatively uncritical. Its free spectral range is still given by eqn 2.38, although manufacturers often (mistakenly) specify the mode spacing ($c/4d$) instead. The spherical and planar geometries share this decoupling between the spectrum and the alignment, but have the disadvantage of being on the edge of the stability region.

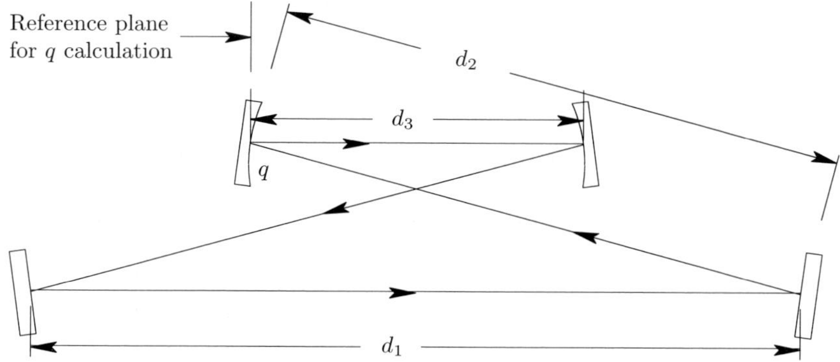

**Fig. 2.6** The symmetrical traveling wave bow-tie cavity.

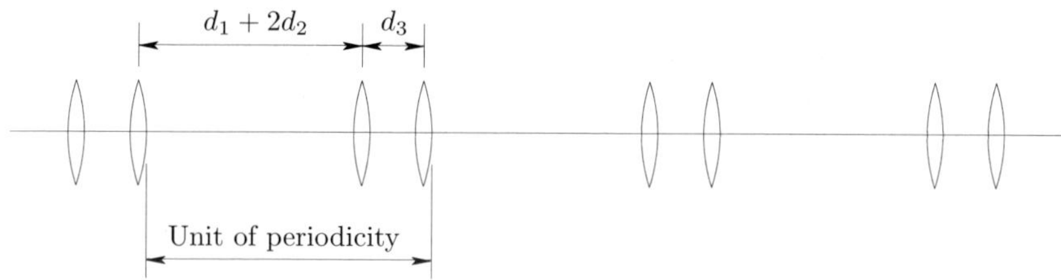

**Fig. 2.7** The lens sequence equivalent of a ring resonator

## 2.7   The traveling wave (ring) cavity

There are a number of ways of constructing a traveling wave cavity; we will describe here the *symmetrical bow-tie configuration* shown in Fig. 2.6. This cavity consists of four mirrors – two curved and two flat – and the mirrors are aligned so that the beam follows the zig-zag path shown in the figure. The curved mirrors have the same radii of curvature, and the configuration is symmetrical about a vertical line halfway between the mirrors. The equivalent sequence of lenses is shown in Fig. 2.7. We will use exactly the same procedure in analyzing the ring cavity as we used for the two-mirror standing wave cavity. If we choose the reference point to be just to the right of the left-hand curved mirror, the round trip ABCD matrix is:

$$
\begin{pmatrix} A & B \\ C & D \end{pmatrix} = \begin{pmatrix} 1 & 0 \\ \frac{-2}{R} & 1 \end{pmatrix} \begin{pmatrix} 1 & d_1 + 2d_2 \\ 0 & 1 \end{pmatrix} \begin{pmatrix} 1 & 0 \\ \frac{-2}{R} & 1 \end{pmatrix} \begin{pmatrix} 1 & d_3 \\ 0 & 1 \end{pmatrix}. \tag{2.39}
$$

From the figures, it should be apparent that there are only two important distances: the long path $d_1 + 2d_2$ and the shorter curved mirror separation, $d_3$. We will again introduce the two dimensionless parameters $g_1$ and $g_2$:

$$
g_1 = 1 - \frac{d_1 + 2d_2}{R}, \tag{2.40}
$$

$$
g_2 = 1 - \frac{d_3}{R}, \tag{2.41}
$$

and write the overall ABCD matrix in terms of them:

$$\begin{pmatrix} A & B \\ C & D \end{pmatrix} = \begin{pmatrix} 1 & 0 \\ \frac{-2}{R} & 1 \end{pmatrix} \begin{pmatrix} 1 & R(1-g_1) \\ 0 & 1 \end{pmatrix} \begin{pmatrix} 1 & 0 \\ \frac{-2}{R} & 1 \end{pmatrix} \begin{pmatrix} 1 & R(1-g_2) \\ 0 & 1 \end{pmatrix}$$

$$= \begin{pmatrix} 2g_1 - 1 & R(-2g_1 g_2 + g_1 + g_2) \\ \frac{-4g_1}{R} & 4g_1 g_2 - 2g_1 - 1 \end{pmatrix}. \tag{2.42}$$

It should come as no surprise that the stability criterion is the same as in the two-mirror case:

$$|A + D| \le 2 \implies |4g_1 g_1 - 2| \le 2 \implies 0 \le g_1 g_2 \le 1. \tag{2.43}$$

Making the indicated substitutions, we get a constraint on both the minimum and maximum value of $d_3$ (unlike the two mirror cavity which has no minimum stable mirror separation):

$$\text{Stability:} \quad R \le d_3 \le \frac{R(d_1 + 2d_2)}{d_1 + 2d_2 - R}. \tag{2.44}$$

The stable range of $d_3$, $\Delta d_3$, can be written down as the following simple expression when $d_1 + 2d_2 \gg R$:

$$\text{Range:} \quad \Delta d_3 \approx \frac{R^2}{d_1 + 2d_2} \quad \text{when} \quad d_1 + 2d_2 \gg R. \tag{2.45}$$

Before carrying out further analysis, we will establish a convention which distinguishes the two waists by using unprimed and primed variables: the unprimed variables refer to the upper waist (between the curved mirrors) and the primed quantities refer to the lower waist. The symmetry of the configuration being studied requires that the two waists be located exactly halfway between the respective pairs of mirrors.

We will use eqn 2.3 to determine the value of $q_1$ ($\equiv 1/q$) at the reference plane. The result is:

$$q_1 = \frac{2g_1(g_2 - 1)}{R(g_1 + g_2 - 2g_1 g_2)} \pm i \frac{2\sqrt{g_1 g_2(1 - g_1 g_2)}}{R(g_1 + g_2 - 2g_1 g_2)}. \tag{2.46}$$

The Rayleigh length, $z_R$, and waist radius, $\omega_0$, for the upper waist are then,

$$z_R = -\frac{\text{Im}\{q_1\}}{|q_1|^2} = \frac{R\sqrt{g_1 g_2(1 - g_1 g_2)}}{2g_1} \tag{2.47}$$

$$\omega_0^2 = \frac{\lambda R\sqrt{g_1 g_2(1 - g_1 g_2)}}{2n\pi g_1}. \tag{2.48}$$

As a consistency check, one can evaluate the distance to the waist $(-Re\{q_1\}/|q_1|^2)$ and confirm that it is indeed $d_3/2$.

The properties of the lower waist are determined by transforming the beam *backwards* past the left curved mirror. The transformation properties of $q_1$ are (eqn 2.1):

$$q_1' = \frac{C + Dq_1}{A + Bq_1} = \frac{2}{R} + q_1, \quad \text{where} \quad \begin{pmatrix} A & B \\ C & D \end{pmatrix} = \begin{pmatrix} 1 & 0 \\ 2/R & 1 \end{pmatrix}. \tag{2.49}$$

Notice that we have used $+R$ instead of $-R$; this is required by the sign conventions for mirror radii when a beam travels *backwards*. The result for $q_1'$ is:

$$q_1' = \frac{2g_2(1 - g_1)}{R(g_1 + g_2 - 2g_1g_2)} \pm i\frac{2\sqrt{g_1g_2(1 - g_1g_2)}}{R(g_1 + g_2 - 2g_1g_2)}. \tag{2.50}$$

The lower Rayleigh length and waist radius are given by:

$$z_R' = -\frac{Im\{q_1\}}{|q_1|^2} = \frac{R\sqrt{g_1g_2(1 - g_1g_2)}}{2g_2} \tag{2.51}$$

$$\omega_0'^2 = \frac{\lambda R\sqrt{g_1g_2(1 - g_1g_2)}}{2n\pi g_2}. \tag{2.52}$$

Again, evaluating $Re\{q_1\}/|q_1|^2$ confirms that the lower waist is a distance $(d_1 + 2d_2)/2$ from the reference plane. An interesting consequence of the preceding calculations is a very simple relation between the two waist radii:

$$\frac{\omega_0}{\omega_0'} = \sqrt{\frac{g_2}{g_1}}. \tag{2.53}$$

Finally, the resonant frequencies are determined by writing down an equation (similar to eqn 2.32) for the round trip phase change in a ring cavity. The result is:

$$\delta = kL - 2(n + m + 1)\left(\tan^{-1}\left(\frac{d_1 + 2d_2}{2z_R'}\right) + \tan^{-1}\left(\frac{d_3}{2z_R}\right)\right). \tag{2.54}$$

Here, $L$ is the total round-trip distance $(= d_1 + 2d_2 + d_3)$. It is interesting (but probably not too surprising) that the two inverse tan functions can be combined to obtain exactly the same result as with the standing wave cavity:

$$\tan^{-1}\left(\frac{d_1 + 2d_2}{2z_R'}\right) + \tan^{-1}\left(\frac{d_3}{2z_R}\right) = \cos^{-1}\pm\sqrt{g_1g_2}, \tag{2.55}$$

with the same treatment of the $\pm$ as in the standing wave case. The resonant frequencies are then,

$$\nu_{nmq} = \left(q + (n + m + 1)\frac{\cos^{-1}\sqrt{g_1g_2}}{\pi}\right)\frac{c}{L}, \tag{2.56}$$

and the *free spectral range* (FSR) is,

$$\text{FSR} = \frac{c}{L}. \tag{2.57}$$

If one uses the appropriate definitions for the $g_{1,2}$ parameters and round-trip distances, all of the important relations for the standing wave and symmetric ring cavities, except for the waist sizes, are identical. This should not be too surprising in view of the similarity of the ABCD matrices for the two cavity types. We should note that the ring cavity symmetry requirement only mandates that the curved mirror radii be the same: all of the results hold for cavities with different diagonal lengths and identical radii provided that one changes the long path distance appropriately (from $d_1 + 2d_2$ to $L - d_3$). Finally, we will summarize the important properties for the two cavity types in Table 2.1.

**Table 2.1** Summary of cavity properties.

| Property | Standing wave | Symmetric ring |
|---|---|---|
| $g_1$ | $1 - \dfrac{d}{R_1}$ | $1 - \dfrac{d_1 + 2d_2}{R}$ |
| $g_2$ | $1 - \dfrac{d}{R_2}$ | $1 - \dfrac{d_3}{R}$ |
| Path length | $2d$ | $L = d_1 + 2d_2 + d_3$ |
| Waist(s) | $\omega^2 = \left(\dfrac{\lambda d}{n\pi}\right)\dfrac{\sqrt{g_1 g_2(1-g_1 g_2)}}{g_1 + g_2 - 2g_1 g_2}$ | $\omega_0^2 = \left(\dfrac{\lambda R}{2n\pi}\right)\dfrac{\sqrt{g_1 g_2(1-g_1 g_2)}}{g_1}$ |
| | | $\omega_0'^2 = \left(\dfrac{\lambda R}{2n\pi}\right)\dfrac{\sqrt{g_1 g_2(1-g_1 g_2)}}{g_2}$ |
| $\nu_{nmq}$ | $\left(q + (n+m+1)\dfrac{\cos^{-1}\pm\sqrt{g_1 g_2}}{\pi}\right)\dfrac{c}{2d}$ | $\left(q + (n+m+1)\dfrac{\cos^{-1}\pm\sqrt{g_1 g_2}}{\pi}\right)\dfrac{c}{L}$ |
| FSR | $\dfrac{c}{2d}$ | $\dfrac{c}{L}$ |

A typical beam profile in a ring cavity is shown in Fig. 2.8. As can be seen, the waist between the curved mirrors is somewhat smaller than that between the flat mirrors. The former waist would normally be the location of a nonlinear crystal when the cavity is used for nonlinear enhancement since the fields will be much larger at the smaller waist. A plot of the waist radii appears in Fig. 2.9. As one approaches the region of instability, the small waist radius decreases very rapidly.

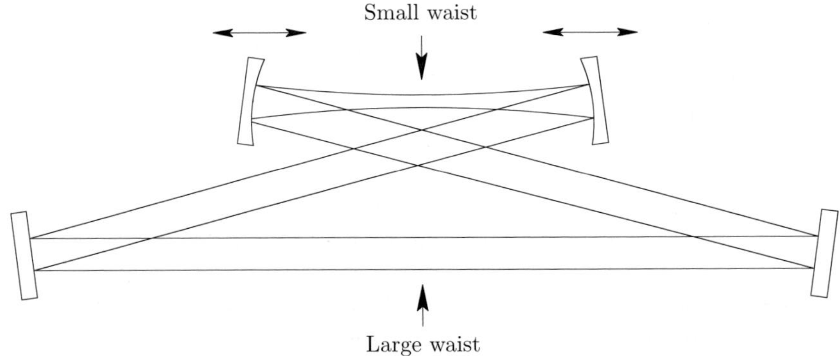


**Fig. 2.8** A typical beam profile in a ring cavity showing the small waist between the curved mirrors and the larger waist between the flat mirrors.

## 2.8 Astigmatism in a ring cavity

The reader will have undoubtedly noticed that we have violated one of our assumptions in analyzing a ring cavity: the ring cavity geometry is no longer cylindrically symmetric. From Fig. 2.6, it can be seen that the only place where this matters is

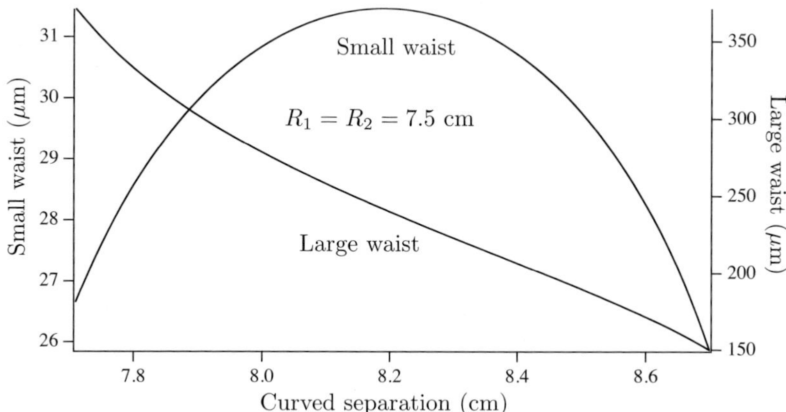



**Fig. 2.9** The small and large waists as a function of the curved mirror separation. The instability region is close to the boundaries of this plot, where the small waist is shrinking rapidly.

where the beam is reflected from the two *curved* mirrors (reflection from the flat mirrors merely *folds* the geometry but introduces no aberrations). The off-axis reflection from the curved mirrors introduces the geometrical aberration called *astigmatism*.

The aberration of astigmatism is illustrated in Fig. 2.10. Astigmatism occurs only for off-axis rays and in describing such rays it is useful to define two planes: the *tangential plane*, which is parallel to the *plane of incidence* (the plane containing the axis of the system and the off-axis ray) and the *sagittal plane*, which is perpendicular to the tangential plane and also contains the off-axis ray. Astigmatism is due to the fact that, for an off-axis bundle of rays, the tangential rays will encounter a different geometry from the sagittal rays.

In the systems described in this book, the causes of astigmatism can always be considered to be optical elements which are tilted away from the symmetry axis of the system so that they break the cylindrical symmetry. We will only discus *orthogonal*, or *simple, astigmatism* in which all of the tilt angles for a given optical system are in one of two orthogonal planes. More general astigmatism is beyond the scope of our treatment.

We treat astigmatism by considering each transverse coordinate separately but use exactly the same procedures as were used for the cylindrically symmetrical case. For orthogonal astigmatism, one can show that the paraxial wave equation (eqn 1.5) will be satisfied when individual paraxial equations for each transverse coordinate, identical in form to the original (with the same value of $k$), are satisfied. Each coordinate will therefore have its own ABCD matrix and its own vale of $q$; the beam profile will now be *elliptical* since the two $\omega$ values are different (the sagittal and tangential wavefront "radii" will also be different). The Gouy phase terms for the transverse coordinates will be different from each other and the phase shift for each coordinate will be $\pi/2$ (the total for both coordinates will still be $\pi$).

We will consider the astigmatism due to both tilted mirrors and tilted crystals (at

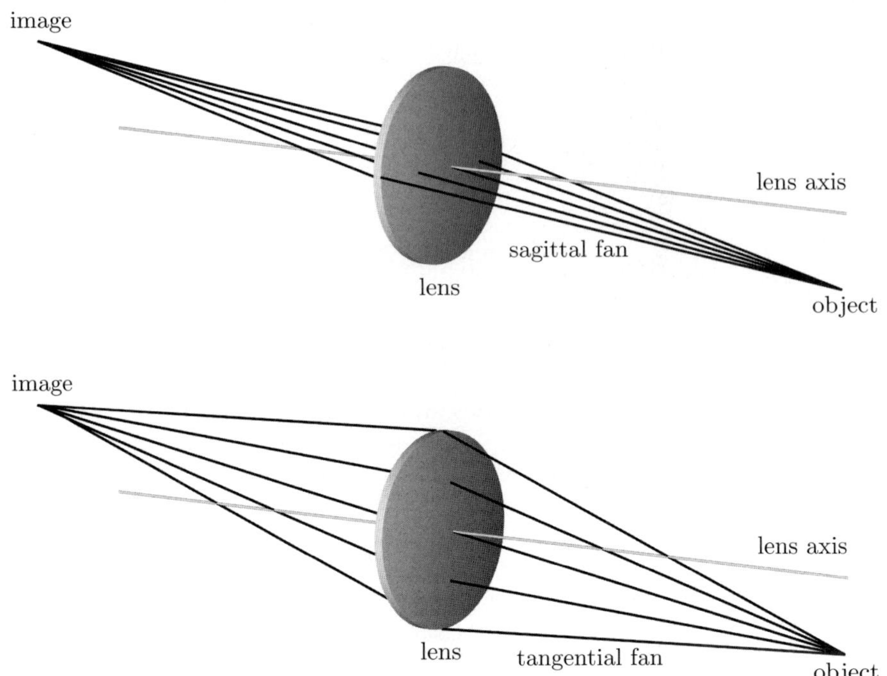


**Fig. 2.10** An illustration of the aberration of *astigmatism*, in which the tangential and sagittal rays need to be treated differently.

Brewster's angle). The tilted mirror with radius of curvature $R$ is shown in Fig. 2.11. It is well known (see, for example, Jenkins and White (1957)) that, for off-axis reflection of a mirror tilted at angle $\theta$, the *effective radius of curvature* of the mirror (and therefore the focal length) seen by a tangential bundle of rays will be different from that seen by a sagittal bundle. The two focal lengths are:

$$\text{Tangential focal length} = \frac{R\cos\theta}{2} \tag{2.58}$$

$$\text{Sagittal focal length} = \frac{R}{2\cos\theta}. \tag{2.59}$$

The tangential and sagittal ABCD matrices are:

$$M_T = \begin{pmatrix} 1 & 0 \\ \frac{-2}{R\cos\theta} & 1 \end{pmatrix}, \qquad M_S = \begin{pmatrix} 1 & 0 \\ \frac{-2\cos\theta}{R} & 1 \end{pmatrix}. \tag{2.60}$$

A crystal of length $l$ and index of refraction, $n$, which is tilted by angle $\theta$ from normal incidence is shown in Fig. 2.12. One can show that the *effective length* of this crystal (the *length* that appears in the ABCD translation matrix, not the *optical length*) is different for tangential and sagittal rays. The resulting ABCD matrices,

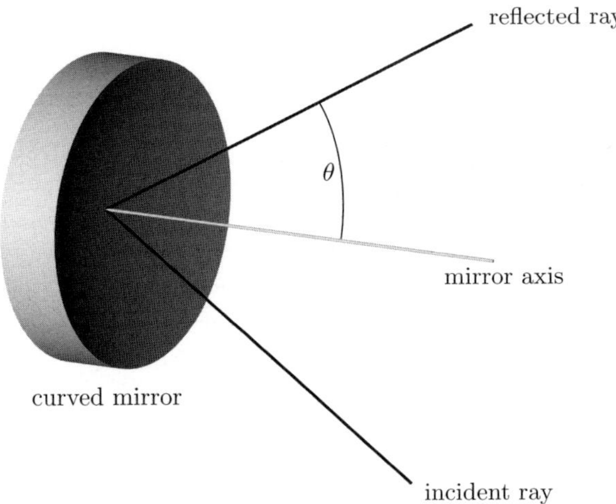



**Fig. 2.11** Off-axis reflection from a spherical mirror.

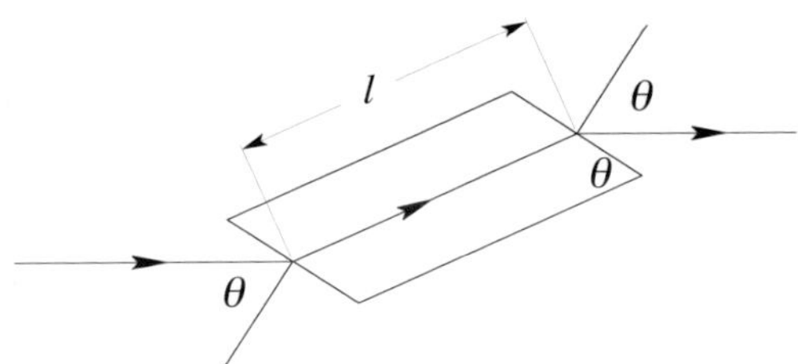



**Fig. 2.12** Off-axis transmission through a crystal.

when $\theta$ is *Brewster's angle*, are:

$$M_T = \begin{pmatrix} 1 & l/n^3 \\ 0 & 1 \end{pmatrix} \tag{2.61}$$

$$M_S = \begin{pmatrix} 1 & l/n \\ 0 & 1 \end{pmatrix}. \tag{2.62}$$

(Note: some treatments yield more complicated expressions since they use the *plate thickness* rather than the *physical length* of the transmitted ray and these differ by a factor equal to the cosine of the angle of the refracted ray.)

Ring cavities are used either for laser feedback or for nonlinear frequency synthesis. In either case, a transparent medium is placed inside the cavity. For reasons which will be discussed in the next chapter, it is highly desirable to reduce the reflection (and absorption) losses of the medium. This can be done in two ways: by tilting the medium so that its surfaces are at Brewster's angle for a particular polarization or by using a medium with antireflection coated surfaces. The Brewster approach allows one to *compensate* the astigmatism which will be an inescapable result of tilted cavity mirrors. The compensation can circularize the cavity mode between flat mirrors, which aids in coupling into the cavity, or it can circularize the mode *inside the crystal*, which might improve the efficiency of the non-linear process for whose enhancement the cavity has been constructed.

In addition to introducing the possible cavity coupling problems mentioned above, astigmatism will narrow the stability range of the cavity. This is shown in Fig. 2.13, where it is evident that the astigmatism results in a different range of stable curved mirror spacings in the $x$ and $y$ directions. The set of mirror spacings which results in both $x$ and $y$ stability is reduced.

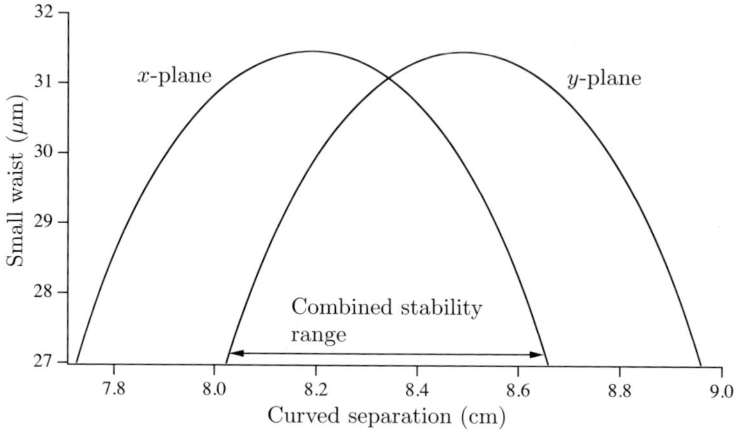


**Fig. 2.13** The reduction in the range of stable mirror spacings due to astigmatism.

## 2.9 Mode matching

It is generally desirable to excite only the lowest (Gaussian) mode in a cavity by judicious choice of the parameters of the input beam. In general, the coupled input beam can be expanded in an infinite series of cavity modes (which form a complete set) and each mode will appear with an amplitude given by the coefficient in the expansion. The problem becomes simplified when the input beam is a Gaussian and the desired cavity mode is the lowest: the internal beam is then completely described by the complex beam parameter and it suffices to match this parameter in the input beam. If one knows the size and location of the cavity mode waist (usually the lower waist in a ring cavity), the mode matching problem becomes one of generating an input beam with the same waist radius and ensuring that the waists are at the same location. This can be done with one or two lenses.

The single lens approach is illustrated in Fig. 2.14. We have two waists: $\omega_1$ is the waist of the input beam and $\omega_2$ is the cavity waist. We first define the quantity $f_0$:

$$f_0 = n\pi\omega_1\omega_2/\lambda = \sqrt{z_{R1}z_{R2}}. \qquad (2.63)$$

We then choose a lens whose focal length $f > f_0$ and adjust the distances $d_1$ and $d_2$ to satisfy:

$$d_1 = f \pm \frac{\omega_1}{\omega_2}\sqrt{f^2 - f_0^2} \qquad (2.64)$$

$$d_2 = f \pm \frac{\omega_2}{\omega_1}\sqrt{f^2 - f_0^2}. \qquad (2.65)$$

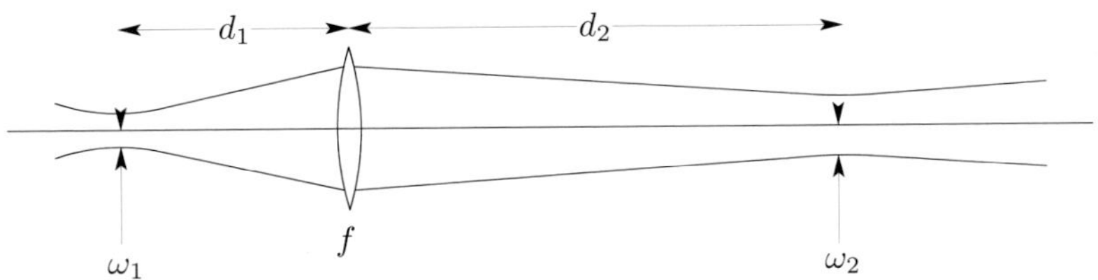

**Fig. 2.14** Mode matching with a single lens.

The disadvantage of this approach is that for a given lens, the distances are fixed and for fixed locations of the two waists, it is not that easy in the usual laboratory environment to find a lens that will satisfy both equations.

A much more flexible approach to mode matching is to use two lenses. This situation is illustrated in Fig. 2.15. One can first evaluate the ABCD matrix up to lens 2:

$$\begin{pmatrix} A & B \\ C & D \end{pmatrix} = \begin{pmatrix} 1 & 0 \\ \frac{-1}{f_2} & 1 \end{pmatrix} \begin{pmatrix} 1 & d_3 \\ 0 & 1 \end{pmatrix} \begin{pmatrix} 1 & 0 \\ \frac{-1}{f_1} & 1 \end{pmatrix} \begin{pmatrix} 1 & d_1 \\ 0 & 1 \end{pmatrix}. \qquad (2.66)$$

Then the complex beam parameter at waist 2 is given by:

$$q_2 = \frac{Aq_1 + B}{Cq_1 + D} + d_2, \qquad (2.67)$$

and the desired beam properties at waist 2 are:

$$R_2 = 1/Re\{1/q_2\} \quad \text{and} \quad \omega_2 = \sqrt{\frac{-\lambda}{n\pi Im\{1/q_2\}}}. \qquad (2.68)$$

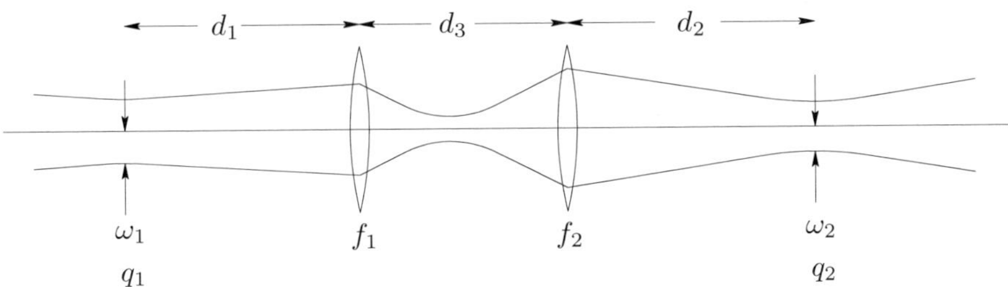

**Fig. 2.15** Mode matching with two lenses.

Although the calculations needed for two-lens mode matching are not too difficult anyway, there is a very simple procedure which can quickly provide an approximate solution which is usually adequate. This approach is illustrated in Fig. 2.16.

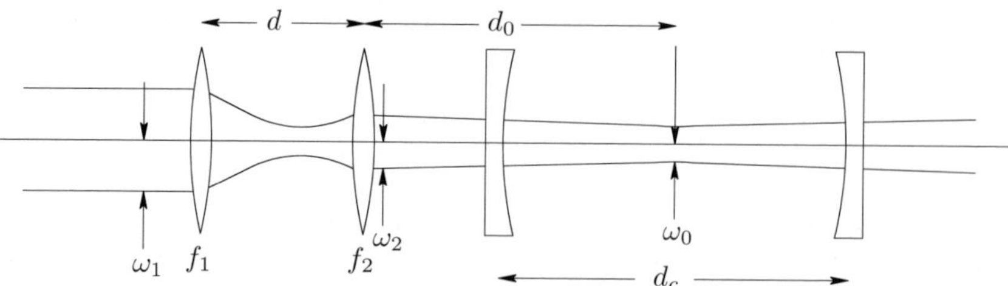

**Fig. 2.16**  Illustration of a simplified approach to two-lens mode matching.

The simplified two-lens procedure is as follows:

**Procedure**                                                             **Example**

1. Determine $\omega_0$ and $z_R$ from the cavity equations.

(Confocal) $d_c = 10$ cm, $\lambda = 1$ $\mu$m, $\omega_0 = 126$ $\mu$m, $z_R = 5$ cm

2. Given $d_0$, the distance from waist to lens 2, determine $\omega_2$ using $\omega_2 = \omega_0\sqrt{1 + (d_0/z_R)^2}$.

$d_0 = 10$ cm, $\omega_2 = 282$ $\mu$m

3. Given input spot size ($\omega_1$), find two lenses whose focal lengths are in ratio $\omega_1/\omega_2$.

$\omega_1 = .05$ cm, ratio $\approx 1.8$, $f_1 = 1.8$ cm, $f_2 = 1$ cm

4. Separate lenses by slightly more than $f_1 + f_2$ so waist is at center of cavity.

$f_1 + f_2 = 2.8$ cm

## 2.10  Beam quality characterization: the $M^2$ parameter

The beams considered in this chapter and the previous one have been idealizations: either a pure Gaussian beam or a single-mode beam described by a Hermite Gaussian function. These idealized beams are rarely found in the real world. A Gaussian beam has the property of being a *minimum uncertainty* function: the product of the beam width and the diffraction angle is the minimum possible. For this reason, a Gaussian beam is in some sense an *ideal, diffraction-limited laser beam*. We seek a simple way of characterizing the departure of the beam characteristics (mainly the divergence) of an arbitrary beam from those of an ideal Gaussian beam. We will show that a very reasonable way of doing this is to use the so-called $M^2$ parameter.

Before describing the $M^2$ parameter, we should note that a measurement of the intensity profile at a *single place* is a completely inadequate way of characterizing a laser beam. There are illustrative examples in the literature of beam profiles which are almost indistinguishable from that of a Gaussian beam but actually contain *zero* TEM$_{00}$ (Gaussian) contribution. Such a beam might look like a Gaussian beam at one place, but it will *diverge much more rapidly* than an actual Gaussian beam. Two conclusions can be drawn from this. First, a beam is properly characterized by *both* its intensity profile and its divergence. Second, if a profile measurement is used for beam characterization, it must be done at *more than one place along the beam path* so that some measure of the divergence is obtained.

To obtain some motivation for the $M^2$ measurement, we will first consider the one-dimensional behavior of two actual beams: the diffracted beam from a plane wave which encounters a rectangular slit and a TEM$_{00}$ laser beam. A slit of width $2a$ will have a *far-field* electric field dependence which is given by

$$\text{Rectangular slit:} \quad E(x, z) \approx \frac{\sin(2\pi a x / z\lambda)}{2\pi a x / z\lambda}, \tag{2.69}$$

where $z$ is the propagation direction and $x$ is the transverse coordinate. It will be useful to examine the $z$-dependence of the *product of the near-field and far-field beam widths*. For this case, this product is

$$\text{Rectangular slit:} \quad \Delta x_0 \times \Delta x(z) = 0.5 \times \lambda z, \tag{2.70}$$

where $\Delta x_0 = a$ is the near-field half-width (the slit half-width) and $\Delta x(z)$ is the distance from $x = 0$ to the first null in the diffracted pattern. The Gaussian beam width as a function of $z$ is given by

$$\text{Gaussian beam:} \quad \omega^2(z) = \omega_0^2 \left( 1 + \left( \frac{z}{z_R} \right)^2 \right), \tag{2.71}$$

where $z = 0$ at the beam waist. The product of the near-field width $\omega_0$ and the far-field width $\omega$ is

$$\text{Gaussian beam:} \quad \omega_0 \times \omega(z) = \frac{1}{\pi} \times \lambda z, \tag{2.72}$$

where we have made the far-field measurement at many Rayleigh lengths from the waist, so that $\omega \gg \omega_0$. For these two cases, the product of the near-field and far-field widths is some constant times $\lambda z$. This numerical constant is a measure of the departure of the beam from the ideal Gaussian behavior: the larger it is, the greater is the departure.

There are a variety of definitions of beam width: the $1/e$ width, the width between first nulls, etc. Without getting too far into the justification of one width over another, we note that the $M^2$ parameter uses the *second moment* or *variance* widths, $\sigma_x$ and $\sigma_y$, defined by

$$\sigma_x^2 = \frac{\int_{-\infty}^{\infty} (x - x_0)^2 I(x, y) dx dy}{\int_{-\infty}^{\infty} I(x, y) dx dy} \tag{2.73}$$

$$\sigma_y^2 = \frac{\int_{-\infty}^{\infty} (y - y_0)^2 I(x, y) dx dy}{\int_{-\infty}^{\infty} I(x, y) dx dy}, \tag{2.74}$$

where $I(x, y)$ is the beam *intensity*, $x, y$ are the transverse coordinates and $x_0, y_0$ are at the *centroid* of the intensity function. It can be shown that the widths defined in this way obey a formula similar to that for a Gaussian beam

$$\sigma_x^2 = \sigma_{0x}^2 + \sigma_{\theta_x}^2 \times z^2 \tag{2.75}$$

$$\sigma_y^2 = \sigma_{0y}^2 + \sigma_{\theta_y}^2 \times z^2, \tag{2.76}$$

where $z = 0$ at the waist and $\sigma_{\theta_{x,y}}$ are the variances of the angular divergence of the beam in the $x$ and $y$ directions. It appears that these equations are rigorously satisfied only for the second moment definition of the beam width. For a Gaussian beam, $\sigma_x = \omega_x/2, \sigma_y = \omega_y/2$. In order to more conveniently compare an arbitrary beam to a *standard* Gaussian beam, the following width definitions are used:

$$W_x \equiv 2\sigma_x \tag{2.77}$$

$$W_y \equiv 2\sigma_y. \tag{2.78}$$

We are now in a position to define the $M^2$ parameter. An arbitrary beam has the divergence properties described by the following equations:

$$W_x^2(z) = W_{0x}^2 \left( 1 + M_x^4 \times \left( \frac{z}{Z_{Rx}} \right)^2 \right) \tag{2.79}$$

$$W_y^2(z) = W_{0y}^2 \left( 1 + M_y^4 \times \left( \frac{z}{Z_{Ry}} \right)^2 \right), \tag{2.80}$$

where $Z_{Rx}$ and $Z_{Ry}$ are *generalized Rayleigh lengths* defined by

$$\begin{aligned} Z_{Rx} &= \frac{n\pi W_{0x}^2}{\lambda} \\ Z_{Ry} &= \frac{n\pi W_{0y}^2}{\lambda} \end{aligned} \tag{2.81}$$

and $W_{0x}, W_{0y}$ are the widths at $z = 0$. In analogy with a Gaussian beam, the near-field far-field products are

$$W_{0x} \times W_x(z) \approx M_x^2 \times \frac{z\lambda}{n\pi} \tag{2.82}$$

$$W_{0y} \times W_y(z) \approx M_y^2 \times \frac{z\lambda}{n\pi}, \tag{2.83}$$

where the far-field measurements are taken many "Rayleigh" lengths from the waist. Note that $M_{x,y}^2 = 1$ for a Gaussian beam and is greater than unity for non-Gaussian beams (since the Gaussian beam has the *minimum uncertainty* property). The $M^2$ parameters are a measure of "how many times diffraction limited" an arbitrary beam is, or, equivalently, how much faster does a beam diverge than an equivalent Gaussian beam with the same "waist" size. Note that $M^2$ has little to do with the apparent shape of a laser beam: an astigmatic beam with an elliptical shape can still have a low $M^2$ since the latter measures the deviation from a diffraction-limited beam in only *a single plane* containing the propagation direction and one transverse coordinate. A practical means for measuring $M^2$ is left to the exercises.

## 2.11   Further reading

This chapter discusses the application of Gaussian beam theory to the geometrical aspects of standing wave and symmetrical ring cavities. Much of the material recommended in the previous chapter will be applicable to the design of cavities, particularly the pioneering paper by Kogelnik and Lee (1966). Some of the ring cavity theory is original to the author of this book, but excellent descriptions of ring cavities can be found in the early papers by members of the Boulder, Colorado NIST group, who used these cavities to enhance the generation of UV light from continuous wave visible lasers. (See, for example, Hemmati (1983).)

## 2.12 Problems

(2.1) Due to manufacturing errors, a pair of mirrors (whose nominal radii should be 200 mm) have radii of 203 mm and 197 mm. There will be a problem if these mirrors are used in a "confocal" cavity with a mirror separation of 200 mm. What is the problem and what are the possible mirror spacings which solve it?

(2.2) You are given a flat plate of glass having index of refraction $n$ and parallel surfaces whose separation is $t$. This object is tilted with respect to the $z$-axis so that the angle between its *normal* and the $z$-axis is $\theta$. As a result of the tilt, the plate exhibits *astigmatism*: the behavior of a paraxial "fan" of rays in the tangential plane (the one containing the normal and $z$-axis) is different from the behavior of such a "fan" in the sagittal plane (the plane which contains the $z$-axis and is perpendicular to the tangential plane). Calculate the ABCD matrices between the entrance and exit points of the plate for both planes. Ignore the parallel displacement of the beam by the tilted plate: assume that the $z$-axis for the ray exiting the plate is displaced from the $z$-axis of the entering ray (and assume the rays make only small angles with respect to both axes).

(2.3) You are given a resonator consisting of a convex mirror ($R_1 < 0$) and a concave mirror ($R_2 > 0$) separated by a distance $d$. Find the values of $d$ for which the resonator is stable both for $|R_1| > R_2$ and $|R_1| < R_2$.

(2.4) Consider a near-planar symmetric resonator made with mirrors of radius $R$ and separation $d \ll R$. Obtain an approximate expression for the spot sizes on the mirrors and at the waist. Use your results to calculate the spot sizes when $R = 8$ m and $d = 1$ m at a wavelength of 514 nm.

(2.5) Consider a near concentric resonator composed of two mirrors of radius $R$ and separation $2R - \Delta d$. Give an approximate expression for the spot sizes on the mirrors and at the waist. Compare this to the result of problem 2.4.

(2.6) Using just a well-corrected lens and a device for measuring the spot size of a beam, describe a method to measure, *very approximately*, the $M^2$ value for an arbitrary beam.

(2.7) The mode-matching techniques for standing wave cavities described in the text did not take account of the refractive power of the input mirror which acts like a weak negative lens. Assume that one has an approximately confocal cavity with mirror separation $d$. Approximately how far is the apparent waist position shifted by refraction from the input mirror? (Treat the input mirror as a thin plano-concave lens with refractive index $n$.) Assume $n = 1.5$ and determine the actual shift in terms of $d$.

# 3

# Energy relations in optical cavities

## 3.1 Introduction

In the previous chapter, we discussed the cavity field distributions which are the eigenmodes of the cavity. We required that the field distribution replicate itself after one round trip but said nothing about how many round trips the propagating field would make before it was attenuated nor did we discuss how the field would be injected into the cavity (or extracted from the cavity). The discussion of cavity eigenmodes depended only on the resonator geometry and the wavelength of the radiation. In discussing the build-up of the field in the cavity, we need to consider everything which affects the power and energy of the internal field. Thus we need to be concerned with both the dissipative losses and the Fresnel reflection losses from the mirrors and intracavity elements. The study of cavity losses is intimately connected with the *frequency resolution* of cavities since it determines how many times the beam can be made to interfere with itself. The discussion of energy relations in cavities is analogous to the study of *dynamics* in classical mechanics, while our previous study of cavity field configurations is analogous to the study of *kinematics*.

We will begin with a simple and elegant derivation of the relation between the field transmission and reflection coefficients based upon time reversal symmetry and, using this, proceed to derive the size and frequency dependence of the internal fields for standing wave and travelling wave cavities. We will then discuss resonator $Q$, finesse, photon lifetime and *impedance matching* into a cavity. Finally we will describe an elegant technique for coupling radiation into a cavity using an intracavity transparent wedge and will conclude with a discussion of some practical matters, such as the experimental determination of certain cavity parameters.

## 3.2 Reflection and transmission at an interface

An elegant derivation of the relation between the field transmission and reflection coefficients was made by the British physicist Sir George Gabriel Stokes (1819–1903). He assumed that there was *no dissipation* at the interface and therefore the process obeyed *time-reversal invariance*. His argument is illustrated in Fig. 3.1, which depicts a wave incident on a transparent surface.

Assume that the amplitude transmission coefficients are $t, t'$ and the reflection coefficients are $r, r'$, where the unprimed quantities are used when one approaches the interface from above and the primed quantities are used when the interface is approached from below. The three waves are shown on the left in the figure. The time-reversed case appears in the center plot. When we reverse time, we need to include three additional waves: the reflection from the bottom wave, and the transmission of both the bottom and top waves. All of the waves are shown in the plot on the right. In order for the right-hand plot to be equivalent to the time-reversed center plot, the two waves in the upper left must sum to $E$ and the two waves in the lower left must sum to zero. Thus,

$$E = Ett' + Err \quad \text{and} \quad 0 = Ert + Er't, \tag{3.1}$$

from which we conclude that

$$tt' + r^2 = 1 \quad \text{and} \quad r = -r.' \tag{3.2}$$

This shows that there is a sign change between the *amplitude* reflection coefficients on opposite sides of an interface and that the *intensity* transmission and reflection coefficients ($T$ and $R$) follow the usual rule (since $t = t'$):

$$R + T = 1, \quad \text{where} \quad R = r^2, T = t^2. \tag{3.3}$$

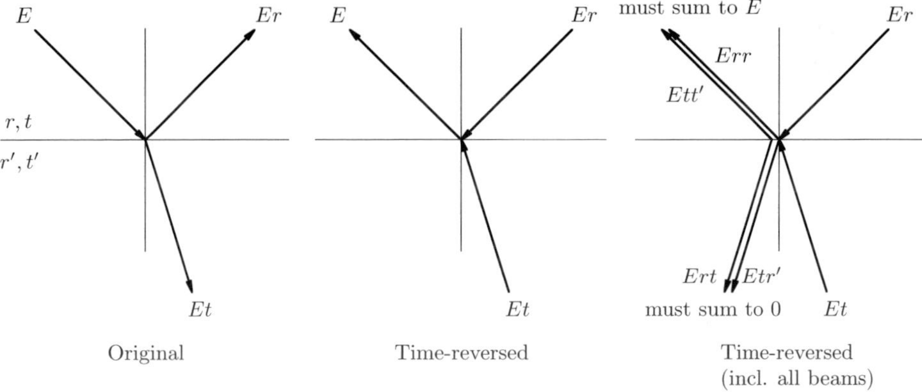


**Fig. 3.1** Original and time reversed behavior at interface.

## 3.3   Reflected fields from standing wave cavity

We will use the above relations to determine the electric field of the wave which is reflected from a two-mirror standing wave cavity. We will assume that the reflection and transmission from the mirrors are free from dissipative losses and will place a potentially lossy medium inside the cavity. The latter will model the Fresnel losses from an intracavity medium (such as a nonlinear crystal) as well as all of the *dissipative* losses of both the mirrors and this medium. The medium will be characterized by the *field* transmission coefficient, $t$. The mirrors have reflection and transmission coefficients, $r_1, r_2$ and $t_1, t_2$.

The total reflected field is calculated by summing the fields from all possible internal passes. The first three passes are shown in Fig. 3.2, where we assume a round trip phase

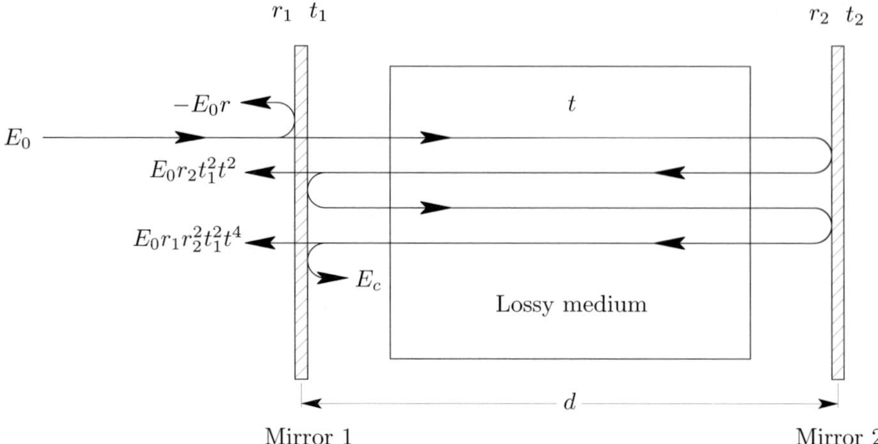



**Fig. 3.2** Reflected fields from cavity due to several round-trips inside the cavity.

delay of $\delta$. We have used eqn 3.2, which requires that the sign from the first (external) reflection be opposite to the signs of the internal reflections. If we sum the first three passes, we obtain a reflected field, $E_r$, given by

$$
\begin{aligned}
E_r &= -E_0 r_1 + E_0 r_2 t_1^2 t^2 e^{-i\delta} + E_0 r_1 r_2^2 t_1^2 t^4 e^{-2i\delta} + \cdots \\
&= -E_0 r_1 + \frac{E_0 t_1^2}{r_1} \left( r_1 r_2 t^2 e^{-i\delta} + (r_1 r_2 t^2 e^{-i\delta})^2 + \cdots \right) \\
&= E_0 \frac{r_2 t^2 e^{-i\delta} - r_1}{1 - r_1 r_2 t^2 e^{-i\delta}},
\end{aligned}
\tag{3.4}
$$

where we have summed the geometrical series and used the relation $r_1^2 + t_1^2 = 1$.

## 3.4 Internal (circulating) field in a standing wave cavity

Resonant cavities are often used to enhance the power incident on an intracavity crystal for nonlinear frequency synthesis. To aid in the analysis of such a system, we will calculate the internal, circulating field to the left of the internal medium in a standing wave cavity; the remaining case of the transmitted field will be left as an exercise for the reader. We are only interested in the fields propagating in *the same direction as the incident wave*, since most applications (such as nonlinear optics enhancement) involving the internal field only work for a *traveling wave*. The actual field, of course, inside the cavity will be a *standing wave*: the superposition of two counter-propagating waves. As we did with the reflected fields, we sum the first three contributions to the internal field and use the formula for an infinite geometrical series to obtain the circulating field, $E_c$:

$$
\begin{aligned}
E_c &= E_0 t_1 + E_0 r_1 r_2 t_1 t^2 e^{-i\delta} + E_0 r_1^2 r_2^2 t_1 t^4 e^{-2i\delta} + \cdots \\
&= E_0 t_1 \left( 1 + r_1 r_2 t^2 e^{-i\delta} + (r_1 r_2 t^2 e^{-i\delta})^2 + \cdots \right) \\
&= \frac{E_0 t_1}{1 - r_1 r_2 t^2 e^{-i\delta}}.
\end{aligned}
\tag{3.5}
$$

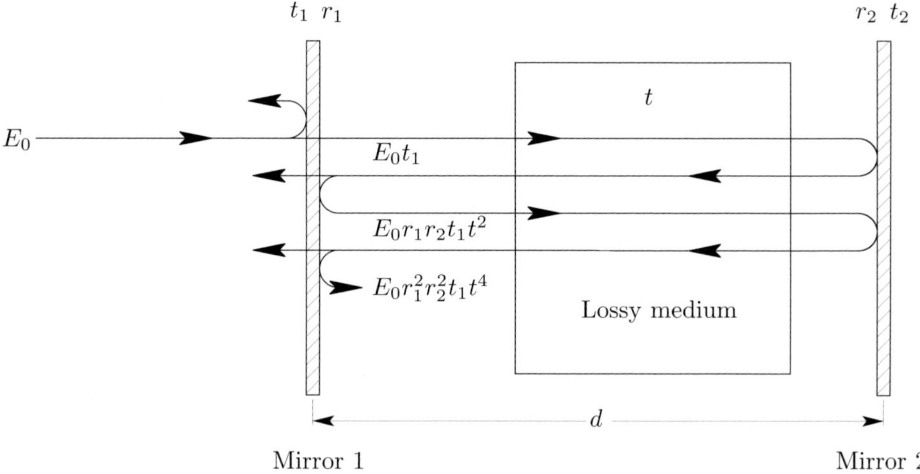


**Fig. 3.3** Internal field from cavity due to several internal round-trips.

## 3.5 Reflected and internal intensities

To calculate the intensities of the reflected and internal waves, we use the well-known quadratic dependence of the intensities upon the fields,

$$\frac{I_{r,c}}{I_0} = \left|\frac{E_{r,c}}{E_0}\right|^2 , \tag{3.6}$$

to obtain

$$\text{Field:} \quad E_r = E_0 \frac{r_m e^{-i\delta} - r_1}{1 - r_1 r_m e^{-i\delta}} \tag{3.7}$$

$$\text{Intensity:} \quad I_r = I_0 \frac{(r_1 - r_m)^2 + 4 r_1 r_m \sin^2 \delta/2}{(1 - r_1 r_m)^2 + 4 r_1 r_m \sin^2 \delta/2} \tag{3.8}$$

for the reflected field and intensity and

$$\text{Field:} \quad E_c = E_0 \frac{t_1}{1 - r_1 r_m e^{-i\delta}} \tag{3.9}$$

$$\text{Intensity:} \quad I_c = I_0 \frac{t_1^2}{(1 - r_1 r_m)^2 + 4 r_1 r_m \sin^2 \delta/2} \tag{3.10}$$

for the circulating field and intensity. In these expressions, we have defined the quantity $r_m \equiv r_2 t^2$ and used the relation $\cos \delta = 1 - 2 \sin^2 \delta/2$.

As mentioned above, the internal intensity is for the wave propagating in the same direction as the incident wave. The internal field will be a standing wave whose *power* will have a $\cos^2 kz$ dependence and whose *average power* will be twice that derived from our expression for the internal intensity, since we are ignoring the wave propagating in the other direction.

## 3.6  The resonant character of the reflected and circulating intensities

We have already found the relation between the round trip phase ($\delta$) and the frequency of the beam (eqn 2.32). In the following, it will be simpler to consider the phase to be the independent variable and to investigate the behavior of the intensities near a *resonance*. We observe that the intensity formulas are periodic functions of $\delta$ with a period of $2\pi$. Thus, in terms of $\delta$, the *free spectral range* is $2\pi$.

The circulating intensity (eqn 3.10) has a maximum when $\delta = 0$ (modulo integral multiples of $2\pi$). The *half maximum intensity* occurs at a phase ($\delta_{1/2}$) which is a solution of:

$$(1 - r_1 r_m)^2 = 4 r_1 r_m \sin^2 \delta_{1/2}/2 \tag{3.11}$$

whose solution for $\delta_{1/2} \ll 1$ is:

$$\delta_{1/2} = 2\sin^{-1}\left(\frac{1 - r_1 r_m}{2\sqrt{r_1 r_m}}\right) \approx \frac{1 - r_1 r_m}{\sqrt{r_1 r_m}}. \tag{3.12}$$

(We will shortly justify our assumption that $\delta_{1/2} \ll 1$.) The full width at half maximum (FWHM), $\Delta\nu_{1/2}$, of the circulating field at resonance is equal to:

$$\text{FWHM} \equiv \Delta\nu_{1/2} = 2\delta_{1/2} = \frac{2(1 - r_1 r_m)}{\sqrt{r_1 r_m}} \tag{3.13}$$

One of the several definitions of the *finesse* (symbolized by $\mathcal{F}$) is that it is the ratio of the free spectral range of the cavity to the full width at half maximum of the *transmitted* (and internal) intensity resonances. Thus,

$$\text{Finesse: } \mathcal{F} = \frac{\text{free spectral range}}{\Delta\nu_{1/2}} = \frac{\pi\sqrt{r_1 r_m}}{1 - r_1 r_m} \quad (\text{FSR} = 2\pi). \tag{3.14}$$

The approximation ($\delta_{1/2} \ll 1$) made in deriving eqn 3.12 is justified by the fact that most useful cavities have a finesse $\gg 1$ and therefore $\delta_{1/2} \ll \pi$. For a lossless, empty cavity ($t = 1$) with identical mirrors ($r_1 = r_2$), eqn 3.14 agrees with the formula for finesse found in many optics textbooks:

$$\mathcal{F} = \frac{\pi\sqrt{r_1^2}}{1 - r_1^2} = \frac{\pi\sqrt{R}}{1 - R} \quad (t = 1, r_1 = r_2, R = r_1^2). \tag{3.15}$$

We conclude this section by obtaining the cavity electric fields and intensities as a function of the frequency of the laser beam. First, we note that the round trip phase change, $\delta$, can be written as:

$$\delta = \frac{4\pi n d}{\lambda_0} + 2\epsilon = \frac{4\pi n d\nu}{c} + 2\epsilon, \tag{3.16}$$

where $\lambda_0$ is the vacuum wavelength of the radiation, $\nu$ is the frequency and $\epsilon$ is the phase change upon reflection from each mirror. Note that we have not included the

Gouy phase term, since it has no frequency dependence and will merely shift all of the axial resonances by the same amount but not change the separation between them. For the same reason, we can set $\epsilon = 0$. From eqns 3.8 and 3.10, it should be obvious that the maximum circulating power and minimum reflected power both occur when $\delta/2 = q\pi$, where $q$ is an integer. From this we get our customary result that the axial resonance frequencies are given by $\nu = qc/2nd$. If we assume that the finesse is very high, we can expand the frequency around any of the resonances and obtain

$$\text{Circulating:} \quad E_c = E_0 \frac{t_1 \mathcal{F} \Delta\nu_{1/2}/\pi}{\Delta\nu_{1/2} + 2i(\nu - \nu_0)} \tag{3.17}$$

$$I_c = I_0 \frac{(t_1 \mathcal{F}\gamma/\pi)^2}{(\Delta\nu_{1/2})^2 + 4(\nu - \nu_0)^2}, \tag{3.18}$$

where $\nu_0$ is one of the resonance frequencies, and the FWHM ($\Delta\nu_{1/2}$) is just the free spectral range ($c/2nd$) divided by the finesse (eqn 3.14). We have assumed that $r_1, r_m \approx 1$ and have removed factors such as $r_1$ or $r_1 r_m$ but have retained differences such as $r_1 - r_m$ and $1 - r_1 r_m$.

The expressions for the reflected fields and intensities can also be written as a function of frequency under the same assumptions as for the internal field and intensity:

$$\text{Reflected:} \quad E_r = E_0 \frac{(r_m - r_1)\mathcal{F}\Delta\nu_{1/2}/\pi - 2i(\nu - \nu_0)}{\Delta\nu_{1/2} + 2i(\nu - \nu_0)} \tag{3.19}$$

$$I_r = I_0 \frac{(r_m - r_1)^2 (\mathcal{F}\Delta\nu_{1/2}/\pi)^2 + 4(\nu - \nu_0)^2}{(\Delta\nu_{1/2})^2 + 4(\nu - \nu_0)^2}. \tag{3.20}$$

## 3.7  Impedance matching

It is well known from electrical network theory that a generator delivers the maximum power to a load when the source impedance is equal to the complex conjugate of the load impedance. Although we haven't defined impedance in the optical realm, inspection of eqn 3.10 indicates that there should be a set of parameters $(r_1, r_m)$ for which the circulating power is a maximum at resonance ($\delta = 0$). By analogy with an electrical network, this optimum coupling is called the *impedance matched* case. We will determine the conditions for impedance matching.

First, it will be useful to state the significance of the parameter $r_m$. This is simply the fraction of the electric field which *remains* after one round trip, *excluding the input coupling*, $t_1$. We will define $L$ to be the *round trip power loss*, again excluding the transmission through the input mirror. From the above,

$$L + r_m^2 = 1 \implies r_m^2 = 1 - L. \tag{3.21}$$

The expressions for the circulating and reflected power at resonance are (from eqns 3.8 and 3.10),

$$\text{Circulating:} \quad I_c = I_0 \frac{t_1^2}{(1 - r_1 r_m)} \tag{3.22}$$

$$\text{Reflected:} \quad I_r = I_0 \frac{(r_1 - r_m)^2}{(1 - r_1 r_m)^2}. \tag{3.23}$$

Differentiating the first with respect to $r_1$ and setting the result equal to zero yields an extremum (maximum) in the circulating power when $r_1 = r_m$. When this is the case, the second equation indicates that there will be *zero reflected power*. Using $r_1^2 + t_1^2 = 1$, the *impedance matched condition* is:

$$\text{Impedance matched:} \quad r_1 = r_m \implies L = t_1^2. \tag{3.24}$$

Stated in words: *A cavity is impedance matched when the intensity transmission coefficient of the input coupling mirror is equal to the sum of all of the other losses, excluding the input coupling.* When this happens, the internal circulating power will be a maximum and the reflection will be zero (at resonance).

It will be useful at this point to investigate the *impedance mismatched* case. We start by observing (from eqn 3.10) that the ratio of $I_c$ under mismatched conditions to $I_c$ under matched conditions is (at resonance):

$$\frac{I_c}{I_{c,matched}} = \frac{(1 - r_1^2)(1 - r_m^2)}{(1 - r_1 r_m)^2}. \tag{3.25}$$

We will define the *mismatch parameter*, $\sigma$, to be:

$$\sigma \equiv \frac{1 - r_1}{1 - r_m} \approx \frac{1 - r_1^2}{1 - r_m^2} = \frac{t_1^2}{L} \quad \text{if } 1 - r_1 \ll 1 \text{ and } 1 - r_m \ll 1. \tag{3.26}$$

Thus, $\sigma = 1$ under matched conditions. From this, eqn 3.25 can be written as

$$\frac{I_c}{I_{c,match}} = \frac{2(1 + r_m)\sigma - (1 - r_m^2)\sigma^2}{(\sigma r_m + 1)^2} \approx \frac{4\sigma}{(\sigma + 1)^2}. \tag{3.27}$$

A plot of the relative circulating power versus the mismatch parameter, $\sigma$, appears in Fig. 3.4. From the figure, it can be seen that an error in the input coupling by a factor of two (in either direction) reduces the circulating power by only 11%, which is fortunate in view of the difficulty in making mirrors with a precise transmission.

The reflected power on resonance will not be zero when the cavity is not impedance matched. Writing the formula for the reflected power (eqn 3.23) in terms of $\sigma$, one gets (when $r_1 \approx 1$):

$$\frac{I_r}{I_0} = \frac{(r_1 - r_m)^2}{(1 - r_1 r_m)^2} = \left(\frac{\sigma - 1}{\sigma r_m + 1}\right)^2 \approx \left(\frac{\sigma - 1}{\sigma + 1}\right)^2. \tag{3.28}$$

Of course, $I_r/I_0 \approx 1$ far from resonance. Plots of the circulating and reflected intensity under non-matched conditions are shown in Fig. 3.5.

A few comments on impedance matching are perhaps in order before proceeding. First, under impedance matched conditions, the *enhancement* ($I_c/I_0$) has a very simple representation and is simply connected to the *finesse*:

$$\text{Enhancement} = \frac{1}{T_1} = \frac{\langle \text{finesse} \rangle}{\pi} \quad \text{(impedance matched)} \tag{3.29}$$

Since $T_1 = L$ in a matched cavity, it is essential to reduce the losses as much as possible in order to obtain a significant enhancement. With a well-designed cavity, it is

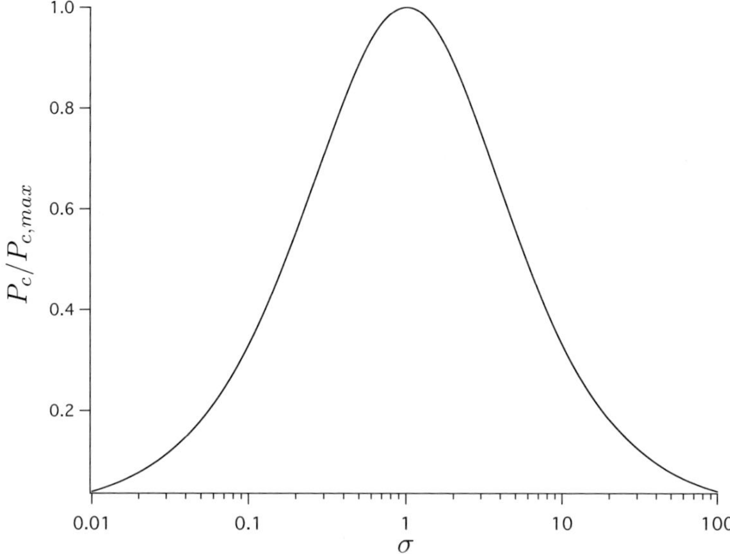

**Fig. 3.4** Relative circulating power versus mismatch parameter, $\sigma$.

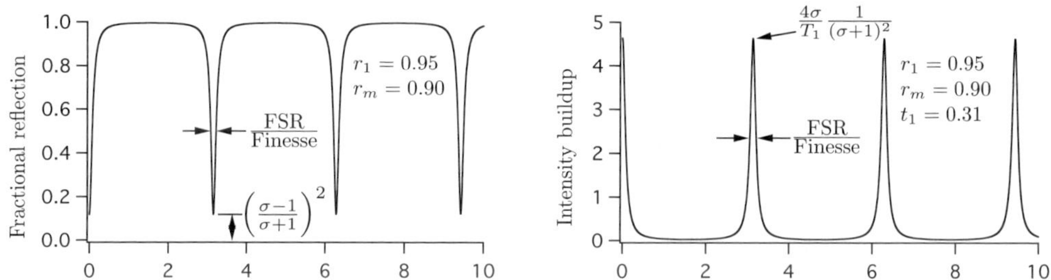

**Fig. 3.5** Spectrum of reflected and circulating power.

possible to obtain an enhancement of a bit greater than 100 when the cavity contains an internal nonlinear crystal. This has enormous implications for certain nonlinear processes: for example, second harmonic generation, which depends quadratically on the fundamental power, can be enhanced by a factor of 10,000. Next, one might think that a practical method for experimentally identifying impedance matching is the absence of reflected intensity on resonance. While this is true, the converse is not: a non-zero reflected power at resonance does not necessarily mean that the cavity is *not* impedance matched. A finite reflection from an impedance-matched cavity can still be obtained when the cavity is not *mode matched*: not all of the incident (perhaps multimode) power is coupled into the single cavity mode and that which is not coupled will be reflected from the input mirror whether or not the cavity is impedance matched.

## 3.8   Fields and intensities in ring cavity

Our analysis of the symmetric ring cavity is similar to that of the standing wave cavity except that we require that all of the mirrors (except the *input coupler*) have nearly 100% reflectivity. The input coupler transmission and reflectivity are (to keep the notation consistent with the standing wave cavity) $t_1$ and $r_1$. If we use the same definition of $r_m$ as with the standing wave cavity, it should be evident that

$$r_m = t. \tag{3.30}$$

A schematic of the cavity appears in Fig. 3.6.

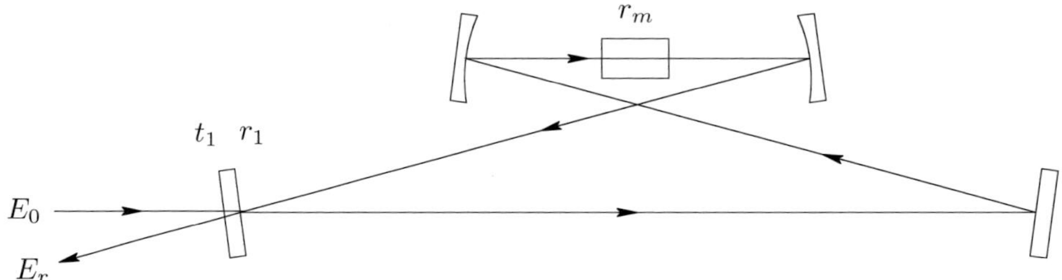


**Fig. 3.6** The ring cavity displaying items needed to calculate internal and reflected fields.

The analysis proceeds in exactly the same fashion as for the standing wave cavity and begins by summing the fields for the first three round trips. The electric field of the reflected wave (with the usual definition of $\delta$ as the round trip phase change) is

$$\begin{aligned}
E_r &= -E_0 r + E_0 r_m t_1^2 e^{-i\delta} + E_0 r_1 r_m^2 t_1^2 e^{-2i\delta} + \cdots \\
&= -E_0 r + E_0 r_m t_1^2 e^{-i\delta} (1 + r_1 r_m e^{-i\delta} + \cdots) \\
&= E_0 \frac{r_m e^{-i\delta} - r_1}{1 - r_1 r_m e^{-i\delta}}.
\end{aligned} \tag{3.31}$$

The circulating field is

$$\begin{aligned}
E_c &= E_0 t_1 + E_0 r_1 r_m t_1 e^{-i\delta} + E_0 r_1^2 r_m^2 t_1 e^{-2i\delta} + \cdots \\
&= E_0 t_1 (1 + r_1 r_m e^{-i\delta} + r_1^2 r_m^2 e^{-2i\delta} + \cdots) \\
&= E_0 \frac{t_1}{1 - r_1 r_m e^{-i\delta}}.
\end{aligned} \tag{3.32}$$

We observe that these are exactly the same results as for the standing wave cavity. This should come as no surprise, since we were performing exactly the same operations and were careful to use the same definition of $r_m$ in both cases (although $r_m$ depends *differently* on the cavity parameters for the two cavity types).

From the preceding analysis, we should be safe in claiming that all of the results for the reflected and circulating fields and intensities of a standing wave cavity as well as all of the impedance matching results should apply unchanged to the ring cavity.

## 3.9   A novel "reflective" coupling scheme using a tilted wedge

The conventional way to couple light into a cavity is to use a partially transmitting mirror as one of the cavity mirrors and to couple light through it. The transmission of this *input coupling mirror* will usually be determined by the impedance matching requirement. Cavities with extremely high finesses will use the slight transmission of ultra-high-reflecting cavity mirrors, but there are situations with medium finesse cavities where one would like to couple an *adjustable* amount of light into the cavity. This might occur with a cavity containing a nonlinear crystal, where it is desirable to maximize the circulating power (by careful impedance matching). We will present an elegant scheme (suggested by N. Yu and J. Torgerson) which uses *reflective* coupling to inject a variable amount of power into a ring cavity.

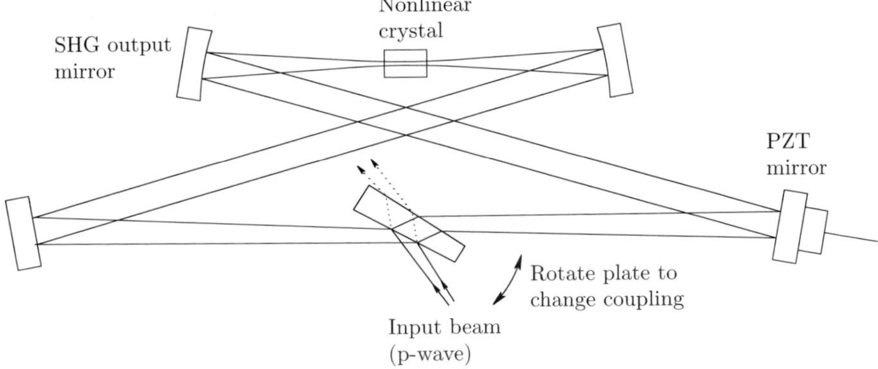


**Fig. 3.7** Diagram of ring cavity using "reflective" coupling.

The scheme is illustrated in Fig. 3.7. A slightly wedged element in inserted into the cavity and the light is coupled into the cavity by reflection from one of the surfaces (the near one in the figure). The loss from the other surface is nearly eliminated by using a polarization in the plane of the ring and ensuring that the angle of incidence on this surface is always near to Brewster's angle. A plot of the reflectivity from both surfaces of a 4° glass wedge appears in Fig. 3.8. For the plate depicted in the plot, the index of refraction is 1.45. As can be seen from the plot, adjusting the tilt angle ±3° around the central angle of 56° will smoothly vary the coupling from 0.5% to 3.5% while keeping the loss at the Brewster surface below 0.1% (a value comparable to other losses in the cavity). One drawback to the scheme is that a change in the tilt angle will require readjustment of *all* of the other mirrors to preserve the cavity alignment. In practice, this has not been found to be a serious problem.

Using this approach, the analogue to the reflected field is the beam transmitted through the wedge. It is left as an exercise to show that all of the theory developed for transmissive coupling holds for reflective coupling, with some simple parameter substitutions.

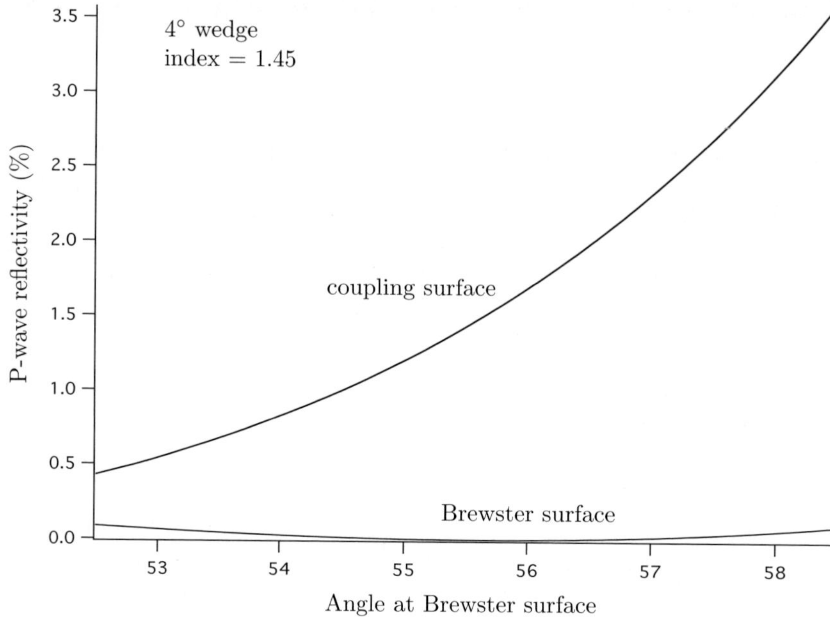



**Fig. 3.8** Reflectivity from "coupling" and "Brewster" surfaces as a function of tilt angle. The "p-wave" is polarized in the plane of the ring.

## 3.10   Photon lifetime

As a result of the finite transmission of the cavity mirrors and the possibility of dissipative losses in the cavity, a wave injected into an optical cavity will decay with time. Although we haven't introduced the concept of *photons* into our discussion, the *photon lifetime*, $t_c$, can be defined to be the average time a photon or injected beam will *live* inside the cavity. If $\mathcal{E}$ is the stored light energy in the cavity, the stored energy will satisfy

$$\frac{d\mathcal{E}}{dt} = -\frac{\mathcal{E}}{t_c},\tag{3.33}$$

whose solution is

$$\mathcal{E} = \mathcal{E}_0 e^{-t/t_c}.\tag{3.34}$$

Introducing the quantity $L_{tot}$, which is the *fractional intensity loss per round trip*, the energy loss per round trip is $L_{tot}\mathcal{E}$ and the fractional energy loss *per unit time* is $(c/l)L_{tot}\mathcal{E}$, where $l$ is the round-trip distance. The energy loss equation can be rewritten as

$$\frac{d\mathcal{E}}{dt} = -\frac{cL_{tot}}{l}\mathcal{E} \implies t_c = \frac{l}{cL_{tot}}.\tag{3.35}$$

Although the relation between the photon lifetime and the cavity resonance width can be obtained by simple Fourier theory, it can also be obtained directly from the relations for the finesse which we derived earlier. When $r_1, r_m \approx 1$, the finesse is approximately (from eqn 3.14):

$$\mathcal{F} \approx \frac{\pi}{1 - r_1 r_m}. \tag{3.36}$$

The round-trip intensity fractional loss ($L_{tot}$) is, using the same approximation,

$$L_{tot} = 1 - r_1^2 r_m^2 = (1 - r_1 r_m)(1 + r_1 r_m) \approx 2(1 - r_1 r_m) \implies \mathcal{F} = \frac{2\pi}{L_{tot}}. \tag{3.37}$$

Using these two relations, the resonance full width at half maximum, $\Delta\nu_{1/2}$, is:

$$\Delta\nu_{1/2} = \frac{\langle \mathrm{fsr} \rangle}{\mathcal{F}} = \frac{c}{l}\frac{1}{\mathcal{F}} = \frac{c}{l}\frac{L_{tot}}{2\pi} = \frac{1}{2\pi t_c}, \tag{3.38}$$

where $\langle \mathrm{fsr} \rangle$ is the free spectral range ($= c/l$). This equation is, of course, just an expression of the *uncertainty relation* in Fourier theory.

## 3.11   The quality factor, $Q$

The quality factor, $Q$, is a familiar quantity from electrical circuit theory and is usually defined as the *frequency resolution* of a resonant circuit. There is an equivalent and much more useful definition of $Q$ which allows it to be applied to *all resonant phenomena*. This definition is:

$$Q \equiv \frac{2\pi \times \text{stored energy}}{\text{energy loss per cycle}}. \tag{3.39}$$

If the loss term is expressed as a *loss rate* instead of a loss per cycle, the definition becomes:

$$Q \equiv \frac{\omega \times \text{stored energy}}{\text{energy loss rate}} = -\frac{\omega \mathcal{E}}{d\mathcal{E}/dt} = \omega t_c, \tag{3.40}$$

where the last identity follows from eqn 3.33 and $\omega$ is the resonant frequency.

Using eqn 3.38, we can now show that the above definitions of $Q$ are equivalent to the familiar notion of $Q$ as a measure of *frequency resolution*. If $\nu = \omega/2\pi$ is the resonant frequency in Hz,

$$Q = 2\pi\nu t_c = 2\pi\nu \left( \frac{1}{2\pi \Delta\nu_{1/2}} \right) = \frac{\nu}{\Delta\nu_{1/2}}. \tag{3.41}$$

## 3.12   Relation between $Q$ and finesse

We first recall that the finesse ($\mathcal{F}$) is defined as

$$\mathcal{F} = \frac{\langle \mathrm{fsr} \rangle}{\Delta\nu_{1/2}} = \frac{c}{l\Delta\nu_{1/2}} = \frac{2\pi c t_c}{l}. \tag{3.42}$$

Using the frequency resolution definition of $Q$,

$$Q = \frac{\nu}{\Delta\nu_{1/2}} = \frac{\nu l \mathcal{F}}{c} = \frac{l\mathcal{F}}{\lambda}. \tag{3.43}$$

In words, the $Q$ is *equal to the finesse times the number of wavelengths in a round-trip* (or the number of *half-waves* between the mirrors in a standing wave cavity).

We saw earlier that the energy in a cavity decays exponentially with a characteristic time equal to $t_c$. Both the *circulating power* and the power coupled out through any cavity mirror will, of course, have the same exponential behavior with time. Thus, after $n$ round-trips, the power will be

$$P = P_0 e^{-t/t_c} = P_0 e^{-\frac{nl}{ct_c}},\tag{3.44}$$

where $P_0$ is the power at $t = 0$. The power is therefore reduced by a factor of $e^{-1}$ when $n = ct_c/l = \mathcal{F}/2\pi$ (eqn 3.42). From this, we see that *the power in the cavity will be reduced by $1/e$ after $\mathcal{F}/2\pi$ round-trips*. From the point of view of interferometry, one can think of a cavity as a device which *causes* $\approx \mathcal{F}/2\pi$ beams to interfere with one other, where the interference is due to the *folding* of the path rather than to the presence of multiple slits or lines, as in a grating. This is similar to multiple beam interference experiments, where the appropriate width is reduced by a factor equal to the *number of beams that interfere*. In a cavity, this number is very roughly equal to the finesse divided by $2\pi$.

## 3.13   Alternative representation of cavity loss

In anticipation of our treatment of *laser gain*, it will be useful to represent the internal cavity loss as being due to a *uniformly distributed absorption coefficient*, $\alpha$, which excludes mirror Fresnel reflection losses. Thus for a standing wave cavity, we require that $\alpha$ satisfy

$$t^4 = e^{-\alpha l}.\tag{3.45}$$

If the round-trip fractional energy loss is $L_{tot}$, the *fraction of the power remaining after one round trip* is $e^{-L_{tot}}$. From this, we obtain

$$e^{-L_{tot}} = r_1^2 r_2^2 t^4 = r_1^2 r_2^2 e^{-\alpha l} \implies L_{tot} = \alpha l - 2\ln(r_1 r_2).\tag{3.46}$$

For the case of a ring cavity, $r_2 = 1$ and the exponent of the $t$ term would be 2, since the beam traverses the internal medium once per round-trip. Equation 3.46 can be considered either as a formula for $L_{tot}$ or a definition of $\alpha$. In the former case, we allow for the intriguing possibility that a *negative* $\alpha$ can make $L_{tot} = 0$, in which case there will be *no damping of the internal field*. This self-sustained cavity field is of course the result of a *coherent laser gain* and will be discussed in detail in Chapter 5.

## 3.14   Experimental determination of cavity parameters

An experimentalist is interested for obvious reasons in measuring the various cavity parameters which have been discussed in this and the preceding chapter. Certain cavity properties, such as the free spectral range, are trivial to obtain, since they depend very simply on the cavity geometry. The *dynamic* properties – those which pertain to the energy relations in a cavity – are a little more difficult to determine. We will briefly discuss techniques used to determine the *enhancement* and *finesse*.

The most straightforward way to measure the finesse is to use its definition as the ratio of the free spectral range to the full width at half maximum of a cavity resonance. This is a reasonable procedure for cavities with low to modest finesse ($\leq 200$), if one has access to a fairly stable tunable laser at the appropriate wavelength. This technique is not easy to apply to cavities with extremely high finesse, since the stability requirement placed on the laser might be too great. Furthermore, it does not measure the enhancement directly, and the relation between enhancement and finesse is simple only when the cavity is impedance matched. We will present two techniques: one which can be used to directly measure the enhancement in a *build-up cavity* and the other which can be used to measure the finesse of a cavity whose finesse is extremely high.

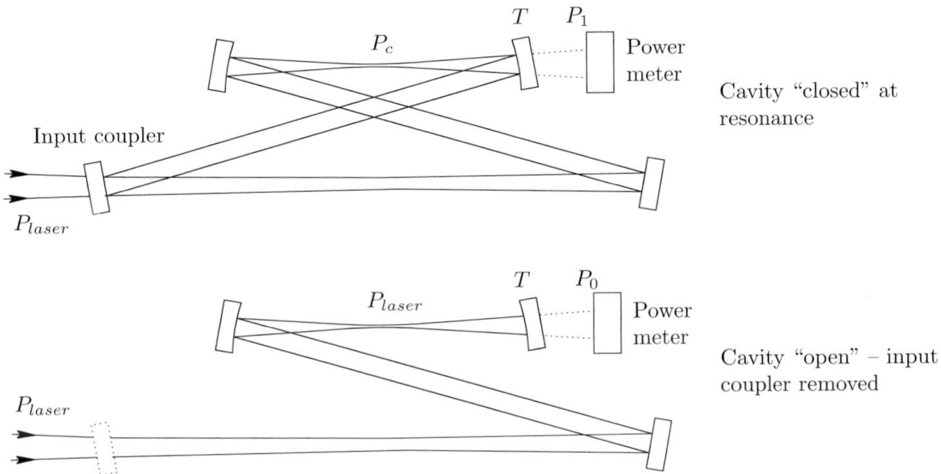


**Fig. 3.9** Technique for measuring cavity enhancement by making two different power measurements of light leaking through cavity mirror, whose (small) transmission is $T$. The enhancement is $P_1/P_0$.

We will discuss the enhancement measurement first, since it is simpler. It is conveniently employed to evaluate a cavity which is used to enhance nonlinear processes such as second harmonic generation or sum frequency mixing. The technique is illustrated in Fig. 3.9. The procedure is to place a sensitive power meter behind one of the cavity mirrors and measure the *leakage* of radiation through this mirror. Two power measurements are made. One, which we will call $P_1$, is done with the cavity at resonance and aligned (it is most convenient if the cavity is *locked* to the laser frequency so that the power can be measured over some modest time period, though it can also be done by *slowly* scanning the laser with the cavity *free-running*). If we assume that $T$ is the *transmission* of the cavity mirror through which the leakage is measured,

$$P_1 = TP_c, \tag{3.47}$$

where $P_c$ is the circulating power. The other power measurement is made by removing the input coupler (keeping the laser power the same). If we call this measured power, $P_0$, and the laser power, $P_{laser}$, then,

$$P_0 = TP_{laser}, \tag{3.48}$$

and we assume that the reflectivity of all of the cavity mirrors is $\approx 1$. The enhancement is simply the ratio of these two power measurements:

$$\text{Enhancement} \equiv \frac{P_c}{P_{laser}} = \frac{P_1}{P_0}. \tag{3.49}$$

The finesse of a cavity can also be determined by applying a very short pulse of light at the resonant frequency of the cavity and observing the exponential decay of the intracavity power as it leaks out through one of the mirrors. A common experimental setup is shown in Fig. 3.10. The laser is tuned near to the cavity resonance and the cavity is allowed to drift into resonance. After the intracavity power reaches a certain threshold (as measured by a detector which samples the leakage through one of the cavity mirrors), a very fast optical switch (usually an *acousto-optic modulator*; see Chapter 14) quickly switches off the beam. A digital sampling oscilloscope stores a curve of the leakage power as a function of time and the finesse is determined by fitting an exponential to the decay curve using,

$$P(t) = P_0 e^{-t/t_c} = P_0 e^{-(2\pi c/l\mathcal{F})t}. \tag{3.50}$$

This approach is called the *ring-down* technique for obvious reasons. It is important

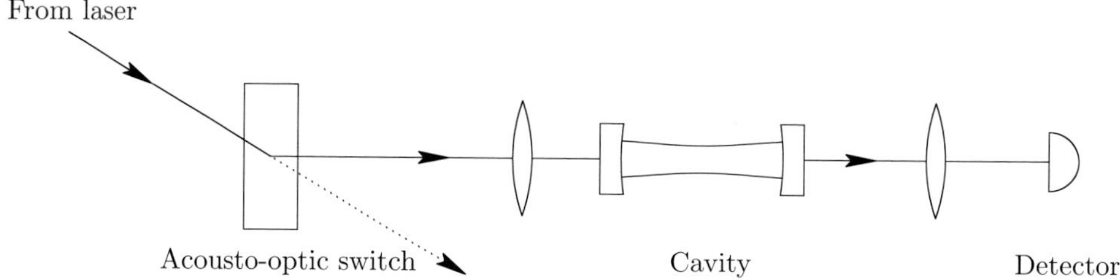


**Fig. 3.10** A commonly used setup for measuring cavity finesse using the *ring-down method*.

that the fall time of the optical switch is short compared to the photon storage time to avoid distorting the exponential decay curve. As one can see, the method works very well for cavities with extremely high finesse (approaching $10^6$ using mirrors with a power loss in the part per million range ) since the storage times are then relatively large and the time delays in the electronics have a smaller effect on the results. A typical curve appears in Fig. 3.11, where the rapid filling is shown for $t < 0$ and the exponential decay is shown for $t > 0$.

## 3.15   Further reading

This chapter discusses the dynamical aspects of optical cavities. The references mentioned in the previous two chapters are still useful as well as a number of excellent treatises on optics, such as Born and Wolf (1980) and Hecht (2002). In addition, a discussion of impedance matching into a cavity occurs in only one place, to the author's knowledge: the paper by Ashkin, Boyd and Dziedzic (1966). The publication on the

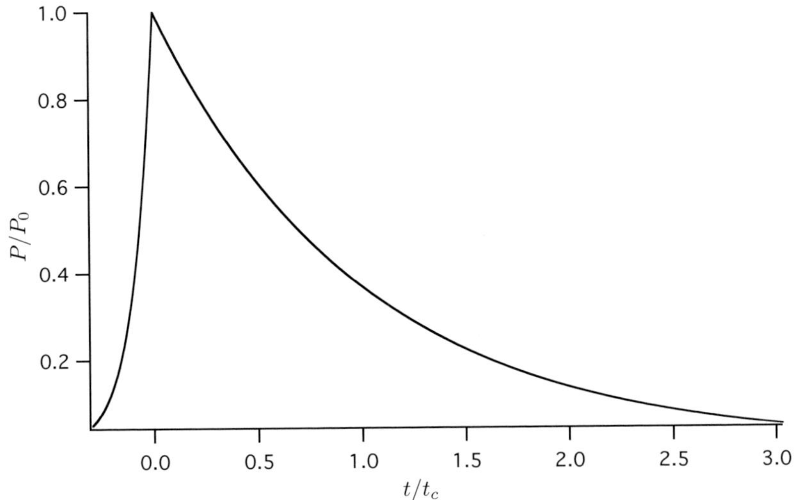

**Fig. 3.11** The signal obtained using the *ring-down* method. The cavity rapidly fills when $t < 0$ and exponentially decays when $t > 0$.

use of a wedge for variable reflective coupling appears in the paper by Torgerson and Nagourney (1999).

## 3.16 Problems

(3.1) A cavity is excited by a 500 nm mode-matched laser beam and has resonances whose separation is 125 MHz and whose width is 2.5 MHz. Determine the following properties of the cavity:

    a The cavity length
    b The cavity finesse
    c The cavity $Q$
    c The photon lifetime

(3.2) Calculate the *transmission* of the electric field and intensity for a cavity having mirrors with reflectivity $r_1, r_2$ (and transmission $t_1, t_2$), separation $d$ and containing a medium with transmission $t$. This is just the case not discussed in the text.

(3.3) For the three-mirror ring cavity depicted below, calculate the free spectral range, finesse, $Q$, and photon lifetime. As shown in the diagram, the three arms are 20 cm long and have the indicated *field* reflectivities.

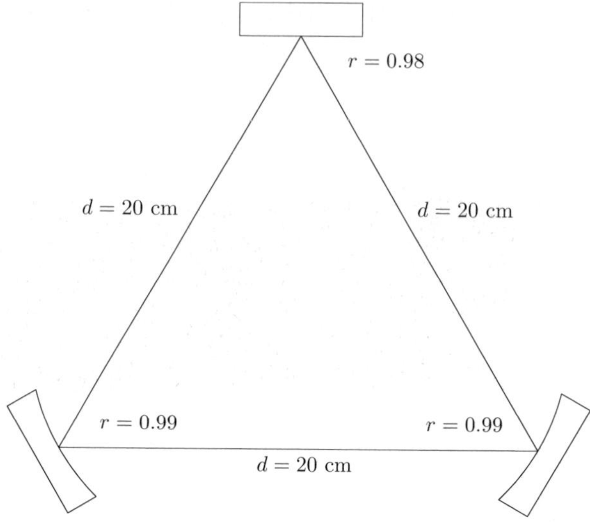



(3.4) A *thin etalon* is sometimes used in a laser cavity to help achieve single-mode operation. Such a device consists of a thin plate of quartz which is operated near normal incidence and tilted about an axis normal to the laser beam for tuning purposes. If the index of refraction is 1.5 and the thickness is 0.2 cm, find the linewidth of the etalon and an expression for the frequency as a function of the tilt angle (this dependence is not necessarily linear). The tilt angle is the angle between the normal to the plate and the input $k$-vector.

(3.5) Calculate the internal and *transmitted* fields and intensities for a ring cavity which uses the *reflective* coupling approach discussed in this chapter. Assume that the mirrors have 100% reflectivity and that there is an internal medium with transmission $r_m$. Neglect the effects of the surface at Brewster's angle and assume that the *coupling* surface has *field* reflectivity and transmission $r$ and $t$. The internal fields and intensities should be calculated just to the left of the coupling surface.

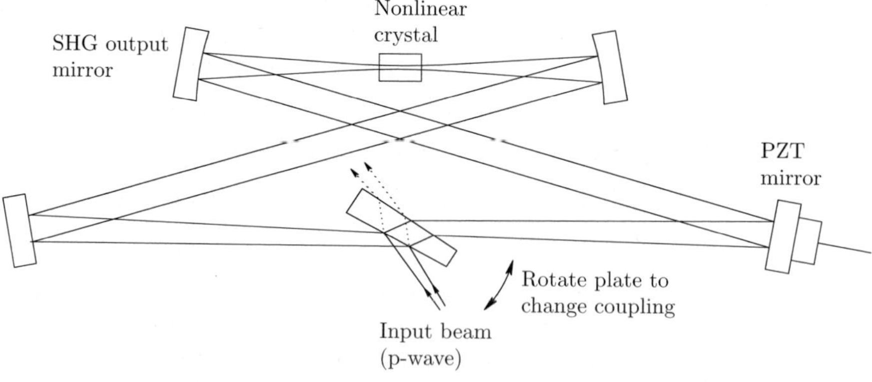

# 4

# Optical cavity as frequency discriminator

## 4.1 Introduction

Earlier in this book, we mentioned some of the uses for optical cavities in an atomic physics laboratory, such as the feedback mechanism in lasers, spectrum analyzers and enhancement devices for nonlinear processes. These applications are fairly simple to understand and do not require additional pieces of apparatus to perform their functions. The use of a cavity as a frequency reference element, however, requires that one have a means of generating an *error signal* which will correct the laser frequency when it differs from that of an an exact cavity resonance. The correction signal is ideally a linear function of the frequency deviation from resonance and is called a *discriminant*. This application requires some additional equipment and warrants some additional discussion. The use of a cavity as a *passive frequency discriminator* will be discussed in this chapter. In a later chapter, we will discuss *active* discriminators derived from atomic or molecular resonances.

Two techniques will be discussed in detail: the Pound–Drever–Hall (PDH) scheme, which uses a frequency-modulated signal reflected from a cavity, and the passive polarization scheme, which analyzes the polarization of the reflected signal from a cavity containing a polarizing element. The former scheme is usually used for locking a laser to a cavity and the latter for the reverse process: locking a cavity to a laser (usually for enhancement of a nonlinear process). We will briefly touch on the use of a cavity in transmission, which historically preceded the PDH method and is still found in some commercial dye or Ti-sapphire laser frequency locking systems. We will illustrate the principles by beginning with a simple "quick and dirty" approach to generate a discriminant using a length-modulated cavity and phase-sensitive detection. We will conclude the discussion by deriving the *frequency response* of the discriminator to a frequency-modulated laser beam; these results are important when one analyzes the servomechanism aspects of a laser which is frequency locked to a cavity.

## 4.2 A simple example

A simple and probably very common way to generate a frequency discriminant when a commercial optical spectrum analyzer is available is to apply a small, low-frequency sinusoidal voltage to the cavity's piezo-electric length-controlling element and detect the transmitted signal using a phase-sensitive detector whose reference is a sample of the modulation voltage. A block diagram of the apparatus appears in Fig. 4.1. The

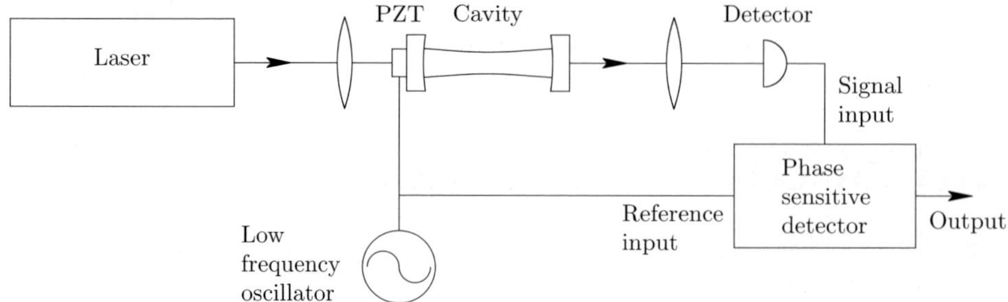


**Fig. 4.1**  Block diagram of "simple" apparatus used to generate a discriminant.

cavity will have its resonance frequency modulated by the applied voltage. As can be seen from Fig. 4.2, when the laser is below resonance, the transmission characteristic of the cavity will convert the modulation of the cavity's *resonant frequency* into amplitude modulation with a certain phase and when the laser is above resonance the converted amplitude modulation will have the opposite phase. At exact resonance, there will be no signal at the modulation frequency (there will be one at *twice* this frequency). Thus, one will get a curve with the appropriate shape for a discriminant when one plots the output of a phase-sensitive detector against the laser frequency – if the modulation *excursion* is small compared to the cavity width, the signal will essentially be proportional to the *first derivative* of the cavity transmission lineshape function. The phase-sensitive detector can be *defined* as a device which produces an output which is jointly proportional to the amplitude of the input and the cosine of the phase difference between the signal and some reference. The operation of *multiplication* by a constant sinusoidal reference followed by low-pass filtering is a reasonable model of a phase-sensitive detector.

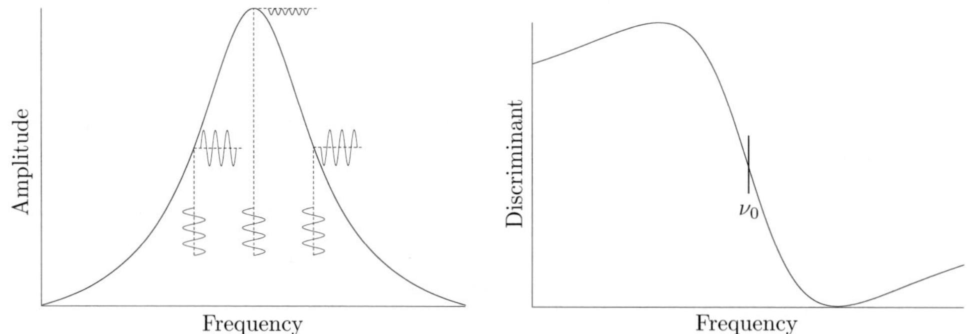


**Fig. 4.2**  Illustration of simple discriminator function. Below resonance (left), the output is in phase with modulation, above resonance the output is 180° out of phase and on resonance the output is at twice the frequency. The output of the phase sensitive detector is the *first derivative* of the resonance curve for small deviation, as shown on the right.

The functional form of the resonance can be easily obtained if we assume that the modulation frequency is much lower than the inverse of the photon storage time –

this allows us to consider the cavity transmission to be a quasi-static process even in the presence of the modulation of its resonance frequency. It can then be described by an expression proportional to that for the circulating intensity (eqn 3.18). If we further assume that the size of the *frequency excursion* due to the modulation is small compared to the cavity FWHM, we can perform a *Taylor expansion* of the function which gives the transmitted intensity in terms of the frequency. The length modulation will cause the cavity resonance to be sinusoidally modulated. From the point of view of the cavity, one can consider the frequency, $\nu(t)$, of the *laser beam* to be modulated by the same amount:

$$\nu(t) = \nu_0 + \delta\nu\sin(\nu_m t), \tag{4.1}$$

where $\delta\nu$ is the maximum deviation of the cavity frequency due to the modulation, $\nu_m$ is the (low) modulation frequency and $\nu_0$ is the laser frequency. Performing the expansion about $\nu_0$, the intensity at the detector ($I_d$) will have following time dependence:

$$I_d(t) = I_d(\nu_0) + \left.\frac{\partial I_d}{\partial \nu}\right|_{\nu_0} (\delta\nu\sin(\nu_m t)). \tag{4.2}$$

The time-dependent intensity will generate a time-dependent photocurrent in the detector and the signal at the output of the phase-sensitive detector will simply be proportional to the first derivative with respect to frequency of the Lorentzian lineshape function given in eqn 3.18 and plotted in Fig. 4.2.

## 4.3 Side of resonance discriminant

Historically, one of the first successful methods of generating an *optical* discriminant was also the simplest. It uses the side of a Lorentzian resonance, which is fairly linear for small excursions from the "lock point". A sample of the laser intensity is subtracted from the signal due to transmission though a cavity to prevent the zero point (lock point) from changing with changes in the laser power. The scheme is illustrated in Fig. 4.3 and the resulting discriminant is plotted in Fig. 4.4.

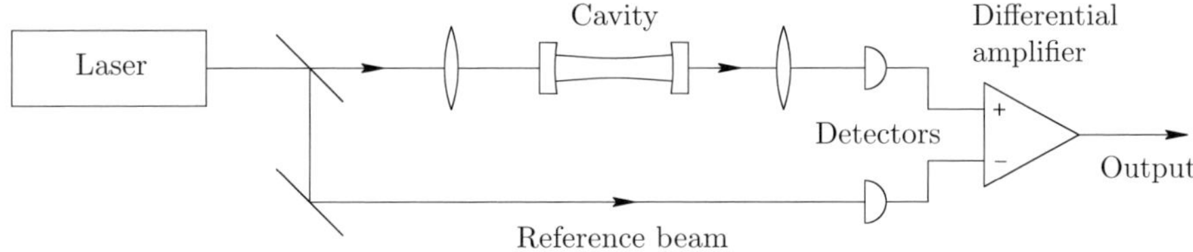


**Fig. 4.3** Apparatus used to generate a discriminant using the side of a resonance curve.

When used as a reference for laser stabilization, this scheme has a number of problems, perhaps the principal one being the time delays resulting from the transmission through the cavity. These delays are on the order of the photon lifetime in the cavity and can greatly limit the bandwidth of the stable servo system used for frequency locking. The successor to this approach – the Pound–Drever–Hall scheme – uses the reflected beam from the cavity and avoids this problem.

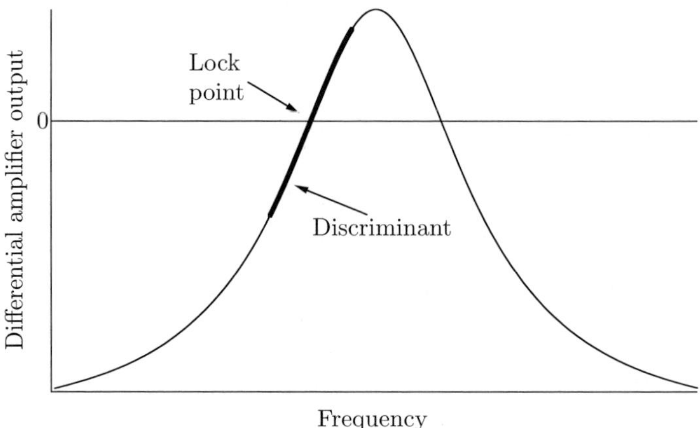


**Fig. 4.4** Plot of discriminant generated using the side of a resonance. Of course, the portion above the Lorentzian resonance could have also been used.

## 4.4　The manipulation of polarized beams: the Jones calculus

We will take a short diversion to summarize a convenient formalism for describing and manipulating *fully polarized* beams since the next two methods used to generate a discriminant make extensive use of polarization-manipulating devices. This formalism is called the *Jones calculus*. We should emphasize that this approach is only applicable to fully polarized waves – partially polarized light is described by other formalisms, such as the *Stokes parameters*.

A polarized beam is represented by a two-element column vector (called a *Jones vector*) giving the Cartesian components of the electric field as a function of time:

$$\boldsymbol{E}(t) = \begin{pmatrix} E_x(t) \\ E_y(t) \end{pmatrix}. \tag{4.3}$$

The components of the field are *complex numbers*. They are often *normalized* to unity complex norm ($\boldsymbol{E} \cdot \boldsymbol{E}^* = 1$), though this is not always done. Usually, common complex factors (including the time dependence of a complex sinusoid) are factored out and removed. Thus, the four common polarizations are (normalized):

$$\text{Horizontal:} \quad \boldsymbol{E}_h = \begin{pmatrix} 1 \\ 0 \end{pmatrix} \tag{4.4}$$

$$\text{Vertical:} \quad \boldsymbol{E}_v = \begin{pmatrix} 0 \\ 1 \end{pmatrix} \tag{4.5}$$

$$\text{Right circular:} \quad \boldsymbol{E}_h = \frac{1}{\sqrt{2}} \begin{pmatrix} 1 \\ -i \end{pmatrix} \tag{4.6}$$

$$\text{Left circular:} \quad \boldsymbol{E}_h = \frac{1}{\sqrt{2}} \begin{pmatrix} 1 \\ i \end{pmatrix}. \tag{4.7}$$

The *Jones matrices* represent the *linear transformation* which a Jones vector undergoes when it passes through an optical device. If the Jones matrix is $\mathcal{A}$ and the incident and transmitted Jones vectors are $\boldsymbol{E}_i$ and $\boldsymbol{E}_t$ respectively, then

$$\boldsymbol{E}_t = \mathcal{A}\boldsymbol{E}_i, \tag{4.8}$$

or writing the elements out,

$$\begin{pmatrix} E_{tx} \\ E_{ty} \end{pmatrix} = \begin{pmatrix} a_{11} & a_{21} \\ a_{12} & a_{22} \end{pmatrix} \begin{pmatrix} E_{ix} \\ E_{iy} \end{pmatrix}. \tag{4.9}$$

The key to the utility of the Jones calculus is the fact that one can represent a complex device as a *sequence of simple devices* and the overall Jones matrix can be obtained by multiplication of the Jones matrices of the simple components. If the beam encounters devices represented by Jones matrices $\mathcal{A}_1, \mathcal{A}_2 \cdots \mathcal{A}_n$, the overall Jones matrix of the sequence is

$$\mathcal{A} = \mathcal{A}_n \mathcal{A}_{n-1} \cdots \mathcal{A}_1, \tag{4.10}$$

where one must write the matrices in the *reverse order* from that in which the devices are encountered by the beam. (One should note the similarity to the treatment of a composite ABCD matrix for cascaded systems.)

The Jones matrices of some common devices are:

$$\text{Horizontal polarizer:} \quad \begin{pmatrix} 1 & 0 \\ 0 & 0 \end{pmatrix} \tag{4.11}$$

$$\text{Vertical polarizer:} \quad \begin{pmatrix} 0 & 0 \\ 0 & 1 \end{pmatrix} \tag{4.12}$$

$$\lambda/4, \text{ vertical fast axis:} \quad \begin{pmatrix} 1 & 0 \\ 0 & i \end{pmatrix} \tag{4.13}$$

$$\lambda/4, \text{ horizontal fast axis:} \quad \begin{pmatrix} 1 & 0 \\ 0 & -i \end{pmatrix}, \tag{4.14}$$

where the last two devices are *quarter-wave plates*, devices which advance or retard the phase of a beam by $\pi/2$ radians.

Since the Jones vector is a conventional two-dimensional vector, it can be evaluated in a *rotated coordinate system* by applying the familiar rotation matrix in the $xy$ plane:

$$\boldsymbol{E}' = S(\theta)\boldsymbol{E}, \tag{4.15}$$

where $\boldsymbol{E}'$ is the Jones vector in the rotated system and $S(\theta)$ is the rotation matrix for an angle of $\theta$:

$$S(\theta) = \begin{pmatrix} \cos\theta & \sin\theta \\ -\sin\theta & \cos\theta \end{pmatrix}. \tag{4.16}$$

In the rotated frame,

$$\boldsymbol{E}'_t = \mathcal{A}'\boldsymbol{E}'_i, \tag{4.17}$$

where $\mathcal{A}'$ is the Jones matrix in the rotated frame.

Since

$$\boldsymbol{E}'_i = S(\theta)\boldsymbol{E}_i \tag{4.18}$$

$$\boldsymbol{E}'_t = S(\theta)\boldsymbol{E}_t, \tag{4.19}$$

multiplying eqn 4.8 on the left by $S$ and using the identity $S^{-1}S = 1$, one has

$$\boldsymbol{E}'_t = S\mathcal{A}S^{-1}\boldsymbol{E}'_i, \tag{4.20}$$

which implies that the *Jones matrix in the rotated system* is

$$\mathcal{A}' = S\mathcal{A}S^{-1}, \tag{4.21}$$

where the inverse of the *orthogonal matrix* $S$ is $S^\dagger$.

Using this, one can immediately write down the Jones matrices for rotated objects, such as

$$\text{Linear polarizer at } 45°: \quad \frac{1}{2}\begin{pmatrix} 1 & 1 \\ 1 & 1 \end{pmatrix} \tag{4.22}$$

$$\text{Linear polarizer at } -45°: \quad \frac{1}{2}\begin{pmatrix} 1 & -1 \\ -1 & 1 \end{pmatrix}. \tag{4.23}$$

$$\tag{4.24}$$

and, in general, the Jones matrix for a linear polarizer at angle $\theta$ to the $x$-axis is:

$$\mathcal{A} = \begin{pmatrix} \cos^2\theta & \sin\theta\cos\theta \\ \sin\theta\cos\theta & \sin^2\theta \end{pmatrix}. \tag{4.25}$$

It should be kept in mind that *a rotation of the coordinate system by angle $\theta$ is equivalent to a rotation of the apparatus by angle $-\theta$.*

## 4.5  The polarization technique

The *polarization* technique generates a discriminant by placing a polarization-sensitive element in the optical cavity and observing the *interference* between two reflected beams with orthogonal polarizations; the orientation of the intracavity element is chosen so that the two polarizations have very different intracavity losses. The signal is actually the *difference* between the intensities of the *two opposite circular polarization components* of the reflected beam. We will show that this does indeed generate a useful discriminant using our cavity theory and some shortcuts from the Jones calculus. Since a potentially lossy element intracavity element is used, the technique is not employed in the highest finesse cavities (where the losses are in the parts per million range) but finds frequent application in build-up cavities used in nonlinear optics where there is often a polarization-sensitive element already in place in the cavity.

The apparatus for generating a polarization signal is shown in Fig. 4.5, where it is used in a ring cavity containing a nonlinear crystal whose faces are at Brewster's angle for polarization in the plane of the figure. Since the *lossless* polarization of

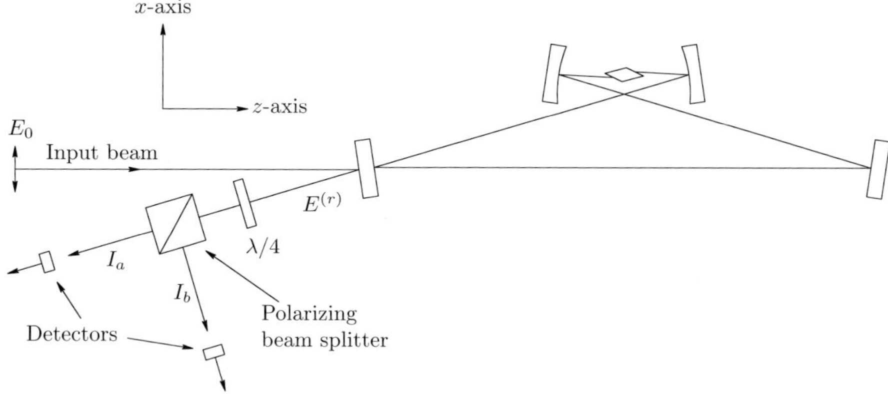

**Fig. 4.5** The polarization technique used with a ring cavity. The $x$-axis is in the plane of the ring; the beam $I_b$ is in the ring plane and the incoming field, $E_0$, makes an angle $\theta$ with the $x$-axis. The $y$-axis is perpendicular to the plane of the ring. The *fast axis* of the quarter wave plate makes an angle of $45°$ to the $x$-axis.

the Brewster surfaces is in the ring plane, the electric field of the incoming beam $(E_0)$ is tilted slightly by an angle of $\theta$ with respect to the $x$-axis. The reflected beam electric field, $\boldsymbol{E}^{(r)}$, has two components: $E_x^{(r)}$ and $E_y^{(r)}$. The *polarizing beam splitter* is just a polarizer with two outputs: polarization along the $x$-axis (beam with intensity $I_a$) passes straight through it without loss, and polarization along the $y$-axis ($I_b$) is reflected by $90°$. The Jones matrix for the quarter wave plate can be obtained from eqn 4.14 after rotating by $45°$ (eqn 4.21). This can be shown to be:

$$\lambda/4 \text{ at } 45° : \quad \mathcal{A} = \frac{1-i}{2}\begin{pmatrix} 1 & -i \\ -i & 1 \end{pmatrix}. \tag{4.26}$$

The Jones vectors for beams $I_a$ and $I_b$ are

$$\text{Beam a: } \boldsymbol{E}_a = \frac{1-i}{2}\begin{pmatrix} 1 & 0 \\ 0 & 0 \end{pmatrix}\begin{pmatrix} 1 & -i \\ -i & 1 \end{pmatrix}\begin{pmatrix} E_x^{(r)} \\ E_y^{(r)} \end{pmatrix} \tag{4.27}$$

$$\text{Beam b: } \boldsymbol{E}_b = \frac{1-i}{2}\begin{pmatrix} 0 & 0 \\ 0 & 1 \end{pmatrix}\begin{pmatrix} 1 & -i \\ -i & 1 \end{pmatrix}\begin{pmatrix} E_x^{(r)} \\ E_y^{(r)} \end{pmatrix}. \tag{4.28}$$

The signal at each detector is obtained by taking the square modulus of the fields:

$$I_a \propto |E_x^{(r)} - iE_y^{(r)}|^2 \tag{4.29}$$

$$I_b \propto |E_x^{(r)} + iE_y^{(r)}|^2, \tag{4.30}$$

where the proportionality constants are the same for $I_a$ and $I_b$. The overall signal $(I_a - I_b)$ is due to an interference between the real part of one beam and the imaginary part of the other.

We will assume that the finesse of the cavity is very low for the $y$-polarized wave. This is justified by the fact that this polarization has a $\approx 20\%$ *Fresnel* loss from the

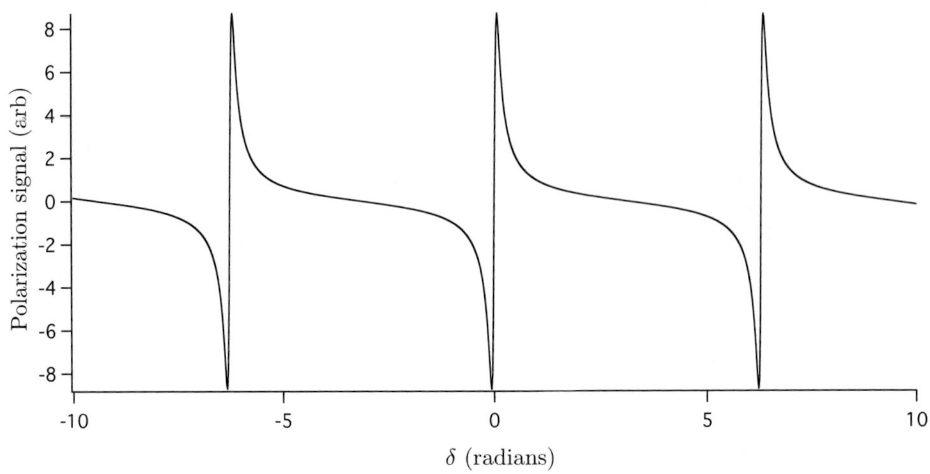

**Fig. 4.6** Plot of polarization signal versus $\delta$.

crystal (for a refractive index of 1.5), so that the finesse of the $y$-polarization in an impedance-matched cavity must be less than about 15 while that of the $x$-polarization can be 300 or more. This assumption allows us to ignore the frequency dependence and write the field reflection of the $y$-component simply as

$$E_y^{(r)} = -rE_{0y}, \tag{4.31}$$

where $r$ is the field reflectivity of the input mirror. Under impedance matched conditions, the $x$-polarization field reflection is (from eqn 3.7):

$$E_x^{(r)} = E_{0x}\frac{r(e^{-i\delta} - 1)}{1 - r^2e^{-i\delta}}. \tag{4.32}$$

If $\theta$ is the angle that the input field makes with the $x$-axis, then the input field components are

$$E_{0x} = E_0 \cos\theta \tag{4.33}$$
$$E_{0y} = E_0 \sin\theta. \tag{4.34}$$

The intensities are then

$$I_{ab} \propto E_0^2 \left| \frac{r(e^{-i\delta} - 1)}{1 - r^2e^{-i\delta}} \cos\theta \pm ir\sin\theta \right|^2, \tag{4.35}$$

where the minus sign is for $I_a$ and the plus is for $I_b$. The desired signal is proportional to $I_a - I_b$:

$$\text{Signal} \propto 4E_0^2 r^2 t^2 \cos\theta \sin\theta \frac{\sin\delta}{(1 - r^2)^2 + 4r^2\sin^2\delta/2}, \tag{4.36}$$

where $t$ is the field transmission of the input mirror. The independent variable, $\delta$, is the round-trip phase change of the intracavity beam. A plot of the discriminant appears in

Fig. 4.6. Normally, one would use the smallest practical $\theta$ in the interests of coupling most of the power into the low-loss polarization where it would be effectively enhanced by the cavity.

## 4.6  Frequency modulation

As an aid in understanding the Pound–Drever–Hall (PDH) technique, we will discuss the frequency modulation of a monochromatic optical field. We will consider the modulated field in both the *frequency* and *time* domains and will start with a graphical aid: the *phasor* model of an oscillating signal. This model will also help in describing the PDH approach to generating a discriminant.

Figure 4.7 represents a *narrowband* frequency-modulated wave using both phasors and the frequency domain (the term *narrowband* will be defined shortly). The phasor

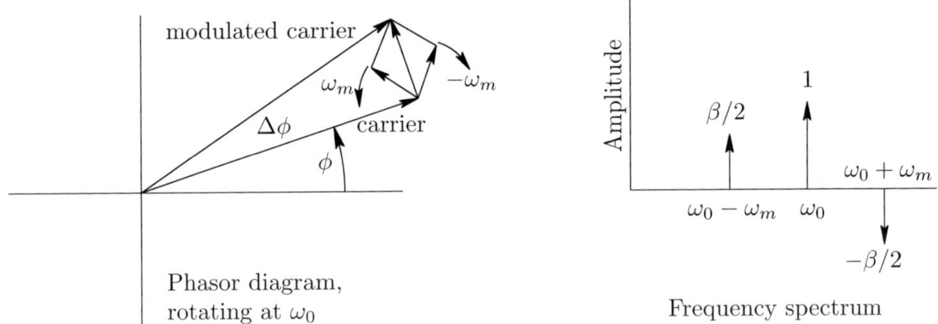



**Fig. 4.7**  Representation of frequency-modulated wave using phasors (left) and the frequency domain (right).

model represents a sinusoidal signal by a uniformly rotating vector with constant length. To aid in visualization, the coordinate system is assumed to rotate at the same angular speed as the wave so that the phasor is fixed in space. The amplitude of the wave is simply the length of the vector and the *instantaneous phase* is the angle it makes with the $x$-axis. At any time, one can generate a physically realistic function by projecting the phasor onto either the $x$- or the $y$-axis. Phasors have the nice property of following the rules of vector addition: one can superpose any number of fields simply by adding the phasors representing them as vectors.

One begins with an unmodulated sinusoidal signal at angular frequency $\omega_0$; this is called the *carrier*. Frequency modulation is obtained by adding two small phasors (called *sidebands*) of equal length and rotating in opposite directions at angular frequency (in the rotating frame) $\omega_m$; they are phased so that they are perpendicular to the carrier phasor when they overlap. As can be seen in the diagram, the resultant of the three phasors is one whose *phase* varies sinusoidally at the rotation speed of the sideband phasors but whose amplitude is essentially constant. This is *phase modulation*, which is identical to frequency modulation for a single sinusoidal modulating signal. If the amplitude of the modulating phasors is small compared to that of the carrier, only two are needed to preserve the constancy of the resultant amplitude. In this case, the size of the phase modulation (called the *phase deviation*) is much less than 1 radian and the modulation is called *narrowband*. The two sidebands are at $\omega_0 + \omega_m$ and $\omega_0 - \omega_m$. Thus, there are three components in the spectral representation of the phase-modulated carrier as shown on the right side of Fig. 4.7. The 180° difference between the two sidebands can be understood by considering the relative sideband phase when both sideband phasors are lined up along the carrier: they are 180° out of phase at that instant of time.

Some further insight into the modulation process can be obtained by considering *amplitude modulation*. Fig. 4.8 displays the phasors and spectrum of an amplitude-modulated carrier. As can be seen from the plots, amplitude modulation is identical to

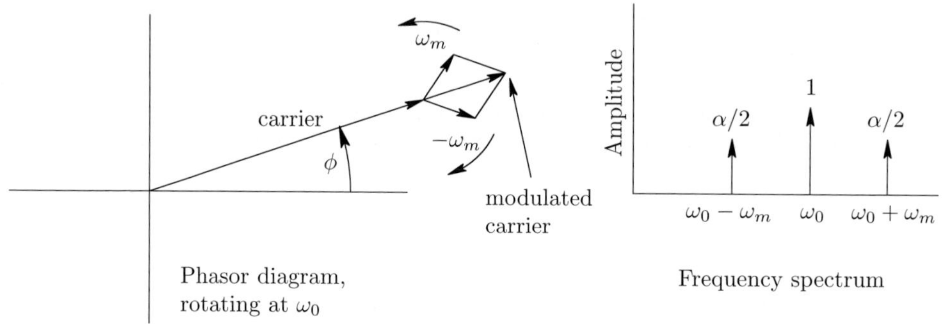


**Fig. 4.8** Representation of amplitude modulated wave using phasors (left) and the frequency domain (right).

frequency modulation except that both sidebands are shifted in phase by 90° relative to the carrier so that they modulate the amplitude of the carrier, keeping its phase constant. In fact, one can convert between amplitude and frequency modulation by using a scheme that rotates the sideband phase by 90° without changing the phase of the carrier (this is the idea behind *phase contrast microscopy*). Both sidebands have the same phase in the spectral representation.

The analytic representation of a phase-modulated carrier is completely straightforward. If we represent an unmodulated monochromatic wave by its time-dependent electric field:

$$E(t) = E_0 \cos(\phi(t)) = E_0 \cos(\omega_0 t), \tag{4.37}$$

the time-dependent phase $\phi(t)$ is just $\omega_0 t$ and increases linearly with time. In order to modulate this phase, we add a time-dependent modulation term $\Delta\phi(t)$. If this modulation is sinusoidal with a maximum of $\beta$ radians, then the phase-modulated field is

$$E(t) = E_0 \cos(\phi(t) + \Delta\phi(t)) = E_0 \cos(\omega_0 t + \beta \sin \omega_m t), \tag{4.38}$$

where the *phase modulation deviation is $\beta$ radians*. The equivalence of this to *frequency modulation* should now be clear if we define the *instantaneous frequency ($\omega(t)$)* to be the *time derivative of the total phase*

$$\omega(t) = \frac{d}{dt}(\phi(t) + \Delta\phi(t)) = \omega_0 + \beta\omega_m \cos\omega_m t. \tag{4.39}$$

The quantity $\beta$ is called the *modulation index* regardless of whether one is describing frequency or phase modulation. The carrier frequency ($\omega_0$) is now being modulated sinusoidally and the maximum frequency shift is $\beta\omega_m$. This latter quantity is called the frequency modulation *deviation*, $\Delta\omega$. The criterion for narrowband modulation is that $\beta \ll 1$ or $\Delta\omega \ll \omega_m$. The narrowband modulated field can be easily expanded using standard trigonometric identities and taking account of the fact that $\beta \ll 1$,

$$E(t) = E_0(\cos\omega_0 t + (\beta/2)\cos(\omega_0 - \omega_m)t - (\beta/2)\cos(\omega_0 + \omega_m)t). \tag{4.40}$$

where the 180° phase difference between the sidebands is now clearly shown. If $\beta > 1$, we have *wideband frequency modulation*, which requires additional sidebands at integral multiples of the modulation frequency. These sidebands are present to essentially preserve the constant amplitude of the modulated carrier when the phase deviation is large. Analytically, the field is given by an expansion in *Bessel functions*:

$$E(t) = E_0 \sum_{-\infty}^{\infty} J_n(\beta)e^{i(\omega_0 + n\omega_m)t}, \tag{4.41}$$

where $J_n(\beta)$ is the $n$th order Bessel function and the real part of the expansion reduces to eqn 4.40 when $\beta \ll 1$.

For completeness, we will give the analytic representation of amplitude modulation:

$$\begin{aligned}
E(t) &= E_0(1 + \alpha\cos\omega_m t)\cos\omega_0 t \\
&= E_0(\cos\omega_0 t + (\alpha/2)\cos(\omega_0 - \omega_m)t + (\alpha/2)\cos(\omega_0 + \omega_m)t, 
\end{aligned} \tag{4.42}$$

where $\alpha$ is the *index of amplitude modulation*. In communication practice, $\alpha$ is sometimes represented as a *per cent*, where 100% amplitude modulation is considered to be the largest index which can be detected without distortion.

## 4.7   The Pound–Drever–Hall approach

The Pound–Drever–Hall method has become the standard method for frequency locking a laser to a very stable cavity and has achieved spectacular performance – lasers with short term linewidths of well under 1 Hz are now routine. It is the optical version of the very old microwave *Pound stabilizer*, which used almost an identical approach in its later incarnations (earlier versions didn't use frequency modulation). The technique is characterized by the use of both the reflected wave from a cavity and frequency modulation to generate the discriminant. A schematic of the apparatus appears in Fig. 4.9.

The laser beam is first frequency (or phase) modulated at a frequency ($\omega_m$) which is somewhat larger than the cavity linewidth, usually with an *electro-optic modulator* (which will be described in a later chapter). A convenient way of efficiently accessing the reflected beam is to use a polarizing beam splitter and quarter-wave plate whose fast axis is at 45° to the plane of the figure. The wave plate converts the light into

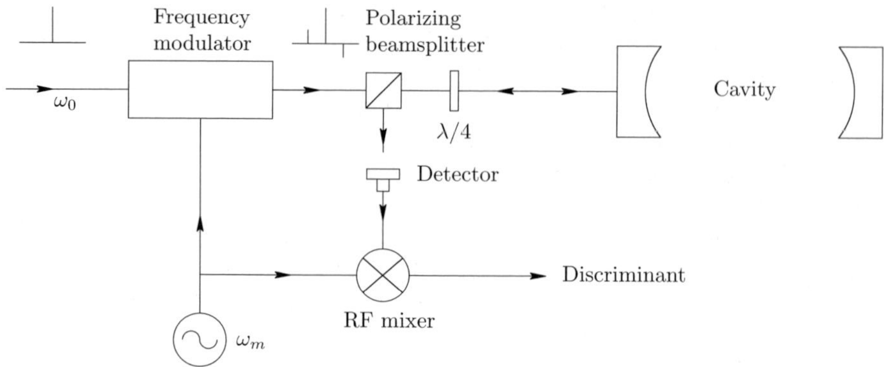

**Fig. 4.9** Schematic of Pound–Drever–Hall apparatus.

circular polarization and the reflected wave is converted back to linear polarization which is orthogonal to the input beam, causing the reflected beam to be deflected by $90°$ by the polarizing beam splitter. The cavity will convert the frequency modulation into amplitude modulation resulting in a signal at $\omega_m$ at the output of the detector. The discriminant is obtained by using the mixer to coherently extract the component of the signal at $\omega_m$. The mixer performs the algebraic operation of *multiplication*, where it *coherently* shifts the frequency of the signal spectrum both up and down by the reference frequency. The up-shifted output (at $2\omega_m$) is removed by a low pass filter and the downshifted DC part is the discriminant.

Some insight into the scheme can be obtained from Fig. 4.10, where we consider the effect of a cavity on the phasors which represent a frequency-modulated beam. When

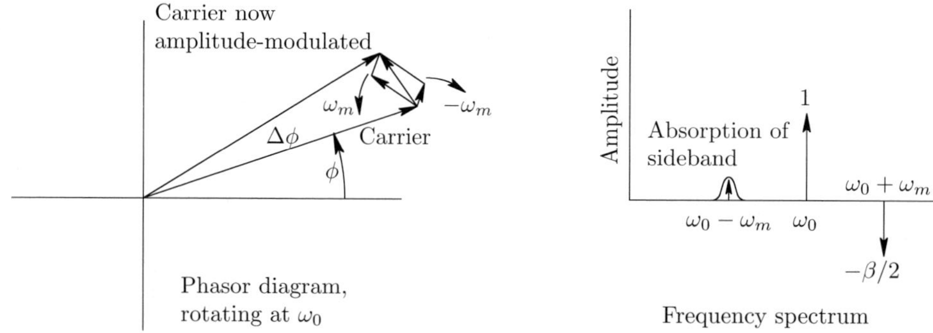

**Fig. 4.10** Influence of cavity on phasor diagram and spectrum of frequency-modulated beam.

the lower sideband is over the cavity resonance, both the amplitude and the *phase* of the reflected component at $\omega_0 - \omega_m$ will be changed. The amplitude reduction is shown on the left side of the figure (the phase change is zero at the exact resonance of the lower sideband). From the phasor diagram, one can easily see that the delicate balance between the sidebands and the carrier has been disrupted and the resultant phasor will now be amplitude modulated at $\omega_m$ as the sidebands rotate in opposite directions.

If one studies the diagram, it should be plausible that the case of the upper sideband being over the cavity resonance will generate an identical amplitude modulation which is 180° out of phase with the first. Thus, one will generate a signal which is of one sign when the laser is below resonance and of the opposite sign when the laser is above resonance: i.e., a discriminant. In summary, the cavity reflection disrupts the amplitude and phase relations among the sidebands and carrier of a frequency modulated beam, causing the frequency modulation to be converted to amplitude modulation which can be extracted by a mixer whose reference is at the modulation frequency.

Before proceeding to the derivation of the signal, it would be useful to say a few words about the polarization manipulation which takes place as the beam traverses the apparatus. Using the Jones calculus, the incident Jones vector undergoes the following transformations up to just before it enters the beam splitter for the second time:

$$\frac{1-i}{2}\begin{pmatrix} 1 & i \\ i & 1 \end{pmatrix}\begin{pmatrix} -1 & 0 \\ 0 & -1 \end{pmatrix}\frac{1-i}{2}\begin{pmatrix} 1 & i \\ i & 1 \end{pmatrix}\begin{pmatrix} 1 & 0 \\ 0 & 0 \end{pmatrix}\begin{pmatrix} E_{0x} \\ E_{0y} \end{pmatrix}. \tag{4.43}$$

The rightmost Jones matrix represents the polarization due to the polarizing beam splitter and the matrix to the left of this one is due to the quarter-wave plate at 45°. The reflection from the cavity is represented by a minus identity matrix, since the (off-resonance) reflection causes a 180° phase shift for both field components. Finally, the leftmost matrix is due to a second pass through the quarter-wave plate. Performing the indicated multiplications,

$$\text{Polarization before second beam splitter pass:} \quad \propto \begin{pmatrix} 0 \\ E_{0x} \end{pmatrix}. \tag{4.44}$$

This is of course the desired polarization for a 90° deflection by the beam splitter. One is thus able to detect the beam reflected from the cavity without loss. The reflected beam can also be extracted using a conventional (dissipative) 50/50 beam splitter, which might have the advantage of being less sensitive to polarization errors.

In deriving the discriminant for the Pound–Drever–Hall method, we start with the frequency-modulated field, where we will consider only two sidebands but with a modulation index ($\beta$) which can be unity or greater. The higher-order sidebands produce discriminant features which are separated from the main resonance by integral multiples of the modulation frequency, but we are only interested in the main resonance. The field is (eqn 4.41)

$$E(t) = E_0 \left\{ J_0(\beta)e^{i\omega_0 t} + J_1(\beta)(e^{i(\omega_0+\omega_m)t} - e^{i(\omega_o-\omega_m)t}) \right\}, \tag{4.45}$$

where we used the fact that $J_{-n}(\beta) = (-1)^n J_n(\beta)$, and the modulation frequency is $\omega_m$.

The complex reflectivity of the cavity is taken from eqn 3.19 under impedance-matched conditions and expressed as a function of angular frequency, $\omega$,

$$E_r \equiv E_0 F(\omega) = -E_0 r\frac{2i(\omega-\omega')/\delta\omega}{1+2i(\omega-\omega')/\delta\omega}, \tag{4.46}$$

where $\omega'$ is the cavity resonant frequency and $\delta\omega$ is the full width at half maximum of the cavity resonance. We will assume that $|r| \approx 1$ in the following. Since there are three frequency components incident on the cavity and reflection from the cavity is a *linear operation*, by the *principle of superposition* the reflected signal is simply the sum of the reflections from each component taken individually:

$$E_r/E_0 = J_0(\beta)F(\omega_0)e^{i\omega_0 t} + J_1(\beta)\left(F(\omega_0 + \omega_m)e^{i(\omega_0 + \omega_m)t} - F(\omega_0 - \omega_m)e^{i(\omega_0 - \omega_m)t}\right).$$
$$(4.47)$$

If we write the detected signal at the output of the photodetector amplifier as $V_{out} = GP_r \propto |E_r|^2$, where $P_r$ is the reflected power and the detector responsivity is $G$,

$$V_{out} = 2GP_0 J_0(\beta)J_1(\beta)\left\{\,\mathrm{Re}\left[F(\omega_0)F^*(\omega_0 + \omega_m) - F^*(\omega_0)F(\omega_0 - \omega_m)\right]\cos\omega_m t \right.$$
$$\left. + \mathrm{Im}\left[F(\omega_0)F^*(\omega_0 + \omega_m) - F^*(\omega_0)F(\omega_0 - \omega_m)\right]\sin\omega_m t \,\right\},$$
$$(4.48)$$

where $P_0$ is the incident power and we have left out the squared terms, which are constant in time, and have also left out the cross terms between the sidebands. It turns out that only the coefficient of the $\sin\omega_m t$ term gives a signal having the desired *dispersion shape* (the $\cos\omega_m t$ term generates a series of *Lorentzians*). The final signal is therefore

$$V_{out} = 2GP_0 J_0(\beta)J_1(\beta)\left\{\mathrm{Im}\left[F(\omega_0)F^*(\omega_0 + \omega_m) - F^*(\omega_0)F(\omega_0 - \omega_m)\right]\right\}\sin\omega_m t.$$
$$(4.49)$$

The mixer will remove the $\sin\omega_m t$ term by shifting it down to zero frequency (DC). Substituting for $F(\omega)$ and defining $\Delta\omega \equiv \omega_0 - \omega'$

$$V'_{out} = \frac{32GP_0 J_0(\beta)J_1(\beta)\omega_m^2}{\delta\omega^3}\frac{\Delta\omega - 4\frac{\Delta\omega(\Delta\omega+\omega_m)(\Delta\omega-\omega_m)}{\delta\omega^2}}{\left(1 + 4\left(\frac{\Delta\omega}{\delta\omega}\right)^2\right)\left(1 + 4\left(\frac{\Delta\omega+\omega_m}{\delta\omega}\right)^2\right)\left(1 + 4\left(\frac{\Delta\omega-\omega_m}{\delta\omega}\right)^2\right)},$$
$$(4.50)$$

where $V'_{out}$ is the signal after being demodulated by a mixer whose local oscillator frequency is $\omega_m$. This expression is plotted in Fig. 4.11. If one is only interested in the central resonance ($\Delta\omega \leq \delta\omega$) and $\omega_m \gg \delta\omega$, this expression can be greatly simplified as

$$V'_{out} = \frac{8GP_0 J_0(\beta)J_1(\beta)}{\delta\omega}\frac{-\Delta\omega}{1 + 4\left(\frac{\Delta\omega}{\delta\omega}\right)^2} \qquad \Delta\omega \leq \delta\omega, \quad \omega_m \gg \delta\omega, \qquad (4.51)$$

which is a conventional *dispersion signal*. The slope at resonance can be easily determined by differentiation:

$$\frac{\partial V'_{out}}{\partial\omega_0} = -8GP_0 J_0(\beta)J_1(\beta)\frac{1}{\delta\omega}.$$
$$(4.52)$$

The positive and negative peaks occur when $\Delta\omega = \pm\delta\omega/2$. The actual slope can be shown to be twice the value obtained by dividing the difference between the maximum and the minimum by $\delta\omega$. Finally, the expression $J_0(\beta)J_1(\beta)$ can be shown to be at its maximum when $\beta = 1.08$.

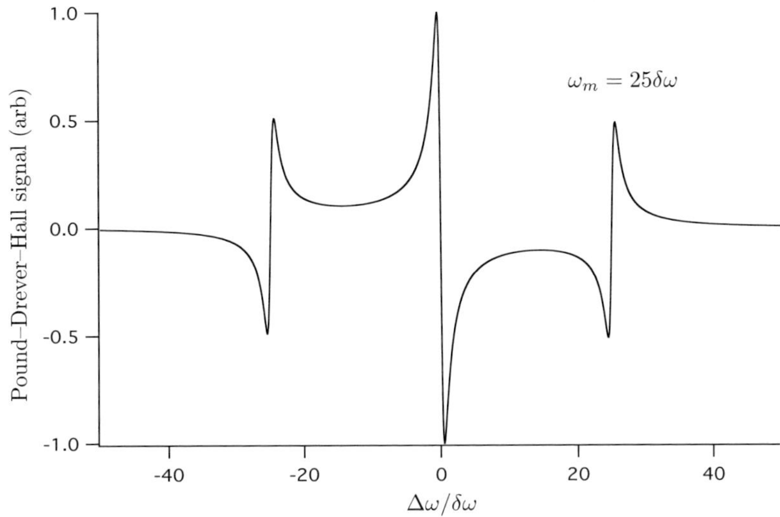


**Fig. 4.11** Signal from Pound–Drever–Hall apparatus plotted for $\beta = 1.08$.

## 4.8   Frequency response of a cavity-based discriminator

In this chapter, we have discussed the use of an optical cavity as a frequency discriminator, with the understanding that it will be used in some kind of frequency servo system. The analysis of the servo loop requires knowledge of the frequency response of each element in the loop, including the discriminator. The discussion of the latter can be a bit confusing, since the independent variable (the item being controlled) is itself a frequency and we are seeking the *frequency response of a frequency*. Operationally, we can apply some frequency modulation at $\omega_N$ to the laser and derive the amplitude and phase of the response of the discriminator as a function of this modulation frequency. Viewing the desired result as a response to a *frequency-modulated laser* should avoid the confusion.

In order to analyze the frequency response of the PDH discriminant, we will first assume that the average laser frequency is at the cavity resonant frequency (probably by frequency locking). We will then generate a small sinusoidal *frequency dither* at $\omega_N$ and determine the amplitude and phase of the discriminator output at $\omega_N$:

$$\text{Frequency dither input signal:}\quad \Delta\omega(t) = \Delta\omega_0 \cos\omega_N t. \tag{4.53}$$

The phase modulation and modulation index of the input signal are

$$\text{Phase modulation:}\quad \Delta\phi(t) = \frac{\Delta\omega_0}{\omega_N}\sin\omega_N t = \alpha\sin\omega_N t \tag{4.54}$$

$$\text{Modulation index:}\quad \alpha = \frac{\Delta\omega_0}{\omega_N}. \tag{4.55}$$

The calculation proceeds in much the same way as the discriminant calculation except that we now have *two phase-modulation inputs*: the one at $\omega_m$ needed to implement the PDH approach and the "noise" input at $\omega_N$. Since the noise signal is very small, we can treat it as a *narrowband* modulation. The doubly modulated field is

$$
\begin{aligned}
E(t) &= E_0 \exp(i\omega_0 t + i\beta \sin\omega_N t + i\alpha \sin\omega_N t) \\
&= E_0 \left[ J_0(\beta)e^{i\omega_0 t} + J_1(\beta)(e^{i(\omega_0+\omega_m)t} - e^{i(\omega_0-\omega_m)t}) \right] \\
&\quad \times \left[ 1 + \frac{\alpha}{2}(e^{i(\omega_0+\omega_N)t} - e^{i(\omega_0-\omega_N)t}) \right] \\
&= E_0 \left[ J_0(\beta)e^{i\omega_0 t} + \frac{J_0(\beta)\alpha}{2}(e^{i(\omega_0+\omega_N)t} - e^{i(\omega_0-\omega_N)t}) \right. \\
&\quad \left. + J_1(\beta)(e^{i(\omega_0+\omega_m)t)} - e^{i(\omega_0-\omega_m)t)}) \right],
\end{aligned}
\tag{4.56}
$$

where we have left out terms at $\omega_m \pm \omega_N$ since they will not contribute to the signal at $\omega_m$. Using the principle of superposition to obtain the reflected field,

$$
\begin{aligned}
\frac{E_r}{E_0} &= \left[ J_0(\beta)e^{i\omega_0 t}F(\omega_0) + \frac{J_0(\beta)\alpha}{2}\left\{ e^{i(\omega_0+\omega_N)t}F(\omega_0+\omega_N) \right. \right. \\
&\quad \left. \left. - e^{i(\omega_0-\omega_N)t}F(\omega_0-\omega_N) \right\} + 2ie^{i\omega_0 t}J_1(\beta)\sin\omega_m t \right],
\end{aligned}
\tag{4.57}
$$

where $F(\omega_0 \pm \omega_m) = 1$ since the sidebands lie well outside the cavity width. Since $\omega_0 = \omega'$ (laser on resonance), we can considerably simplify this expression using $F(\omega_0) = 0$ and $F(\omega_0 + \omega_N) = F^*(\omega_0 - \omega_N)$ to obtain

$$
\begin{aligned}
\frac{E_r}{E_0} &= \left[ iJ_0(\beta)\alpha e^{i\omega_0 t}\mathrm{Im}\{e^{i\omega_N t}F(\omega_0+\omega_N)\} \right. \\
&\quad \left. + 2ie^{i\omega_0 t}J_1(\beta)\sin\omega_m t \right].
\end{aligned}
\tag{4.58}
$$

Proceeding as before to determine $V_{out}$, we calculate the square modulus and consider only the terms with a $\sin\omega_m t$ time dependence since the mixer and low-pass filter will isolate these terms:

$$
\begin{aligned}
V_{out}(\omega_m) &\propto 4\alpha J_0(\beta)J_1(\beta)\mathrm{Im}\{F(\omega_0+\omega_N)e^{i\omega_N t}\}\sin\omega_m t \\
&= 4\alpha J_0(\beta)J_1(\beta)|F(\omega_0+\omega_N)|\cos(\omega_N t + \phi)\sin\omega_m t,
\end{aligned}
\tag{4.59}
$$

where

$$
\phi = -\tan^{-1}\frac{\mathrm{Re}\{F(\omega_0+\omega_N)\}}{\mathrm{Im}\{F(\omega_0+\omega_N)\}}
\tag{4.60}
$$

and the slightly unusual expression for the phase is chosen so that the response is a cosine rather than a sine for easier comparison to the cosine "noise" input signal ($\Delta\omega(t)$). The output can be seen as a "beating" between the reflected sidebands at $\omega_m$ and those at $\omega_N$ (the reflected carrier is zero on resonance). Finally, dropping the constant multiplicative terms (except for $\alpha$, which is proportional to the *amplitude of the input signal*), the output of the mixer is

$$V_{mixer} \propto |F(\omega_0 + \omega_N)|\alpha \cos(\omega_N t + \phi)$$
$$= \frac{2\omega_N/\delta\omega}{\sqrt{1 + 4\left(\frac{\omega_N}{\delta\omega}\right)^2}}\alpha \cos(\omega_N t + \phi). \tag{4.61}$$

This can be rewritten as follows to emphasize the *transfer function* aspect of the equation:

$$V_{mixer} \propto \left(\frac{1}{\sqrt{1 + 4\left(\frac{\omega_N}{\delta\omega}\right)^2}}\right)\Delta\omega \cos(\omega_N t + \phi) \tag{4.62}$$

$$\phi = -\tan^{-1}\frac{2\omega_N}{\delta\omega}. \tag{4.63}$$

Thus, an input frequency signal $\Delta\omega \cos\omega_N t$ is multiplied by the bracketed factor to obtain the output amplitude, and its phase is shifted by $\phi$ to obtain the output phase. The *transfer function* of the discriminator is identical to that for a Butterworth low-pass filter consisting of a series $R$ and shunt $C$ having time constant $\tau = RC$. It is easy to show that the transfer function of this circuit is:

$$G(\omega) = \frac{1}{1 + i\omega\tau}. \tag{4.64}$$

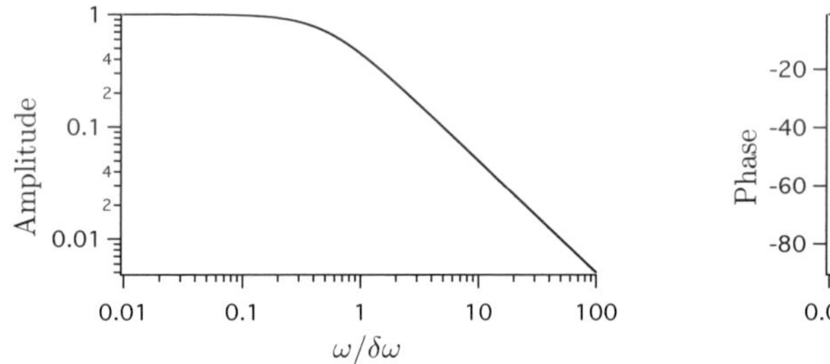
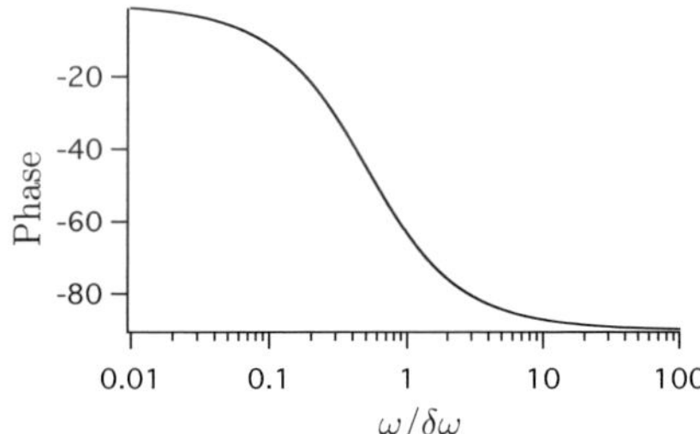

**Fig. 4.12** Magnitude and phase of the PDH transfer function. The $-3$ db point occurs at $\delta\omega/2$.

The magnitude and phase of $G(\omega)$ are:

$$|G(\omega)| = \frac{1}{1 + \omega^2\tau^2} \tag{4.65}$$

$$\phi = -\tan^{-1}\omega\tau. \tag{4.66}$$

If we let $\tau = 2/\delta\omega$ the equivalence is established.

A plot of the magnitude and phase of the transfer function appears in Fig. 4.12. The plot displays the characteristic amplitude and phase characteristic of a single-pole low-pass filter. For the purposes of servo theory, the transfer function has one *pole* (zero of the denominator) which occurs at the $-3$ db point of the amplitude plot and the point where the phase lags by $-45°$. This point occurs at

$$\text{Pole location } (-3 \text{ db point}): \quad \omega_N = \frac{\delta\omega}{2}. \tag{4.67}$$

Below this frequency the output of the discriminator is more or less proportional to the input signal; above it, the discriminator behaves like an integrator with its characteristic 90° phase lag and 6 db/octave amplitude roll-off. This lag can be compensated in the electronics by placing a *zero* (a *differentiator*) at the same frequency.

## 4.9   Further reading

The best references on the three methods for generating a discriminant appear in the original papers describing the techniques. The side-of-line approach was described in the paper by Helmcke, Lee and Hall (1973); the Pound–Drever–Hall (PDH) scheme was originally described in the paper by Drever, Hall et al. (1983) and the polarization approach was described in the paper by Hansch and Couillaud (1980). There is a very large literature on the PDH method due to its use in the LIGO project, whose purpose is the detection of gravitational waves. Most papers on the PDH technique don't discuss its frequency response; one of the few exceptions is the paper by Day, Gustafson and Byer (1992).

## 4.10   Problems

(4.1) Determine the Jones matrix for a polarizing beam-splitter whose transmission axis is rotated (about the propagation direction) by angle $\theta$ from the $x$-direction. A polarizing beam-splitter is a device (in the form of a cube) which transmits one component of light and reflects the orthogonal component (see figure below).

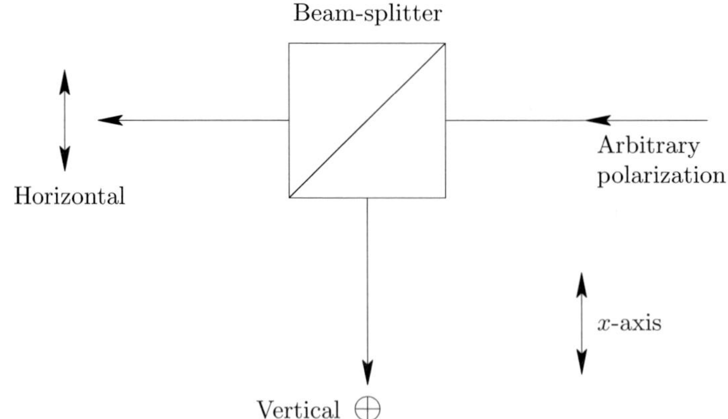

(4.2) Show, using Jones matrices, that a circular analyzer's behavior is independent of rotations about the propagation axis. A circular analyzer consists of a quarter-wave plate and polarizer with the fast axis of the wave plate at $45°$ to the transmission polarization direction of the polarizer. It transmits one type of circular polarization (right or left) and rejects the other.

(4.3) All of the discriminants discussed in this chapter can have a problem with offsets: any *additive* constant signal at the discriminator output will shift the *lock point* (the zero crossing) away from the exact cavity resonance. Calculate the size of the lock point shift (in frequency units) as a function of the cavity width and peak fringe height for the Pound–Drever–Hall discriminant.

(4.4) The offsets mentioned in the previous problem can be generated by inadvertent *amplitude modulation* from the electro-optic modulator which is used to phase modulate the beam. Calculate the approximate size of the offset due to a small amplitude modulation whose index is $\alpha$.

# 5
# Laser gain and some of its consequences

## 5.1 Introduction

The crucial element common to all lasers is the ability of a specially prepared medium to amplify optical radiation. We will describe this phenomenon from two viewpoints. First, we will derive the *optical Bloch equations*, based upon a semiclassical treatment of a two-level atomic system and a radiation field. As a visualization aid, we will discuss the formally identical *classical Bloch equations*, which describe the behavior of a magnetic moment in a combined static and rotating magnetic field. It will be shown that laser gain is proportional to the imaginary part of the linear susceptibility, which will be obtained from the semiclassical treatment. The second viewpoint will use a slightly modified form of Einstein's A and B coefficients adapted to describing interactions between atoms and *monochromatic* fields. Various damping and broadening mechanisms will be introduced in a reasonable but non-rigorous fashion and the crucial distinction between *homogeneous* and *inhomogeneous* broadening will be introduced.

Our approach will be reasonably complete but not as exhaustive as other books on the subject. The goal is not to prepare the reader for a career in the design of lasers but rather to develop an appreciation in the reader of the several laser systems described in later chapters.

## 5.2 The wave equation

We will use the *semiclassical* approach to the interaction of an atom with radiation: the atom will be treated quantum mechanically but the radiation will be treated classically as a time-dependent term in the Hamiltonian. We start with the Schrödinger equation

$$i\hbar\frac{\partial\psi}{\partial t} = (H_0 + H_I(t))\psi, \tag{5.1}$$

where $H_0$ is the Hamiltonian of the atom *in the absence* of radiation and $H_I(t)$ is the time-dependent term which completely describes the interaction of the atom with radiation.

The atom has two *non-degenerate* levels labeled 1 (*ground state*) and 2 (*excited state*) with level 2 having the larger energy. The wavefunction, $\psi(\boldsymbol{r}, t)$, is expanded in eigenstates of the unperturbed Hamiltonian $H_0$ with time-dependent expansion coefficients. The rapidly varying time dependence of the eigenstates is factored out so that the coefficients $c_1(t)$ and $c_2(t)$ vary slowly with time. Thus,

$$\psi(\boldsymbol{r},t) = c_1(t)|1\rangle e^{-i\omega_1 t} + c_2(t)|2\rangle e^{-i\omega_2 t}, \qquad |c_1|^2 + |c_2|^2 = 1, \tag{5.2}$$

where

$$H_0|1\rangle = E_1|1\rangle \tag{5.3}$$
$$H_0|2\rangle = E_2|2\rangle \tag{5.4}$$
$$\omega_{1,2} = E_{1,2}/\hbar. \tag{5.5}$$

Substituting $\psi(\boldsymbol{r},t)$ (eqn 5.2) into Schrödinger's equation (eqn 5.1) and making use of the *orthonormality* of $|1\rangle$ and $|2\rangle$ yields two coupled equations:

$$\begin{aligned} i\hbar\dot{c}_1 &= \langle 1|H_I(t)|2\rangle e^{-i\omega_0 t} c_2 \\ i\hbar\dot{c}_2 &= \langle 2|H_I(t)|1\rangle e^{i\omega_0 t} c_1, \end{aligned} \tag{5.6}$$

where $\omega_0 = \omega_2 - \omega_1$ and we have used $\langle 1|H_I|1\rangle = \langle 2|H_I|2\rangle = 0$ for states of definite parity, since the presumed (electric dipole) interaction term, $H_I$, has *odd parity*.

## 5.3 The interaction term

In the presence of an electromagnetic field, the Hamiltonian is:

$$H = \frac{1}{2m}(\boldsymbol{p} - e\boldsymbol{A})^2 + e\phi + V = H_0 + H_I, \qquad H_0 = \frac{\boldsymbol{p}^2}{2m} + V, \tag{5.7}$$

where $\boldsymbol{p}$ is the canonical momentum, $V$ is the potential energy exclusive of the radiation field, and $\phi$ and $\boldsymbol{A}$ are the scalar and vector potentials defined by:

$$\boldsymbol{E} = -\nabla\phi - \frac{\partial \boldsymbol{A}}{\partial t} \tag{5.8}$$
$$\boldsymbol{B} = \nabla \times \boldsymbol{A}. \tag{5.9}$$

Due to the way they are defined, the potentials $\phi$ and $\boldsymbol{A}$ are somewhat arbitrary. The ability to change the potentials without changing the fields is called *gauge invariance*, and the constraints on a particular choice of potentials is called a *gauge*. In the absence of sources (charges and currents), one can use the *Coulomb gauge*, which is defined by the equation

$$\text{Coulomb gauge:} \quad \nabla \cdot \boldsymbol{A} = 0. \tag{5.10}$$

Since there are no sources, we also have

$$\phi = 0. \tag{5.11}$$

Then, the interaction term is

$$H_I = \frac{e}{m}\boldsymbol{A} \cdot \boldsymbol{p} + \frac{e^2}{2m}\boldsymbol{A}^2 \approx \frac{e}{m}\boldsymbol{A} \cdot \boldsymbol{p}, \tag{5.12}$$

where it can be shown that the $\boldsymbol{A}^2$ term is generally much smaller than the $\boldsymbol{A}\cdot\boldsymbol{p}$ term.

We will assume that the field is given by a plane traveling wave whose vector potential is:

$$A(r, t) = A_0 \cos(k \cdot r - \omega t). \tag{5.13}$$

The electric field is

$$E = E_0 \sin(k \cdot r - \omega t), \qquad E_0 = \omega A_0. \tag{5.14}$$

The interaction term is currently expressed as a function of $p$; we can use the following identity to convert matrix elements of $p$ into matrix elements of $r$:

$$\langle i | p | j \rangle = i m \omega_{ij} \langle i | r | j \rangle, \tag{5.15}$$

where $\omega_{ij} = (E_i - E_j)/\hbar$. This identity is easily obtained by taking matrix elements of the commutator $[H_0, r]$ between energy eignestates $i$ and $j$ and using the identity $[r_s, p_t] = i\hbar \delta_{st}$. From eqns 5.12 and 5.15, the matrix elements of $H_I$ are

$$\langle i | H_I | j \rangle = i e E_0 \cdot \langle i | r \cos(k \cdot r - \omega t) | j \rangle, \tag{5.16}$$

where we are assuming that we are in the vicinity of a *resonance* and thus substitute $\omega$ for $\omega_{ij}$. In evaluating the matrix element of the cosine term, one integrates over the atom, whose spatial extent is $\approx 0.1$ nm. The spatial period of the cosine function is the wavelength of optical radiation, $\approx 500$ nm. Thus, one can make the *dipole approximation* and assume that $k \cdot r \ll 1$ over the atom. Then, ignoring a constant phase term ($i$), the desired *dipole matrix element* is

$$\langle i | H_I | j \rangle = \langle i | e E_0 \cdot r | j \rangle \cos(\omega t). \tag{5.17}$$

## 5.4   The rotating wave approximation

The interaction matrix element can be written in terms of a *Rabi frequency*, $\Omega$, which is defined as:

$$\Omega \equiv \frac{1}{\hbar} \langle 1 | e r \cdot E_0 | 2 \rangle = \frac{e}{\hbar} \int \psi_1^*(r) r \cdot E_0 \psi_2(r) d^3 r. \tag{5.18}$$

The coupled equations (eqns 5.6) are now

$$\begin{aligned} i\dot{c}_1 &= \Omega \cos(\omega t) e^{-i\omega_0 t} c_2 \\ i\dot{c}_2 &= \Omega^* \cos(\omega t) e^{i\omega_0 t} c_1. \end{aligned} \tag{5.19}$$

These equations depend upon two (very high) optical frequencies, $\omega$ and $\omega_0$, and upon two slowly varying quantities, $c_1$ and $c_2$. These disparate time dependencies can be resolved if we can find a way of eliminating the rapidly oscillating terms. This can be done by expanding the cos() in exponentials and throwing away the very high frequency $e^{\pm i(\omega + \omega_0)t}$ terms. The coupled equations then become

$$\begin{aligned} i\dot{c}_1 &= \frac{\Omega}{2} e^{i\delta t} c_2 \\ i\dot{c}_2 &= \frac{\Omega^*}{2} e^{-i\delta t} c_1, \end{aligned} \tag{5.20}$$

where $\delta = \omega - \omega_0$. Ignoring the high frequency terms is called the *rotating wave approximation*. It is equivalent to replacing the *oscillating* term $\cos \omega t$ with a (rotating) complex exponential, $\cos \omega t \pm i \sin \omega t \ (= e^{\pm i \omega t})$, where the sign is chosen to make all of the factors in the equations slowly varying. The validity of this will be discussed later on and we will see what this means *graphically* when we consider the *classical Bloch equations*. Finally, we eliminate all of the explicit time dependence on the right-hand side by transforming the coefficients $c_{1,2}$ into coefficients $\tilde{c}_{1,2}$ using a matrix:

$$\begin{pmatrix} \tilde{c}_1 \\ \tilde{c}_2 \end{pmatrix} = \begin{pmatrix} e^{-i\delta t/2} & 0 \\ 0 & e^{i\delta t/2} \end{pmatrix} \begin{pmatrix} c_1 \\ c_2 \end{pmatrix}. \tag{5.21}$$

The coupled equations are now:

$$\begin{aligned} i\dot{\tilde{c}}_1 &= \frac{1}{2}(\delta\tilde{c}_1 + \Omega\tilde{c}_2) \\ i\dot{\tilde{c}}_2 &= \frac{1}{2}(\Omega\tilde{c}_1 - \delta\tilde{c}_2). \end{aligned} \tag{5.22}$$

## 5.5   Density matrix of two-level system

There are three approaches to solving the original coupled equations (eqns 5.6). One can use a perturbation approach, where the field is so weak that $c_2$ never has any appreciable amplitude and integrate the second equation assuming that $c_1 \approx 1$ (this is done *after* damping terms are added to the equations). The time integral of a function of the form $\cos(\omega t)e^{\pm i\omega_0 t}$ has two terms: one with a resonant denominator and one without. Invoking the *rotating wave approximation* consists in eliminating the term with the non-resonant denominator.

The perturbation approach is not appropriate when studying lasers because the excited state amplitude can be quite large. A solution for arbitrary excitation can be obtained by combining the two equations into a single second-order differential equation. An alternative approach which is also valid for strong excitation is to take the transformed equations (eqns 5.22) and form a *density matrix*. We will discuss the density matrix in this section.

Density matrices are generally used in quantum statistics where one has a statistical *ensemble* and can assign a probability to each possible wave function of the ensemble. In our case, we have a *pure state* whose wave function is known. However, there are some advantages to using a density matrix even then. The density operator, $\rho$, for a pure state $\psi$ is defined formally as

$$\rho = |\psi\rangle\langle\psi| \tag{5.23}$$

and is obviously Hermitian. The matrix representation (also denoted by $\rho$) of the density operator can be easily seen to be the following *outer product*:

$$\rho = \begin{pmatrix} c_1 \\ c_2 \end{pmatrix} \begin{pmatrix} c_1^* & c_2^* \end{pmatrix} = \begin{pmatrix} |c_1|^2 & c_1 c_2^* \\ c_2 c_1^* & |c_2|^2 \end{pmatrix}. \tag{5.24}$$

The reason for the absence of the exponential time factors is that the density matrix is being evaluated in a system which is *rotating* at angular rate $\omega$. When we discuss the classical Bloch equations, we will also perform a rotation at the frequency of the rotating magnetic field. (In the *laboratory frame*, the diagonal elements will be the same as above but the off-diagonal elements will acquire a time dependence at optical frequencies).

The diagonal and off-diagonal elements of $\rho$ have interesting interpretations. The diagonal elements $\rho_{ii}$ give the *probability* of finding the atom in state $i$. These are sometimes called the *populations* of the state $i$, though this term is perhaps more appropriate when one multiplies by the number density of atoms. The off-diagonal elements are the *coherences*, which are used when one takes the expectation value of some non-diagonal operator such as the dipole moment of the atom. The populations vary slowly with time and the coherences have a time-dependence at optical frequencies (in the *laboratory* frame).

Using the $\tilde{c}_i$, the density matrix $\tilde{\rho}$ is

$$\tilde{\rho} = \begin{pmatrix} |\tilde{c}_1|^2 & \tilde{c}_1\tilde{c}_2^* \\ \tilde{c}_2\tilde{c}_1^* & |\tilde{c}_2|^2 \end{pmatrix}. \tag{5.25}$$

From the definition of $\tilde{\rho}_{ij}$ together with the coupled differential equations (eqns 5.22), one can obtain the following differential equations for $\tilde{\rho}$:

$$\begin{aligned}
\frac{d\tilde{\rho}_{12}}{dt} &= \frac{d\tilde{\rho}_{21}^*}{dt} = -i\delta\tilde{\rho}_{12} + \frac{i\Omega}{2}(\tilde{\rho}_{11} - \tilde{\rho}_{22}) \\
\frac{d\tilde{\rho}_{22}}{dt} &= -\frac{d\tilde{\rho}_{11}}{dt} = \frac{i\Omega}{2}(\tilde{\rho}_{21} - \tilde{\rho}_{12}).
\end{aligned} \tag{5.26}$$

These equations can be rewritten in a very compact form if we define the following quantities:

$$\begin{aligned}
u &\equiv \tilde{\rho}_{12} + \tilde{\rho}_{21} & &= 2\mathrm{Re}\{\tilde{\rho}_{12}\} & &(5.27) \\
v &\equiv -i(\tilde{\rho}_{12} - \tilde{\rho}_{21}) & &= 2\mathrm{Im}\{\tilde{\rho}_{12}\} & &(5.28) \\
w &= \tilde{\rho}_{11} - \tilde{\rho}_{22} & &= \text{Population difference} & &(5.29) \\
\boldsymbol{R} &= u\hat{x} + v\hat{y} + w\hat{z} & &= \text{Bloch vector} & &(5.30) \\
\boldsymbol{W} &= \Omega\hat{x} + \delta\hat{z} & &= \text{Effective field.} & &(5.31)
\end{aligned}$$

Substituting these into eqns 5.26 yields the *optical Bloch equations (without damping)*:

$$\dot{u} = \delta v \tag{5.32}$$

$$\dot{v} = -\delta u + \Omega w \tag{5.33}$$

$$\dot{w} = -\Omega v. \tag{5.34}$$

These equations can be combined into a single vector equation:

$$\frac{d\boldsymbol{R}}{dt} = \boldsymbol{R} \times \boldsymbol{W}. \tag{5.35}$$

It should be kept in mind that the optical Bloch equation is just the Schrödinger equation (in the so-called *interaction representation*) for an irradiated two-level system expressed in terms of quantities (*populations* and *coherences*) which are more familiar to the experimentalist than is the wavefunction. We will discuss the significance of this equation by deriving a formally identical *classical* equation which describes the behavior of a magnetic dipole in a combined static and rotating magnetic field (the *classical Bloch equation*).

## 5.6   The classical Bloch equation

Consider a *magnetic moment* $\boldsymbol{\mu}$ in a vertical magnetic field, $\boldsymbol{H}_0 = H_0\hat{\boldsymbol{z}}$ (Fig. 5.1). Equating the time rate of change of the angular momentum, $\boldsymbol{L}$, to the torque,

$$\frac{d\boldsymbol{L}}{dt} = \gamma \boldsymbol{L} \times \boldsymbol{H}_0, \tag{5.36}$$

where $\boldsymbol{\mu} = \gamma \boldsymbol{L}$, and $\gamma$ is the *gyromagnetic ratio* (ratio of the magnetic moment to the angular momentum). It is straightforward to show that the magnetic moment

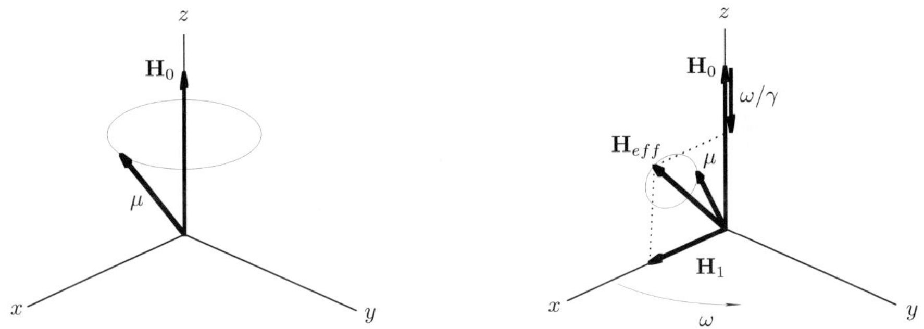

**Fig. 5.1** Magnetic moment precessing in magnetic field (left). The addition of a rotating $\boldsymbol{H}_1$ field in a frame rotating with $\boldsymbol{H}_1$ shortens the static field (in the rotating frame) and the moment now precesses around $\boldsymbol{H}_{eff}$ (right).

precesses around the static field at an angular rate equal to $\omega_L = \gamma H_0$, called the *Larmor frequency*.

We now add a small magnetic field, $\boldsymbol{H}_1$, *rotating* in the $xy$-plane at angular rate $\omega$. It is very convenient to transform to a *frame which is rotating at the same rate*, so that $\boldsymbol{H}_1$ is fixed in the new frame (along the $x$-axis, $\boldsymbol{H}_1 = H_1\hat{\boldsymbol{x}}$). We use the relationship between the time derivative of a vector in a *fixed frame* and one in a *frame rotating* at $\omega$ (this can be easily derived by performing an infinitesimal rotation and using the definition of the derivative):

$$\frac{d}{dt} = \frac{\partial}{\partial t} + \boldsymbol{\omega} \times \tag{5.37}$$

where $d/dt$ is in the fixed frame and $\partial/\partial t$ is in the rotating frame. Thus,

$$\frac{d\boldsymbol{L}}{dt} = \frac{\partial \boldsymbol{L}}{\partial t} + \boldsymbol{\omega} \times \boldsymbol{L} \tag{5.38}$$

and, substituting into eqn 5.36,

$$\frac{\partial \boldsymbol{L}}{\partial t} = \gamma \boldsymbol{L} \times \left( \left( H_0 - \frac{\omega}{\gamma} \right) \hat{\boldsymbol{z}} + H_1 \hat{\boldsymbol{x}} \right) \tag{5.39}$$

is the equation for the rate of change of $\boldsymbol{L}$ in the *rotating frame*. The quantity in the outer brackets is called the *effective field* since it is the field in the rotating frame about which the magnetic moment precesses. The *classical Bloch equation* is obtained by identifying this quantity as $\boldsymbol{W}/\gamma$:

$$\frac{\partial \boldsymbol{L}}{\partial t} = \boldsymbol{L} \times \boldsymbol{W}, \tag{5.40}$$

which is the same equation as the one derived earlier for a two-level quantum mechanical system.

The interpretation of the classical equation is straightforward. With no $\boldsymbol{H}_1$ field, the magnetic moment precesses about $\boldsymbol{H}_0$ at the Larmor frequency, $\omega_L$. When one adds a *fixed* $\boldsymbol{H}_1$ field, the precession is now around the combined fields $H_0\hat{\boldsymbol{z}} + H_1\hat{\boldsymbol{x}}$. If one allows $\boldsymbol{H}_1$ to rotate at angular rate $\omega$, the time derivative transformation causes $\boldsymbol{H}_0$ to shorten by $(\omega/\gamma)\hat{\boldsymbol{z}}$. When the rotation is at the Larmor frequency, the $z$-component of the magnetic field will be reduced to zero (in the rotating frame) and the magnetic moment will precess around the horizontal $\boldsymbol{H}_1$ field at the *Rabi frequency*, $\gamma H_1$. If one is measuring the $z$-component of the magnetic moment, it will have a sinusoidal time dependence at the Rabi frequency. This is the *resonant* behavior that is sought after in magnetic resonance experiments, where the Rabi frequency was first introduced.

We now make the following *equivalences* between the classical quantities and those used in the quantum mechanical optical Bloch equation,

| Classical | | Quantum | |
|:---:|:---:|:---:|:---|
| $\gamma H_1$ | $\Longleftrightarrow$ | $\Omega$ | Rabi frequency |
| $\gamma H_0$ | $\Longleftrightarrow$ | $\omega_0$ | Resonance frequency |
| $\boldsymbol{L}$ | $\Longleftrightarrow$ | $\boldsymbol{R}$ | Bloch vector |
| $\boldsymbol{W}/\gamma$ | $\Longleftrightarrow$ | $\boldsymbol{W}$ | Effective field |
| $\delta = \omega_0 - \omega$ | $\Longleftrightarrow$ $\delta = \omega - \omega_0$ | | Frequency detuning, |

$$\tag{5.41}$$

with the result that $\boldsymbol{W} = \delta\hat{\boldsymbol{z}} + \Omega\hat{\boldsymbol{x}}$ in both the classical and quantum mechanical cases. The reason for calling $\boldsymbol{W}$ the *effective field* in the optical Bloch equation should now be clear.

The interpretation of the two-level atom irradiated by a monochromatic field follows from the above assignments. The Rabi frequency is the rate at which the atom cycles between the two levels when the optical frequency is exactly resonant with the energy separation between the levels. The resonance frequency of the two-level system is identified with the Lamor frequency of the magnetic moment (with no $\boldsymbol{H}_1$ field). Although the Rabi frequency is strictly only defined at resonance, the off-resonance precession frequency is often referred to, and is called the *generalized Rabi frequency*. It is equal to $|W|$:

$$\text{Generalized Rabi frequency} = |W| = \sqrt{\delta^2 + \Omega^2}. \tag{5.42}$$

The precessing vector is called the *Bloch vector*. From the Bloch equation, it should be obvious that the length of $\boldsymbol{L}$ (or $\boldsymbol{R}$) is fixed but its orientation will change with time depending upon the value of $\boldsymbol{W}$ (and initial conditions). The projection of the vector on the $z$-axis is the $z$-component of the magnetic moment in magnetic resonance experiments and the *population inversion*, $\rho_{11} - \rho_{22}$, in a two-level atom. The population inversion will play a central role in laser theory: it is just the excess probability of finding the atom in the ground state over the probability of finding it in the excited state. Thus, a Bloch vector pointing up will correspond to an atom in the ground state and a Bloch vector pointing down will correspond to an atom in the excited state. (From the way we have defined the population inversion, a *negative* inversion corresponds to the excited state population being *greater* than that of the ground state.) The $x$- and $y$- components of the Bloch vector are proportional to the real and imaginary parts of the coherence element, $\rho_{12}$. We will show that these components are proportional to the quadrature components of an *oscillating dipole moment* induced in the atom by the optical field. In the classical case, the horizontal projection of the Bloch vector is the horizontal component of the oscillating magnetic dipole moment.

The classical model also provides a rationale for the *rotating wave approximation*. In most magnetic resonance experiments, one uses an *oscillating field* rather than a rotating one. An oscillating field can be written as the sum of two counter-rotating fields:

$$\boldsymbol{H}(t) = H_0\hat{\boldsymbol{x}}\cos\omega t = \boldsymbol{H}_+(t) + \boldsymbol{H}_-(t)$$

$$\boldsymbol{H}_+(t) = \frac{1}{2}H_0\left(\hat{\boldsymbol{x}}\cos\omega t + \hat{\boldsymbol{y}}\sin\omega t\right) \tag{5.43}$$

$$\boldsymbol{H}_-(t) = \frac{1}{2}H_0\left(\hat{\boldsymbol{x}}\cos\omega t - \hat{\boldsymbol{y}}\sin\omega t\right),$$

where it should be clear that $\boldsymbol{H}_+$ and $\boldsymbol{H}_-$ rotate in opposite senses. At resonance, one of these fields will rotate exactly in step with the Lamor precession of the magnetic moment and the rotating frame will rotate so that this component is fixed. In the rotating frame, the other component will rotate at $2\omega_L$ and will be very far off resonance and will therefore only slightly perturb the magnetic moment (the idea of *far off resonance* will be clarified when we discuss radiative damping and a *resonance width*). In the two-level atom, the rotation is not in coordinate space but is in the complex plane: it corresponds to the use of complex exponentials ($e^{\pm i\omega t}$) rather than sines and cosines to represent the oscillatory time dependence of the fields. Just as in the magnetic resonance case, the oscillatory field can be resolved into two counter-rotating complex exponentials and the *non-resonant* (rapidly oscillating) term will have very little effect on the Bloch vector (the off-resonance perturbation is sometimes called an *AC Stark shift* or a *Bloch-Siegert shift*).

## 5.7   Radiative and collision-induced damping

The semiclassical treatment so far can actually describe most of the aspects of a laser, in particular the *coherent amplification* which results from *stimulated emission*. Although spontaneous decay can be patched into the semiclassical theory using the *Fermi golden rule*, a satisfactory treatment requires the quantization of the radiation field. We will use a *phenomenological approach* to introduce spontaneous decay in a possibly ad hoc (but reasonable) manner: we will add *damping terms* to the differential equations satisfied by the density matrix. It turns out that it is much easier to do this in a natural way to the density matrix equations than to the differential equations for the coefficients, $c_i(t)$, since the behavior of the populations and coherences under various damping mechanisms is easier to describe. We will also modify the Bloch equations to account for *collision-induced damping* in a low-pressure gas.

We will begin with the observation that, in the absence of radiation, an atom with an excited state *population* $(\tilde{\rho}_{22})$ will spontaneously decay at a rate $(\gamma)$ which is proportional to the population:

$$\frac{d\tilde{\rho}_{22}}{dt} = -\gamma\tilde{\rho}_{22}, \tag{5.44}$$

whose solution is

$$\tilde{\rho}_{22}(t) = \tilde{\rho}_{22}|_0 e^{-\gamma t}. \tag{5.45}$$

Since we call level 1 the "ground state", we will assume that it does not decay.

What is the corresponding effect on the coherence elements, $\tilde{\rho}_{12}$ and $\tilde{\rho}_{21}$? We can answer this by using the following relation, which follows from the definition of the density matrix:

$$\tilde{\rho}_{12}(t) = \sqrt{\tilde{\rho}_{11}(t)\tilde{\rho}_{22}(t)}. \tag{5.46}$$

We have already established that $\tilde{\rho}_{22}(t)$ decays exponentially at rate $\gamma$, and $\tilde{\rho}_{11}$ does not decay. Thus, in the absence of a radiation field,

$$\tilde{\rho}_{12}(t) = \tilde{\rho}_{12}|_0 e^{-\gamma t/2}, \tag{5.47}$$

and therefore $\tilde{\rho}_{12}$ decays at a rate $\gamma/2$. We will display the most recent density matrix equations of motion here for reference:

$$\dot{u} = \delta v$$
$$\dot{v} = -\delta u + \Omega w$$
$$\dot{w} = -\Omega v.$$

Recalling that

$$u = 2Re\{\tilde{\rho}_{12}\}$$
$$v = 2Im\{\tilde{\rho}_{12}\}$$
$$w = \tilde{\rho}_{11} - \tilde{\rho}_{22},$$

the following modified density matrix equations will have the correct damping properties when the field is turned off:

$$\dot{u} = \delta v - \frac{\gamma}{2}u$$
$$\dot{v} = -\delta u + \Omega w - \frac{\gamma}{2}v \tag{5.48}$$
$$\dot{w} = -\Omega v - \gamma(w - 1).$$

The term in brackets in the last equation is due to the fact that $w$ relaxes to $w = 1$ (ground state), while the coherences $(u, v)$ relax to zero.

The steady-state solution of eqns 5.48 can be obtained by setting all of the time derivatives equal to zero and solving the three resulting linear equations. The results are:

$$u = \frac{\Omega\delta}{\delta^2 + \Omega^2/2 + \gamma^2/4} \tag{5.49}$$

$$v = \frac{\Omega\gamma/2}{\delta^2 + \Omega^2/2 + \gamma^2/4} \tag{5.50}$$

$$w = \frac{\delta^2 + \gamma^2/4}{\delta^2 + \Omega^2/2 + \gamma^2/4}. \tag{5.51}$$

The value of $\rho_{22}$ comes up in the study of *fluorescence*, which is proportional to $\rho_{22}$:

$$\rho_{22} = \frac{1 - w}{2} = \frac{\Omega^2/4}{\delta^2 + \Omega^2/2 + \gamma^2/4}. \tag{5.52}$$

This is a simple *Lorentzian*. We see here the presence of the *resonance width*: all of these expressions have denominator $\delta^2 + \Omega^2/2 + \gamma^2/4$ and, in particular, the full width at half maximum of the expression for $\rho_{22}$ vs the detuning, $\delta$, is just $\gamma$ (when $\Omega \ll \gamma$). We also notice, in these equations, the phenomenon of *Saturation*: when $\Omega \geq \gamma$ the resonance width, $\gamma$, of $\rho_{22}$ will increase:

$$\text{Full width at half maximum} = 2\sqrt{\Omega^2/2 + \gamma^2/4} = \sqrt{2\Omega^2 + \gamma^2}. \tag{5.53}$$

Finally, the unsaturated excited state population on resonance is $(\Omega/\gamma)^2$ while the saturated population on resonance approaches $1/2$. We will have more to say about saturation later.

The treatment of collisions in a low-pressure gas can be quite involved; we will touch on the main points. We will start with the result from *kinetic theory* that the probability, $p(t)dt$, that an atom has a period $t$ which is free from collisions is:

$$p(t)dt = (1/\tau_{col}) \exp(-t/\tau_{col})dt, \tag{5.54}$$

where $dt$ is a very small observation time and $\tau_{col}$ is the *average time between collisions* (which depends upon the *type of collision*).

There are two kinds of collisions: *elastic*, which conserve energy in the atom, and *inelastic*, which do not. Elastic collisions therefore *do not result in transitions* but instead *randomize* to some extent the *phase* of the atomic wavefunction. Inelastic collisions result in transitions, almost always to a lower energy level. It turns out that inelastic collisions have the same effect on the Bloch vector as spontaneous decay, and the *rate for inelastic collisions can be added to that for spontaneous decay*. Thus, the total damping rate, $\gamma_{total}$, is

$$\gamma_{total} = \gamma_{spont} + \gamma_{inelast}. \tag{5.55}$$

Elastic collisions are observed far more frequently and their treatment is slightly more complicated. We need to find a relationship between the collision rate, $1/\tau_{col}$, and the damping parameter, $\gamma_{elast}$, for elastic collisions. Our model will be the radiation *emitted by an ensemble of $n$ atoms*, where we assume that the effect of collisions on absorption will be the same as on emission. A model of the emission process which is consistent with the semiclassical theory is that the atom emits a series of wavetrains whose average length is $c\tau_{coll}$ and whose phase is completely random. We will evaluate the *autocorrelation function* of the emitted electric field and use the *Wiener–Khintchine* theorem to obtain the *power spectrum*, whose frequency dependence will give us our desired result.

The electric field is

$$E(t) = E_1(t) + E_2(t) + \cdots + E_n(t)$$
$$= E_0 e^{-i\omega_0 t}\left\{ e^{i\phi_1(t)} + e^{i\phi_2(t)} + \cdots + e^{i\phi_n(t)} \right\}, \tag{5.56}$$

where the $\phi_i(t)$ are the *random* phases of the wavetrains. The autocorrelation function is then

$$\langle E(t)^* E(t+\tau)\rangle = E_0^2 e^{-i\omega_0 \tau}\langle \sum_i^n \sum_j^n e^{i(\phi_i(t)-\phi_j(t))}\rangle \tag{5.57}$$

$$= E_0^2 e^{-i\omega_0 \tau}\sum_i^n \langle e^{i(\phi_i(t+\tau)-\phi_i(t))}\rangle \tag{5.58}$$

$$= n\langle E_i(t)^* E_i(t+\tau)\rangle, \tag{5.59}$$

where the second equation is a result of the *cross terms* averaging to zero and the third is due to the fact that all of the wavetrains are *statistically* equivalent. The autocorrelation function on the last line is that due to a *single atom*.

The salient feature of our model of elastic collisions is that the *phase of the emitted wavetrain is completely randomized by the collision*. Thus, after each collision, the contribution of an atom drops to zero. The single atom autocorrelation function, $\langle E_i(t)^* E_i(t+\tau)\rangle$, is therefore proportional to the probability that the time between collisions is *longer than $\tau$* (collision times shorter than $\tau$ yield a zero autocorrelation since the two electric field factors are then *totally uncorrelated*). Using the probability distribution (eqn 5.54), we therefore have

$$\langle E_i(t)^* E_i(t+\tau)\rangle = E_0^2 e^{-i\omega_0 \tau}\langle e^{i[\phi_i(t+\tau)-\phi_i(t)]}\rangle \tag{5.60}$$

$$= E_0^2 e^{-i\omega_0 \tau}\int_\tau^{-\infty} p(t)dt \tag{5.61}$$

$$= E_0^2 e^{-i\omega_0 \tau - \tau/\tau_{coll}}, \tag{5.62}$$

and the autocorrelation function for all $n$ atoms is

$$\langle E(t)^* E(t+\tau)\rangle = n E_0^2 e^{-i\omega_0 \tau - \tau/\tau_{coll}}. \tag{5.63}$$

The *power spectrum* is obtained by invoking the Wiener–Khintchine theorem, which states: *the power spectrum of a signal is equal to the Fourier transform of its autocorrelation function*. Thus, the power spectrum, $F(\omega)$, is proportional to

$$F(\omega) \propto \frac{1}{(\omega - \omega_0)^2 + (1/\tau_{coll})^2}. \tag{5.64}$$

The *linewidth* (full width at half maximum) of this function is $2/\tau_{coll}$. We observe that the power spectrum has the same Lorentzian frequency dependence as the population provided that we use the following relation between the *elastic collision width* and the collision time:

$$\gamma_{elast} = \frac{2}{\tau_{coll}}. \tag{5.65}$$

The next question is how to incorporate $\gamma_{elast}$ into the density matrix equations. Since only the *coherences* and not the *populations* are involved in elastic collisions, it should be plausible that the damping be included *only* in the $\rho_{12}$ (i.e., $u$ and $v$) equation (once more demonstrating the utility of the density matrix since it is not obvious how one would do this using the coefficients, $c_i(t)$). We will define the total damping (due to radiation and collisions of both types) as $\gamma'$, where

$$\gamma' = \gamma_{total} + \gamma_{elast} = \gamma_{spont} + \gamma_{inelast} + \gamma_{elast}. \tag{5.66}$$

Thus, the density matrix equations are

$$\dot{u} = \delta v - \frac{\gamma'}{2} u \tag{5.67}$$

$$\dot{v} = -\delta u + \Omega w - \frac{\gamma'}{2} v \tag{5.68}$$

$$\dot{w} = -\Omega v - \gamma_{total}(w - 1), \tag{5.69}$$

where it will be recalled that $\gamma_{total}$ is the combined inelastic collision and radiative damping rate. The steady-state solutions to these equations are

$$u = \frac{\Omega \delta}{\delta^2 + \frac{1}{4}\gamma'^2 + \frac{1}{2}(\gamma'/\gamma_{total})\Omega^2} \tag{5.70}$$

$$v = \frac{\frac{1}{2}\Omega\gamma'}{\delta^2 + \frac{1}{4}\gamma'^2 + \frac{1}{2}(\gamma'/\gamma_{total})\Omega^2} \tag{5.71}$$

$$w = \frac{\delta^2 + \frac{1}{4}\gamma'^2}{\delta^2 + \frac{1}{4}\gamma'^2 + \frac{1}{2}(\gamma'/\gamma_{total})\Omega^2} \tag{5.72}$$

and

$$\rho_{22} = \frac{\frac{1}{4}(\gamma'/\gamma_{total})\Omega^2}{\delta^2 + \frac{1}{4}\gamma'^2 + \frac{1}{2}(\gamma'/\gamma_{total})\Omega^2}. \tag{5.73}$$

It is perhaps worth noting that the two kinds of collision-induced relaxations (elastic and inelastic) were first observed in magnetic resonance experiments where they were associated with the relaxation times $T_2$ and $T_1$. The relaxation due to elastic collisions is sometimes called *transverse relaxation* and it occurs with characteristic time $T_2$ ($= \tau_{coll}$). It is also occasionally referred to as *spin-spin relaxation*, since in magnetic resonance it is due to an interaction between spins which causes the component perpendicular (transverse) to the magnetic field to randomize while the longitudinal component is not affected. Conversely, *longitudinal relaxation* in magnetic resonance occurs with characteristic time $T_1$ ($= 1/\gamma_{inelast}$) and is also called *spin-lattice* relaxation since the component of the spin along the magnetic field (and therefore the energy) relaxes, due (in a solid) to the interaction between the spin and the surrounding lattice. Using the optical Bloch model, longitudinal relaxation is seen to be the relaxation of the population (the $z$-component of the Bloch vector) while transverse relaxation is due to the randomization of the atomic wavefunction's *phase* (the $x$- and $y$- components of the Bloch vector).

## 5.8   The atomic susceptibility and optical gain

We will now calculate the *atomic susceptibility* for a classical wave traveling through a linear, isotropic medium whose polarizability will be described using the semiclassical theory from the first part of this chapter. This will lead in a fairly natural way to *laser gain*.

We will start with the *material relations* among the quantities $\boldsymbol{D}$ (electric displacement), $\boldsymbol{E}$ (electric field) and $\boldsymbol{P}$ (polarization):

$$\boldsymbol{D} = \epsilon_0 \boldsymbol{E} + \boldsymbol{P} \tag{5.74}$$

$$\boldsymbol{P} = \epsilon_0 \chi \boldsymbol{E}. \tag{5.75}$$

We will consider the polarization to consist of two parts: that due to the atomic transition and that from all other sources:

$$\begin{aligned}
\boldsymbol{D} &= \epsilon_0 \boldsymbol{E} + \boldsymbol{P} + \boldsymbol{P}_{transition} \\
&= \epsilon \boldsymbol{E} + \epsilon_0 \chi \boldsymbol{E} \\
&= \epsilon' \boldsymbol{E},
\end{aligned} \tag{5.76}$$

where

$$\epsilon'(\nu) = \epsilon \left[ 1 + \frac{\epsilon_0}{\epsilon} \chi(\omega) \right] \tag{5.77}$$

and $\chi(\omega)$ is due *only* to the resonant processes in the atoms.

A monochromatic wave propagating in the $z$-direction through the medium is described by

$$\boldsymbol{E}(z,t) = \boldsymbol{E}_0 e^{i(\omega t - k' z)}, \tag{5.78}$$

where $k'$ is the propagation constant in the medium and is given by:

$$k' = \frac{\omega}{c'} = \omega \sqrt{\mu \epsilon'} \approx k \left( 1 + \frac{\epsilon_0}{2\epsilon} \chi \right), \qquad |\chi| \ll 1, \tag{5.79}$$

where $c'$ is the speed of light in the medium. The frequency-dependent *atomic susceptibility* is a complex quantity:

$$\chi = \chi' - i\chi'', \tag{5.80}$$

where the minus sign is used to be consistent with other treatments. The propagation constant is now

$$k' \approx k \left[ 1 + \frac{\chi'(\omega)}{2n^2} \right] - i \frac{k\chi''(\omega)}{2n^2}, \tag{5.81}$$

where $n = \sqrt{\epsilon/\epsilon_0}$ is the *index of refraction* of the medium (exclusive of the resonant processes). Substituting this into the equation for the traveling wave,

$$\boldsymbol{E}(z,t) = \boldsymbol{E}_0 e^{i(\omega t - kz)} \left\{ e^{(i\Delta k + \gamma(\omega)/2)z} \right\}, \tag{5.82}$$

where

$$\Delta k = -\frac{k\chi'(\omega)}{2n^2} \tag{5.83}$$

$$\gamma(\nu) = -\frac{k\chi''(\omega)}{n^2}. \tag{5.84}$$

The electric field of the wave will undergo a *phase change* given by $\Delta k z$ and will be fractionally *attenuated* (or amplified) by $\gamma/2$ per unit length (the factor of 2 is present since $\gamma$ is actually an *intensity* absorption coefficient). We see that the attenuation or gain of the wave is given by an expression which is proportional to the imaginary part of the atomic susceptibility, which we will now derive in terms of the atomic quantities.

We will first modify the density matrix equations to allow for an *equilbrium population* which is different from that of our original treatment, where we assumed that the atom will relax into the ground state ($w = 1$). We now allow for the possibility that some external agency (called a *pumping mechanism*) can maintain an arbitrary (and probably *non-thermal*) population difference, $w_0$, in the absence of radiation. The density matrix equations are now

$$\dot{u} = \delta v - \frac{\gamma'}{2} u \tag{5.85}$$

$$\dot{v} = -\delta u + \Omega w - \frac{\gamma'}{2} v \tag{5.86}$$

$$\dot{w} = -\Omega v - \gamma_{total}(w - w_0), \tag{5.87}$$

where in the third equation it should be evident that, in the absence of radiation, the equilibribrium value of $w$ is $w_0$. The steady-state solutions are

$$u = \frac{\Omega \delta w_0}{\delta^2 + \frac{1}{4}\gamma'^2 + \frac{1}{2}(\gamma'/\gamma_{total})\Omega^2} \tag{5.88}$$

$$v = \frac{\frac{1}{2}\Omega\gamma' w_0}{\delta^2 + \frac{1}{4}\gamma'^2 + \frac{1}{2}(\gamma'/\gamma_{total})\Omega^2} \tag{5.89}$$

$$w = \frac{(\delta^2 + \frac{1}{4}\gamma'^2)w_0}{\delta^2 + \frac{1}{4}\gamma'^2 + \frac{1}{2}(\gamma'/\gamma_{total})\Omega^2}. \tag{5.90}$$

Since we seek the atomic susceptibility, we must calculate the atomic polarization induced by the radiation. The polarization of each atom is obtained by taking the expectation value of the polarization operator, $p = -er$. If we use the original wavefunction (eqn 5.2),

$$
\begin{aligned}
\langle p \rangle &= \left( c_1^* e^{i\omega_1 t} \langle 1| + c_2^* e^{i\omega_2 t} \langle 2| \right) p \left( c_1 e^{-i\omega_1 t} |1\rangle + c_2 e^{-i\omega_2 t} |2\rangle \right) \\
&= p_{12}(\rho_{12} e^{i\omega_0 t} + \rho_{21} e^{-i\omega_0 t}) \\
&= p_{12}(\tilde\rho_{12} e^{i\omega t} + \tilde\rho_{21} e^{-i\omega t}) \\
&= 2p_{12}(\mathrm{Re}\{\tilde\rho_{12}\}\cos\omega t - \mathrm{Im}\{\tilde\rho_{12}\}\sin\omega t) \\
&= p_{12}(u\cos\omega t - v\sin\omega t),
\end{aligned}
\tag{5.91}
$$

where we assume that $p_{12} \equiv p$ is real (we will henceforth ignore the vectorial nature of the dipole moment). The last equation confirms our earlier claim that the coherences are proportional to the components of an oscillating (actually rotating) induced atomic dipole moment. Using this result, we can now write down the *macroscopic dipole moment*, $P = N\langle p \rangle$, where $N$ is the number of atoms per unit volume:

$$
P = -pN\Omega w_0 \frac{-\frac{1}{2}\gamma'\sin\omega t + \delta\cos\omega t}{\delta^2 + \frac{1}{4}\gamma'^2 + \frac{1}{2}(\gamma'/\gamma_{total})\Omega^2} E_0,
\tag{5.92}
$$

where we use an overall minus sign since $p < 0$ for an electron. We extract the imaginary component of $\chi$ by first considering the time-dependent polarization,

$$
\begin{aligned}
P(t) &= \mathrm{Re}\{\epsilon_0 \chi E_0 e^{i\omega t}\} \\
&= E_0\epsilon_0(\chi'\cos\omega t + \chi''\sin\omega t),
\end{aligned}
\tag{5.93}
$$

from which we obtain

$$
\chi''(\omega) = \frac{p\Omega\gamma'\Delta N_0}{2\epsilon_0 E_0} \frac{1}{\delta^2 + \frac{1}{4}\gamma'^2 + \frac{1}{2}(\gamma'/\gamma_{total})\Omega^2},
\tag{5.94}
$$

where we define the zero-field population difference $Nw_0 \equiv \Delta N_0$. Using the expression for the Rabi frequency (which we consider to be intrinsically positive), we can simplify this equation:

$$
\chi''(\omega) = \frac{p^2\gamma'\Delta N_0}{2\hbar\epsilon_0} \frac{1}{\delta^2 + \frac{1}{4}\gamma'^2 + \frac{1}{2}(\gamma'/\gamma_{total})\Omega^2}.
\tag{5.95}
$$

The population difference of the ensemble of atoms is called the *inversion* and is represented by $\Delta N$, which is obtained by multiplying the equation for $w$ by $N$:

$$
\Delta N = Nw = \frac{(\delta^2 + \frac{1}{4}\gamma'^2)\Delta N_0}{\delta^2 + \frac{1}{4}\gamma'^2 + \frac{1}{2}(\gamma'/\gamma_{total})\Omega^2}.
\tag{5.96}
$$

From this, we can further simplify the equation for $\chi''$:

$$
\chi''(\omega) = \frac{p^2\gamma'\Delta N}{2\hbar\epsilon_0} \frac{1}{\delta^2 + \frac{1}{4}\gamma'^2}.
\tag{5.97}
$$

We note that $\Delta N$ (unlike $\Delta N_0$) is the inversion *in the presence of the optical field*. We further note that this expression for $\chi''$ seems to have lost the term in the denominator which is responsible for saturation. However, the saturation behavior is now due to the factor $\Delta N$, which clearly changes from the field-free value $\Delta N_0$ when $\Omega$ becomes comparable to $\gamma'$. Finally, the $\chi''$ equation can be written in terms of a *normalized lineshape function*, $g(\omega)$, whose integral over all frequencies is unity:

$$\chi''(\omega) = \frac{p^2 \Delta N \pi}{\hbar \epsilon_0} g(\omega),  \qquad (5.98)$$

where

$$g(\omega) = \frac{\gamma'/2\pi}{(\omega - \omega_0)^2 + \frac{1}{4}\gamma'^2}.  \qquad (5.99)$$

The absorption coefficient, $\gamma(\omega)$ is obtained by substituting the expression for $\chi''$ into eqn 5.84:

$$\gamma(\omega) = -\frac{kp^2 \Delta N \pi}{n^2 \hbar \epsilon_0} g(\omega),  \qquad (5.100)$$

where it is hoped that the distinction between the damping constant $\gamma'$ and the absorption coefficient $\gamma(\omega)$ is clear, despite the confusing notation. The effect of $\gamma(\omega)$ on a traveling wave is

$$\boldsymbol{E}(z) = \boldsymbol{E}(0)e^{\gamma(\omega)z/2}.  \qquad (5.101)$$

From the way that $\Delta N$ is defined ($N(\tilde{\rho}_{11} - \tilde{\rho}_{22})$), it will be *positive* when the atom is in thermal equilibrium at a modest temperature and most of the population is in the ground state. Thus, there will be *absorption* of the light as it passes through the medium. However, if the excited state population is greater than that of the ground state, we have a *population inversion* and $\Delta N$ will be *negative* with the result that there will be *coherent optical gain*: the amplitude of the wave will *increase* as the wave traverses the medium.

We can further simplify eqn 5.100 by replacing $p^2$ with an expression derived from the *rate of spontaneous emission* (see, for example, Loudon (1983)):

$$\gamma_{spont} = \frac{1}{\tau_{spont}} = \frac{\omega^3 n^3 D^2}{3\pi \epsilon_0 \hbar c^3},  \qquad (5.102)$$

where $D^2 = |\langle 1|ex|2\rangle|^2 + |\langle 1|ey|2\rangle|^2 + |\langle 1|ez|2\rangle|^2$. Since the three matrix elements in $D^2$ are equal (there is no preferred direction), $D^2 \Rightarrow 3|\langle 1|ex|2\rangle|^2 = 3p^2$, and the expression for $\gamma_{spont}$ is

$$\gamma_{spont} = \frac{\omega^3 n^3 p^2}{\pi \epsilon_0 \hbar c^3}.  \qquad (5.103)$$

Substituting the value of $p^2$ into the expression for $\gamma(\omega)$ yields the formula for *laser gain*:

$$\text{Laser gain:} \qquad \gamma(\omega) = -\frac{\lambda^2 \Delta N}{4n^2 \tau_{spont}} g(\omega) = -\frac{\lambda^2 \Delta N}{8\pi n^2 \tau_{spont}} g(\nu),  \qquad (5.104)$$

where the last equality comes from $g(\omega) = g(\nu)/2\pi$. A *laser amplifier* consists of a medium in which $\Delta N < 0$ with the result that a traveling wave will increase in amplitude as it traverses the medium. Since $\Delta N$ *depends upon the intensity of the wave*, the dependence of the field amplitude (and intensity) on $z$ will not necessarily be a simple exponential. One must go back to the differential equation for the intensity, $I(z)$,

$$\frac{dI(z)}{dz} = \gamma(\omega)I(z), \tag{5.105}$$

from which we get the *laser gain equation*:

$$\text{Laser gain equation:} \quad \frac{dI(z)}{dz} = -\frac{\lambda^2 \Delta N}{8\pi n^2 \tau_{spont}} g(\omega)I(z). \tag{5.106}$$

## 5.9  The Einstein A and B coefficients

The most important result of the recent sections, the laser gain equation, can also be derived from the phenomenological treatment due to Einstein. According to this theory, there are three possible processes which an atom can undergo in the presence of radiation. These are: *spontaneous emission, absorption and stimulated emission.* If the *energy spectral density* (energy per unit volume per unit frequency interval) is $\rho(\nu)$, the three processes are described by three parameters, $A_{21}, B_{21}$ and $B_{12}$, in three rate equations:

$$\text{Spontaneous emission:} \quad \frac{dN_2}{dt} = -A_{21}N_2$$

$$\text{Absorption:} \quad \frac{dN_2}{dt} = -\frac{dN_1}{dt} = +B_{12}N_1\rho(\nu) \tag{5.107}$$

$$\text{Stimulated emission:} \quad \frac{dN_2}{dt} = -\frac{dN_1}{dt} = -B_{21}N_2\rho(\nu),$$

where it is assumed the atom has two levels, labeled 1 and 2, with number densities in each level $N_1$ and $N_2$. The photon emitted by stimulated emission is assumed to have the same values of $\mathbf{k}$, polarization and phase as the incident photon. The three processes are shown schematically in Fig. 5.2.

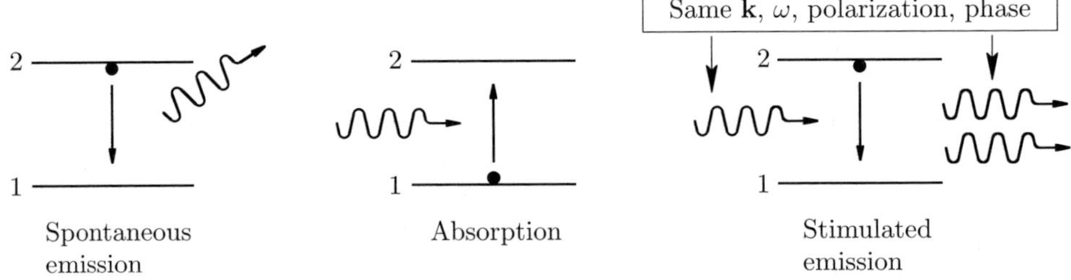


**Fig. 5.2** The three possible radiative interactions in a two-level atom according to the Einstein theory.

Before solving these rate equations, it will be useful to derive the *Planck formula* for the spectral density, $\rho(\nu)$, of radiation which is in thermal equilibrium with a *black body* at temperature, $T$. Planck's hypothesis was that the electromagnetic field is *quantized* – the energy in the radiation field is carried by *photons*, whose energy is $h\nu$, where $\nu$ is the frequency. Since the radiation is in thermal equilibrium with a thermal reservoir (the black body) at temperature $T$, the probability of finding $n$ photons at frequency $\nu$ is given by a *Boltzmann factor*:

$$\text{Probability of } n \text{ photons} \propto e^{-nh\nu/k_B T}. \tag{5.108}$$

From this, the average number of photons at frequency $\nu$ is

$$\bar{n} = \frac{\sum_n n e^{-nh\nu/k_B T}}{\sum_n e^{-nh\nu/k_B T}}. \tag{5.109}$$

This is easily evaluated if we define

$$X \equiv \sum_n e^{-h\nu/k_B T}. \tag{5.110}$$

Then, the value of $\bar{n}$ is

$$\bar{n} = \left(\frac{-k_B T}{h}\right) \frac{1}{X} \frac{\partial X}{\partial \nu} = \frac{1}{e^{h\nu/k_B T} - 1}. \tag{5.111}$$

This is the number of photons per *mode* at $\nu$. Since we are seeking a *spectral density*, we need to multiply $\bar{n}$ by the *number of modes per frequency interval*. In order to do this, we need to model the modes of the radiation field. There are two useful models. One can consider the field to be a series of *standing waves* in a rectangular cavity of dimensions $L_x \times L_y \times L_z$ or to be a series of *traveling waves* satisfying *periodic boundary conditions*. Both models yield the same results; we will choose the cavity since it is based upon slightly more familiar concepts. In order to satisfy the boundary conditions at the walls, the possible values of $\boldsymbol{k}$ are

$$\boldsymbol{k} = \left(\frac{\pi l}{L_x}, \frac{\pi m}{L_y}, \frac{\pi n}{L_z}\right). \tag{5.112}$$

Each mode occupies a volume of $k$-space given by:

$$\text{Volume of } k\text{-space per mode} = \frac{\pi^3}{L_x L_y L_z} = \frac{\pi^3}{V}, \tag{5.113}$$

where $V$ is the volume of the cavity. Since it turns out that the result will be independent of $V$, we can let the dimensions of the cavity become infinite after the calculation is completed; this might avoid conceptual problems with the apparent arbitrariness of the cavity dimensions. The modes would then form a *continuum*, which is, of course, what is actually observed.

In order to calculate the number of modes between $k$ and $k + dk$, we consider a spherical shell in *the first octant of $k$-space*:

$$\text{Volume of spherical shell} = \tfrac{1}{8} \times 4\pi k^2 dk. \tag{5.114}$$

The factor of $\tfrac{1}{8}$ is there in order to avoid counting the modes twice: changing the sign of any component of $\boldsymbol{k}$ yields a value *from the same mode* (since the modes are *standing waves*; this would not be the case with *traveling waves*). The number of modes between $k$ and $k + dk$ is obtained by dividing the volume of the shell by the volume per mode and multiplying the result by 2, since there are *two possible polarizations per mode*:

$$\text{Number of modes between } k \text{ and } k + dk = \frac{V k^2 dk}{\pi^2}. \tag{5.115}$$

We can convert this to the number of modes between $\nu$ and $\nu + d\nu$ (using $k = 2\pi\nu/c$):

$$\text{Number of modes between } \nu \text{ and } \nu + d\nu = V \frac{8\pi\nu^2 d\nu}{c^3}. \tag{5.116}$$

The spectral density is given by the product of three factors: the number of photons per mode, the energy per photon ($h\nu$) and the number of modes per frequency interval. Evaluating this product and dividing by $V$ (since we want the energy density) yields the Planck formula:

$$\rho = \frac{8\pi h \nu^3}{c^3} \frac{1}{e^{h\nu/k_B T} - 1}. \tag{5.117}$$

The rate equations (eqn 5.107) are solved by adding together the three sources for $dN_2/dt$ and setting the sum equal to zero (since the populations are at *equilibrium*). The result is:

$$\text{At equilibrium:} \quad 0 = \frac{dN_2}{dt} = -\frac{dN_1}{dt} = -A_{21}N_2 + N_{12}N_1\rho(\nu) - B_{21}N_2\rho(\nu). \tag{5.118}$$

Since the levels are in *thermal equilibrium* with the radiation, the ratio $N_2/N_1$ is equal to a Boltzmann factor:

$$\frac{N_2}{N_1} = \frac{g_2}{g_1}e^{-h\nu/k_B T}, \tag{5.119}$$

where $g_i$ is the *multiplicity* of the $i$th state: the number of levels having the same energy. Solving eqn 5.118 for $N_2/N_1$ and setting it equal to the Boltzmann factor, one can obtain an equation for $\rho(\nu)$:

$$\rho(\nu) = \frac{A_{21}}{B_{21}} \frac{1}{\frac{B_{12}g_1}{B_{21}g_2}e^{h\nu/k_B T} - 1}. \tag{5.120}$$

This equation is formally similar to the Planck formula (eqn 5.117) and is identical to it if

$$g_2 B_{21} = g_1 B_{12} \tag{5.121}$$

$$\frac{A_{21}}{B_{21}} = \frac{8\pi h \nu^3}{c^3}. \tag{5.122}$$

Using these very important relationships among $A_{21}$, $B_{21}$, and $B_{12}$, one can obtain the rate that the level 2 population changes due to absorption and stimulated emission (ignoring spontaneous emission):

$$\frac{dN_2}{dt} = (B_{12}N_1 - B_{21}N_2)\rho(\nu) = A_{21}\frac{c^3}{8\pi h \nu^3}\left(\frac{g_2}{g_1}N_1 - N_2\right)\rho(\nu). \tag{5.123}$$

Before proceeding further, we note that we are attempting to use Einstein's theory with *monochromatic* radiation instead of the very broadband radiation for which the theory was originally intended. To adapt the theory to monochromatic radiation, we first observe that the rates involving the $B$ coefficient are written as:

$$\langle \text{Transition rate} \rangle = B \times \langle \text{spectral energy density} \rangle. \tag{5.124}$$

When we derived the transition rates using quantum mechanics, the parameter which we now call the $B$ coefficient had a spectral width due to damping processes. The quantity $\rho(\nu)$ also has a spectral width: in the classical use of the theory, the width of the $B$ coefficient is much *smaller* than that of the radiation and the latter can be ignored. For monochromatic radiation, the quantity $\rho(\nu)$ can be represented as a *delta function*:

$$\rho(\nu) = \rho_{\nu'}\delta(\nu - \nu'). \tag{5.125}$$

In order to allow both spectrally broad and spectrally narrow radiation in a single theory, we therefore modify the theory so that the transition rates are given by the *frequency integral* of the product of three factors:

$$\text{Transition rate} = \int Bg(\nu)\rho(\nu)d\nu, \tag{5.126}$$

where we have factored the $B$ coefficient into a constant factor and a lineshape function, $g(\nu)$, whose integral over all frequencies is unity. Thus, in the classical case of a broad $\rho$ and narrow $B$, the integral is $B\rho$, as required by the original Einstein theory. For the case of monochromatic radiation, we have

$$\text{Transition rate} = Bg(\nu)\rho_{\nu}. \tag{5.127}$$

Note that $\rho_{\nu}$ is no longer a spectral density, but instead is simply an *energy density* of the radiation (energy per unit volume). For a plane wave, the intensity, $I_{\nu}$, is

$$I_{\nu} = c\rho_{\nu}. \tag{5.128}$$

We can now write the rate equation for $N_2$ in terms of $I_{\nu}$ and the wavelength $\lambda$:

$$\frac{dN_2}{dt} = A_{21}\frac{\lambda^2}{8\pi}g(\nu)\left(\frac{g_2}{g_1}N_1 - N_2\right)\frac{I_{\nu}}{h\nu}, \tag{5.129}$$

where we note that the last factor $(I_{\nu}/h\nu)$ is the *photon flux*.

This equation is an expression for a rate which is equal to some quantity times a photon flux. The multiplier of the flux and population is therefore a *cross section*, $\sigma(\nu)$:

$$\sigma(\nu) = A_{21}\frac{\lambda^2}{8\pi}g(\nu). \tag{5.130}$$

For *radiative damping only*, the normalized lineshape function $g(\nu)$ is

$$g(\nu) = \frac{\Delta\nu/2\pi}{(\nu - \nu_0)^2 + \frac{1}{4}(\Delta\nu)^2}. \tag{5.131}$$

The Einstein $A_{21}$ coefficient is $1/\tau$, where $\tau$ is the radiative lifetime. Therefore, $\Delta\nu = A_{21}/2\pi$ and at *resonance* $g(\nu_0) = 4/A_{21}$. The result is that the resonant cross section is

$$\text{At resonance:}\quad \sigma(\nu_0) = \frac{\lambda^2}{2\pi}. \tag{5.132}$$

Interpreting $\sigma$ as a cross-sectional area across which a photon will cause an excitation, this result means that an atom's influence extends out about one wavelength, despite the atomic diameter being more than 1000 times smaller (the same is true of radio antennas).

If we multiply eqn 5.129 by $h\nu$, we obtain the *rate of change of the internal energy density of the atoms* as a result of their interaction with the radiation. Thus,

$$\frac{dW}{dt} = A_{21}\frac{\lambda^2 \Delta N g(\nu)}{8\pi} I_\nu \qquad \Delta N \equiv \frac{g_2}{g_1} N_1 - N_2, \tag{5.133}$$

where the atoms' internal energy density is $W$. In the absence of *sources of radiation,* one can write a *continuity equation* for the intensity (this actually is a direct result of the *Poynting theorem*):

$$\nabla \cdot \boldsymbol{S} + \frac{\partial\rho}{\partial t} = 0, \tag{5.134}$$

where $\boldsymbol{S}$ is the Poynting vector (the vector of the intensity) and $\rho$ is the electromagnetic energy density. If there is a source,

$$\nabla \cdot \boldsymbol{S} + \frac{\partial\rho}{\partial t} = -\frac{dW}{dt}, \tag{5.135}$$

where the minus sign comes from the fact that an *increase* in the energy of the atoms comes about as a result of a *decrease* in the energy of the radiation. Substituting for $dW/dt$ and setting $d\rho/dt = 0$ (since we are interested in the equilibrium situation),

$$\frac{dI_\nu}{dz} = -A_{21}\frac{\lambda^2 \Delta N g(\nu)}{8\pi} I_\nu, \tag{5.136}$$

where we are assuming wave propagation in the $z$-direction. This is the same as the quantum mechanical result (since $A_{21} = 1/\tau_{spont}$ and we assumed $n = 1$ in the derivation).

## 5.10   Doppler broadening: an example of inhomogeneous broadening

In earlier sections, we discussed two broadening mechanisms: radiative (natural) broadening and collision (elastic and inelastic) broadening. These two mechanisms have this in common: *at any time, the distribution of possible absorption or emission frequencies is the same for all atoms.* This type of broadening is called *homogeneous.* Conversely, if *at any time the distribution of absorption/emission frequencies is different for different atoms,* the broadening is called *inhomogeneous.* The presence of two categories of broadening has profound implications for the spectrum of the light emitted by a laser, as we shall discover later.

There are a number of possible sources of inhomogeneous broadening. Perhaps the most obvious is in a crystal, where, due to a random distribution of impurities or other inhomogenieties, different atoms will have slightly different local environments and therefore will have different transition frequencies. The most common inhomogeneous broadening in gas lasers is due to the distribution of atomic velocities and consequent distribution of transition frequencies because of the *Doppler shift*. We will discuss *Doppler broadening* in this section.

A collection of atoms in thermal equilibrium at temperature $T$ in an atomic vapor will have a *Maxwellian* distribution of velocities, where the probability $(dN/N)$ of the atom's velocity along the $z$-axis being between $v_z$ and $v_z + dv_z$ is

$$\frac{dN}{N} = \left(\frac{M}{2\pi k_B T}\right)^{1/2} e^{-\frac{Mv^2}{2k_B T}} dv_z, \tag{5.137}$$

where $M$ is the atomic mass and $k_B$ is the Boltzmann constant. As a result of the motion of the atoms, their transition frequencies will experience a first-order Doppler shift:

$$\nu = \nu_0 \left(1 - \frac{v_z}{c}\right), \tag{5.138}$$

where $\nu_0$ is the rest frequency of the atom and we assume that the laser beam is directed in the $z$-direction. Combining these two equations, we get a normalized Doppler line shape, $g(\nu)$:

$$g(\nu) = \left(\frac{4\ln 2}{\pi}\right)^{1/2} \frac{1}{\Delta\nu_D} e^{-4\ln 2\left(\frac{\nu - \nu_0}{\Delta\nu_D}\right)^2}, \tag{5.139}$$

where the full width at half maximum, $\Delta\nu_D$, is given by:

$$\Delta\nu_D = \left(\frac{8k_B T \ln 2}{Mc^2}\right)^{1/2} \nu_0$$

$$\approx 7.2 \times 10^{-7} \sqrt{\frac{T}{A}} \nu_0 \tag{5.140}$$

and $A$ is the atomic mass in amu.

A plot of a Gaussian and Lorentzian line with the same width and height appears in Fig. 5.3. The lineshapes are very similar within one half width of resonance but the much longer *tail* of the Lorentzian line should be evident. Since homogeneous broadening usually results in a Lorentzian, and inhomogeneous broadening usually results in a Gaussian, one might say that *all broadening is homogeneous sufficiently far from resonance.*

When there is more than one broadening mechanism, the *combined* lineshape is the *convolution* of the two component lineshapes:

$$g_{combined}(\nu) = \int_{-\infty}^{\infty} g_1(\nu - \nu')g_2(\nu')d\nu', \tag{5.141}$$

where $g_1(\nu)$ and $g_2(\nu)$ are the two original lineshape functions. For similar broadening mechanisms, the combined linewidth is a simple function of the original linewidths.

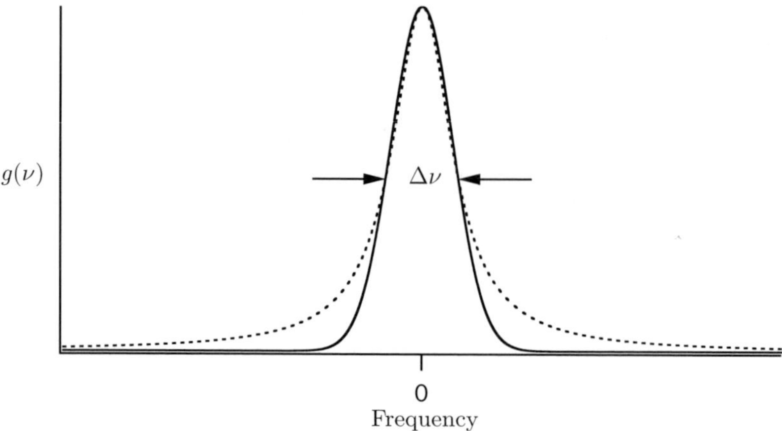


**Fig. 5.3** A Lorentzian (dashed) and Gaussian (solid) with the same linewidth and amplitude.

We saw earlier that for radiative and collision broadening (or two types of collision broadening), the rates and therefore the widths simply add:

$$\text{Homogeneous (Lorentzian):} \quad \Delta\nu_{total} = \Delta\nu_1 + \Delta\nu_2 \qquad (5.142)$$

and the combined lineshape function is also a Lorentzian. It is well known that the convolution of a Gaussian with another Gaussian is also a Gaussian whose width is the quadratic sum of the original widths:

$$\text{Inhomogeneous (Gaussian):} \quad \Delta\nu_{total} = \sqrt{\Delta\nu_1^2 + \Delta\nu_2^2}. \qquad (5.143)$$

The case of *different types* of broadening mechanisms is a little more complicated. When the linewidth of one is much greater than the other, one can simply use the dominant function, although far from resonance the Lorentzian always dominates. When the widths are similar, one obtains a *Voigt profile*, which is not surprisingly similar to both a Lorentzian and a Gaussian. (See, for example, Liu (2001) for a discussion of the Voigt profile together with a simple analytical expression for evaluating it.)

## 5.11  Comments on saturation

Now that we have touched on the notion of saturation and have discussed the difference between homogeneous and inhomogeneous broadening, we would like to illustrate this difference using a simple "thought" experiment. The experiment introduces the phenomenon of "hole burning", which we will describe in the limiting case of an inhomogeneous width which is much larger than the homogeneous width. First, we will derive some simple consequences of saturation.

Earlier in this chapter, we saw that the linewidth of a transition would increase when the Rabi frequency, $\Omega$, was comparable to or greater than the homogeneous linewidth, $\gamma'$. We will examine the effects of a large Rabi frequency by first writing the gain constant (eqn 5.104) in terms of $\Delta N_0$ instead of $\Delta N$ in order to display clearly the dependence on the Rabi frequency:

$$\gamma(\omega) = -\frac{\lambda^2 \gamma' \Delta N_0}{16\pi^2 \tau_{spont}} \frac{1}{(\omega - \omega_0)^2 + \frac{1}{4}\gamma'^2 + \frac{1}{2}(\gamma'/\gamma_{total})\Omega^2}. \tag{5.144}$$

In order to simplify the discussion, we will assume that there is only radiative damping ($\gamma_{total} = \gamma' = \gamma$). When $\Omega \geq \gamma$, there will be two consequences. First, as already mentioned, the width will increase:

$$\Delta\omega = \Delta\omega_0 \sqrt{1 + 2(\Omega/\gamma)^2}, \tag{5.145}$$

where $\Delta\omega_0 = \gamma$ is the linewidth in the limit of zero $\Omega$. The peak gain (or absorption) at *resonance* will also be reduced:

$$\gamma(\omega_0) = \gamma(\omega_0)|_{\Omega=0} \frac{1}{1 + 2(\Omega/\gamma)^2}. \tag{5.146}$$

It would be instructive at this point to observe the behavior of the inversion ($\Delta N$) as a function of the Rabi frequency by rewriting the equation for $\Delta N$ (eqn 5.96):

$$\Delta N = \Delta N_0 \frac{(\omega - \omega_0)^2 + \frac{1}{4}\gamma^2}{(\omega - \omega_0)^2 + \frac{1}{4}\gamma^2 + \frac{1}{2}\Omega^2}. \tag{5.147}$$

From this, we see clearly that $\Delta N$ varies from $\Delta N_0$ when $\Omega = 0$ to something approaching zero when $\Omega \gg \gamma$. The fact that, in the absence of pumping, $\Delta N$ is never negative leads to the general principle that *in the absence of an external pumping mechanism, it is impossible to invert a two-level population by continuously irradiating the atoms at or near their resonant frequency.* As the intensity increases, the populations will tend to be *equalized* and the population difference (the inversion) will approach zero.

We can rewrite eqns 5.146 and 5.145 as:

$$\gamma(\omega_0) = \gamma(\omega_0)|_{\Omega=0} \frac{1}{1 + \frac{I}{I_s}} \tag{5.148}$$

$$\Delta\omega = \Delta\omega_0 \sqrt{1 + \frac{I}{I_s}}, \tag{5.149}$$

where $I$ is the intensity and $I_s$ is the *saturation intensity* at resonance and is equal to

$$I_s = \frac{\pi \hbar \omega_0}{\lambda^2 \tau_{spont}}, \tag{5.150}$$

where we used $I = \epsilon_0 c |E|^2 / 2$ and eqn 5.103 to obtain this result and assumed that $n = 1$. The saturation intensity can also be obtained by recognizing that the onset of saturation occurs when *the excitation rate is equal to the spontaneous decay rate:*

$$\text{Saturation:} \quad \left(\frac{I_s}{\hbar\omega_0}\right)\sigma = \left(\frac{I_s}{\hbar\omega_0}\right)\left(\frac{\lambda^2}{2\pi}\right) = \frac{1}{\tau_{spont}} \implies I_s = \frac{2\pi\hbar\omega_0}{\lambda^2 \tau_{spont}}, \tag{5.151}$$

which a similar to eqn 5.150.

Equation 5.148 gives the gain reduction due to saturation at *resonance*; we can obtain the off-resonance result by factoring eqn 5.144 into the product of a Lorentzian and a second intensity-dependent factor

$$\gamma(\omega) = \gamma(\omega)|_{\Omega=0}\frac{1}{1 + \frac{I}{I_s}\bar{g}(\omega - \omega_0)}, \tag{5.152}$$

where $\bar{g}(\omega - \omega_0)$ is a Lorentzian normalized to unity at resonance:

$$\bar{g}(\omega - \omega_0) \equiv \frac{\gamma'^2}{\gamma^2 + 4(\omega - \omega_0)^2}. \tag{5.153}$$

The population (eqn 5.147) can also be factored in a similar fashion:

$$\Delta N = \Delta N_0 \frac{1}{1 + \frac{I}{I_s}\bar{g}(\omega - \omega_0)}. \tag{5.154}$$

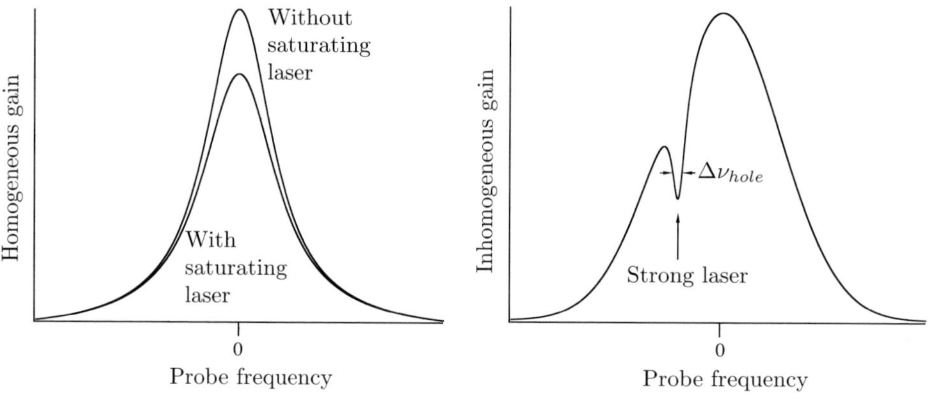


**Fig. 5.4** On the left are two *homogeneously* broadened lines observed with a weak tunable probe beam both with and without the saturating beam. On the right is an *inhomogeneously* broadened line with a *hole* at the frequency of the saturating beam.

With these preliminaries, we will now discuss the phenomenon of *hole burning*, which will be illustrated by a simple experiment. Assume that one has an inhomogeneously broadened medium with two laser beams passing through it. One beam is very intense and is tuned slightly to one side of the resonance. The second is tunable and weak enough to not alter the populations of the atoms. The former is called the *saturating* beam and the latter is called the *probe* beam. If one has a means of measuring the *gain* of the probe as one tunes it over the inhomogeneous line, one will obtain a curve similar to the one on the right of Fig. 5.4. The saturating beam will *eat a hole out of the inhomogeneous profile*; the hole is located at the frequency of the strong beam.

The reason for this behavior is that the inhomogeneously broadened medium can be divided up into a large number of *independent, spectrally narrow packets*, each of which is homogeneously broadened and whose center frequencies cover the entire inhomogeneous width. When a strong saturating field appears at some frequency within the inhomogeneous width, only the packet resonant with the field will be saturated; the other packets will not be affected since they are not resonant with the field. Thus, the saturated packet will have a smaller $\Delta N$ and therefore a smaller gain and a *hole* in the inhomogeneous gain curve will appear.

An approximate analytical expression can be obtained for the gain experienced by the probe beam in the presence of the saturating beam by assuming that the inhomogeneous broadening is due to the Doppler effect. For atoms moving with speed $v$ along the $z$-axis, the Doppler shift for the saturating laser (which propagates in the $z$-direction) is

$$\Delta\omega'_{Doppler} = \frac{v}{c}\omega' = k'v, \tag{5.155}$$

where $k'$ is the wave vector of the fixed frequency saturating laser and $\omega'$ is its frequency. The population difference induced by the saturating laser is therefore (from eqn 5.154)

$$\Delta N(v) = \Delta N_0 \frac{1}{1 + \frac{I}{I_s}\bar{g}(\omega' - \omega_0 - k'v)}, \tag{5.156}$$

where $I$ is the intensity of the saturating laser and $I_s$ is the saturation intensity. The velocity distribution of the atoms is given by $N(v)$, where $N(v)dv$ is the number of atoms whose velocity component along the laser beam is between $v$ and $v + dv$. Since the unsaturated population difference, $\Delta N_0$, is proportional to $N(v)$, the saturated population difference is given by:

$$\text{For atoms with speed } v:\quad \Delta N(v) \propto N(v)\frac{1}{1 + \frac{I}{I_s}\bar{g}(\omega' - \omega_0 - k'v)}. \tag{5.157}$$

From eqn 5.104, we see that the gain of the atoms for the probe laser is proportional to $\Delta N(v) \times g(\omega - \omega_0 - kv)$, where $\omega$ and $k$ are the frequency and wave vector of the probe laser.

$$\text{For atoms with speed } v:\quad \gamma(\omega) \propto N(v)g(\omega - \omega_0 - kv)\frac{1}{1 + \frac{I}{I_s}\bar{g}(\omega' - \omega_0 - k'v)} \tag{5.158}$$

To include all of the atoms, we *sum (integrate)* over $v$:

$$\gamma(\omega) \propto \int_{-\infty}^{\infty} N(v)g(\omega - \omega_0 - kv)\frac{1}{1 + \frac{I}{I_s}\bar{g}(\omega' - \omega_0 - k'v)}\,dv$$

$$\propto N((\omega - \omega_0)/k)\frac{1}{1 + \frac{I}{I_s}\bar{g}(\omega - \omega')}, \tag{5.159}$$

where we have treated the $g(\omega - \omega_0 - kv)$ factor as a delta function, selecting velocities which satisfy $\omega - \omega_0 = kv$. (In the argument of the lineshape function, $\bar{g}(\omega' - \omega_0 - k'v)$, very little error is introduced by setting $k'$ equal to $k$.) A simple expression can be obtained in the limit that $I$ is somewhat smaller than $I_s$ by expanding the second factor,

$$\gamma(\omega) \propto N((\omega - \omega_0)/k) \left\{ 1 - \frac{I}{I_s} \bar{g}(\omega - \omega') \right\}. \tag{5.160}$$

When $\omega - \omega' \gg \gamma'$, we have a simple unmodified Doppler lineshape proportional to $N((\omega - \omega_0)/k)$. At frequencies near $\omega'$, the second term in the curly brackets will produce a *hole* in the population difference and therefore a hole in the gain expression, as shown in Fig. 5.4.

If one performs this experiment with a *homogeneously broadened* medium, the results will be very different. This case is illustrated by the curves on the left side of of Fig. 5.4. Here, the effect of the strong saturating beam is to lower the gain over the *entire line* rather than over a small portion of it as in the inhomogeneous case. These two very different behaviors will alter the spectrum of light emitted by lasers whose amplifying media are subject to the two types of broadening.

## 5.12  Further reading

The best book on the semiclassical laser theory is, in the author's opinion, Loudon's *The Quantum Theory of Light* (1983). Virtually all of the previously mentioned books on laser theory and Gaussian beams use an approach similar to ours; perhaps one of the most readable is Yariv's book on quantum electronics (1989). Finally, an excellent atomic physics text which describes the optical Bloch equations is Foot (2005); we used this text for much of our treatment of this topic.

## 5.13  Problems

(5.1) For a system in thermal equilibrium, calculate the temperature at which the spontaneous and stimulated emission rates are equal for $\lambda = 500$ nm and the wavelength at which these rates are equal for a temperature of 4000 K.

(5.2) Derive the *Wien law* by determining the product of the temperature and wavelength at the peak of Planck's formula for the spectral density of black body radiation. This product should depend upon $h$ and $k$ (Boltzmann constant). Make reasonable approximations if necessary.

(5.3) A HeNe laser produces light at 632.8 nm due to transitions in neon atoms. Assuming pure $^{20}$Ne at a temperature of $\approx 300°$C, answer the following questions.

    a What is the Doppler linewidth (FWHM) of the transition?

    b How many *blackbody* modes per unit volume couple to this transition?

    c How many $\text{TEM}_{00q}$ modes are within the Doppler full width of this transition if the mirror spacing is 1 m?

(5.4) All resistors generate thermal noise, called "Johnson noise", whose RMS voltage is given by $V_{Johnson} = \sqrt{4k_B T R \Delta \nu}$ where $R$ is the resistance, $T$ is the absolute temperature, $k$ is Boltzmann's constant and $\Delta \nu$ is the electrical bandwidth of the system. This noise can be considered to be blackbody radiation in a one-dimensional "cavity" where $h\nu \ll k_B T$. Derive this formula using the following steps.

- Use the procedure for determining the spectral density of blackbody radiation that we derived in the text. (Three factors need to be multiplied together.)
- The number of modes can be determined using the following model. Assume that the resistor $R$ is connected to a transmission line (a *one-dimensional* enclosure) whose characteristic impedance is also $R$. The only things you need to know about characteristic impedance are that a transmission line terminated in it will *not experience any reflections* and will have a resistance at the input equal to the characteristic impedance, $R$. Thus, the transmission line's length is irrelevant (which aids the model, since the length does not appear in the result).
- Once you get the correct spectral density of the energy, calculate the *power* by determining the *time* for the energy emitted by one resistor to travel down the line and be absorbed by the other. Then use the remaining known parameters $(R, \Delta\nu)$ to obtain the voltage.
- The following diagram displays the appropriate model. One needs a resistor $R$ at *both* ends to avoid reflections (both resistors have Johnson noise).

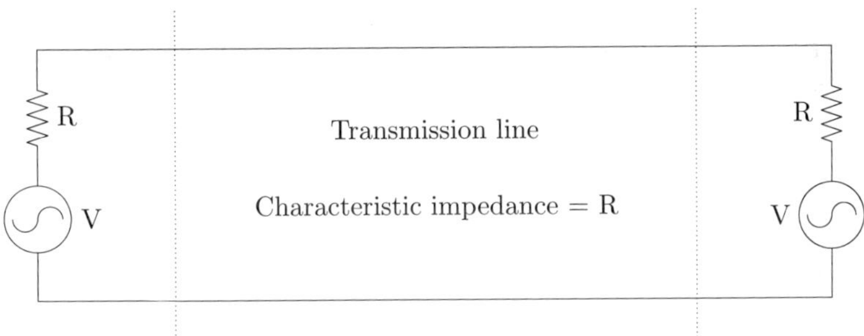

# 6

# Laser oscillation and pumping mechanisms

## 6.1 Introduction

This chapter will combine the results of Chapters 3 and 5 by placing an amplifying medium in a resonant cavity and evaluating the behavior of the resulting *laser oscillator*. First, we will determine the conditions for laser oscillation and the expected power output of the laser. Next, we will discuss optical *pumping* techniques for three-level and four-level lasers. Finally, the spectral characteristics of laser radiation will be considered. This will include the *frequency pulling* of the empty cavity modes, the multimode behavior of an inhomogeneously broadened laser (including *spatial hole burning*) and the ultimate linewidth (Schawlow–Townes limit) of a laser. Although it is slightly off topic, the remarkable behavior of the photon occupation number above threshold will also be discussed. The fairly subtle distinction between the photon statistics of a coherent laser beam and that of thermal radiation will be treated in some detail. For most of the chapter, the laser will be modeled by a gain medium inside a *traveling wave cavity* since the spatially nonuniform field in a standing wave cavity is somewhat more difficult to treat.

## 6.2 The condition for laser oscillation

In Chapter 3, the round-trip fractional intensity loss in a cavity was represented by the quantity $L_{tot}$ so that the *intensity remaining after one round trip* is $e^{-L_{tot}}$. The internal intensity transmission is $t^2$ for a traveling-wave cavity and $t^4$ for a standing-wave cavity. The internal loss was considered to be distributed throughout the cavity and was therefore described by an absorption coefficient, $\alpha$. The intensity remaining after one round trip due to absorption is $e^{-\alpha l}$, where $l$ is the round-trip distance. Equating $e^{-L_{tot}}$ to the combined absorption effects and Fresnel reflectivity, one has:

$$e^{-L_{tot}} = r_1^2 r_2^2 e^{-\alpha l}, \tag{6.1}$$

where $r_1$ and $r_2$ are the field reflectivities of the cavity mirrors. Although most lasers using either type of cavity have only one partially transmitting mirror, we will continue to allow for the possibility of two partially transmitting mirrors. The fractional round-trip loss is

$$L_{tot} = \alpha l - 2\ln(r_1 r_2). \tag{6.2}$$

In Chapter 5, the fractional intensity gain (or loss) per unit length due to resonant atomic processes was represented by $\gamma(\omega)$, which satisfies the equation

$$\frac{dI_\nu}{dz} = \gamma(\omega)I_\nu. \tag{6.3}$$

In a *laser amplifier*, the intensity increase after one pass can be substantial and the $z$-dependence of $\gamma(\omega)$ is therefore not exponential, since $\gamma(\omega)$ depends upon the intensity via the quantity $\Delta N$. The single-pass gain of a *laser oscillator* is usually much smaller and we will assume that the intensity has an exponential dependence on $z$:

$$I_\nu = I_{\nu 0} e^{\gamma(\omega)z}, \tag{6.4}$$

where we are assuming the gain medium is in a *traveling wave cavity*. This expression is formally the same as that responsible for absorption in a cavity containing an absorbing medium; we will thus include the possibility of gain by modifying $\alpha$:

$$\alpha \Longrightarrow \alpha - \gamma(\omega) \tag{6.5}$$

so that the expression for $L_{tot}$ is now

$$L_{tot} = (\alpha - \gamma(\omega))l - 2\ln(r_1 r_2). \tag{6.6}$$

When $L_{tot} = 0$, any field inside the cavity will be indefinitely sustained, since there are no net losses. We thus have the condition for *laser oscillation:*

$$\text{Laser oscillation condition:} \quad L_{tot} = 0 \Longrightarrow \gamma(\omega) = \alpha - \frac{2}{l}\ln(r_1 r_2), \tag{6.7}$$

where, of course, this happens *only at the resonant frequencies of the cavity*. Essentially, this equation tells us that the gain must compensate for both the distributed loss, $\alpha$, and the losses in cavity intensity due to the radiation leaking out through the partially transmitting cavity mirrors. It is important to realize that the compensation must be *exact:* if it is too small the oscillations will die out and if it is too large the power will increase indefinitely, an obviously unphysical condition. As the gain constant is increased (via an external pumping mechanism), the laser will reach *threshold* when the gain is exactly equal to the losses. This is called the *threshold gain*, $\gamma_t(\omega)$. The gain can not be larger than this: it is *clamped* to $\gamma_t$. Of course, by increasing the pumping, the power output will increase, but the gain will remain the same. This seeming paradox will be resolved in the next section.

## 6.3    The power output of a laser

In the last section, we showed that self-sustained laser oscillation will occur when the gain is equal to the loss:

$$\gamma_t = \alpha - \frac{2}{l}\ln(r_1 r_2). \tag{6.8}$$

As the zero-field inversion, $\Delta N_0$, is made more negative (via increased pumping), the zero-field gain ($\gamma_0$) will increase, but the saturated gain will remain clamped to $\gamma_t$ (eqn 5.148):

$$\gamma_t = \gamma_0 \frac{1}{1 + \frac{I_\nu}{I_s}}. \tag{6.9}$$

Again, we assume a *traveling wave* cavity, since then $I_\nu$ (the internal intensity) will be spatially uniform. We see that as the unsaturated gain, $\gamma_0$, increases, $I_\nu$ will be increased just enough to keep $\gamma_t$ constant. Therefore, the output power will increase with increased pumping. This equation also explains how laser oscillation is *initiated*. One increases the pumping until $\gamma > \gamma_t$. Then, any noise (from spontaneous emission) or other disturbance that has Fourier components at the laser frequency will be amplified and the resulting intensity will grow until *saturation reduces the gain to* $\gamma_t$, at which point the intensity will be constant. Thus, the *clamping mechanism of the gain is saturation.*

Combining these equations and solving for $I_\nu$,

$$I_\nu = I_s \left( \frac{\gamma_0}{\gamma_t} - 1 \right) = I_s \left( \frac{\gamma_0 l}{\alpha l - 2\ln(r_1 r_2)} - 1 \right). \tag{6.10}$$

Assume that only one mirror (the *output coupler*, mirror 1) has any appreciable transmission. Then,

$$-2\ln(r_1 r_2) = -\ln r_1^2 = -\ln(1 - T) \approx T, \tag{6.11}$$

where $T$ is the (intensity) transmission of the output coupler and we assume that $T \ll 1$. If we call the round-trip loss (excluding transmission, $T$) $L_i$, then $L_i = \alpha l$ and

$$I_\nu = I_s \left( \frac{g_0}{L_i + T} - 1 \right), \tag{6.12}$$

where $g_0 \equiv \gamma_0 l$ is the *unsaturated gain per round trip*.

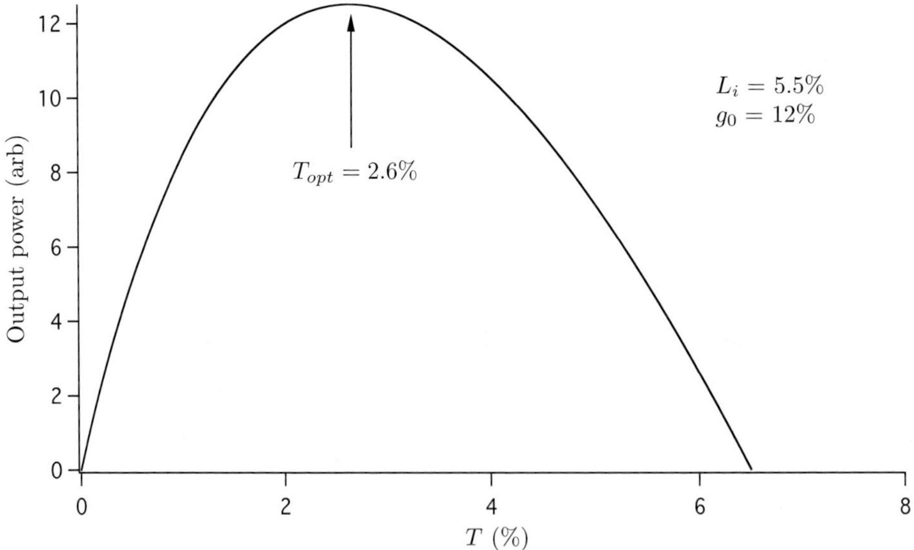


**Fig. 6.1** Output power versus output coupling.

The quantity $I_\nu$ is the internal circulating intensity, some of which will be absorbed by the atoms and some of which will stimulate emission. In order to determine the power output, we need to multiply $I_\nu$ by the *saturated gain*, $\gamma_t$. This will yield the *net* *power per unit volume, $P_e$, provided by the amplifying medium.* The *internal power* is obtained by multiplying this by the *mode volume, $V_m$*, which is the volume overlap between the amplifying medium and the radiation. Finally, the *external power* is obtained by multiplying the internal power by $T/(L_i + T)$. Putting all of these factors together, the useful output power, $P_0$, is

$$P_0 = P_e V_m \frac{T}{L_i + T} = \gamma_t I_\nu V_m \frac{T}{L_i + T} = \left( \frac{L_i + T}{l} \right) I_\nu V_m \left( \frac{T}{L_i + T} \right), \qquad (6.13)$$

where the last step comes from equating the gain $(\gamma_t l)$ to the total loss $(L_i + T)$ and solving for $\gamma_t$. Finally, substituting for $I_\nu$, yields

$$\text{Available power from a laser:} \quad P_0 = \frac{V_m I_s}{l} \left( \frac{g_0}{L_i + T} - 1 \right) T. \qquad (6.14)$$

We should note that we could have also obtained this result by multiplying the internal intensity, $I_\nu$ (eqn 6.12), by the effective *cross-sectional area* $(V_m/l)$ and the mirror transmission, $T$.

The only conveniently available parameter which can be adjusted to maximize the power output is the output coupler transmission, $T$. One can determine the optimum value of $T$ by differentiating and setting the derivative equal to zero. The result is:

$$\text{Optimum output coupling:} \quad T_{opt} = -L_i + \sqrt{g_0 L_i}. \qquad (6.15)$$

A plot of the output power versus $T$ appears in Fig. 6.1, where $L_i = 5.5\%$ and $g_0 = 12\%$. Unlike the analogous case of impedance matching into an empty cavity, a factor of 2 error in $T$ will result in a reduction of the output power by approximately a factor of 2 (the analogous reduction in the circulating intensity in the impedance matching case is about 11%).

## 6.4 Pumping in three-level and four-level laser systems

In order for a medium to amplify radiation, a method for maintaining a field-free inversion $(\Delta N_0 < 0)$ needs to be provided. This is called a *pumping mechanism*; we will discuss *optical* pumping mechanisms in this section. From our earlier discussion (Chapter 5), we established that it is impossible to invert a two-level system using optical radiation alone. Thus, one needs three or more levels to produce an inversion. Despite the presence of these additional levels, the optical amplification is still provided by only *two levels*, and the theory which we have developed is still applicable with at most minor changes.

We will consider laser media with three and four levels and determine the population inversion due to pumping using the *rate equations*, which describe the relationships among the *populations* and their rates of change. As with the Einstein $A$ and $B$ coefficients, a theory which uses the rate equations needs additional ad hoc assumptions, which only appear naturally using quantum mechanics. The added terms are intuitively obvious and fairly easy to evaluate using the theory already developed.

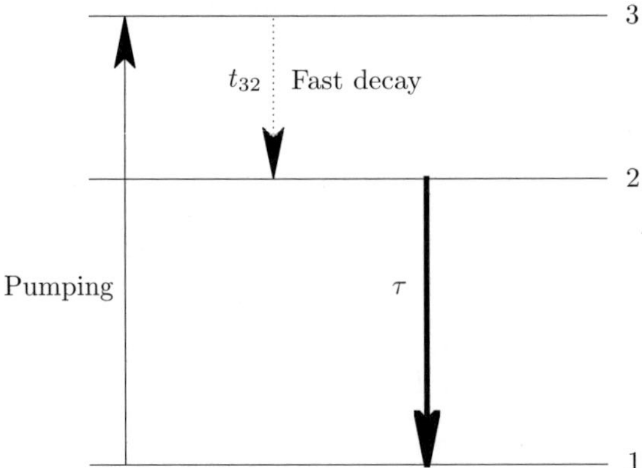



**Fig. 6.2** Level structure of a three-level laser.

Figure 6.2 depicts a three-level atomic system. For simplicity, we assume that all of the levels have a *multiplicity* of 1. The excitation rate of level 2 due to the laser radiation is given by eqn 5.129

$$\frac{dN_2}{dt} = A_{21}\frac{\lambda^2}{8\pi}g(\nu)(N_1 - N_2)\frac{I_\nu}{h\nu} = W(\nu)(N_1 - N_2), \tag{6.16}$$

which defines the quantity $W(\nu)$ as the *stimulated transition rate*. The rate equations can be created for each level, $N_i$, by equating $dN_i/dt$ to an expression in which one *adds* the rates of the *sources* to $N_i$ and subtracts the rates of the *sinks* from $N_i$. Thus, an exponential decay (at rate $1/\tau$) from level $i$ to level $j$ requires a term $-N_i/\tau$ in the equation for $N_i$ and a term $N_i/\tau$ in the equation for $N_j$. We will represent the *optical pumping rate* by $R$, so that level 1 is depleted at rate $-RN_1$ and level 2 is increased at rate $RN_1$. Level 3 is just an intermediate in the pumping process and does not need a rate equation as long as $t_{32} \ll \tau$ so that the pumping can be considered to be directly to level 2. The rate equations for levels 1 and 2 are,

$$\frac{dN_1}{dt} = -RN_1 + \frac{N_2}{\tau} + (N_2 - N_1)W(\nu) \tag{6.17}$$

$$\frac{dN_2}{dt} = RN_1 - \frac{N_2}{\tau} - (N_2 - N_1)W(\nu) \tag{6.18}$$

$$N_T = N_1 + N_2, \tag{6.19}$$

where $N_T$ is the *total* number density of atoms. These equations can be solved in the *steady-state* by setting the derivatives equal to zero and solving the resulting algebraic equations. Noting that the equations for $N_1$ and $N_2$ are the same, we have only *two* independent equations which can be solved for the inversion. The result is

$$\Delta N = \frac{1/\tau - R}{1/\tau + R + 2W(\nu)}N_T. \tag{6.20}$$

We are mainly interested in the inversion ($\Delta N_0$) at *threshold*, where $W(\nu) = 0$:

$$\text{Three-level laser:} \quad \Delta N_0 = \frac{1/\tau - R_t}{1/\tau + R_t} N_T, \tag{6.21}$$

where $R_t$ is the *threshold pumping rate*. We see that with the addition of level 3, we can now invert levels 1 and 2 simply by making $R > 1/\tau$. The ground state still retains a considerable population (the Boltzmann factor $\approx 1$) and one needs to pump *fairly hard* to obtain a significant inversion. If we add a level somewhat *below* level 1, we can avoid this and greatly increase the ease of obtaining an inversion. This is the *four-level* laser scheme.

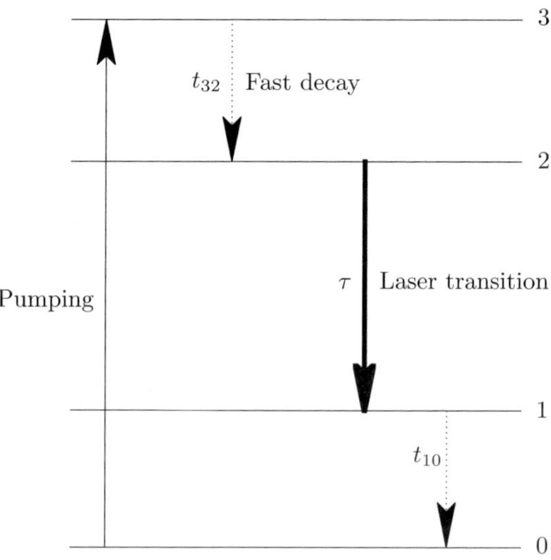


**Fig. 6.3** Level structure of a four-level laser

A four-level laser system is shown in Fig. 6.3. We have taken the three-level system and added level 0 below the lower amplification level (level 1) with an energy gap sufficient to make the Boltzmann factor for level 1 much less than one. As in the three-level system, the decay time from level 1 to level 0 is much less than $\tau$. The three rate equations are (excluding one for level 3 for the same reasons as for the three-level system),

$$\frac{dN_0}{dt} = -RN_0 + \frac{N_1}{t_{10}} \tag{6.22}$$

$$\frac{dN_1}{dt} = -\frac{N_1}{t_{10}} + \frac{N_2}{\tau} + (N_2 - N_1)W(\nu) \tag{6.23}$$

$$\frac{dN_2}{dt} = RN_0 - \frac{N_2}{\tau} - (N_2 - N_1)W(\nu) \tag{6.24}$$

$$N_T = N_0 + N_1 + N_2. \tag{6.25}$$

Solving for $\Delta N_0$ at threshold in the steady state (again realizing that the equations are not all independent),

$$\text{Four-level laser:} \quad \Delta N_0 = -\frac{R_t}{R_t + 1/\tau} N_T, \tag{6.26}$$

where we used $t_{10} \ll \tau$. Notice that at the pumping rate needed to barely reach threshold in a three-level system ($R = 1/\tau$), the four-level system is already significantly inverted ($\Delta N_0 = -\frac{1}{2} N_T$). If one solves for the threshold pumping rates, the ratio of the pumping rates for a threshold inversion for the two laser systems is

$$\text{At threshold:} \quad \frac{(R_t)_{four-level}}{(R_t)_{three-level}} = \frac{\Delta N_0}{N_T + \Delta N_0} \ll 1. \tag{6.27}$$

Thus, we see that by adding a level somewhat below the lower amplification level we have tremendously eased the pumping requirements.

## 6.5  Laser oscillation frequencies and pulling

Laser oscillation occurs only at the resonant frequencies of the enclosing cavity, since these are the frequencies at which the considerations of loss and gain apply (the *feedback* necessary for oscillation is not available at other frequencies). When we derived the quantum mechanical gain constant, we found that it is proportional to the *imaginary part* of the complex susceptibility, $\chi$. We also found that the *real part* of the susceptibility produced a *phase shift*. This phase shift will alter the cavity resonances from their values in the absence of the amplifying medium. To summarize the results from Chapter 5, a wave propagating in the amplifying medium will be described by

$$\vec{E}(z,t) = \vec{E}_0 e^{i(\omega t - kz)} \left\{ e^{(i\Delta k + \gamma(\omega)/2)z} \right\}, \tag{6.28}$$

where

$$\Delta k = -\frac{k\chi'(\omega)}{2n^2} \tag{6.29}$$

$$\gamma(\nu) = -\frac{k\chi''(\omega)}{n^2}. \tag{6.30}$$

In Chapter 2, we found that the resonant frequencies of an empty cavity (one not containing an amplifying medium) are determined by

$$nkl - 2(m + n + 1)\cos^{-1} \pm\sqrt{g_1 g_2} = 2q\pi, \tag{6.31}$$

where $l$ is the round-trip distance, $g_{12}$ are geometrical parameters, $q$ is an integral *axial mode index* and the $n$ before the $k$ is the index of refraction. The integers $m$ and $n$ multiplying the $\cos^{-1}$ are transverse mode indices. This expression is valid for both standing wave and traveling wave cavities. In the presence of an amplifying medium, this equation is changed to

$$nkl \left[ 1 + \frac{\chi'}{2n^2} \right] - 2(m + n + 1)\cos^{-1} \pm\sqrt{g_1 g_2} = 2q\pi, \tag{6.32}$$

due to the $\Delta k$ term in the susceptibility. The resonant frequencies are

$$\nu = \left( q + (n + m + 1)\frac{\cos^{-1}\sqrt{g_1 g_2}}{\pi} \right) \frac{c}{nl} \left( 1 + \frac{\chi'}{2n^2} \right)^{-1}. \tag{6.33}$$

The above equation can be used to determine the resonant frequencies, but it is more useful to express them in terms of the empty cavity resonance frequencies. This change in the resonant frequencies from those of the empty cavity is called *frequency pulling*. We begin by rewriting the above equation in terms of the *empty cavity resonance frequency*, $\nu_{qnm}$,

$$\nu \left[ 1 + \frac{\chi'}{2n^2} \right] = \nu_{qnm}, \tag{6.34}$$

where the empty cavity is still described by a refractive index, $n$. A *dispersion relationship* exists between $\chi'$ and $\chi''$

$$\chi' = \frac{2(\nu_0 - \nu)}{\Delta\nu}\chi'', \tag{6.35}$$

where $\Delta\nu$ is the width of the atomic resonance whose center frequency is $\nu_0$. This relation is a general result but can be derived from eqns 5.92 and 5.93. We now have

$$\nu \left[ 1 - \frac{\nu_0 - \nu}{\Delta\nu}\frac{\gamma(\nu)}{k} \right] = \nu_{qnm}, \tag{6.36}$$

where the expression for $\gamma(\omega)$ (eqn 6.30) was used. If the pulling effect is small and $\nu$ is very close to $\nu_{qnm}$,

$$\nu \approx \nu_{qnm} - (\nu_{qnm} - \nu_0)\frac{c\gamma(\nu_{qnm})}{2\pi n \Delta\nu}. \tag{6.37}$$

At threshold, we have $\gamma_t(\nu_{qnm})l = L_{tot}$ (round-trip gain equals loss) and from eqn 3.37 the finesse is $2\pi/L_{tot}$. Thus,

$$\frac{c\gamma(\nu_{qnm})}{2\pi n} = \frac{c}{\mathcal{F}l} = \Delta\nu_{1/2}, \tag{6.38}$$

where $\Delta\nu_{1/2}$ is the *empty cavity linewidth*. Putting these equations together, we have the *frequency pulling relationship*

$$\text{Frequency pulling:} \quad \nu = \nu_{qnm} - (\nu_{qnm} - \nu_0)\left(\frac{\Delta\nu_{1/2}}{\Delta\nu}\right). \tag{6.39}$$

Thus, the resonance frequencies will be shifted from their empty cavity values by the product of the ratio of the cavity linewidth to the gain linewidth (usually small) and the detuning from the gain resonance frequency. The pulling will always be toward $\nu_0$.

## 6.6   Inhomogeneous broadening and multimode behavior

When we placed an amplifying medium in a resonant cavity, we discovered that under certain conditions a self-sustained field is possible: the laser oscillates. The frequencies at which oscillation is possible are the resonant frequencies of the *cavity*, modified (*pulled*) by the presence of the amplifying medium. The actual cavity mode(s) at which oscillation takes place is determined by the requirement that the gain is equal to the loss. The question arises: is it possible for a laser to oscillate *simultaneously* at more than one cavity mode frequency? The answer depends upon the nature of the broadening in the amplifying medium.

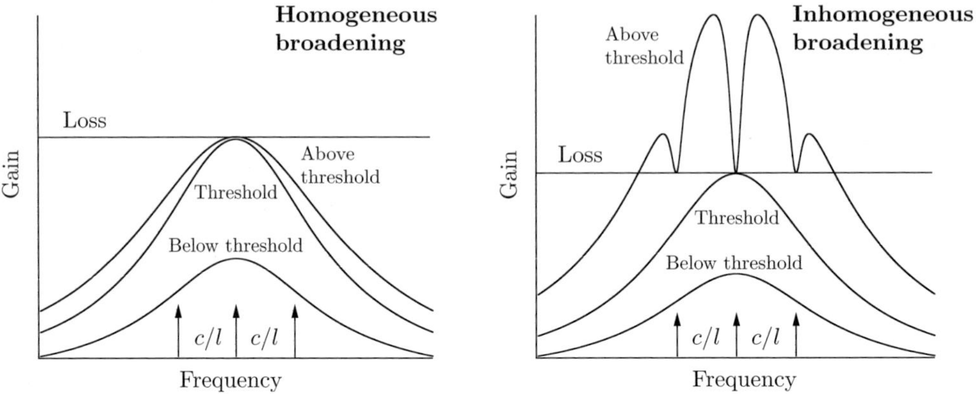


**Fig. 6.4** Gain curves for homogeneously and inhomogeneously broadened lasers for three pumping rates: below threshold, at threshold and above threshold.

A set of gain versus frequency curves appears in Fig. 6.4 for both homogeneously and inhomogeneously broadened lasers. The behavior of the curves for different pumping conditions illustrates how an inhomogeneously broadened laser can oscillate in more than one cavity mode simultaneously. In the figure, the cavity is adjusted so that one mode is always at the peak of the gain curve.

The curve on the left in the figure shows a homogeneously broadened gain curve at three pumping rates: below threshold, exact threshold and above threshold. As the pumping increases, the gain curve will increase in both amplitude and width until *threshold* is reached where the peak gain just equals the loss. The peak gain is *clamped* at this level and the laser will oscillate at the frequency corresponding to the peak. Further increases in the pumping rate will cause the gain curve to widen, but the peak will still be clamped to $\gamma_t$ and the unique oscillation frequency will be the same. All of this is a consequence of the behavior of the homogeneous gain curve under various levels of saturation, as discussed in Chapter 5. One can therefore state the general principle that *at any particular time, a homogeneously broadened laser is inherently single-mode.* There might be *technical noise* (mechanical fluctuations in the medium or cavity causing frequency fluctuations of a non-fundamental nature) and this might cause a *mode hop*, but the laser will still oscillate in only one mode at a time.

Below threshold, the inhomogeneously broadened gain curve behaves in a similar fashion to the homogeneously broadened curve, as shown on the right side of the figure. Again, at exact threshold, the laser will oscillate at the peak of the gain curve. Above threshold the behavior is very different from the homogeneous case. Since the model of the inhomogeneously broadened medium is a series of homogeneously broadened packets distributed in frequency over the inhomogeneous linewidth, each *cavity* mode can oscillate independently of the others provided that the gain at the mode frequency is greater than or equal to the loss. The plot shows a number of *holes* at several cavity modes with the bottom of the hole clamped to the loss as required by the condition for laser oscillation. For all frequencies *between cavity modes*, the gain can be greater than the loss since no clamping can occur at frequencies not corresponding to a cavity mode (possible laser oscillation can only occur at a cavity mode). It is as though a *large number of independent lasers exist* whose peak gain curve frequencies are distributed over the inhomogeneous linewidth. Thus, the general principle is that *an inhomogeneously broadened medium can sustain laser oscillation at a number of modes within the inhomogeneous linewidth.*

## 6.7 Spatial hole burning

Before leaving the topic of inhomegeneous broadening, a very important form of inhomogeneous broadening in a *standing wave laser* should be mentioned. It is due to the *spatial inhomegeniety* which is caused by the standing wave intensity distribution inside the gain medium. Because of the dependence of the field distribution between the mirrors on the mode frequency, the laser can oscillate simultaneously in more than one mode. The phenomenon is called *spatial hole burning* and is illustrated in Fig. 6.5.

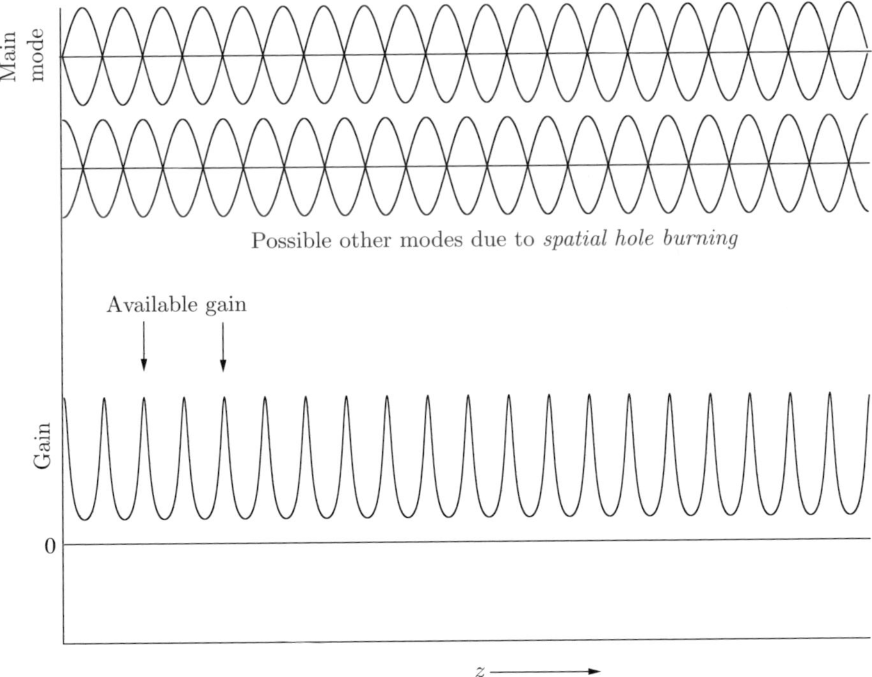


**Fig. 6.5** Illustration of *spatial hole burning*. The lowest curve is the saturated gain due to the upper standing wave field. The lower standing wave field has a different spatial distribution and can also oscillate due to the availability of gain.

In a standing wave cavity, if the intensity of the internal field in *each direction* is $I_\nu$, the gain saturation relationship is

$$\gamma(z) = \gamma_0 \frac{1}{1 + \frac{2I_\nu \sin^2 kz}{I_s}}, \tag{6.40}$$

which explicitly shows the $z$-dependence of the saturated gain. The presence of a spatially periodic gain is clearly depicted in Fig. 6.5 and is the reason that this phenomenon is called *spatial hole burning*. (The hole burning formerly discussed is called *spectral hole burning*.) In a standing wave, the clamping of the gain to the loss *at each point* is no longer a reasonable condition for laser operation. Instead, one requires that the rate of energy gain by the field due to stimulated emission be equal to the rate of energy loss *integrated over the entire length of the amplifying medium*. The intensity increase due to the gain is

$$\text{Intensity increase:} \quad \int_0^l I\gamma(z)dz = 2\gamma_0 I_\nu \int_0^l \frac{\sin^2(kz)dz}{1 + (2I_\nu/I_s)\sin^2(kz)}, \tag{6.41}$$

where $I = 2I_\nu \sin^2 kz$ is the total intensity in a standing wave. The power output in a standing wave laser can be calculated by equating the intensity increase to the integrated loss and performing a fair bit of algebra. One will find that the output power is as much as 30% smaller than in a traveling wave cavity due to the standing wave intensity pattern.

The possibility of multimode oscillation in a standing wave laser containing a homogeneously broadened medium can be explained by Fig. 6.5. The fundamental mode (at the top of the figure) gives rise to the periodic gain curve at the bottom of the figure. The maximum gain points are located at the *nodes* in the standing wave pattern and the minimum gain points are at the *antinodes* of the field. The figure also shows a different cavity mode whose nodes and antinodes are displaced considerably from those of the fundamental mode. As a result of this *different field distribution*, the second cavity mode has *gain available at places where its field is strong* and can therefore oscillate at the same time as the fundamental mode. Thus, we have possible multimode behavior due to *spatial inhomogenieties* as well as *spectral inhomogenieties*. This form of multimode behavior is most easily cured by using a traveling wave cavity instead of a standing wave cavity: the resulting laser is usually called a *ring laser*. Another *treatment* is to place the gain medium *very close* to one of the standing wave cavity mirrors. This will force the different field distribution to be *displaced considerably in frequency* from the fundamental mode where it can be more easily suppressed by intracavity etalons or other frequency-discriminating elements.

## 6.8   Some consequences of the photon model for laser radiation

The semiclassical theory treats the radiation field classically and is capable of describing the most important aspects of laser oscillation, such as the gain which is due to stimulated emission. There are, however, some aspects of laser operation which are best described using a *photon model*, which requires a fully quantum mechanical treatment of the field. We will postulate the existence of photons in a somewhat ad hoc manner and derive a number of important laser beam properties from this model. The first of the three topics using the photon model is an examination of the occupation number, $\bar{n}$, above threshold.

The determination of the occupation number requires the steady-state solution of two rate equations: one for the atoms and one for the photons. We first encountered photons when we derived the *Planck formula* for the spectral density of blackbody radiation. We postulated that the energy in the radiation field is an integral multiple of $h\nu$ or, alternatively, the field at frequency $\nu$ is composed of an integral number of *photons*, each of which has energy $h\nu$. From this, the intensity, $I_\nu$, of a unidirectional wave propagating inside the amplifying medium is

$$I_\nu = c\frac{nh\nu}{V_m}, \tag{6.42}$$

where $V_m$ is the *mode volume* of the field and $n$ is the number of photons. The rate equation for the upper amplifying level of a four-level laser is

$$\frac{dN_2}{dt} = RN_0 - \frac{N_2}{\tau} - (N_2 - N_1)\left(\frac{\lambda^2}{8\pi}\frac{I_\nu}{h\nu}A_{21}g(\nu)\right)$$

$$= RN_0 - \frac{N_2}{\tau} - (N_2 - N_1)\frac{K}{V_m}n \tag{6.43}$$

$$\text{where} \quad K = \frac{\lambda^3}{4\pi^2}A_{21}\frac{\nu}{\Delta\nu}, \tag{6.44}$$

and we have defined $K$ so that $KnN_2$ is the *stimulated emission rate at resonance*. We will assume that $t_{10}$ is so small that level 1 has negligible population ($N_1 \approx 0$). Then, multiplying by $V_m$

$$V_m\frac{dN_2}{dt} = RV_mN_0 - \frac{N_2V_m}{\tau} - KnN_2. \tag{6.45}$$

The rate equation for the photon number, $n$, requires one additional assumption: the decay rate for *spontaneous emission* is the same as that for stimulated emission with one photon per mode. Thus, the stimulated rate $KnN_2$ becomes $K(n+1)N_2$ when we add spontaneous decay. If the *photon decay rate* (due to cavity losses) is $1/t_c$, the photon number rate equation is

$$\frac{dn}{dt} = K(n+1)N_2 - \frac{n}{t_c}. \tag{6.46}$$

These two rate equations are readily solved for $n$ in the steady state (they are quadratic equations since $nN_2$ appears in both):

$$n = \frac{V_m}{2K\tau}\left\{\left(\frac{R}{R_t}-1\right) + \left[\left(\frac{R}{R_t}-1\right)^2 + \frac{4R}{R_t}\frac{K\tau}{V_m}\right]^{1/2}\right\}, \tag{6.47}$$

where $R_t$ is the threshold pumping rate

$$\text{At threshold:} \quad R_t = \frac{1}{N_0K\tau t_c}. \tag{6.48}$$

A plot of the photon occupation number from below to above threshold appears in

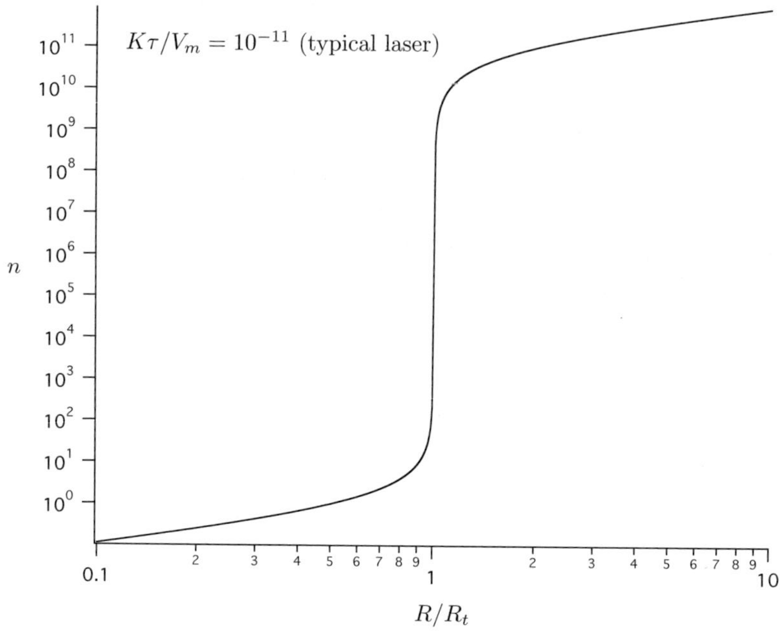


**Fig. 6.6** Photon occupation number, $n$, both above and below threshold.

Fig. 6.6. The remarkable jump in $n$ above threshold is akin to a *phase transition*. Since photons are bosons, it is a *Bose–Einstein condensation* of photons: the sudden appearance of an enormous number of particles in the *same quantum state*.

The energy spectral density of a *thermal* (blackbody) field is described by the Planck formula. The number of photons per mode (occupation number) is:

$$\bar{n} = \frac{1}{e^{h\nu/k_BT} - 1}. \tag{6.49}$$

We see that in order to approach one photon per mode, the temperature of the radiation source would need to be about 5000 K (the temperature of the surface of the sun). If the photon field from a laser were described by the same statistics as a thermal field (this is *not* an appropriate description, as we shall see later), the temperature of the source would need to be about $10^{10}$ K, clearly a stupendous temperature. This is one of the several remarkable features of the radiation from a laser; we will discuss others in the following sections.

## 6.9 The photon statistics of laser radiation

Before considering the statistics of laser radiation, we will briefly discuss *temporal coherence*. The temporal coherence of a wave is determined with the help of the *normalized first-order correlation function*, $g_1(\tau)$, which is defined as

$$g_1(\tau) = \frac{\langle E^*(t)E(t+\tau)\rangle}{\langle E^*(t)E(t)\rangle} \tag{6.50}$$

$$\text{where} \quad \langle E^*(t+\tau)E(t)\rangle = \frac{1}{T}\int_0^T E^*(t+\tau)E(t)dt, \tag{6.51}$$

where $T$ is a large but finite averaging time. Essentially, $g_1(\tau)$ is a measure of the correlation between a wave and the same wave delayed by $\tau$, or, equivalently, the correlation between the wave and the same wave a distance $c\tau$ from the original position along the propagation direction. Its magnitude can be obtained from a *Michelson interferometer* (Fig. 6.7). The intensity at the detector, $I_d$, is related to that at the

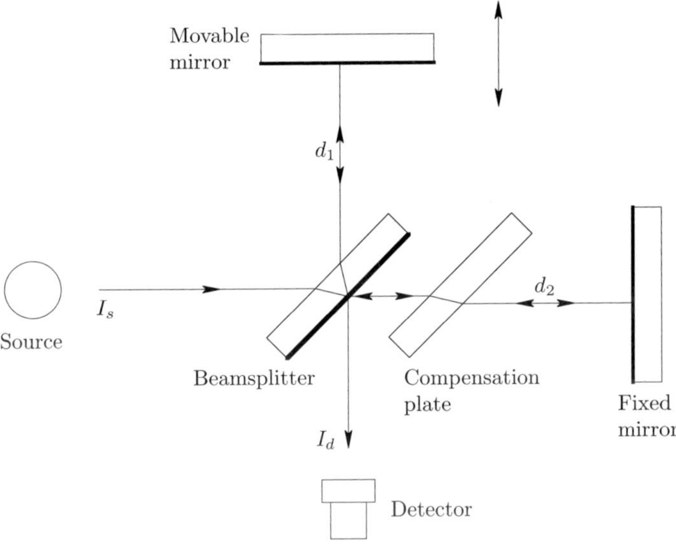


**Fig. 6.7** A Michelson interferometer used to measure temporal coherence.

source, $I_s$, by

$$I_d = \tfrac{1}{2}I_s[1 + \mathrm{Re}\{g_1(\tau)\}], \tag{6.52}$$

where $\tau = 2(d_1 - d_2)/c$. The *fringe visibility*, $V(\tau)$, is

$$V(\tau) \equiv \frac{I_{max} - I_{min}}{I_{max} + I_{min}} = |g_1(\tau)|. \tag{6.53}$$

We define the *coherence time*, $\tau_c$, to be *the value of $\tau$ at which $g_1$ and therefore the fringe visibility in a Michelson interferometer is reduced to one half of its maximum* (the maximum occurs when $\tau = 0$). We can relate $\tau_c$ to the *bandwidth*, $\delta\nu$, of the radiation using the following approximate argument:

Define $\Delta \equiv |d_1 - d_2| = c\tau/2$. The fringe peaks will appear when $\Delta = n\lambda$, where $\lambda$ is the average wavelength of the radiation. If there is a finite bandwidth, $\delta\nu$, of the radiation, the fringe peaks will be smeared by $\delta\Delta = -\Delta(\delta\nu/\nu)$. The fringes will disappear when $\delta\Delta \approx \lambda/2$ which implies that $-\Delta(\delta\nu/\nu) = -(c\tau_c/2)(\delta\nu/\nu) = \lambda/2$ and therefore $\tau_c \approx 1/\delta\nu$. A more careful calculation yields the following result:

$$\text{Coherence time:} \quad \tau_c = \frac{1}{2\pi\delta\nu} \tag{6.54}$$

$$\text{Coherence length:} \quad l_c \equiv c\tau_c. \tag{6.55}$$

We will show later on that the bandwidth of laser radiation can be well below 1 Hz; thus the coherence time can be 1 s or more and the coherence length can be *greater than* 300,000 km!

Light sources can be divided into two distinct classes, depending upon the photon statistics of the radiation they emit. Conventional light sources based upon a thermal emitter or a discharge lamp are said to emit *chaotic light*. Included in this category is a laser below threshold where the spontaneous emission will be amplified by the gain medium. The other type of source is a *laser* above threshold whose radiation is said to be *coherent*. The photon statistics of chaotic radiation is very different from that of coherent radiation.

We will begin our photon statistics study by assuming that one has an apparatus which counts photons during fixed time intervals of duration $T$. The quantity $P_n(t,T)$ is the probability of counting $n$ photons in an interval from $t$ to $t+T$ and $p(t)dt$ is the probability of counting *one* photon in a very small interval, $dt$ (it is assumed there is negligible probability of counting *two or more* photons in this interval). We can obtain a differential equation for $P_n(t,T)$ by considering the two *uncorrelated* ways of counting $n$ photons during the interval from $t$ to $t+T+dt$:

1. $n-1$ photons in $T$ and 1 photon in $dt$
2. $n$ photons in $T$ and 0 photons in $dt$

Adding the two probabilities, we obtain,

$$P_n(t,T+dt) = P_{n-1}(t,T)p(t)dt + P_n(t,T)(1-p(t))dt. \tag{6.56}$$

We can rearrange the terms to make the left side a derivative:

$$\frac{P_n(t,T+dt) - P_n(t,T)}{dt} = \frac{dP_n(t,t')}{dt'} = (P_{n-1}(t,T) - P_n(t,T))p(t). \tag{6.57}$$

If $\bar{I}$ is the photon emission rate, averaged over one optical cycle, and $\alpha$ is the *quantum efficiency* of the photodetector, then

$$p(t)dt = \alpha\bar{I}dt \tag{6.58}$$

and

$$\frac{dP_n(t,t')}{dt'} = (P_{n-1}(t,T) - P_n(t,T))\alpha\bar{I}(t). \tag{6.59}$$

It can be shown (by *substitution*) that the solution to this equation is

$$P_n(t,T) = \frac{[X(t,T)T]^n e^{-X(t,T)T}}{n!}, \tag{6.60}$$

where

$$X(t,T) = \frac{\alpha}{T} \int_t^{t+T} \bar{I}(t')dt' \qquad (6.61)$$

is the photon *counting rate* averaged over $T$. The usual practice is to average a large number of measurements over time. Then, the *average* value of $P_n(t,T)$ is called $P_n(T)$ and is given by

$$P_n(T) = \left\langle \frac{[X(t,T)T]^n e^{-X(t,T)T}}{n!} \right\rangle_t, \qquad (6.62)$$

where the subscript $t$ denotes averaging over $t$.

The key to the difference between the two types of statistics lies in the behavior with time of the quantity $\bar{I}(t)$. If $\bar{I}(t)$ is independent of $t$, then $X(t,T)$ is independent of both $t$ and $T$ and

$$X(t,T)T = \alpha \bar{I} T = \bar{n}. \qquad (6.63)$$

The calculation of the time average is trivial (since there is *no* time dependence), yielding the *Poisson distribution*

$$\text{Poisson distribution (constant } \bar{I}): \quad P_n(T) = \frac{\bar{n}^n e^{-\bar{n}}}{n!}. \qquad (6.64)$$

From the way that $\bar{I}(t)$ is defined, $\bar{n}$ is clearly the *average number of photons* registered by the counting apparatus during interval $T$. This result is valid for a classical stable wave and also for the so-called quantum mechanical *coherent state* of the radiation field (indeed, the latter is just a sum over $n$-photon states with coefficients equal to the square root of $P_n$, so Poisson statistics are an immediate consequence when one squares the coefficient). It is also valid for a *chaotic* source if *one counts for a period greater than the coherence time*, $t_c$, since the fluctuations will average out and $\bar{I}(t)$ can be considered constant. Thus, the distinction between the different statistics depends upon *the behavior of $\bar{I}(t)$ over counting periods which are short compared to the coherence time*.

A chaotic wave will be described by an $\bar{I}(t)$ which depends upon time for counting periods less than $t_c$. In order to calculate the average over $t$ as required by eqn 6.62, we need to know the *distribution of intensities*, $p[\bar{I}(t)]$, which is the probability that a measurement will yield a value between $\bar{I}(t)$ and $\bar{I}(t) + d\bar{I}(t)$. This can be obtained from the following considerations. We will model the chaotic source as a group of similar atoms emitting waves whose electric fields have the same frequency and amplitude but random phases:

$$\begin{aligned} E(t) &= E_1(t) + E_2(t) + \cdots + E_n(t) \\ &= E_0 e^{-i\omega_0 t}\left(e^{i\phi_1(t)} + e^{i\phi_2(t)} + \cdots + e^{i\phi_n(t)}\right) \qquad (6.65) \\ &= E_0 e^{-i\omega_0 t}\beta(t)e^{i\phi(t)}. \end{aligned}$$

The quantities $\beta(t)$ and $\phi(t)$ are the time-varying amplitude and phase which result from the addition of $n$ complex numbers with unity magnitude and random phases. The intensity is

$$\bar{I}(t) = \tfrac{1}{2}\epsilon_0 c |E(t)|^2 = \tfrac{1}{2}\epsilon_0 c \beta^2(t) E_0^2. \qquad (6.66)$$

The distribution of intensities reduces to finding the distribution of the quantity $\beta(t)$. The appropriate (normalized) probability can be obtained from *random walk* theory and is given by

$$p[\beta(t)] = (1/\pi n)e^{-\beta^2(t)/n}, \tag{6.67}$$

since we are adding a series of vectors of fixed length and random direction, which is equivalent to a random walk. From eqn 6.66, one can convert this into a distribution of intensities,

$$p[\bar{I}(t)] = (1/\bar{I})e^{-\bar{I}(t)/\bar{I}}, \qquad \bar{I} = \tfrac{1}{2}\epsilon_0 c E_0^2 n. \tag{6.68}$$

The appropriate value of $X(t,T)$ for chaotic light is

$$\text{Chaotic light:} \quad X(t,T) = \alpha\bar{I}(t), \tag{6.69}$$

since, for $T \ll t_c$, the intensity is essentially *constant* during the counting interval, but varies from one counting interval to the next. The value of $P_n(T)$ for chaotic light is obtained by integrating the product of the distribution of intensities and $P_n(t,T)$,

$$P_n(T) = \frac{1}{\bar{I}} \int_0^\infty e^{-\bar{I}(t)/\bar{I}} \frac{[\alpha\bar{I}(t)T]^n e^{-\alpha\bar{I}(t)T}}{n!} d\bar{I}(t). \tag{6.70}$$

This integral is actually easier to evaluate than it appears and the result is

$$\text{Chaotic source (Bose–Einstein distribution):} \quad P_n(T) = \frac{\bar{n}^n}{(1+\bar{n})^{1+n}}, \tag{6.71}$$

where $\bar{n}$ is again $\alpha\bar{I}T$.

Our procedure for calculating $P_n(T)$ for chaotic sources was very general and was based upon some simple photon counting considerations together with the assumption that the source intensity was not constant for times much shorter than the coherence time. For the *particular case* of a thermal (blackbody) source, it is very straightforward to obtain the same result. The probability of observing $n$ photons in thermal equilibrium with the source is just the normalized Boltzmann factor

$$P_n = \frac{e^{-nh\nu/k_B T}}{\sum_m e^{-mh\nu/k_B T}}, \tag{6.72}$$

where we have dropped the $T$ dependence in $P_n$. The numerator can be evaluated from the expression for $\bar{n}$ obtained during the derivation of Planck's formula and the denominator is just a geometrical series which can also be expressed as a function of $\bar{n}$. The result is

$$\text{Blackbody radiation:} \quad P_n = \frac{\bar{n}^n}{(1+\bar{n})^{n+1}}. \tag{6.73}$$

Finally, we will evaluate the *variance* for the two distributions. The variance, $\Delta n^2$, is defined as the average of $(n-\bar{n})^2$ and is given by

$$\text{Variance:} \quad \langle \Delta n^2 \rangle = \langle n^2 \rangle - \bar{n}^2. \tag{6.74}$$

It will be left as an exercise for the reader to verify that the variances of the two distributions are:

$$\text{Chaotic source:} \quad P_n = \frac{\bar{n}^n}{(1+\bar{n})^{n+1}}, \qquad \langle \Delta n^2 \rangle = \bar{n}^2 + \bar{n} \tag{6.75}$$

$$\text{Laser source:} \quad P_n = \frac{\bar{n}^n e^{-\bar{n}}}{n!}, \qquad \langle \Delta n^2 \rangle = \bar{n}. \tag{6.76}$$

We thus have the familiar result that the RMS fluctuation in count rates is equal to $\sqrt{\bar{n}}$ for almost all sources encountered in the laboratory. Most chaotic sources have bandwidths of many MHz and would require counting times of less than a microsecond to observe their departure from Poisson statistics. Figure 6.8 displays an exception: it shows the difference in the distributions for $\bar{n} = 20$. These two very different distributions were observed in the laboratory for a laser both above and below threshold (Freed 1966).

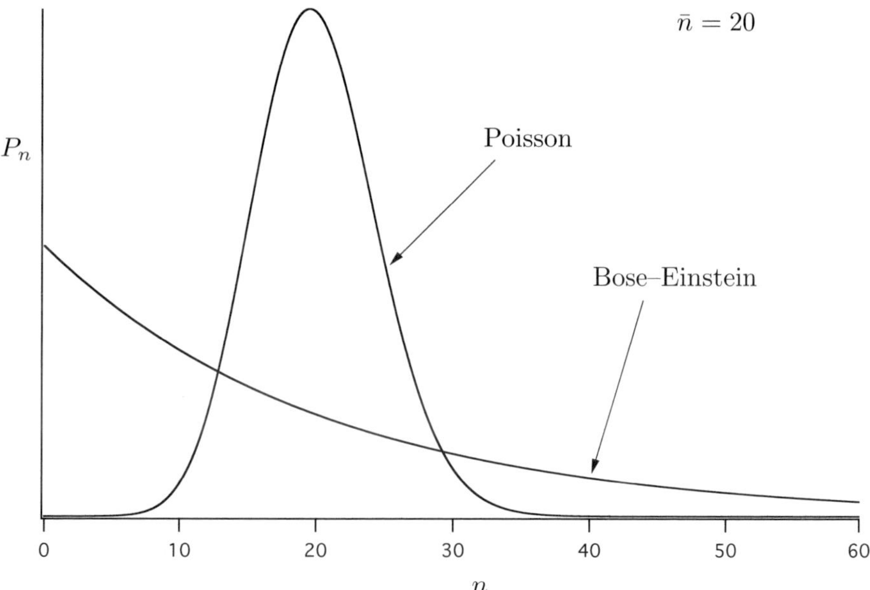


**Fig. 6.8** Comparison of coherent (Poisson) and chaotic (Bose–Einstein) distributions for $\bar{n} = 20$. These are theoretical curves but the two very different distributions have been observed in the laboratory (Freed 1966).

## 6.10 The ultimate linewidth of a laser

From our calculation of the occupation number $(\bar{n})$ of the radiation from a laser above threshold, we discovered that $\bar{n}$ is extremely high and therefore the radiation will be an excellent approximation to a classical, stable, monochromatic wave. Our approach ignored the effects of *spontaneous emission*, since this process does not fit naturally into the framework of a semiclassical treatment. When one includes spontaneous emission, the randomness of the spontaneous photons will add some uncertainty to the phase and frequency of the electric field, and the laser radiation will acquire a finite linewidth. We will give here a simplified derivation of the *Schawlow–Townes* expression for the ultimate linewidth of a laser.

Our approach will be to consider the atoms and radiation to be *one composite quantum system*, described by a quality factor, $Q$, which is defined in terms of the *stored energy* and *energy loss rate* (eqn 3.39). The stored energy will be in the *field*.

We begin with the rate equation for the photon number, which is reproduced here

$$\frac{dn}{dt} = K(n+1)N_2 - \frac{n}{t_c}, \qquad K = \frac{\lambda^3}{4\pi^2} A_{21} \frac{\nu}{\Delta\nu}, \qquad (6.77)$$

where $N_2$ is the upper laser level population. Let us assume that we have a four-level laser and that the lower laser level has negligible population. In the steady state $(dn/dt = 0)$ and well above threshold $(n \gg 1)$, $N_2$ is

$$\text{Steady state, well above threshold:} \quad N_2 = \left(\frac{n}{n+1}\right)\frac{1}{Kt_c} \to \frac{1}{Kt_c}. \qquad (6.78)$$

From the rate equation, the *spontaneous rate* (the "1" in the brackets) into a mode is

$$\text{Spontaneous rate} = KN_2 = \frac{1}{t_c} \qquad (6.79)$$

from the equation immediately above and the *spontaneous power into the mode* is obtained by multiplying by $h\nu$:

$$\text{Spontaneous power} = \frac{h\nu}{t_c}. \qquad (6.80)$$

If we assume that the *output coupling* is the only loss, the output power, $P_o$, is

$$P_o = \frac{nh\nu}{t_c}, \qquad (6.81)$$

since $t_c$ is the rate that the energy leaks out through the output coupler. The *stored mode energy* is $nh\nu$:

$$\text{Stored energy} = nh\nu = t_c P_o. \qquad (6.82)$$

The $Q$ factor is given by

$$Q = \frac{\nu}{\delta\nu} = \omega \times \frac{\text{stored energy}}{\text{spontaneous energy emission rate}} = \frac{(2\pi\nu)t_c P_o}{(h\nu/t_c)} = \frac{2\pi t_c^2 P_o}{h}. \qquad (6.83)$$

Solving for $\delta\nu$,

$$\delta\nu \approx \frac{2\pi(\Delta\nu_{1/2})^2 h\nu}{P_o}, \qquad (6.84)$$

where we used the cavity expression, $t_c = 1/2\pi\Delta\nu_{1/2}$. Other treatments include a factor $N_2/\Delta N$ which is unity in our case. The important parameter is the *cavity linewidth* ($\Delta\nu_{1/2}$), which can be quite large in the extremely short cavities found in semiconductor lasers whose ultimate linewidths can therefore be in the tens of MHz (there is actually another *linewidth-enhancing* factor in semiconductor lasers). At the other extreme, a conventional laboratory continuous wave dye or Ti-sapphire laser can have an ultimate linewidth of less than one Hz. It should be stressed again that the Schawlow–Townes linewidth is a *lower bound*: other (usually remediable) factors such as mechanical vibrations and thermal effects will invariably make the observed linewidth much larger than that derived from this expression. The increased linewidth from the latter factors is called *technical noise*. When we discuss *laser frequency stabilization*, we will show how one can remove the technical frequency noise and even reduce the laser linewidth to below the Shawlow–Townes limit.

## 6.11 Further reading

Much of the discussion of the photon statistics is based upon Loudon's book on the quantum theory of light (1983). The discussions of the laser oscillation condition and pulling comes largely from Yariv (1989) and the discussions of the Schawlow–Townes limit and photon statistics are similar to those in Milonni and Eberly (1988).

## 6.12 Problems

(6.1) The *threshold* pumping power is measured with two different output couplers whose power transmissions are $T_1$ and $T_2$. If the corresponding threshold pumping powers are $P_1$ and $P_2$ and everything else is the same, write down a formula for the *internal losses* of the laser (assume that all the other mirrors are totally reflecting). For simplicity, assume that the laser is a ring laser with very low losses.

(6.2) Calculate the shift due to frequency pulling in a He-Xe laser operating at the 3.51 $\mu$m transition of Xe. Assume that the transition is Doppler broadened with a FWHM of 200 MHz and that the laser is a standing wave cavity with single-pass logarithmic loss $\rho$ (defined as $\rho = -\ln(1-T)$) of 0.5 and length of 30 cm. Estimate the pulling when the cavity resonance is detuned 50 MHz from the gain curve center. (Use the expression for *homogeneous broadening* derived in the text; this is less than a factor of two different from the *inhomogeneous* broadening result.)

(6.3) Determine the quantum limit to the linewidth for the following lasers:
    (a) A He-Ne laser ($\lambda = 632.8$ nm) with a cavity length of 30 cm, total loss per pass of 1% (including output coupling) and power output of 1 mW.
    (b) A GaAs/GaAlAs diode laser ($\lambda = 850$ nm) with uncoated facets (refractive index 3.5), a cavity length of 120 $\mu$m, loss coefficient ($\alpha$) of 10 cm$^{-1}$ and power output of 5 mW.
    (c) A ring dye laser ($\lambda = 590$ nm) with a total path length of 75 cm, loss factor ($\alpha l_{tot}$) of 0.05%, output coupling of 3% and power of 100 mW.

(6.4) (a) Estimate the bandwidth and coherence time of white light, assuming that it comprises wavelengths from 400 nm to 700 nm.

    (b) Show that the coherence length of white light is on the order of the wavelength.

    (c) Using a Michelson interferometer, is it possible to observe "white-light" fringes (that is, an observable fringe pattern with the white light defined above)?

(6.5) Calculate the coherence time and the coherence length (both of these are related to the "longitudinal coherence") for the following sources (this comparison should illustrate one of the main differences between *thermal* light and laser light):

    (a) a mercury source emitting at 546 nm with a bandwidth of $\Delta\lambda \approx 0.01$ nm

    (b) a YAG laser at 1064 nm with a bandwidth $\Delta\nu \approx 10$ kHz

(6.6) Show that the average intensity ($\langle I(P)\rangle$) observed at the detector in a Michelson interferometer depends upon the intensity at the source ($\langle I(R)\rangle$) and the complex degree of coherence at the source ($\gamma(R, R, \tau)$) and is given by the following relationship:
$$\langle I(P)\rangle = \frac{1}{2}[1 + \mathrm{Re}\{\gamma(R, R, \tau)]$$

(6.7) Show that the variance ($\langle \Delta n^2\rangle$) of the photon number for a Poisson distribution is given by the following equation:
$$\langle \Delta n^2\rangle \equiv \langle (n - \bar{n})^2\rangle = \bar{n}$$

(6.8) Show that the variance ($\langle \Delta n^2\rangle$) of the photon number for a thermal (Bose–Einstein) distribution is given by the following equation:
$$\langle \Delta n^2\rangle \equiv \langle (n - \bar{n})^2\rangle = \bar{n}^2 + \bar{n}$$

# 7
# Descriptions of specific CW laser systems

## 7.1  Introduction

In this chapter, we will describe several continuous wave laser systems, aided by the theory formulated in the previous two chapters. Due to its dominance, the ubiquitous diode laser warrants at least one chapter of its own. Several of the laser systems described in this chapter are currently being replaced by simpler and less expensive diode and diode-pumped solid state systems.

In the category of lasers whose days are probably numbered in the atomic physics laboratory are argon ion lasers and dye lasers. Some commercial realizations of these lasers will be described. The diode-pumped frequency-doubled Nd:YVO$_4$ laser and the Ti-sapphire laser (pumped by the former) will be discussed. Finally, the formerly ubiquitous He-Ne laser (currently supplanted by red diode lasers) will be described, in part for historical reasons and also because it employs an interesting and unique excitation transfer mechanism for pumping. We will start with the He-Ne laser.

## 7.2  The He-Ne laser

A schematic of an early helium neon (He-Ne) laser appears in Fig. 7.1. The laser

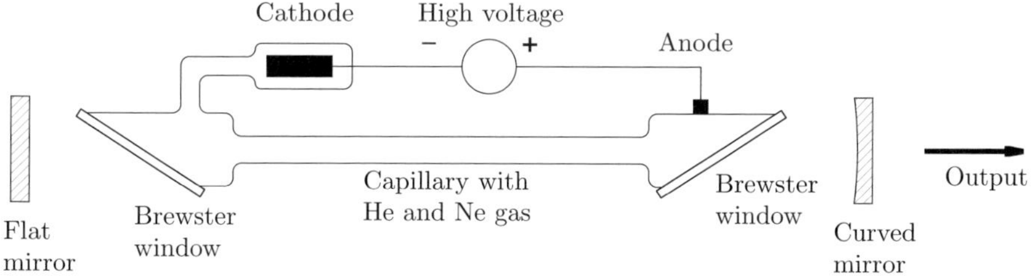


**Fig. 7.1** Schematic of helium-neon laser.

consists of a glass capillary filled with a mixture of helium and neon gases and with windows inclined at Brewster's angle on both ends. A DC glow discharge is maintained by a high voltage power supply connected to an anode and cathode placed inside the tube. We shall shortly demonstrate that there is an optimum current which produces the maximum output power and the discharge current is set to this value (the voltage–current characteristic depends on the pressure of the gases and the capillary dimensions). It has been found that the optimum performance is obtained when the helium partial pressure is about five times that of neon and the product of the total pressure and capillary diameter is about 4 Torr-mm. Thus, for a typical laser with a 2 mm diameter capillary, about 2 Torr of helium and 0.4 Torr of neon is used. The laser emits radiation at a number of wavelengths and the optimum parameters have a slight wavelength dependence.

The pumping mechanism can be understood by reference to Fig. 7.2, which shows the energy diagrams of both neutral helium and neutral neon. The principal pumping

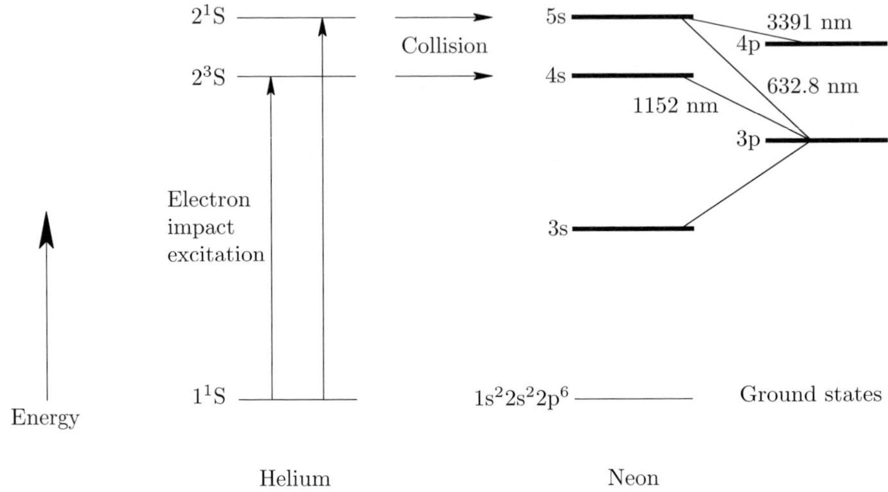


**Fig. 7.2** Level diagram of helium and neon as an aid in explaining the pumping mechanism in He-Ne laser. The excited levels in neon are actually *groups* of levels (composed of different *terms*) whose outer electron state is specified in the diagram. The *s*-states have four levels and the *p*-states have ten.

mechanism is a two-step process which relies on the near resonance between the $2^3S$ and $2^1S$ levels in helium and the $4s$ and $5s$ levels in neon. The energetic electrons in the discharge excite the helium atoms into the *metastable* $2^3S$ and $2^1S$ levels, where they remain for $10^{-4}$ seconds and $5 \times 10^{-6}$ seconds respectively. These lifetimes are much shorter than expected for the $2^3S$ and $2^1S$ states since they are fairly rapidly quenched by the *discharge*.

Once the helium atoms are excited to the metastable states, they transfer their excitation, *via collisions*, to the $4s$ and $5s$ states of neon. This process is greatly enhanced by the accidental near resonance between these levels and the metastable levels in helium. This step *pumps* the neon atoms into their *upper lasing level*. From this point, the neon atoms behave like a four-level laser system, since the $S$ to $P$ transition rate ($\approx 10^{-7}$ /s) responsible for laser amplification is somewhat smaller than the rate ($\approx 10^{-8}$ /s) of the $3p, 4p$ to $3s$ transitions, which deplete the lower laser level. The figure displays three possible laser frequencies; the most common one is 632.8 nm (there is another one at 543 nm which is not shown; it is due to a transition from the $5s$ state to a different term in the $3p$ configuration).

The reason that there is an optimum discharge current is the following. Let the ground state population in helium be $N_1$, the mestatable population be $N_2$ and the discharge current be $I$. If $\alpha_1$ and $\alpha_2$ are constants, the helium metastable state *excitation rate* is $N_1\alpha_1 I$ and the *de-excitation rate* is $N_2(\alpha_2 + \alpha_3 I)$, where the constant term is due to other de-excitations, such as wall collisions. At equilibrium, the two rates are equal and the metastable state helium population is

$$N_2 = N_1 \frac{\alpha_1 I}{\alpha_2 + \alpha_3 I}. \tag{7.1}$$

We assume that, via collisions, the excited state neon populations will be proportional to the helium metastable populations. Thus, the upper laser level (in neon) will *saturate* with increasing current. However, the lower lasing level is increasingly populated as the current increases, due to electron bombardment. The result will be a *peak* in the population difference and, therefore, gain. This phenomenon is depicted in Fig. 7.3 which displays the populations and gain in neon as a function of the current. The peak

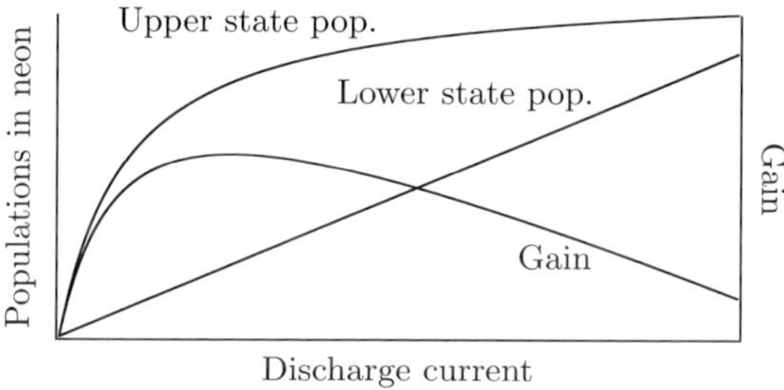


**Fig. 7.3** Saturation of upper state population and gain maximum in neon in a He-Ne laser.

gain occurs at a discharge current of some tens of milliamps in a He-Ne laser, thus restricting the maximum power of such lasers to several milliwatts at most. Despite the much lower efficiency of an argon laser, the latter can produce 1000 times more power since the gain does not saturate with increased current.

## 7.3 The argon ion laser

The argon ion laser is another system which is pumped by an electrical discharge which both ionizes the argon gas and populates the upper laser level in the ion. It is an extraordinarily inefficient system but it can generate much more power than a He-Ne laser on a number of wavelengths in the blue and UV regions of the spectrum. A simplified cross-sectional drawing of a commercial argon ion laser tube appears in Fig. 7.4.

We will shortly show that the gain in an argon ion laser is proportional to the *square* of the current density in the discharge and therefore a considerable effort is made to confine the discharge as much as possible. An ingenious technique for accomplishing this uses a series of perforated tungsten disks whose outer edge is in contact with a

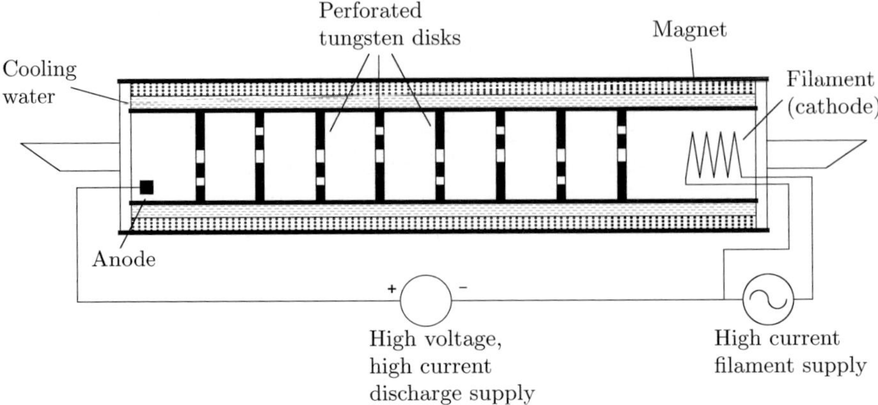


**Fig. 7.4** Simplified cross-sectional view of argon ion laser tube with associated power supplies.

cooling water jacket. The central hole in the disks confines the discharge to a column whose diameter is equal to that of the hole and the excellent thermal conductivity of the tungsten removes the considerable heat generated by the discharge. The additional holes in the disks serve as a gas return path, required to prevent large axial pressure gradients due to the *pumping* from the discharge. The holes are deliberately misaligned between adjacent disks (shown as *staggered* in the figure) to provide a larger path and ensure that the discharge uses only the central holes. A filament is used to increase the area of the cathode and reduce its work function, thereby increasing the ionization rate and protecting the cathode. Most commercial argon lasers have an external argon reservoir which is provided with an automatically operated solenoid valve to keep the argon pressure relatively constant (the discharge will remove argon from the tube over time). A relatively large magnetic field (about 1000 gauss), generated by a solenoid surrounding the tube, will cause the electrons to spiral around the field lines and more effectively ionize the argon atoms.

The level diagram of $Ar^+$ appears in Fig. 7.5. The *zero* of the energy scale is set equal to the energy of the neutral argon ground state. This shows more clearly the total energy needed to excite the upper laser level of the ion. Unlike in the He-Ne laser, the laser levels in $Ar^+$ are directly excited by electron bombardment in the discharge. Due to the exponential dependence of the electron energy distribution, which is proportional to $e^{-E/k_B T}$, a two-step pumping process is much faster than a single-step process, even for the relatively *hot* electrons in a discharge. A single-step process would require an electron with more than 35 eV while the two-step process needs two $\approx$ 18 eV electrons. The first electron ionizes the neutral argon and the second electron excites the ion to the upper laser level. There are actually three ways that the second excitation can proceed. The ion can be directly excited to the upper laser level (from the ion ground state), the ion can be excited to higher energy levels followed by radiative decay to the upper laser level and the ion can be excited to a metastable level, where a collision with another electron will transfer the ion to the

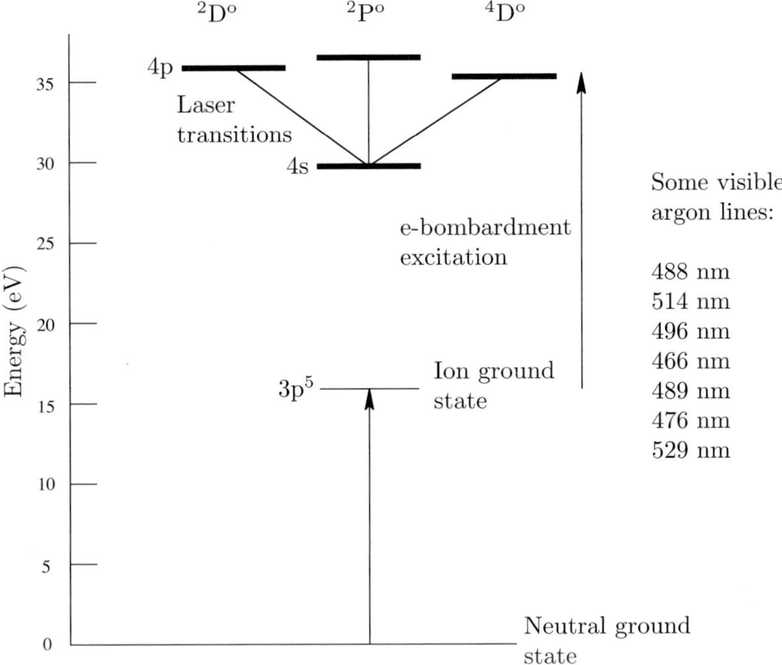


**Fig. 7.5** Levels of $Ar^+$ refered to ground state of neutral argon.

upper laser level. The first two mechanisms proceed at a rate which is proportional to the electron density, $N_e$, in the discharge. Since the ionization rate is also proportional to the electron density, the total excitation rate is proportional to the electron density squared:

$$\frac{dN_2}{dt} \propto N_e^2 \propto J^2, \tag{7.2}$$

where $J$ is the current density (which is proportional to $N_e$). From this, we see why argon ion laser designers go to great lengths to increase the current density, for example, by using perforated disks and strong magnetic fields as discussed above. Since the excitation rate to level 2 is proportional to the current squared and the quenching rate from either laser level is linear in the current, the pumping rate does not *saturate*, as it does in the He-Ne laser.

As is the case with neon, the lifetime of the upper laser level ($\approx 10^{-8}$ s) is about 10 times longer than that of the lower laser level, so one can apply four-level laser theory. One can define the laser *quantum efficiency* as the ratio of the energy of the stimulated photon to the energy needed to excite the upper laser level. The quantum efficiency of an argon laser is very low, about 7%. However, since the argon pumping scheme does not saturate, one can obtain a significant amount of power by using a very large discharge current. Typical argon lasers can operate with currents of 60 A or more while He-Ne lasers are limited to currents in the tens of milliamps range. The actual overall efficiency of an argon laser is much less than its quantum efficiency. At a discharge current of 60 A, the discharge voltage is about 500 V while the power output is only about 20 W; this yields an overall efficiency of less than 0.001 and explains why water cooling is absolutely necessary in *large frame* argon ion lasers.

## 7.4  The continuous wave organic dye laser

It is often desirable in the laboratory to have a laser which is tunable over a wide range of wavelengths. In addition to the obvious utility of such systems, they also make very good candidates for *mode-locked* operation, as we shall see later. There are two practical widely tunable systems in the visible and near infrared: those using an organic dye as the active medium and those using a titanium-sapphire crystal. Although they have some superficial similarities, they will be considered separately.

The active medium in a dye laser is an organic dye dissolved in a solvent (which actually has a surprisingly important role in determining the laser characteristics). The dye molecules have a complicated level structure which can be modeled as a simple diatomic molecule, as shown in Fig. 7.6. The figure displays the *vibrational* levels in

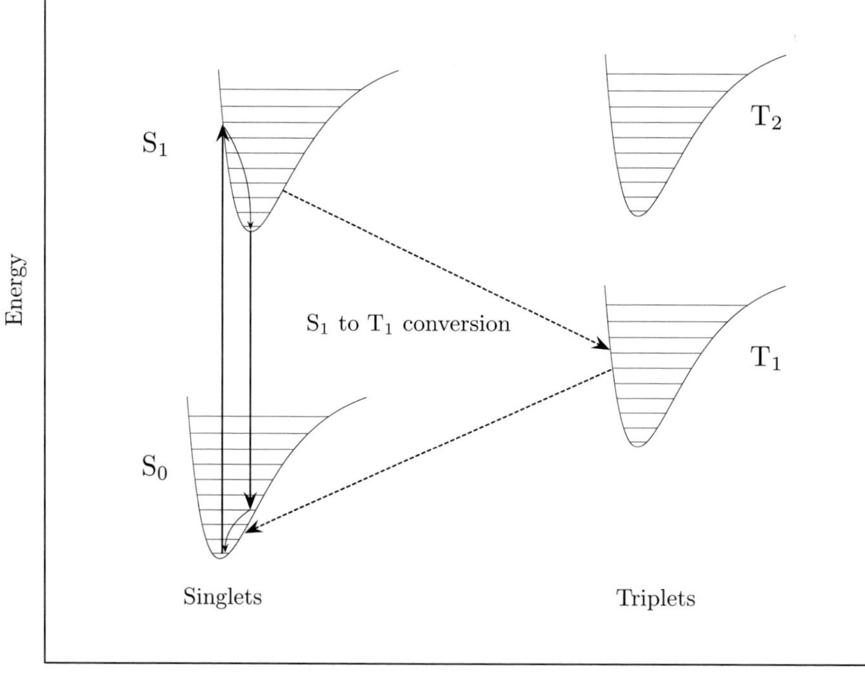


**Fig. 7.6** Simplified model of organic dye level structure.

a diagram whose horizontal coordinate is the intranuclear distance. Each potential energy curve corresponds to a different *electronic* level and the vibrational levels have a fine structure of rotational levels, which are not shown. The minima of the potential energy curves for the different *electronic* levels are at different intranuclear distances;

this is necessary for laser behavior, as we will see below. The wide tunability of the dye laser is due to the rich vibration-rotational level structure which is further broadened by the presence of the liquid solvent.

Four sets of *electronic levels* are shown in the diagram: two *singlets* ($S_1$ and $S_2$) and two *triplets* ($T_1$ and $T_2$). The laser transitions are between the two singlet levels and they are very fast (nanosecond lifetimes) since the transitions are *allowed*. The transitions between singlets and triplets are *forbidden* and therefore have very long *radiative* lifetimes. When the dye is excited to a higher electronic level, the various vibrational levels *very quickly* thermalize (in picoseconds), with the maximum population appearing at the potential minimum. Any molecule which ends up in a triplet state will be removed from the laser amplification process until it decays back to the singlet levels (which can take a good fraction of a second). An important rule for determining the behavior of the vibrational levels during an electronic transition is the *Franck–Condon principle* which states that electronic transitions will proceed along a *vertical line* in diagrams such as Fig. 7.6. The reason for this is that the electronic transition time is much shorter than the vibrational period and therefore the intranuclear distance will not change very much during an electronic transition.

With these preliminaries, it should be clear that the dye level structure is a good approximation to a four-level laser. The following sequence of transitions will take place when an optical pumping source is turned on. First there will be a transition from the bottom of $S_0$ to an excited vibrational state in $S_1$ (due to the Franck–Condon principle). Then the $S_1$ levels will quickly thermalize to populate the upper laser level (near the bottom of $S_1$). The laser transition will be between this level and an excited vibrational level in the lower singlet system, $S_0$. Due to the *different intranuclear distances* in $S_0$ and $S_1$, the laser transition will be very distinct from the pumping transition and a four level-laser system will be realized.

The absorption and emission curves for a popular dye are shown in Fig. 7.7. The absorption curve is due to the pumping transitions and the emission curve is due to the laser transitions. The figure clearly shows the shift between the two curves, which corroborates our model, and the $\approx 70$ nm breadth of the emission curve, which allows good tunability. The unfortunate partial overlap between the absorption and emission curves will harm the shorter wavelength performance of the dye. A curve of the triplet absorption is also shown; this can also reduce the power output of the laser. The peak of the absorption curve is near 500 nm, which is ideal for pumping with the very strong 488 nm or 514 nm lines from an argon ion laser.

The long triplet decay lifetime is a very serious problem for continuous wave operation of a dye laser and requires that the liquid dye solution circulate in a closed loop so that the triplets are given an ample time to decay. Although the triplet to singlet transitions are radiatively forbidden, non-radiative transitions promoted by the solvent can quickly populate the triplet levels and the dye needs to flow fairly rapidly to prevent this from significantly reducing the laser power. The commonly adopted scheme is to produce a thin *dye stream* (about 0.1 mm thickness) by forcing the dissolved dye through a jet under pressure from a circulating pump (Fig. 7.8). The stream is positioned at the waist between two concave laser cavity mirrors and oriented at Brewster's angle to minimize the Fresnel loss. With a typical dye stream speed of 10 m/s

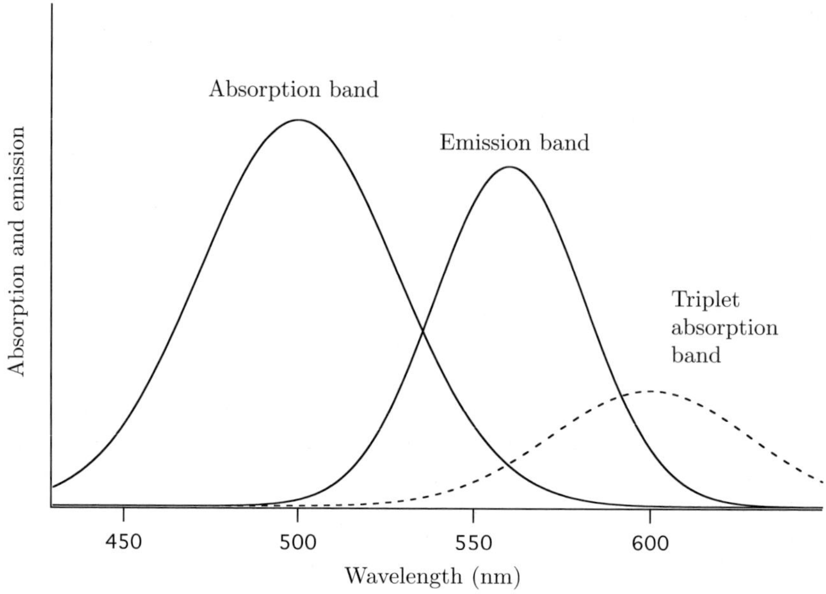



**Fig. 7.7** Absorption and emission curves for a common dye.

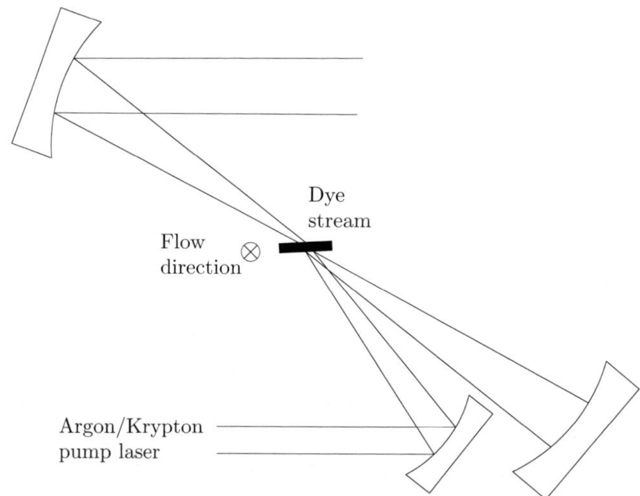



**Fig. 7.8** Detail of dye jet and associated optics. Both standing wave and ring lasers employ essentially the identical system for circulating the dye.

and spot size of 10 $\mu$m, a dye molecule spends only about 1 $\mu$s in the laser cavity; this is usually shorter than the time for significant triplet build-up. A disadvantage of such large dye stream velocities is that a bubble will cause disturbances in $\approx 1$ $\mu$s and therefore generate frequency noise in the MHz region, which requires very fast servo systems to remove.

A complete commercial ring dye laser system is shown in Fig. 7.9. The mirror

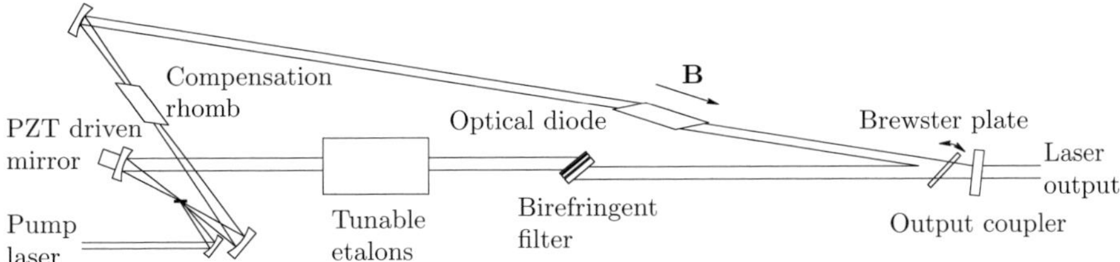


**Fig. 7.9** Schematic of ring dye laser with electronic frequency controls.

configuration is essentially a distorted version of the bow-tie ring discussed earlier with three instead of two curved mirrors. The third curved mirror generates a second small waist which can be used for intracavity frequency doubling (the smallest waist is at the dye stream). The cavity astigmatism is compensated partly by the dye jet and mostly by a *compensation rhomb* (or a *doubling crystal* in place of the rhomb) whose surfaces are at Brewster's angle. The tendency of the laser to operate in both directions is eliminated with an *optical diode*, which uses the *Faraday effect* to rotate the polarization in a manner which makes one direction much more lossy than the other (this will be elaborated on later).

Since the organic dye medium is *homogeneously broadened* and a ring configuration is used (eliminating spatial hole burning), the ring dye laser should be intrinsically single mode. While this is true *at any time*, there is a tendency for the laser to *mode hop* due to mechanical disturbances, especially those due to hydrodynamical fluctuations in the dye stream. This tendency is increased by the spectrally broad gain curve which allows several modes to have very nearly the same gain. To increase the frequency selectivity and reduce this tendency to mode hop, additional frequency-selective elements are placed in the cavity. These also allow electronic control of the laser frequency to some extent.

Coarse frequency control is provided by a *birefringent filter*, which is essentially a three-plate *Lyot filter*. Each element consists of a plate of birefringent material which converts the linearly polarized light to *elliptical polarization* at all but a series of equally spaced frequencies, whose spacing is dependent on the thickness of the plate. Since laser operation is possible only at frequencies which are *not* converted to elliptical polarization, each plate *allows* a set of possible laser frequencies separated by a *free spectral range* and having a *finite width*. The overall free spectral range of the three-plate filter is very large since the three plates have different, incommensurate free spectral ranges. The *fast axes* of the elements are in the plane of the plates, and the frequency of the filter is changed by rotating the plates about an axis normal to their surface.

Fine frequency control is provided by changing the cavity length using two intracavity elements: the piezo-electric (PZT) controlled mirror can provide relatively rapid frequency changes (at a rate of up to about 15 kHz) and a galvanometrically driven plate at Brewster's angle (the *Brewster plate*) can change the path length at a somewhat lower rate by rotating, as shown in the figure. Mode hops are reduced using a pair of *etalons* (simple Fabry–Perot filters having low finesse) with considerably different widths. The etalon with the smaller width is electronically locked to the laser frequency and acts like a *tracking filter*. The laser can be frequency locked to an external cavity using these electronic frequency controls, and its stabilized linewidth is about 1 MHz. Although a given dye might be tunable only over a range of 50 to 100 nm, a large number of dyes are available covering the entire visible and near infrared region of the spectrum and generating from several milliwatts to more than a watt of narrow-band radiation.

Dye lasers have numerous difficulties. Perhaps their greatest problem is their sheer inefficiency when one includes the pump laser. One needs from 30 kW to 50 kW from the mains (with all of its attendant problems such as the need for water cooling and the increasing cost of electrical power) to generate as little as a few milliwatts of narrow-band radiation (this inefficiency becomes even more stark when the visible light is *frequency doubled* into the UV). The finite lifetime and high replacement cost of the pumping laser tubes are well known to most atomic physicists. The finite lifetime of dyes is another nuisance, together with the need to periodically handle potentially toxic dyes during a dye change. Finally, a complete dye laser and argon pump laser can cost more than \$200,000. Frequency-doubled and diode-pumped solid-state lasers for dye pumping are more efficient and have less costly maintainance than argon lasers, though their initial cost is similar. The next laser to be discussed – the titanium-sapphire laser – avoids some of the problems of a liquid dye solution, but real advances in the utility of laboratory laser systems probably depend upon the rapidly increasing sophistication of diode and fiber lasers.

## 7.5   The titanium-sapphire laser

The titanium-sapphire laser is continuously tunable from about 700 nm to 1020 nm using a single crystal, making it the most widely tunable laser available. It is usually pumped with the 532 nm line of a diode-pumped, frequency-doubled CW Nd:YVO$_4$ laser and can deliver more than one watt of narrow-band radiation over most of its range. The active medium is a crystal of sapphire (aluminum oxide, Al$_2$O$_3$) *doped* with about 1% by weight of TiO$_3$. The titanium atoms replace the aluminum atoms at some sites and behave as though they were triply ionized (Ti$^{3+}$).

The level structure of Ti$^{3+}$ appears in Fig. 7.10. The Ti$^{3+}$ ion has a single active 3d electron available for laser gain; the other electrons participate in the bonding with the nearby oxygen atoms. Each Ti$^{3+}$ ion is surrounded by six oxygen ions and the 3d level is split by the crystal field into a number of levels in two groups, labeled $^2$E and $^2$T$_2$. The laser transitions take place between these two groups of electronic states. The electronic states are split into a large number of vibrational (phonon) levels and one can draw potential energy diagrams as a function of the titanium–oxygen separation; these resemble the curves for an organic dye. The resemblance to the dye is further enhanced by the strong coupling between the electronic states ($^2$E and $^2$T) and the Ti–O distance (crystal lattice), resulting in a horizontal shift between the lower and upper laser levels. This shift is much larger than in a dye and the absorption and emission curves have very little overlap (Fig. 7.11). The analysis of the laser transitions in a titanium-sapphire system is very much like that in a dye molecule

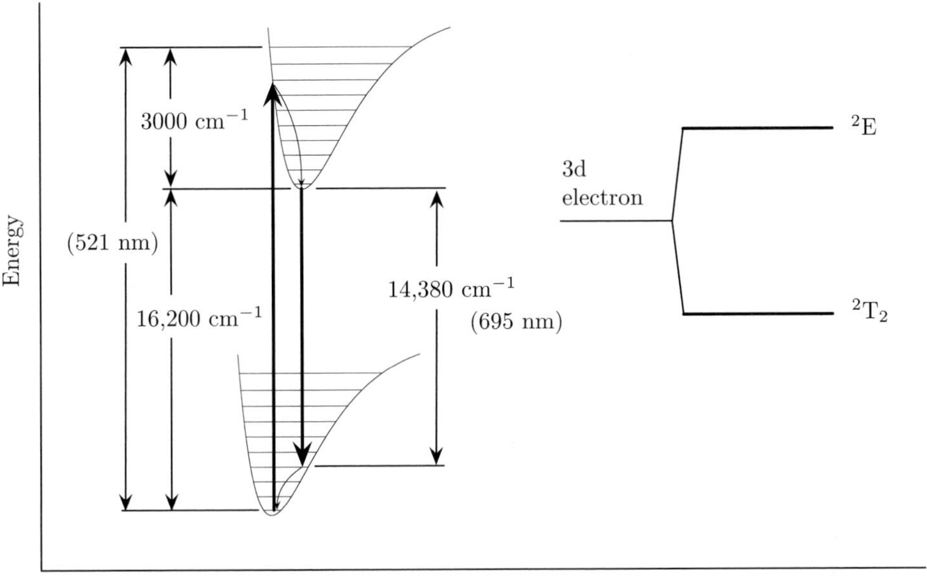

Fig. 7.10 Level structure of active $Ti^{3+}$ ions in sapphire crystal. The heavy lines on the right represent several levels.

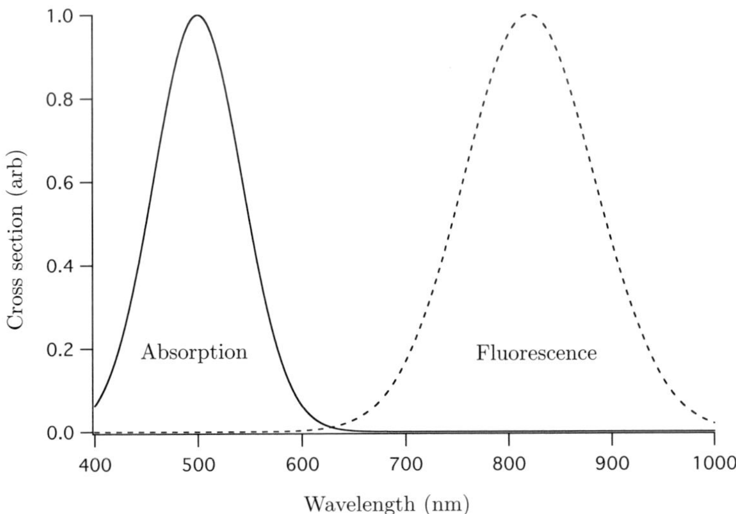

Fig. 7.11 Absorption and emission in a titanium sapphire crystal.

and the Franck–Condon principle is still valid. The displacement between the ground and excited state manifolds and the extremely short (picosecond) thermalization times within each manifold allow four-level laser theory to be used. The lifetime of the upper laser level is about 3.9 $\mu$s and, unlike in the dye, there are no nonradiative transitions to contend with (there is no analogue to the dye laser *triplet* states in a titanium sapphire laser). As can be seen in Fig. 7.10, the pumping wavelength should be about 521 nm. Although current practice is to use the 532 nm line from a high-power frequency-doubled Nd:YVO$_4$ laser as a pump, one can also use the 514 nm or 488 nm lines from an argon ion laser.

A rough schematic of a commercial titanium-sapphire laser appears in Fig. 7.12. The laser is very similar to the ring dye laser discussed earlier. One significant difference

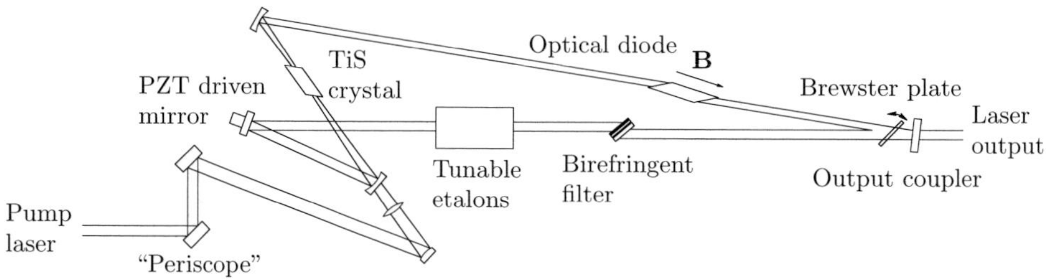


**Fig. 7.12** Commercial realization of a titanium-sapphire laser.

is the presence of the titanium-sapphire crystal at approximately the place of the compensation rhomb in the dye laser. Since the gain of titanium-sapphire is much less than that of most dyes, the titanium-sapphire crystal is much longer than the dye film thickness ($\approx$ 200 times longer) and the crystal is *collinearly* pumped. The pumping optics consist of a *periscope* to steer the pump beam, a lens and convex mirror to focus it and a dichroic cavity mirror (transmissive at the pump wavelength, reflective at the laser wavelength) to allow the beam to impinge on the crystal. The frequency control elements are exactly the same as in the ring dye laser.

Although titanium-sapphire lasers are very useful in spectroscopy (and their high power allows a practical extension of their range via frequency doubling), their principal current use is in extremely narrow pulse (femtosecond) *mode locked* systems. This application is facilitated by the very large width of the gain curve. This topic will be revisited in a later chapter.

## 7.6 The CW neodymium-yttrium-aluminum-garnet (Nd:YAG) laser

A very common continuous wave laser uses a crystal of *yttrium-aluminum-garnet* (Y$_3$Al$_5$O$_{12}$, abbreviated *YAG*) doped with about 1% Nd$^{3+}$, which replaces the Y and is responsible for the laser gain. Since all of the Nd$^{3+}$ ions experience similar fields from the host lattice, the laser transition is only slightly inhomogeneously broadened (unlike the Nd:glass laser, in which the host is amorphous glass and the inhomogeneous broadening is very great). At room temperature, the laser has two principal lines: 1064 nm (the stronger) and 946 nm. A level diagram of $Nd^{3+}$ in a YAG host appears in Fig. 7.13.

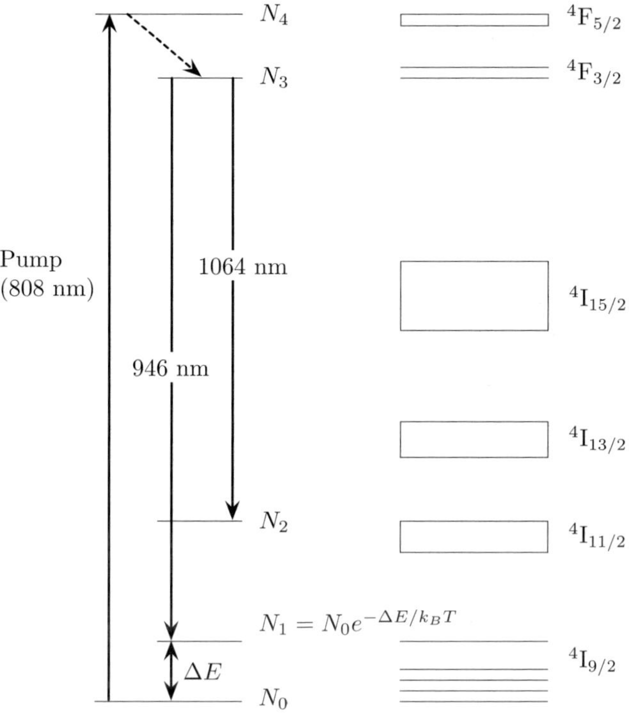


**Fig. 7.13** Level structure of the Nd:YAG system.

The YAG laser is pumped with 808 nm radiation which excites the $Nd^{3+}$ from the ground state to the $^4F_{5/2}$ level where it decays very quickly to the $^4F_{3/2}$ upper laser level via a non-radiative transition. The laser transitions to the lower-lying $^4I$ levels are forbidden in isolated $Nd^{3+}$ ions by the selection rules for dipole transitions ($\Delta J = 0, \pm 1$), but the perturbations from the crystal fields allow these transitions to take place. The lifetime of the $^4F_{3/2}$ state is about 230 $\mu$s. The lower laser level for the 1064 nm line ($^4I_{11/2}$) has a lifetime of about 30 ns and the thermal population is very small since the energy is about 10 times $k_B T$. Thus, the laser follows classical four-level theory for this line.

The laser can also emit radiation at 946 nm using the $^4I_{9/2}$ state as the lower laser level. Unfortunately, this level has appreciable population at room temperature and the level structure for this transition is a *quasi-three-level system*. The cross section for this transition is also about 10 times smaller than that for the 1064 nm line. For these reasons, the 946 nm line is somewhat weaker than the 1064 nm line.

The CW YAG laser can be pumped with a linear discharge lamp. This lamp is optically coupled to the YAG crystal by placing the lamp and crystal at the two foci of an *elliptical* cavity. To remove the considerable heat generated by the lamp, cooling water is allowed to flow between the lamp and YAG rod (both of which are sealed to avoid contact with the water). The water is usually deionized to reduce its tendency to coat the reflecting surface of the elliptical cavity. These lamp-pumped lasers can generate more than 10 W of CW power at 1064 nm. The YAG laser can also be pumped with the 808 nm radiation from a high-power laser diode, and two such lasers will be discussed in the next two sections.

## 7.7 The YAG non-planar ring oscillator: a novel ring laser geometry

There are two interesting variants of the continuous wave Nd:YAG laser. One, which will be discussed in the next section, is the diode-pumped high-power frequency-doubled Nd:YVO$_4$ laser, which is increasingly replacing argon ion lasers as pumping sources, particularly for titanium sapphire lasers. The other is the so-called *monolithic non-planar ring oscillator*, which is an exceptionally simple and frequency stable source of up to 1 W at 1064 nm (and about 500 mW at 946 nm).

The YAG non-planar ring oscillator consists of a single crystal of Nd:YAG material, cut in such a manner that it supports an *internal* circulating beam which does not lie in a single plane. A drawing of the laser showing the internal path of the beam appears in Fig. 7.14. The beam is confined by total internal reflection on all of the

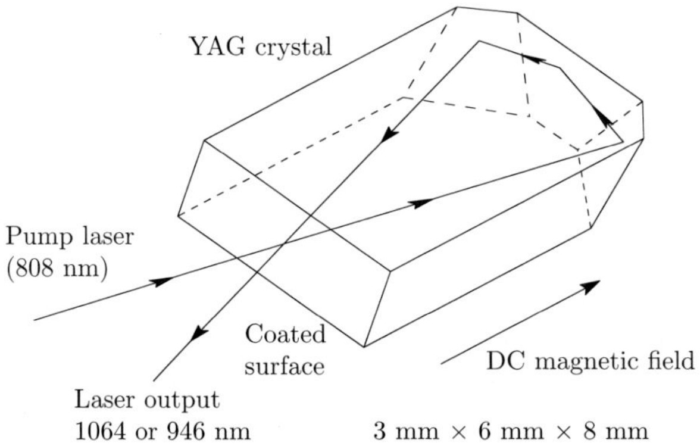


**Fig. 7.14** The Nd:YAG non-planar ring oscillator.

facets except the output face, where the output coupling is determined by the reflection from a surface with a multilayer coating. The pump beam at 808 nm is directed at the point where the laser beam emerges; the pump is quickly absorbed. The direction and beam parameters of the pump beam are adjusted to ensure that only the lowest mode (TEM$_{00}$) is excited.

Unidirectional operation is obtained using the Faraday effect, produced by a magnetic field. The magnetic field is chosen to be largely along the internal propagation direction of the circulating beam. In general, optical diodes based upon Faraday rotation have three elements: a *reciprocal* polarization rotator, a *non-reciprocal* rotator (using the Faraday effect) and a polarizer. The rotations produced by the reciprocal and non-reciprocal elements are approximately equal and add for one beam direction and cancel for the other (desired) direction. With the addition of a polarizer, the two directions have different losses, and oscillation will take place in the direction with lower loss. In the NPRO, the reciprocal rotation is provided by the *non-planar* path inside the YAG crystal and the non-reciprocal rotation is due to the Faraday effect. The multilayer coating on the output coupling facet serves as the (partial) polarizer. The inventors of the YAG NPRO used the Jones calculus to analyze the integral optical diode and the calculations are quite complicated.

The key advantage of the monolithic NPRO is its extraordinary freedom from technical frequency noise. This is entirely due to the laser resonator being constructed from a single piece of Nd:YAG material, which is intrinsically very rigid. The single direction ring oscillation eliminates the possibility of spatial hole burning, facilitating single-frequency operation. The *free-running* (i.e., without active frequency stabilization) short-term linewidth of the NPRO was measured by beating two independent lasers together and is about 3 kHz. This is still much greater than the Shawlow–Townes linewidth (less than 1 Hz) but is much less than the free-running linewidth of most single-frequency lasers constructed from discrete components.

The YAG NPRO is also quite efficient, due to its large quantum efficiency and the high efficiency of the high-power laser diodes used for pumping. A commercial realization of the laser produces about 1 W (at 1064 nm) from a pair of pumping sources which together provide 3 W. The pump sources are about 50% efficient, so the overall electrical efficiency is very good.

## 7.8   Diode-pumped solid-state (DPSS) YAG lasers

In earlier times, CW YAG lasers were pumped with the broadband light from a high-pressure discharge lamp. Although ample power can be obtained with this arrangement, there is a great deal of waste since much of the lamp's output contributes little to the pumping since it is at wavelengths other than the optimum pumping wavelength (808 nm). With the advent of high-power laser diodes, whose electrical efficiency is $\approx 50\%$ and which can be made to operate at 808 nm, laser manufacturers began providing neodymium-based *diode-pumped solid-state* (DPSS) lasers with high power and high efficiency. Their output can be extended into the visible with little loss in output power (or efficiency) by placing a frequency doubling crystal inside the cavity. Neodymium-based frequency-doubled sources are currently available whose output power in the green (532 nm) is greater than 15 W and these are rapidly supplanting argon-ion lasers for certain laser pumping applications (particularly for pumping titanium-sapphire lasers). These systems draw less than 2 kW from the mains, which is about 20 times less than a comparable argon-ion laser.

A schematic of a commercial frequency-doubled DPSS laser appears in Fig. 7.15. This laser uses $Nd^{3+}$ in an yttrium orthovanadate ($YVO_4$) crystal. The level structure is similar to that of Nd:YAG with a 1064 nm line and a few other lines at somewhat different wavelengths from the analogous lines in YAG. The $Nd:YVO_4$ medium has a greater gain and greater pump bandwidth but provides similar performance to YAG at the high powers of the DPSS laser shown in the figure (it has a lower threshold than

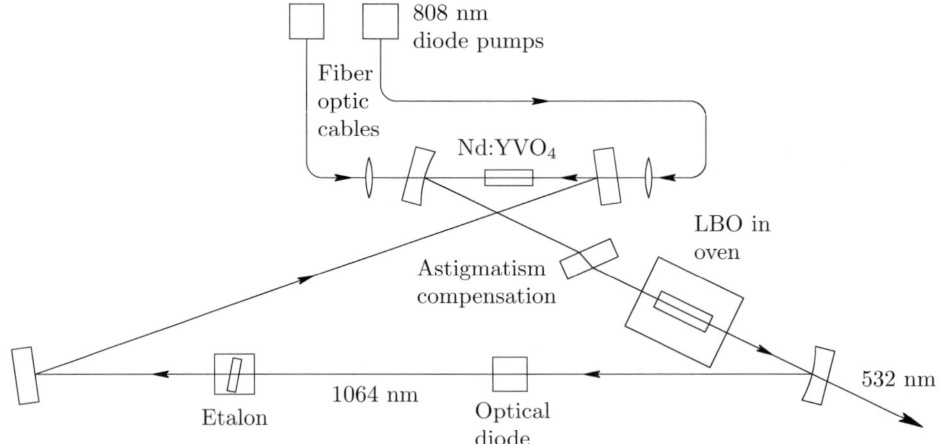


**Fig. 7.15** The Nd:YVO$_4$ diode-pumped solid-state laser with intracavity frequency doubling.

Nd:YAG providing better low-power performance). Its principal advantage over YAG is a larger pump bandwidth, making the laser performance less sensitive to drifts in the pump diode wavelength.

A ring configuration is used to prevent spatial hole burning, and a frequency doubling crystal (lithium triborate, LBO) is placed at a waist between the two curved mirrors. Two high-power diode sources collinearly pump the laser crystal from both ends. The pump lasers are arrays arranged in *diode bars* and the pump radiation is delivered to the laser crystal using fiber optic cables and focusing lenses. The pump diodes together with their cooling apparatus are located in a remote power supply to reduce the cooling requirements for the laser cavity. All but one of the cavity mirrors are *dichroic*, transmitting at the pump wavelength (such as 808 nm or 532 nm) and reflecting at 1064 nm. A conventional optical diode, consisting of a reciprocal rotator and a Faraday rotator, is used to ensure single-direction operation.

## 7.9 Further reading

Excellent descriptions of actual lasers can be found in Svelto (2004), Verdeyen (1995), Yariv (1989) and Milonni and Erberly (1988). A brief description of the YAG NPRO appears in an article in *Science* by Byer (1988) and a detailed theory (using the Jones calculus) is in the paper by Nilsson, Gustafson and Byer (1989). It goes without saying that a great deal of current information about commercial laser systems is available on the worldwide web, where one can obtain the original patents and occasionally useful sales brochures.

# 8
# Laser gain in a semiconductor

## 8.1  Introduction

There is little doubt that the most common laser system is the semiconductor laser diode, which appears in CD and DVD players, laser pointers, supermarket scanners, fiberoptic communications, etc. The analogy between semiconductor lasers and transistors is quite illuminating, and the ultimate replacement of most traditional discharge-pumped lasers (analogous to vacuum tubes) by diode lasers is assured. Almost all of the important laser sources currently used in atomic physics laboratories are either semiconductor diodes or are pumped by diodes.

The theory of semiconductor laser gain will be given here. The most important results of the solid state physics of semiconductors will be presented first, as a background. Obviously, the topic of laser diode physics could easily fill a book several times as thick as this one. Following our earlier guideline, just enough theory will be provided to give the reader an appreciation of the several kinds of diode laser likely to be encountered in the laboratory. The next chapter will be devoted to a description of specific semiconductor lasers and their characterictics.

## 8.2  Solid state physics background

When we derived the Planck formula, we obtained a *distribution function* which gives the number of photons per mode as a function of the photon energy (or frequency). Although this function was derived using a model of discrete, equally spaced energy levels, it could also have been derived entirely from the *behavior of the photon wavefunction under an interchange of identical particles*. Particles with *integral spin*, such as photons, are called *Bosons* and have wavefunctions that are *symmetrical upon exchange of identical particles*. One can then use *purely statistical arguments* to derive the *Bose–Einstein distribution function*. Electrons, on the other hand, are *Fermions* (particles with half-integral spin), whose wavefunctions are *antisymmetrical under particle exchange*. Using similar statistical methods, one can derive the probability, $f(E)$, of finding an electron with energy $E$. This probability is given by the *Fermi–Dirac distribution function*:

$$f(E) = \frac{1}{e^{(E-E_f)/k_B T} + 1},\tag{8.1}$$

where $E_f$ is called the *Fermi energy* and is essentially a chemical potential. The functional form of this distribution resembles that of the Bose–Einstein distribution (with a plus sign instead of a minus sign in the denominator), although there is no Fermi energy in the Bose–Einstein distribution, since the *particle number* is not conserved in a photon gas in thermal equilibrium. However, the Fermi–Dirac distribution differs in one very important respect: $f(E) \leq 1$. This is a concise statement of the *Pauli exclusion principle*, which allows *at most one electron in a quantum state at a time*. Plots of $f(E)$ at absolute zero ($T = 0$) and at a small finite temperature appear in Fig. 8.1. From the plots, we see that the Fermi energy is just the maximum energy

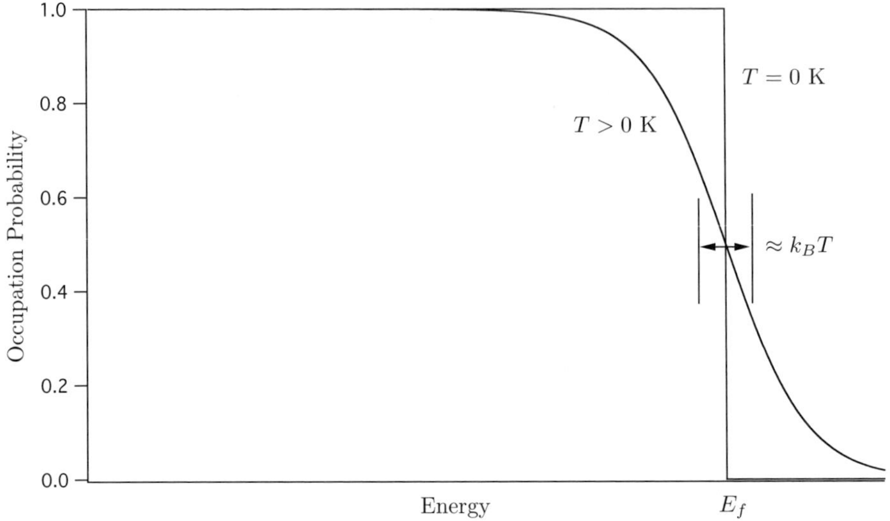


**Fig. 8.1** Fermi–Dirac distribution function.

of the electrons at absolute zero, when all of the possible levels are filled. The Fermi energy can also be defined at non-zero temperatures as *the energy for which the occupation probability is one half*. At non-zero temperature, the hard edge will have a finite breadth which is approximately equal to $k_B T$.

The *distribution of electron energies* is obtained by multiplying $f(E)$ by the *number of electron states per frequency interval*, $\rho(E)$. To derive the density of states, we will proceed in a similar fashion to our approach with photons and model the electron states as plane waves whose allowed values of $k$ are determined by the boundary conditions. A common practice in solid state physics is to take a cubic sample of the crystal and replicate it infinitely in all six directions. We then use *periodic boundary conditions* and require that the plane wave solutions repeat themselves as one crosses each crystal boundary. If the edge length of the cube is $L$, then the possible values of $\boldsymbol{k}$ are

$$\boldsymbol{k} = \left( \frac{2\pi l}{L}, \frac{2\pi m}{L}, \frac{2\pi n}{L} \right), \qquad l, m, n \text{ integers.} \tag{8.2}$$

Each cell in $k$-space has a volume of $8\pi^3/L^3 = 8\pi^3/V$, where $V$ is the crystal volume. Thus, the number of states in a *shell* of radius $k$ and thickness $dk$ is

$$\text{Number of states} = 2 \left( \frac{V}{8\pi^3} \right) (4\pi k^2 dk) = \frac{V k^2 dk}{\pi^2}, \tag{8.3}$$

where the factor of two is due to the two possible electron *spin states*. Note that we do *not* divide by 8 as we did with photons, since changing the sign of a component of $\boldsymbol{k}$ yields a *different* state when using *periodic* boundary conditions. If we express the number of states as a function of energy, we use the dispersion relation for a *free particle*

$$E = \frac{(\hbar k)^2}{2m} \tag{8.4}$$

$$\Rightarrow k = \sqrt{\frac{2mE}{\hbar^2}}, \quad dk = \frac{1}{2\hbar}\sqrt{\frac{2m}{E}}\,dE, \tag{8.5}$$

with the result that the *density of states* (number of states per energy interval per unit volume), $\rho(E)$, is

$$\rho(E) = \frac{1}{2\pi^2}\left(\frac{2m}{\hbar^2}\right)^{3/2}\sqrt{E}. \tag{8.6}$$

The *density of electron energies* is $\rho(E)f(E)$ and is plotted in Fig. 8.2. One can derive

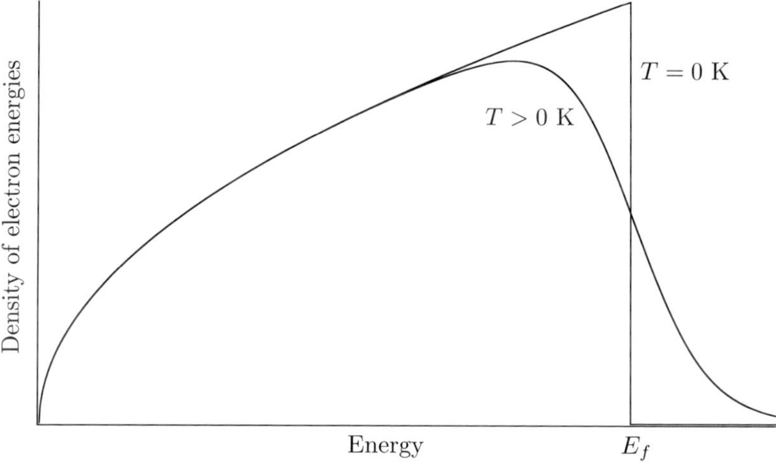


**Fig. 8.2** Density of electron energies as a function of energy.

a number of useful results (such as heat capacities of metals, etc.) solely from this model (called the Sommerfeld model), where the electrons are treated as free particles subject to the Fermi–Dirac distribution. However, all of the phenomena of interest to us will need to include the influence of a *periodic crystal lattice*.

The electronic wavefunction in a crystal is considered to be a plane wave *modulated* by a spatial function, $u_k(\boldsymbol{r})$, whose periodicity is that of the crystal lattice. The wavefunction is called a *Bloch function*:

$$\text{Bloch function:} \quad \psi(\boldsymbol{r}) = u_k(\boldsymbol{r})e^{i\boldsymbol{k}\cdot\boldsymbol{r}}. \tag{8.7}$$

The *Bloch theorem* states that this is a solution of the Schrödinger equation with a potential whose periodicity is the same as that of $u_k(\boldsymbol{r})$. Because of the lattice, the energy spectrum will be a series of discrete *bands* of allowed energies, between which are forbidden regions. We can provide some motivation for this statement by considering a one-dimensional crystal in the delta-function limit of the *Kronig–Penney model*. The time independent Schrödinger equation is

$$\frac{d^2\psi}{dx^2} + \frac{2m}{\hbar^2}(E - V(x)) = 0, \tag{8.8}$$

where $E$ is the total energy and $V(x)$ is an infinite series of delta functions whose spacing is $a$. The delta functions are derived from a rectangular function of width $b$ and height $V_0$ in the limit of infinite height and infinitesimal width with the *area* kept constant. The *strength* of the potential is proportional to the area and is represented by the parameter, $P$,

$$P = \lim_{\substack{b \to 0 \\ \beta \to \infty}} \frac{\beta^2 ab}{2}, \tag{8.9}$$

where

$$\beta = \frac{1}{\hbar}\sqrt{2m(V_0 - E)}. \tag{8.10}$$

One can show that the condition for the existence of a solution to Schrödinger's equation with this periodic potential is

$$P\frac{\sin \alpha a}{\alpha a} + \cos \alpha a = \cos ka, \tag{8.11}$$

where

$$\alpha = \frac{1}{\hbar}\sqrt{2mE}. \tag{8.12}$$

The left-hand side of eqn 8.11 is plotted in Fig. 8.3. Clearly, for this equation to have a *real* solution, the value of the left-hand side evaluated at $\alpha a$ must lie between the two dotted lines in the figure since the range of $\cos ka$ is between $+1$ and $-1$. The energy eigenvalues are proportional to $\alpha^2$ and the spectrum consists of a series of *bands* of allowed energies with *disallowed* energies in the *gaps* between the bands. One can numerically solve the transcendental equation for $\alpha a$ as a function of $ka$ to display the band structure more clearly. The solution for $P = 3\pi/2$ (a fairly strong potential) appears in Fig. 8.4, where the electron energy ($\propto (\alpha a)^2$) is plotted as a function of $ka$ from 0 to $4\pi$.

When $P = 0$, $\alpha = k$, and we have a free electron, whose energy is

$$\text{Free electron:} \quad E = \frac{(\hbar k)^2}{2m}. \tag{8.13}$$

We have plotted the free electron energy ($(\alpha a)^2$ in our units) as a dashed line and one can see the resemblance between the energy of a free electron and that of an electron in a crystal lattice. A quadratic dependence of the energy near the top and bottom of the bands is strongly suggested by the plot. As $P$ becomes larger, the bands shrink until they become points whose values are

$$\text{Very strong potential:} \quad E = \frac{n^2\hbar^2}{8ma^2}, \quad n = 1, 2\ldots\infty. \tag{8.14}$$

These are the eigenvalues for a particle in a box of width $a$ with impenetrable walls.

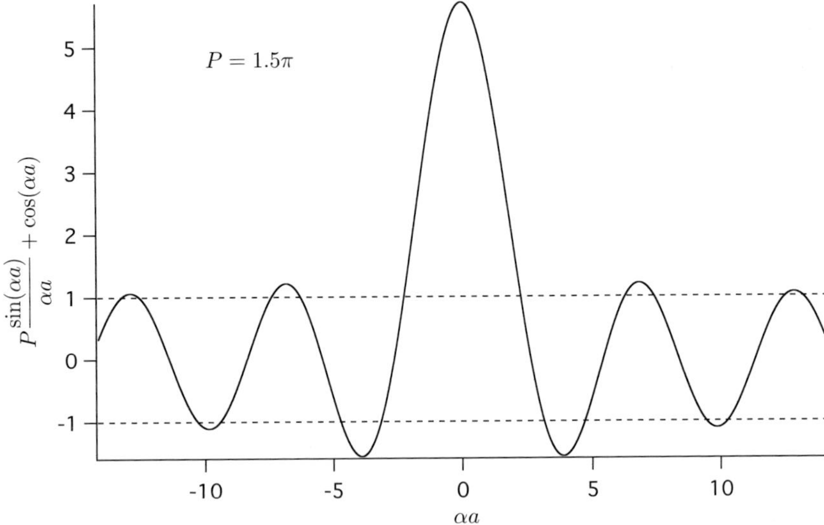

**Fig. 8.3** Condition for the existence of solution to periodic delta function potential problem. Only those values of $k\alpha$ for which the function lies between the dotted lines can yield a valid solution.

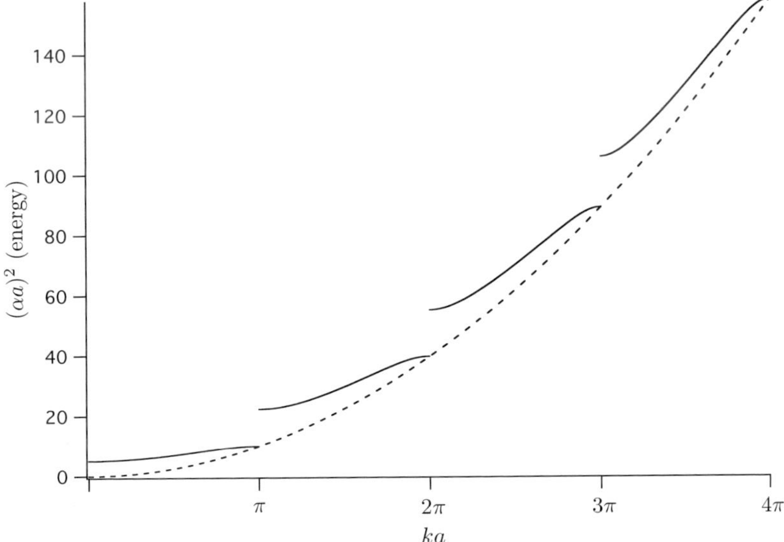

**Fig. 8.4** Energy eigenvalues for a periodic delta function potential with $P = 3\pi/2$. The band structure is clearly visible. A free-particle energy curve (dashed line) is shown for comparison.

The band theory (together with the Pauli exclusion principle) is very useful in explaining some familiar properties of materials, such as the reason that some substances are conductors and others are insulators. If there are $N$ atoms in a sample, it is easy to show that there will be $2N$ electronic states in each band. Thus, if there is one valence electron (as in the alkali metals), the band will be half full and the electrons are free to move and occupy unfilled states. These substances are therefore *conductors*. Materials composed of atoms with two valence electrons will have filled bands, since the number of electrons is equal to the number of states. If the bands are non-overlapping, the electrons will not be able to move without violating the Pauli exclusion principle and these substances are *insulators*. Some substances with two valence electrons and *overlapping bands*, such as the alkaline earth metals, are also conductors (the two overlapping bands behave like a *single partially filled band*).

We have seen from the Bloch theorem that the electron wavefunctions are *modulated* plane waves. The electrons are actually *wave packets* made up of a large number of states having a range of $k$ values. These wave packets will have a *group velocity*,

$$\text{Group velocity:} \quad v_g = \frac{d\omega}{dk} = \frac{1}{\hbar}\frac{dE}{dk}. \tag{8.15}$$

If we apply an electric field, $E_0$, the work $dW$ done in time $dt$ is

$$dW = eE_0 v_g dt. \tag{8.16}$$

But,

$$dW = \frac{dE}{dk}dk = \hbar v_g dk \implies dk = \frac{eE_0}{\hbar}dt \implies \frac{dk}{dt} = \frac{eE_0}{\hbar}. \tag{8.17}$$

The acceleration is

$$\frac{dv_g}{dt} = \frac{1}{\hbar}\frac{d^2 E}{dk^2}\frac{dk}{dt}. \tag{8.18}$$

Thus, we have

$$\frac{dv_g}{dt} = \frac{d^2 E}{dk^2}\frac{eE_0}{\hbar^2}. \tag{8.19}$$

The classical equivalent of this is

$$\text{Classically:} \quad \frac{dv}{dt} = \frac{eE_0}{m}. \tag{8.20}$$

In order to allow one to use the classical equation when an external field is applied, we will endow the electron with an *equivalent mass*, $m^*$, which is due to the interaction with the lattice. With this, we have in a crystal

$$\frac{dv_g}{dt} = \frac{eE_0}{m^*}, \tag{8.21}$$

$$\text{where} \quad m^* = \hbar^2 (d^2 E/dk^2)^{-1}. \tag{8.22}$$

A semiconductor is a material which has a very small amount of room temperature conductivity even though the lowest band is filled at $T = 0$. This conductivity can be explained with the aid of Fig. 8.5, which displays the *energy curves* for the two lowest

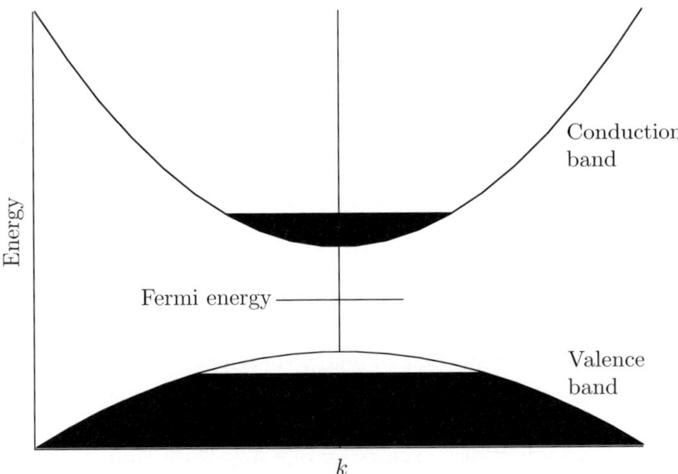


**Fig. 8.5** The energy bands in an intrinsic semiconductor

bands. We assume that the energy depends quadratically on $k$ and that the bottom
of a band opens upward while the top of a band opens downward (as in the Kronig–
Penney model). An idealized semiconductor which is completely free from impurities is
called an *intrinsic semiconductor*. At absolute zero, the lowest band (which we call the
*valence band,* since it contains electrons attached to the lattice centers) will be filled and
there can be no intrinsic conductivity. At temperatures above absolute zero, there will
be some *thermal* excitation of the valence electrons into the next higher band, which is
empty at absolute zero. This partially filled band is called the *conduction band,* since
it can support electron motion. The *band gap* (minimum energy difference between the
bands) is about 1 eV in many semiconductors, though it can vary from 0.1 eV to 2 eV.
Thus, the thermal excitation at room temperature is very small (room temperature
corresponds to 0.026 eV). Using the band model, the classification of a substance as
a conductor, semiconductor or insulator can be viewed as a measure of the *size of the
band gap.* An insulator has a large band gap, a semiconductor has a modest-size gap
and a conductor has either no band gap or overlapping bands. It should be kept in
mind that an intrinsic semiconductor is an idealization, since the conductivity in even
the purest materials will still probably be dominated by impurities, particularly at low
temperatures.

The flow of current from the electrons in the conduction band is fairly easy to
understand, since it is similar to conduction in a metal. What is a little unusual in a
semiconductor is that the electrons in the *valence band* can *rearrange themselves* when
a field is applied and will also contribute to the current flow. Since the valence band is
mostly filled, we can model this contribution to the current as the motion of positively
charged *vacancies,* called *holes,* in the valence band. In the following sequence, we
show how this is possible by writing down an expression for the current density in the
valence band:

$$\boldsymbol{J}_{valence} = \frac{1}{V} \sum_{valence} (-e)\boldsymbol{v}_i = \frac{1}{V} \sum_{all} (-e)\boldsymbol{v}_i - \frac{1}{V} \sum_{empty} (-e)\boldsymbol{v}_i = \frac{1}{V} \sum_{empty} e\boldsymbol{v}_i, \quad (8.23)$$

where the second summation vanishes since no current is possible in a filled band. Thus we see how conduction in the valence band can be considered to be due to *positive* charge carriers (holes). One can show by equating the number of empty states in the valence band to the number of filled states in the conduction band that the Fermi energy in an intrinsic semiconductor lies approximately halfway between the bottom of the conduction band and the top of the valence band.

The number of free electrons and the equal number of holes in an intrinsic semiconductor is very small, since there is little thermal excitation into the conduction band. One can control the type and number of charge carriers by *doping* the semiconductor. Doping is the deliberate addition of a controlled amount of some impurity to a semiconductor in order to control its physical properties (mainly conductivity). The *dopant* usually has a valence which is one greater or one less than that of the semiconductor. In the former case, the dopant yields a *free electron* and the material is called *n-type*, since the *majority carriers* are electrons. When the valence is one less, the material is called *p-type* and the majority carriers are holes. The situation is illustrated in Fig. 8.6 for the case of a silicon semiconductor (valence 4) doped with arsenic (valence 5) or boron (valence 3). It has been experimentally established that the dopant atoms actually *replace* the original atoms rather than filling the interstitial spaces between them. The silicon atoms make *covalent* bonds with their four neighbors, and the substituted

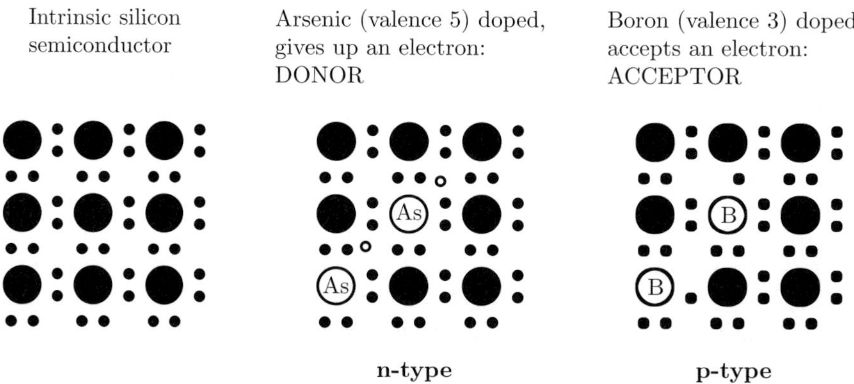


**Fig. 8.6** Doping of silicon by arsenic and boron to produce n-type and p-type semiconductors.

arsenic *donor* atom has one electron left over, which is loosely bound and can easily be *ionized* into the conduction band. In similar fashion, the boron atom has a site which can easily accept an additional electron, forming a hole in the valence band.

One can determine the density of the majority carriers using the kinetic form of the *law of mass action*, which states that the rate of a chemical reaction is proportional to the product of the concentrations of the reactants. In our case, the reactants are electrons and holes and the two possible *reactions* are the creation of a photon when an electron *annihilates* a hole and the creation of a *hole–electron pair* when a photon is absorbed (the photons are part of the *blackbody radiation* field). If $n$ is the density of electrons and $p$ is the density of holes, the two rates are

$$e + h \longrightarrow \text{photon:} \quad \text{Rate} = B(T)np \tag{8.24}$$

$$\text{photon} \longrightarrow e + h: \quad \text{Rate} = A(T), \tag{8.25}$$

where $A(T)$ and $B(T)$ are similar to the Einstein coefficients for the two processes and $T$ is included to show the strong temperature dependence of these reactions. At equilibrium, the two rates are *equal*,

$$A(T) = B(T)np \implies np = \frac{A(T)}{B(T)}. \tag{8.26}$$

The $A$ and $B$ coefficients are independent of the doping. Thus, for an intrinsic semiconductor, $n = p = n_i$, where $n_i$ is the intrinsic electron (or hole) density, and for a doped semiconductor, we have

$$np = n_i^2. \tag{8.27}$$

This rule breaks down in so-called *degenerate semiconductors*. These are so heavily doped that the Fermi level lies inside the conduction or valence band and the conductivity is therefore similar to that of a metal. The rule also breaks down when the system is not in *thermal equilibrium* (for example, when a current flows). For significant electron doping,

$$n \gg p \implies n \approx \rho_n, \tag{8.28}$$

where $\rho_n$ is the *dopant concentration*. An analogous equation holds for p-type semiconductors. Therefore, the general rule is that the *majority carrier concentration in a doped semiconductor is approximately equal to the dopant concentration*. (We will shortly show that most of the dopant atoms are *ionized* in a semiconductor.)

A very useful model for analyzing the behavior of donor electrons in an n-type semiconductor is the *Bohr model for donor electrons*. This is illustrated in Fig. 8.7. If the donor concentration is not too large, one can consider the donated electron to be in an approximate *Bohr orbit* around the positively charged core. The well-known energy of the lowest Bohr orbit of the hydrogen atom is

$$E = -\left(\frac{1}{4\pi\epsilon_0}\right)^2 \frac{me^4}{2\hbar^2} = -13.6 \text{ eV}. \tag{8.29}$$

To account for the electrical polarization of the semiconductor material, we replace $\epsilon_0$ by $\epsilon$, the actual permittivity of the medium. The electron mass $m$ is replaced with the effective mass, $m^*$. These changes can be carried out by multiplying $-13.6$ eV by two factors:

$$E = -\left(\frac{\epsilon_0}{\epsilon}\right)^2 \left(\frac{m^*}{m}\right)(13.6 \text{ eV}) = -0.025 \text{ eV}, \tag{8.30}$$

where we used a dielectric constant $(\epsilon/\epsilon_0)$ of 11.8 and an effective mass of $0.26 \times m_{electron}$ for silicon. Thus, room temperature thermal energy (about 0.026 eV) is

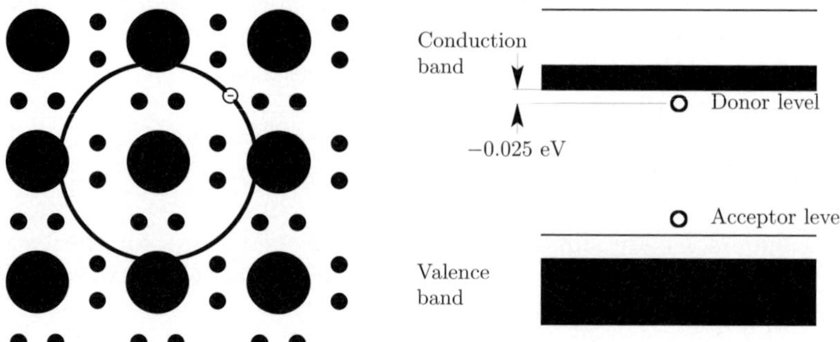

**Fig. 8.7** Illustration of Bohr model for calculating the energy levels of donor and acceptor states.

sufficient to ionize a large fraction of the donor atoms. Although the arguments are much less transparent, one can analyze the acceptor sites in a similar fashion and obtain a similar result (although the acceptor states will be *slightly above* the top of the valence band).

We saw from eqn 8.22 that the effective mass is *inversely proportional to the curvature of the E vs k (dispersion) function.* To simplify our discussion, we will mostly consider semiconductors with fairly simple energy curves: the minima and maxima of the curves will be at $k = 0$. This is an example of a *direct semiconductor*, whose valence band maximum and conduction band minimum occur at the same value of $k$. Our $E$ vs $k$ plots will be qualitatively similar to those in the Kronig–Penney model: the conduction band plot is an upward opening parabola and the valence band plot is a downward opening parabola. One might wonder whether this implies that the effective mass of a hole is negative, since it is inversely proportional to $\partial^2 E/\partial k^2$. The negative curvature of the valence band energy plot is for *electrons* in the valence band. When one considers holes (electron *vacancies*) the mass *changes sign*. Using this model, both holes and electrons behave like free particles, but with a mass which is different from that of a free electron. The energies of the electrons and holes, $E_n$ and $E_p$, are given with respect to the *bottom of the conduction band, $E_c$, and top of the valence band, $E_v$, respectively,* and are

$$\text{electrons:} \quad E_n = E_c + \frac{(\hbar k)^2}{2m_e^*}$$
$$\text{holes:} \quad E_p = E_v - \frac{(\hbar k)^2}{2m_h^*}. \tag{8.31}$$

It was stated (but not proved) that the Fermi energy of an *intrinsic* semiconductor is approximately centered in the forbidden area between the valence and conduction bands. What, then, is the location of the Fermi energy in a doped (*extrinsic*) semiconductor? The result can be obtained by integrating the density of electron energies over all energies and setting the result equal to the number of particles. As the dopant level is increased, the Fermi energy moves toward the band corresponding to the *type* of dopant: the conduction band for donors and the valence band for acceptors. Thus, for moderate to heavy doping in n-type material, $E_f$ is near the bottom of the conduction band and in p-type material it is near the top of the valence band.

In our analysis of p-n junctions, we need to examine the behavior of the Fermi levels of two different materials in contact. We first assert that the Fermi level is a *chemical potential*, $\mu$, which is defined by the following expression of the first law of thermodynamics for a system in which the number of particles can vary:

$$dU = T\, dS - p\, dV - \mu\, dN. \tag{8.32}$$

If we consider a system composed of two phases containing $n_1$ and $n_2$ particles whose chemical potentials are $\mu_1$ and $\mu_2$, then the change in the *Helmholtz free energy* ($F = U - TS$) is

$$dF = -p\, dV - S\, dT - \mu_1\, dn_1 - \mu_2\, dn_2. \tag{8.33}$$

For the two-phase system, the total number of particles is conserved

$$dn_1 = -dn_2. \tag{8.34}$$

Therefore, at constant volume and temperature, the *equilibrium* condition is

$$dF = \mu_1\, dn_2 - \mu_2\, dn_2 = 0, \tag{8.35}$$

which implies that

$$\mu_1 = \mu_2. \tag{8.36}$$

Thus, if there is no infinite barrier to the flow of electrons between the n and p materials, the Fermi energies will be the same. This analysis is true for *equilibrium situations*; we will introduce a slightly different approach when we discuss optical gain in a semiconductor in the presence of an electrical current.

With these preliminaries, we can discuss our final topic in this background section: the behavior of a p-n junction. Figure 8.8 shows the behavior of pieces of n-doped and p-doped material under four different conditions: before contact, immediately after contact, after contact when thermal equilibrium is established and forward biased. Note that the figures display the levels as a function of *position* ($x$) instead of $k$ and the levels are therefore *flat*. Before contact, the two elements have their own Fermi levels which are close to the edges of the appropriate bands. *Immediately after contact*, electrons will flow from the n material to the p material and holes will flow in the opposite direction, driven by the unequal chemical potentials (Fermi levels) in the two media. Two things will happen when the current flow stops: the Fermi levels will be equal (thermal equilibrium) and there will be *charged regions* on either side of the junction. The charged region is lacking in *free carriers* (since they have *recombined*) and is called the *depletion region*. Two equal and opposite currents will flow at equilibrium: the *diffusion current* due to *majority carriers* diffusing across the junction and the *drift current* due to minority carriers which are driven by the electric field from the separated positively and negatively charged regions. If an *external* emf is applied to the junction, the band structures (including the Fermi levels) of the two elements will *shift relative to each other* by the amount of the emf. A *forward bias* is shown in the figure (positive terminal of the battery attached to p material and

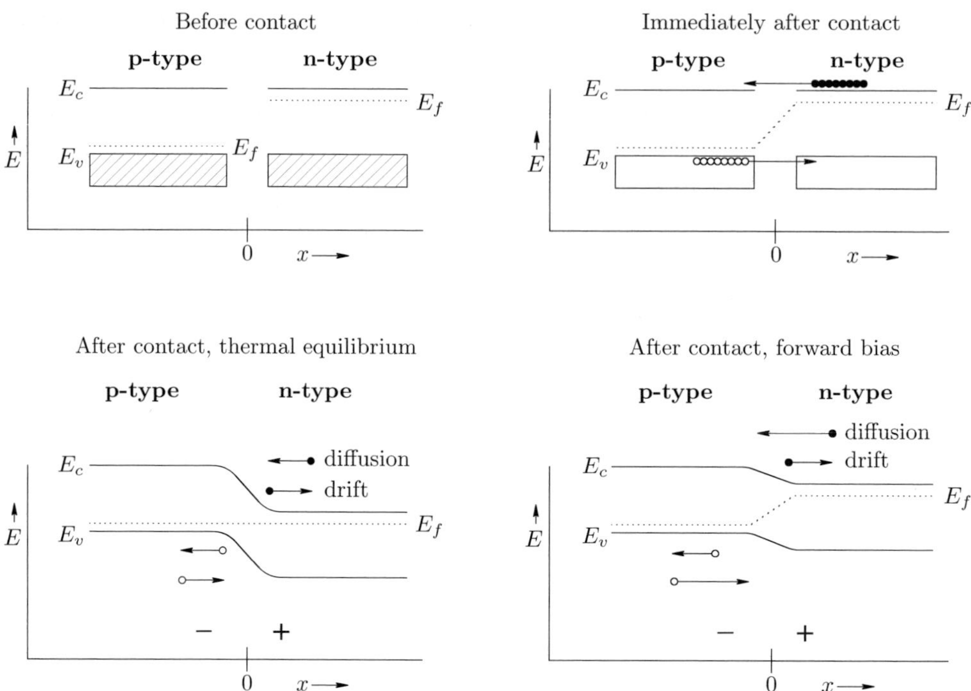


**Fig. 8.8** Behavior of p and n material before contact, after contact and forward biased. Filled circles are *electrons* and empty circles are *holes*.

negative connected to the n material). As can be seen, the forward bias reduces the barrier to the *majority carriers* (diffusion current), allowing the current to increase. The *minority carriers* are not affected, since they experience *no barrier* regardless of the emf and are always swept to the other side of the junction by the electric field. *Rectification* is due to the asymmetry of this process: reverse bias will increase the barrier to the majority carriers, reducing the current. The spatially dependent Fermi level in the composite pn junction with an applied emf is an example of the breakdown of the utility of a single Fermi level when there is no longer thermal equilibrium.

## 8.3 Optical gain in a semiconductor

Despite the obvious differences between free atoms and bulk semiconductors, the optical processes which can take place in the latter are remarkably similar to those that occur in the former. As is the case in atoms, radiative processes in semiconductors act on *electrons*, promoting them from one level to another. In a semiconductor, the levels which are involved in radiative transitions reside in the valence and conduction *bands*. The three processes in a semiconductor are

$$\text{Spontaneous emission:} \quad e + h \longrightarrow \text{photon} \tag{8.37}$$

$$\text{Absorption:} \quad \text{photon} \longrightarrow e + h \tag{8.38}$$

$$\text{Stimulated emission:} \quad \text{photon} + e + h \longrightarrow \text{photon} + \text{photon}. \tag{8.39}$$

The hole/electron model actually slightly obscures the fact that these processes are simply *interband transitions* due to optical radiation. Thus, spontaneous emission is due to an electron dropping from the conduction band to the valence band with the emission of a photon whose energy is equal to the difference in the electron energies.

A diagram of the conduction and valence band energies versus $k$ is shown in Fig. 8.9. Transitions are indicated by the vertical arrow. To analyze the optical behavior, one

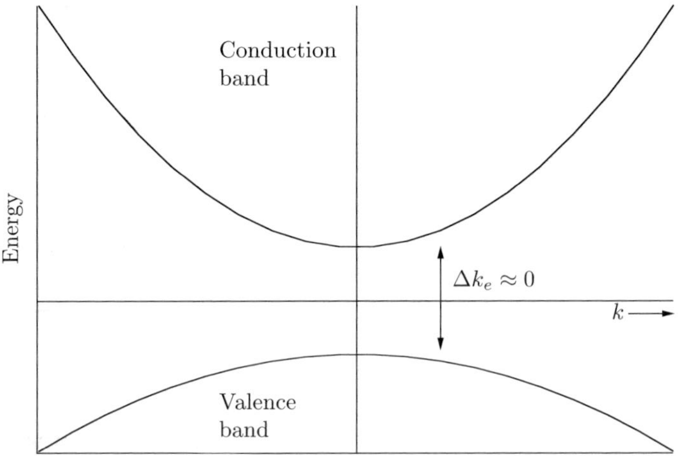


**Fig. 8.9** Energy diagram showing an optical transition in a doped semiconductor.

needs to know the change in the *electron* wave vector, $k_e$, during a transition. This can be determined by the requirement of *momentum conservation* during the transition. In a *direct semiconductor*, the bottom of the conduction band is at the same value of $k$ as the top of the valence band. Since the states at room temperature are very near the band edges, there is no need for a change in the lattice vibration state (phonon change) when a transition takes place. In the absence of phonon changes, the conservation of momentum reduces to:

$$\Delta k_e = k'_e - k_e = k_{photon}, \tag{8.40}$$

where the primed quantity refers to the electron momentum after the transition. At typical diode laser wavelengths (about 1 $\mu$m),

$$k_{photon} \approx 6 \times 10^4 \text{ cm}^{-1}. \tag{8.41}$$

In the conduction band, $k_e$ is

$$k_e = m_e^* v_e / \hbar \approx (3k_B T_e m^*)^{1/2} / \hbar \approx 3 \times 10^6 \text{ cm}^{-1} \tag{8.42}$$

for $T_e = 300$ K. Thus we see that

$$k_{photon} \ll k_e \tag{8.43}$$

and

$$\text{Conservation of momentum} \implies \Delta k_e \approx 0. \tag{8.44}$$

The *selection rule* is that $k$ doesn't change during a transition: transitions are vertical in the energy vs $k$ diagram. A second selection rule is that the *spin state does not change* during a transition since the field does not act on the spin. Summarizing the selection rules for a *direct* semiconductor:

$$\text{Conservation of energy:}\ \ \Delta E_e = h\nu$$
$$\text{Conservation of momentum:}\ \ \Delta k_e = 0 \tag{8.45}$$
$$\text{Conservation of spin:}\ \ \Delta S = 0,$$

where $\Delta E_e$ is the change in the electron energy.

In order for a semiconductor to to be useful in a laser (or light emitting diode), the following process most occur with a reasonable probability:

$$e + h \longrightarrow h\nu, \tag{8.46}$$

where the photon energy, $h\nu$, must be greater than the band gap, $E_g$. This "reaction" easily conserves energy and momentum in a *direct* semiconductor, as we showed earlier. In an *indirect* semiconductor, the momentum conservation becomes a problem, since a much greater momentum than that carried by the photon must be supplied. This additional momentum can come from phonons (crystal vibration modes) and this process is therefore much less probable than electron–hole annihilation in a direct semiconductor. Thus, direct semiconductors, such as gallium arsenide, are invariably used in semiconductor lasers (and indirect semiconductors, such as silicon or germanium, are rarely used).

From our earlier definition of the Fermi level, a serious problem for laser transitions should be evident. In order for the three optical process discussed above to occur, there must be holes and electrons *in the same place*, or, equivalently, there must be electrons in the conduction band and vacancies in the valence band at the same point in space. One can see that the existence of *a single* Fermi level prevents this from happening. An n-doped semiconductor will have its Fermi level near the conduction band and will therefore have *no vacancies* in the valence band. An analogous problem exists for p-doped material.

This problem is resolved by realizing that the Fermi level is only meaningful under conditions of *thermal equilibrium*. When a current flows, the populations *do not* obey the Fermi–Dirac distribution since the material is *no longer in thermal equilibrium*. The conditions can undoubtedly reach a *steady state*, but this is not the same thing as being in thermal equilibrium. The solution is to define *two separate Fermi levels, one for the conduction band and one for the valence band*. These are called *quasi-Fermi levels* and are justified by the fact that the thermal equilibration time *within* the bands ($\approx 10^{-12}$ s) is much shorter than the equilibration time *between* the bands ($\approx 10^{-9}$ s).

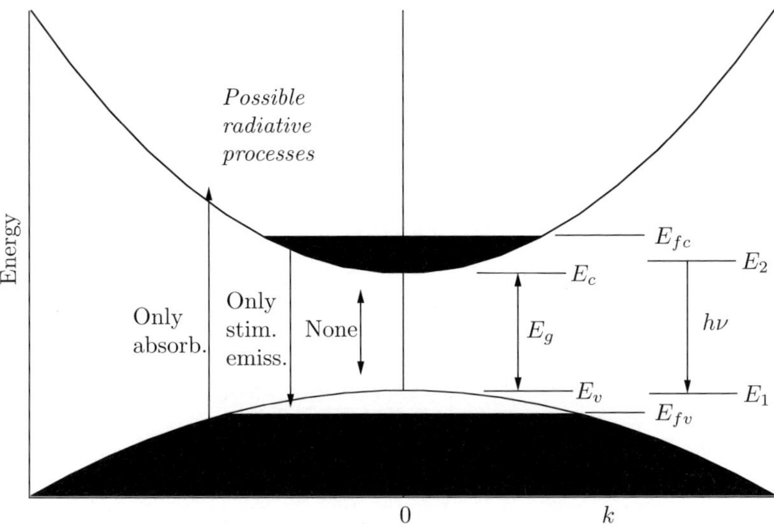


**Fig. 8.10** Band structure of a direct semiconductor showing quasi-Fermi levels. There is sufficient injected current for the $E_{fc}$ to be above the conduction edge and $E_{fv}$ to be below the valence edge in this drawing; this is the condition for the existence of optical gain.

The levels in a direct semiconductor with quasi-Fermi levels ($E_{fc}$ and $E_{fv}$) are shown in Fig. 8.10. The quasi-Fermi levels can be on *either side of the band edges*, depending upon the carrier concentration. If $E_2$ is an energy in the conduction band and $E_1$ is an energy in the valence band, the two distribution functions are:

$$f_c(E_2) = \frac{1}{e^{(E_2 - E_{fc})/k_B T} + 1} \tag{8.47}$$

$$f_v(E_1) = \frac{1}{e^{(E_{fv} - E_1)/k_B T} + 1}. \tag{8.48}$$

The *nature* of the optical process will be determined by the comparison between the photon energy and the difference in quasi-Fermi energies. Consulting Fig. 8.10, it should be apparent that when $h\nu < E_g$, no optical processes can take place. When $E_g < h\nu < E_{fc} - E_{fv}$, there is only *stimulated emission*, since absorption would violate the Pauli principle. This is the range in which laser behavior takes place. When $h\nu > E_{fc} - E_{fv}$, only photon absorption can occur. (These conclusions are only strictly valid at absolute zero, but are good approximations to actual behavior at non-zero temperatures.)

We will use the Einstein A and B coefficients to derive the gain for a semiconductor. The two important processes, absorption and stimulated emission, occur at the following rates in the presence of blackbody radiation whose spectral density is $\rho(\nu)$:

$$\text{Absorption:} \quad R_{12} = B_{12} \times \rho(\nu) \times \rho_j(\nu) \times [f_v(E_1)(1 - f_c(E_2)] \tag{8.49}$$

$$\text{Stimulated emission:} \quad R_{21} = B_{21} \times \rho(\nu) \times \rho_j(\nu) \times [f_c(E_2)(1 - f_v(E_1)], \tag{8.50}$$

where $E_1$ is the energy of the lower laser level and $E_2$ is the energy of the upper laser level. The first two factors are familiar from the Einstein phenomenological approach discussed in Chapter 5. The factor in brackets in the absorption expression is just the product of the probability of finding an electron at $E_1$ in the valence band and the probability of finding *an unoccupied state* in the conduction band at $E_2$. The bracketed factor in the stimulated emission expression has an analogous explanation. The factor $\rho_j(\nu)$ is the *joint density of states* for the transition at frequency $\nu$. This is equal to the number of states that begin in one band and end in the other with an energy difference $h\nu$. Only one density of states is used to determine $\rho_j(\nu)$ since each level in one band involves *a single* level in the other band. To calculate $\rho_j(\nu)$, we start with expressions for $k_c$ and $k_v$:

$$k_c = \sqrt{\frac{2m_e^*(E_2 - E_c)}{\hbar^2}} \tag{8.51}$$

$$k_v = \sqrt{\frac{2m_h^*(E_v - E_1)}{\hbar^2}}. \tag{8.52}$$

We equate these two expressions, since the wave vectors are equal according to the momentum selection rule, obtaining

$$E_2 - E_c = \frac{m_h^*}{m_e^*}(E_v - E_1), \tag{8.53}$$

from which we get the differential form:

$$dE_2 = -\frac{m_h^*}{m_e^*}dE_1. \tag{8.54}$$

Consulting the energy level diagram, we can see that $(E_v - E_1) + (E_2 - E_c) + E_g = h\nu$, from which we obtain

$$E_2 - E_c = \frac{m_h^*}{m_h^* + m_e^*}(h\nu - E_g)$$
$$E_1 - E_v = -\frac{m_e^*}{m_h^* + m_e^*}(h\nu - E_g). \tag{8.55}$$

The value of $k$ is obtained by substituting this expression for $E_2 - E_c$ into eqn 8.51

$$k = \frac{1}{\hbar}\left\{\frac{2m_e^* m_h^*}{m_e^* + m_h^*}(h\nu - E_g)\right\}^{1/2}. \tag{8.56}$$

We arbitrarily take one of the state densities and evaluate it subject to the requirement that $E = h\nu = E_2 - E_1$ and $dE = dE_2 - dE_1$. Since spin is conserved, we divide by two (only one *final* spin state is involved),

$$\rho_j(E)dE = \frac{1}{2\pi^2}k^2 dk \implies \rho_j(E) = \frac{k^2}{2\pi^2}\left(\frac{dk}{dE}\right)_{E=E_2-E_1}. \tag{8.57}$$

The derivative can be evaluated by subtracting the two expressions in eqns 8.31 to obtain

$$dE = dE_2 - dE_1 = \hbar^2 \left( \frac{1}{m_e^*} + \frac{1}{m_h^*} \right) k\,dk \implies \frac{dk}{dE} = \frac{m_r^*}{\hbar^2 k}, \tag{8.58}$$

where $m_r^*$ is the *reduced mass* defined by

$$\frac{1}{m_r^*} = \frac{1}{m_e^*} + \frac{1}{m_h^*}. \tag{8.59}$$

Using eqns 8.56, 8.57, and 8.58, the result for $\rho_j$ is

$$\rho_j(\nu) = \frac{h}{4\pi^2} \left( \frac{2m_r^*}{\hbar^2} \right)^{3/2} (h\nu - E_g)^{1/2}, \tag{8.60}$$

where $\rho_j(\nu) = h\rho_j(h\nu)$.

The remainder of the gain calculation is identical to our approach in Chapter 5 using the Einstein A and B coefficients. The rate of change of the upper laser level population is obtained by subtracting the stimulated emission rate from the absorption rate (compare eqn 5.123)

$$\begin{aligned}
\frac{dN_2}{dt} &= \{B_{12}[f_v(E_1)(1 - f_c(E_2))] - B_{21}[f_c(E_2)(1 - f_v(E_1))]\}\,\rho_j(\nu)\rho(\nu) \\
&= B\rho_j(\nu)\rho(\nu)[f_v(E_1) - f_c(E_2)],
\end{aligned} \tag{8.61}$$

where $B_{12} = B_{21} = B$ since the multiplicities are the same ($g_{1,2} = 2$, due to the electron spin). We next convert the radiation density to a monochromatic intensity and use the relationship between the $B$-coefficient and the $A$-coefficient to obtain

$$\frac{dN_2}{dt} = \int_0^\infty Ag(\nu)\frac{\lambda^2}{8\pi n^2}\rho_j(\nu)[f_v(E_1) - f_c(E_2)]\frac{I_\nu}{h\nu}\,d\nu \tag{8.62}$$

$$= A\frac{\lambda^2}{8\pi n^2}\rho_j(\nu)[f_v(E_1) - f_c(E_2)]\frac{I_\nu}{h\nu}, \tag{8.63}$$

where we include the *lineshape function*, $g(\nu)$, and integrate over all frequencies, as required (see Chapter 5) when using monochromatic radiation ($n$ is the index of refraction). The principal broadening in semiconductors is due to electron–phonon dephasing collisions whose collision time in GaAs is about 0.1 ps, which corresponds to a Lorentzian width of about 1.6 THz. We will shortly see that the statistical factors in the gain have a width which is about 10 times larger than this. We therefore consider $g(\nu)$ to be a delta function (a very broad one!) when we evaluate the integral. Finally, we multiply by $h\nu$ to convert $N_2$ to the *energy stored in the upper state* and recognize the factor multiplying $I_\nu$ on the right-hand side as the gain:

$$\gamma(\nu) = A\frac{\lambda^2}{8\pi n^2}\rho_j(\nu)[f_v(E_1) - f_c(E_2)]. \tag{8.64}$$

The last factor is roughly equivalent to the *inversion*, $\Delta N$, in our earlier discussion of laser gain (Chapter 5). We note that, despite the fact that the $f$ terms depend upon

$E_1$ and $E_2$, they can be readily expressed as functions of $h\nu$ and the fixed band and Fermi energies using eqns 8.55.

In order to calculate the gain for a given electron and hole density, one needs to know the location of the quasi-Fermi levels. This requires the solution of two integral equations for $E_{fc}$ and $E_{fv}$:

$$n = \int_0^\infty \rho_c(E_2) f_c(E_2) dE_2 \tag{8.65}$$

$$p = \int_0^\infty \rho_v(E_1) f_v(E_1) dE_1, \tag{8.66}$$

where $n$ and $p$ are the electron and hole densities while $\rho_c$ and $\rho_v$ are the densities of states for the conduction and valence bands. Both of these equations are of the form

$$F = \int_0^\infty \frac{x^{1/2}}{e^{x-\eta} + 1} dx, \tag{8.67}$$

where the unknown is $\eta$. There are a number of approximate solutions to these equations; one fairly simple approach (Suhara (2004) and Joyce, et al. (1977)) is to express the difference between the logarithm of the Fermi–Dirac distribution and the logarithm of the Boltzmann distribution as a power series in the carrier density. We first define $N_c$ and $N_v$:

$$N_c = 2 \left( 2\pi k_B \frac{T}{h^2} \right)^{3/2} (m_e^*)^{3/2} \tag{8.68}$$

$$N_v = 2 \left( 2\pi k_B \frac{T}{h^2} \right)^{3/2} ((m_{hh}^*)^{3/2} + (m_{lh}^*)^{3/2}), \tag{8.69}$$

where $m_{hh}^*$ and $m_{lh}^*$ are the "heavy hole" and "light hole" effective masses. Then,

$$\frac{E_{fc} - E_c}{k_B T} = \ln \left( \frac{n}{N_c} \right) + \sum_{i=1}^4 A_i \left( \frac{n}{N_c} \right)^i$$

$$\frac{E_v - E_{fv}}{k_B T} = \ln \left( \frac{p}{N_v} \right) + \sum_{i=1}^4 A_i \left( \frac{p}{N_v} \right)^i, \tag{8.70}$$

where

$$A_1 = 3.53553 \times 10^{-1} \tag{8.71}$$

$$A_2 = -4.95009 \times 10^{-3} \tag{8.72}$$

$$A_3 = 1.48386 \times 10^{-4} \tag{8.73}$$

$$A_4 = -4.42563 \times 10^{-6}. \tag{8.74}$$

The quasi-Fermi levels for gallium arsenide are plotted in Fig. 8.11 using these formulas.

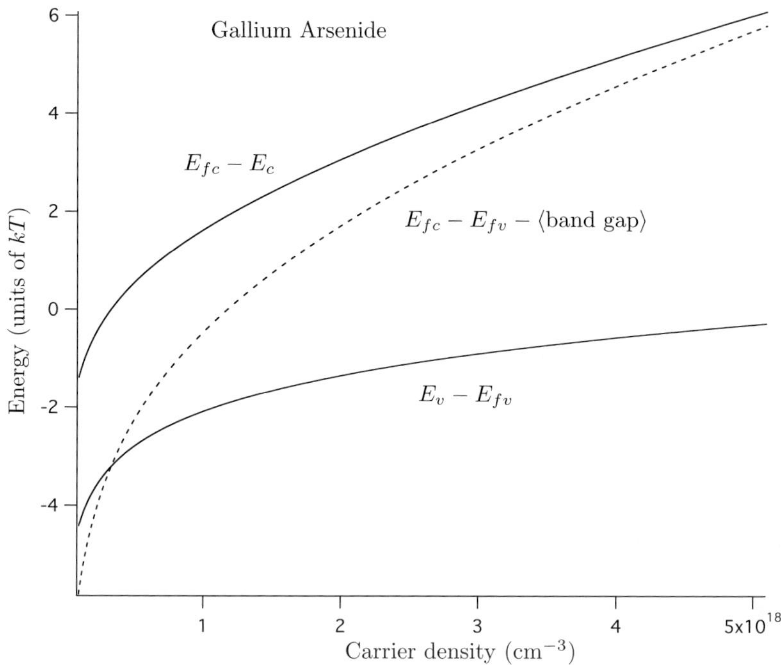


**Fig. 8.11** Quasi-Fermi levels relative to band edges in GaAs. The dashed curve is the quasi-Fermi level difference minus the band gap; the *transparency* condition occurs at the zero of this curve.

We haven't discussed actual pumping mechanisms, although we implicitly assumed that, by some means, majority carriers could be injected into the bands so that the condition for gain, $E_{fc} - E_{fv} > E_g$, could be realized. As can be seen in Fig. 8.11, $E_{fc}$ can be *below* the conduction band and $E_{fv}$ can be *above* the valence band for low carrier concentrations. As the carrier concentration increases, $E_{fc}$ increases and $E_{fv}$ is reduced. The *threshold* for gain occurs when $E_{fc} - E_{fv} = E_g$. We call this the *transparency condition* and the corresponding carrier density is $N_{tr}$. When an injection current is applied, *charge neutrality* requires that

$$n + N_A^- = p + N_D^+, \tag{8.75}$$

where $N_A^-$ and $N_D^+$ are the number densities of *ionized* acceptors and donors. If we assume that most acceptors and donors are ionized, it can be shown that $n, p \gg N_A^-, N_D^+$ at reasonably high injection levels and one can approximate $n \approx p$. From this, we see from the figure that the transparency density for GaAs is about $1.2 \times 10^{18}$ cm$^{-3}$, which agrees with the actual value.

Although the actual expression for the gain as a function of the carrier density, $N$, is quite involved, we will assume that the relation is approximately *linear* (a good approximation for bulk semiconductors but not for quantum wells). Then, the gain can be written as

$$\gamma = \sigma(N - N_{tr}). \tag{8.76}$$

Thus, $N_{tr}$ and $\sigma$ are two of the several useful parameters which characterize a particular semiconductor for use in lasers or LEDs (the others are the electron–hole recombination time, equivalent masses and the band gap). The parameter $\sigma$ is sometimes called the *differential gain constant*.

A plot of the gain versus $h\nu$ appears in Fig. 8.12. It is interesting to note that the expression for $\gamma(\nu)$ is also an expression for *absorption* when $h\nu > E_{fc} - E_{fb}$. The parameters for the plot are:

$$E_g = 1.42 \text{ eV}$$
$$E_{fc} - E_c = 0.071 \text{ eV}$$
$$E_v - E_{fv} = 0.039 \text{ eV}$$
$$m_e^* = 0.067 \times m_{electron}$$
$$m_{hh}^* = 0.46 \times m_{electron}$$
$$m_{lh}^* = 0.08 \times m_{electron}.$$

The quasi-Fermi levels are taken from eqns 8.70 for an electron and hole density of $1.75 \times 10^{18} \text{cm}^{-3}$. Of course, when $h\nu < E_g$ there is neither gain nor absorption.

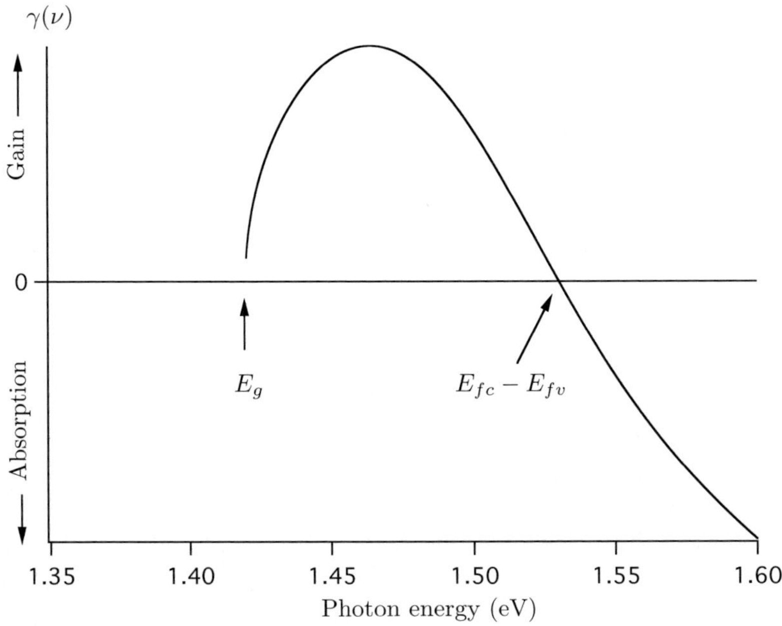


**Fig. 8.12** Gain versus $h\nu$ in GaAs.

Using very similar arguments to those involved in the gain calculation, the *spontaneous decay* rate, $R(\nu)$, can be expressed as

$$R(\nu) = Af_c(E_2)(1 - f_v(E_1))\rho_j(\nu). \tag{8.77}$$

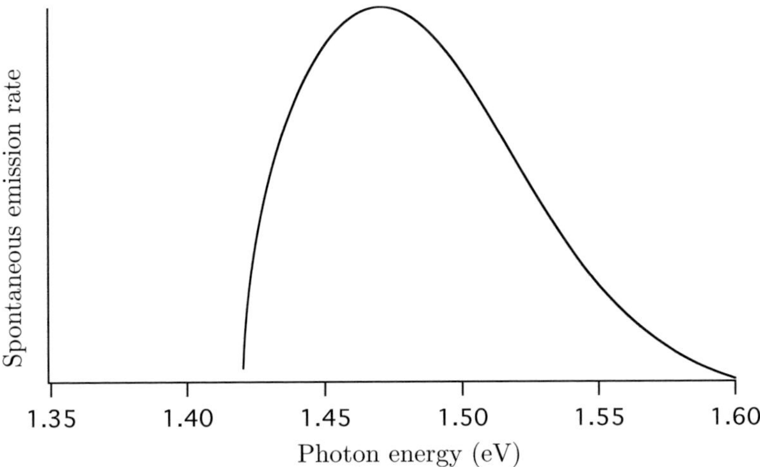

**Fig. 8.13** Spontaneous emission rate versus $h\nu$ in GaAs.

The factors after the $A$ are the probability that there is a conduction electron at $E_2$ times the probability that there is a vacancy at $E_1$ times the number of ways this can happen while conserving momentum and yielding a photon of energy $h\nu$. We have made the same approximation as in the gain calculation that the lineshape factor, $g(\nu)$, is a bit narrower than the other frequency-dependent factors and used this fact to evaluate the integral. A plot of the spontaneous emission for the same parameters as the gain appears in Fig. 8.13. It is interesting that, unlike in the atomic case, the spontaneous decay breadth in semiconductors is somewhat larger than the breadth of the gain curve.

## 8.4 Further reading

The books on lasers mentioned at the end of the last chapter all have excellent treatments of semiconductor laser theory. Two very useful additional books devoted to semiconductor lasers are *Semiconductor Laser Fundamentals* by Suhara (2004) and *Semiconductor Lasers I* by Kapon (1999). Two excellent textbooks on solid state physics are the books by Kittel (2004) and Ashcroft & Mermin (1976).

## 8.5 Problems

(8.1) Show that the Fermi level in an *intrinsic* semiconductor is halfway between the bottom of the conduction band and the top of the valence band when the hole and electron masses are the same. This is demonstrated by equating the number of free electrons to the number of holes, which implies that

$$\int_{E_g}^{\infty} \rho_e(E - E_g) f_e(E) dE = \int_{-\infty}^{0} \rho_h(-E) f_h(E) dE, \qquad (8.78)$$

where the electron and hole distribution functions satisfy $f_e + f_h = 1$ (since conduction electrons are due to valence *vacancies*) and one can approximate the distribution functions as $f_{e,h} = e^{\pm(E_f - E)/k_B T}$ (since we anticipate that $E - E_f \gg k_B T$). The quantity $E_g$ is the band gap energy and it is convenient for the energy zero to be at the top of the valence band.

(8.2) Calculate the gain of GaAs at 855 nm (near the peak of the gain curve) using the parameters given in the text for Fig. 8.12. Assume that $A = 10^{-9}$ s.

(8.3) Show that the *joint density of states* can be written as

$$\rho_j(h\nu) = \frac{1}{2}\left[\frac{1}{\rho_c(E_2)} - \frac{1}{\rho_v(E_1)}\right], \qquad (8.79)$$

where $h\nu = E_2 - E_1$ and $\rho_{c,v}$ are the densities of states in the conduction and valence bands.

(8.4) The determination of the quasi-Fermi levels is difficult at finite temperatures; at absolute zero it can be accomplished by using an integral similar to that found in problem 1 with $f$ equal to a step function. Determine the location of the quasi-Fermi levels in GaAs at $T = 0$ assuming an electron density in the conduction band of $5\times10^{18}$ cm$^{-3}$ and an equal hole density in the valence band.

# 9

# Semiconductor diode lasers

## 9.1 Introduction

The previous chapter provided the theoretical background for semiconductor lasers; this chapter will describe some specific systems and discuss their characteristics. We will proceed in historical order, starting with the homojunction laser and introducing each new development up to the strained junction quantum well laser. We will describe longitudinal mode control with distributed feedback and Bragg reflector lasers and the more experimentally ubiquitous external cavity diode laser. The properties of all of these lasers will be discussed in some detail. We will end with a description of semiconductor laser amplifiers and injection locked diode lasers. These latter techniques together with the external cavity diode laser provide cost-effective and efficient solid-state competitors for dye and Ti-sapphire (CW) lasers and should eventually help to render them obsolete.

## 9.2 The homojunction semiconductor laser

The earliest semiconductor lasers were made from a pn junction in a direct semiconductor (usually gallium arsenide). Since there is only one junction between materials of the same type (but differently doped), these lasers are called *homojunction lasers*. They have serious shortcomings and are no longer in use; we describe them both for historical reasons and also because they have a very simple structure.

**Table 9.1** Parameters for gallium arsenide.

|  | GaAs |
| --- | --- |
| $\lambda$ (nm) | 850 |
| $E_g$ (eV) | 1.424 |
| $m_e/m_{electron}$ | 0.067 |
| $m_{hh}/m_{electron}$ | 0.46 |
| $m_{lh}/m_{electron}$ | 0.08 |
| $N_{tr}$ ($10^{18}$ cm$^{-3}$) | 1.2 |
| $\sigma$ ($10^{-16}$ cm$^2$) | 1.5 |
| $\tau$ (ns) | 3 |

The important parameters for GaAs are tabulated in Table 9.1. Gallium arsenide is a *binary* III-V semiconductor: it is composed of equal amounts of a group III element (gallium) and a group V element (arsenic). The valence bands in GaAs are derived from atomic p-states, which have three possible levels ($l = 1, \quad 2l + 1 = 3$), excluding electron spin. In the crystalline form, there are three possible valence bands, with different curvatures of $E_v(k)$ (i.e., different equivalent masses). Since the density of states is proportional to $m^{*3/2}$, the most important valence band is the one with the least curvature (and greatest equivalent mass): the so-called *heavy hole* whose mass ($m_{hh}$) is given in the table.

A schematic of a homojunction laser is shown in Fig. 9.1 and a flat band diagram of the levels (forward biased) appears in Fig. 9.2. The reason for using a pn junction

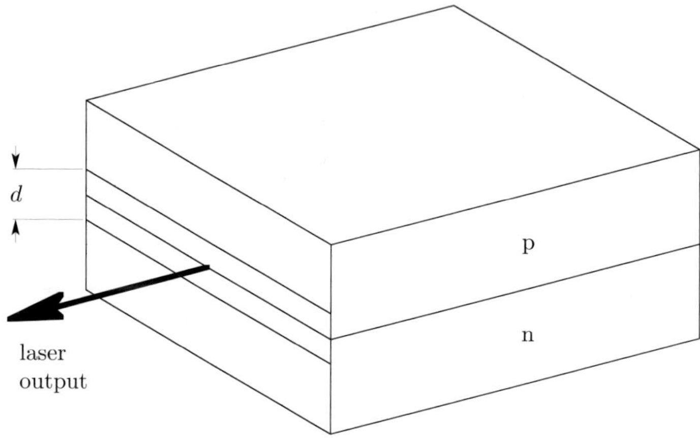


**Fig. 9.1**  Schematic of homojunction laser.

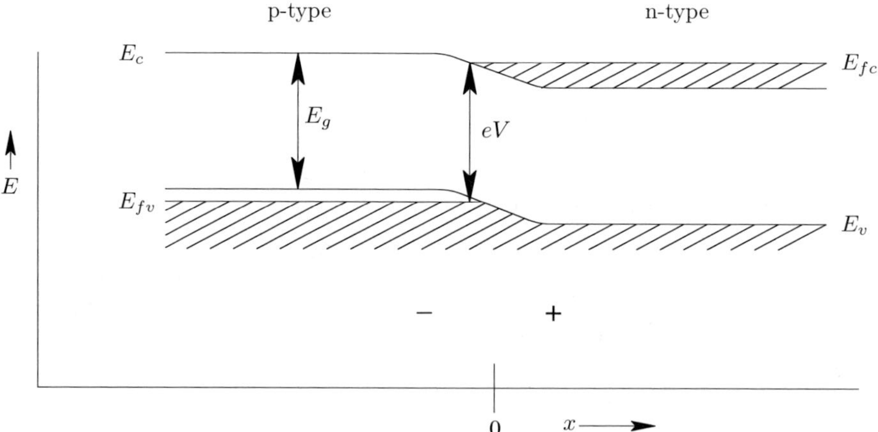


**Fig. 9.2**  Energy bands in a forward-biased (with voltage $V$) homojunction laser.

is that it is a convenient place for greatly enhanced electron–hole recombination (necessary for laser action) which is confined to a fairly small region. The size of the region can be determined from the *diffusion equation*. The electron diffusion constant, $D_e$, is defined by:

$$n\boldsymbol{v}_e = -D_e\nabla n, \tag{9.1}$$

where $n$ is the electron density and $\boldsymbol{v}_e$ is its diffusion velocity. From the *continuity equation*,

$$\nabla \cdot (n\boldsymbol{v}_e) + \frac{\partial n}{\partial t} = -\langle\text{recombination rate per unit volume}\rangle = -n/\tau, \tag{9.2}$$

one obtains the diffusion equation:

$$-D_e\nabla^2 n + \frac{\partial n}{\partial t} = -n/\tau, \tag{9.3}$$

where $\tau$ is the electron–hole recombination time. We assume that the electrons are *generated elsewhere*, so the generation rate doesn't appear on the right-hand side of the continuity equation. In the steady state ($\partial n/\partial t = 0$), the equation in one dimension is

$$D_e\frac{d^2 n}{dx^2} - n/\tau = 0. \tag{9.4}$$

Using a trial solution of the form $n = n_0 e^{-x/L}$, we obtain

$$n = n_0 e^{-x/\sqrt{D_e\tau}}, \tag{9.5}$$

where $L = \sqrt{D_e\tau}$ is called the *diffusion length*. Using $D_e \approx 10$ cm$^2$/s and $\tau = 3$ ns, $L = d \approx 1$ $\mu$m. (This should be distinguished from the length of the *depletion region*, which is calculated in a completely different way and is about ten times smaller in GaAs.) The size of the active region is therefore about 1 $\mu$m in a GaAs homojunction laser.

Semiconductor lasers are almost always pumped by *injection current*: the electron current in a forward biased p-n junction. The current density, $\boldsymbol{J}$, is given by

$$\boldsymbol{J} = n_0 e\boldsymbol{v}_e. \tag{9.6}$$

Assuming the current is flowing in the $x$-direction and neglecting the small $y$ and $z$ dependence of $\boldsymbol{J}$, the continuity equation (eqn 9.2) in the steady state ($\partial n/\partial t = 0$) is

$$v_e\frac{dn}{dx} = -n/\tau, \tag{9.7}$$

where we assume that the diffusion velocity, $v_e$, is spatially constant. We have already derived the spatial dependence of $n = n_0 e^{-x/d}$. Substituting this into the above equation and using the expression for $\boldsymbol{J}$, we obtain

$$v_e n_0 = J_x/e = n_0 d/\tau \implies J_x = e n_0 d/\tau. \tag{9.8}$$

One can estimate the current density needed for laser operation by starting with the *transparency density*, which is $1.2 \times 10^{18}$ cm$^{-3}$ in GaAs. This electron density is generated by the injected current. Using $\tau = 3$ ns and $d = 1$ $\mu$m, the needed current density is

$$J_x = \frac{n_e e d}{\tau} \approx 6.4 \times 10^3 \text{ A/cm}^2. \tag{9.9}$$

This is the minimum current density for *laser gain*; the *threshold* current density will be a bit larger. We see that a fairly significant current density is needed in a homojunction laser and that to reduce this current density we need to reduce $d$.

A second problem with the homojunction design is due to the cavity mode in the p-n junction. As can be seen from the drawing (Fig. 9.1), the end facets of the laser material are used as the cavity mirrors. One can coat the facets to increase the reflectivity, but the high gain in the semiconductor medium allows the laser to work adequately with uncoated facets. The reflectivity of the facets is given by the Fresnel formula for normal incidence

$$\text{Power reflectivity} = \left(\frac{n-1}{n+1}\right)^2 = 0.32. \tag{9.10}$$

where the index of refraction, $n$, is about 3.6 for GaAs. The size of the cavity mode can be estimated from our work with standing wave cavities. The waist radius in a *confocal* cavity of length $L$ is $\sqrt{\lambda L/2\pi n}$ which is about 4.7 $\mu$m in a 500 $\mu$m long cavity when $\lambda = 1$ $\mu$m. The actual waist size will be larger since the facets are closer to being flat than concave. The mode size is much larger than the *active region* ($d$) and the field outside the gain region will be attenuated. Some method for confining the cavity mode is needed to reduce the physical extent of the field.

As a result of these two deficiencies, a very substantial current is needed to obtain useful radiation from a homojunction laser and these lasers are normally operated either in pulse mode at room temperature or CW at 77 K (liquid nitrogen temperature) due to the heating from this large current.

## 9.3    The double heterostructure laser

It should be clear from the discussion of the homojunction laser that the threshold current would be greatly decreased if one could reduce the size of both the active region and the cavity mode. The first successful attempt to reduce these quantities employed the *heterostructure* configuration in which there are one or more junctions involving materials with *different compositions* and dopings (unlike the GaAs homojunction or silicon or germanium p-n junctions which consist of a *single* material with different dopings). Semiconductor lasers constructed in this way alway have *two junctions* and are called *double heterostructure lasers*. The importance of this technology was acknowledged by the 2000 Nobel prize in physics, which was awarded to three scientists for "developing semiconductor heterostructures".

When we discussed the homojunction laser, we introduced the material gallium arsenide, which differs from the previously discussed *single-element* materials (silicon and germanium) in being a *compound*. Certain families of crystalline compounds have a much greater probability of being *direct-bandgap* semiconductors, in which radiative interactions occur with high probability. In addition to *binary* compounds, such as GaAs, direct semiconductors suitable for laser construction include *ternary* compounds, such as $Al_x Ga_{1-x}As$, and *quaternary* compounds, such as $In_{1-x}Ga_x As_y P_{1-y}$, where $x$ and $y$ are the *molar fractions* of the constituents. From the appearance of the subscripts, it should be clear that the ternary and quaternary compounds have similar structures to GaAs since the total number of atoms from group III and group V are independent of the values of $x$ or $y$. For example, one can consider $Al_x Ga_{1-x}As$ to be GaAs with $x$ fraction of Ga atoms replaced with Al atoms. Changing $x$ and $y$ changes the band gap, the index of refraction and the *physical size* of the crystal cells. As $x$ is increased in $Al_x Ga_{1-x}As$, the band gap increases and the refractive index decreases.

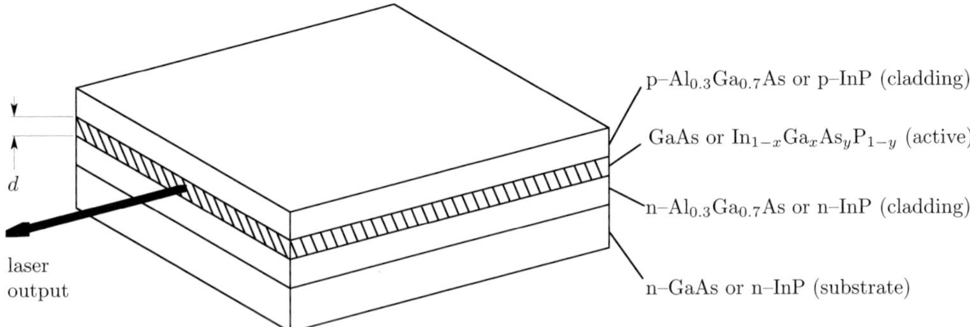


**Fig. 9.3** Schematic of a double heterostructure laser.

The only constraint on the materials on either side of the junctions is that they have essentially the same *lattice constants* (dimensions of lattice), a condition called *lattice matching*. Most semiconductor lasers are composed of atoms from group III and group V (so-called III-V compounds) of the periodic table and all of these compounds form crystals with *cubic* geometry. If the lattice spacings of the materials differ by more that 0.1%, there can be *strains* which will often have adverse effects on the interface (though strains are deliberately used in some *quantum well* lasers). The use of ternary and quaternary compounds increases the number of degrees of freedom so that lattice matching can be obtained without adversely affecting other desirable properties of the materials. A diagram of a common double heterojunction laser using two possible sets of compounds appears in Fig. 9.3.

The use of the double heterostructure confines both the carriers (electrons and holes) and the cavity mode to the *active region*, of width $d$. Since $d$ can be 10 times smaller than its value in the homojunction laser, the threshold is reduced by more than an order of magnitude. An aid to understanding why this can occur appears in Fig. 9.4, where a band diagram of a forward-biased double heterostructure laser is shown together with a plot of the refractive index as a function of position.

As can be seen from the band diagram, with a sufficient forward bias, there is a *well* in the active GaAs region which is energetically favorable for the confinement of both electrons and holes (until they recombine). The electrons are injected from the left and the holes from the right and both carriers are exposed to forces (from the band energy gradients) which confine them in the active region where laser action can take

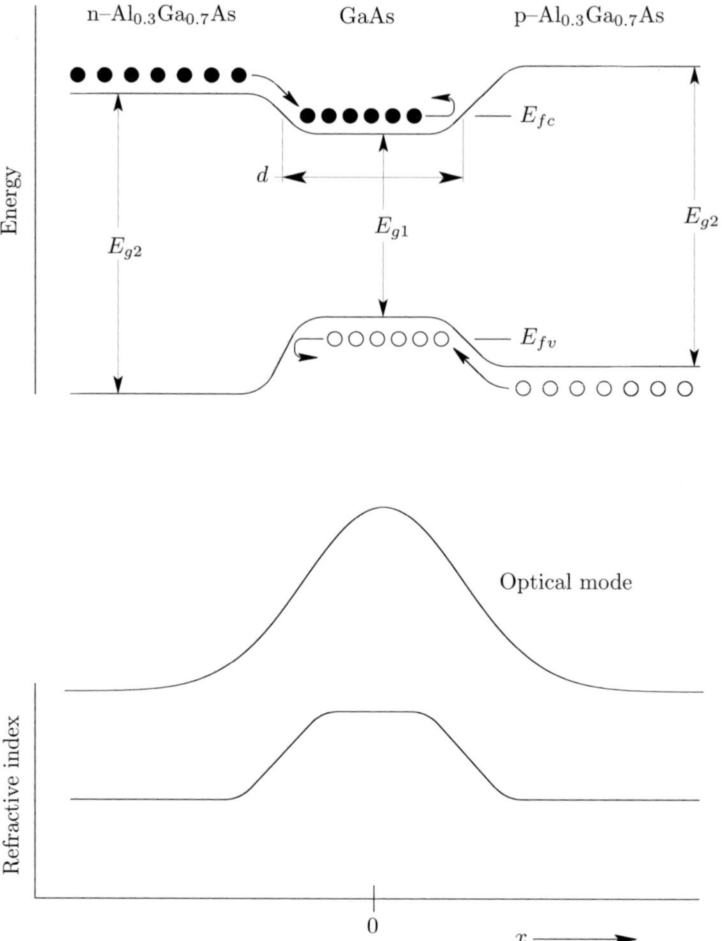



**Fig. 9.4** Plots of band energies and refractive indices in double heterostructure.

place. Thus, unlike the case of the homojunction laser where we relied on uncontrolled diffusion and recombination to define the active region, we can now "engineer" the active region to be in a much smaller layer of material whose size is under our control (within limits).

Photon confinement is due to the *lower* index of refraction on either side of the active region. The heterostructure forms an *optical waveguide* which confines the field, much like an optical fiber. The lower index material is called the *cladding*. The *tail* of the field distribution, which propagates in the cladding, has very little absorption since the photon energy (slightly above $E_{g1}$) likely satisfies the condition for transparency in the cladding ($h\nu < E_{g2}$). The simultaneous occurrence of both carrier and field confinement is due to the fortuitous *inverse relationship* between the band gap and refractive index in III-V crystals: as the band gap increases, the refractive index decreases. Thus, the smaller band gap of the GaAs active region, which results in carrier confinement, is also a region of greater index of refraction, which results in field confinement.

If we assume that the active region is well-defined and the field distribution is known (from optical waveguide theory), we can make some estimates of the threshold gain and injection current density. The region between $x = -d/2$ and $x = d/2$ is characterized by a threshold gain, $\gamma_t$, and absorption, $\alpha_a$, and the regions in the n-type and p-type material are described by absorptions $\alpha_n$ and $\alpha_p$. The gain and losses are weighted by the intensity distributions using the $\Gamma$ quantities:

$$\Gamma_a = \frac{\int_{-d/2}^{d/2} |E|^2 dx}{\int_{-\infty}^{\infty} |E|^2 dx} \tag{9.11}$$

$$\Gamma_n = \frac{\int_{-\infty}^{-d/2} |E|^2 dx}{\int_{-\infty}^{\infty} |E|^2 dx} \tag{9.12}$$

$$\Gamma_p = \frac{\int_{d/2}^{\infty} |E|^2 dx}{\int_{-\infty}^{\infty} |E|^2 dx}, \tag{9.13}$$

where

$$\Gamma_a + \Gamma_n + \Gamma_p = 1, \tag{9.14}$$

and the subscripts $a, n$ and $p$ refer to the active region, the $n$-region and the $p$-region, respectively. These parameters allow one to take into account the imperfect *overlap* between the internal field and the active region of the laser. If the gain and losses are fairly uniform within each of these three regions, the weighted gain and loss are approximately

$$\text{gain} = \gamma \Gamma_a \tag{9.15}$$

$$\text{loss} = \alpha_a \Gamma_a + \alpha_n \Gamma_n + \alpha_p \Gamma_p. \tag{9.16}$$

In order to calculate the threshold gain, we need the *laser oscillation condition* and the *semiconductor gain expression*, which we will repeat here (at threshold):

$$\text{Laser oscillation condition:} \quad \gamma_t = \alpha - \frac{2}{l} \ln(r_1 r_2) \tag{9.17}$$

$$\text{Semiconductor gain:} \quad \gamma_t = \sigma(N_t - N_{tr}). \tag{9.18}$$

From the gain formula,

$$N_t = N_{tr} + \frac{\gamma_t}{\Gamma_a \sigma}, \tag{9.19}$$

where we multiplied the right-hand side of the gain expression by the *weight function*, $\Gamma_a$. The threshold gain is obtained from the laser oscillation condition (replacing $\alpha$ by the total weighted loss):

$$\gamma_t = -\frac{2}{l} \ln(r_1 r_2) + \alpha_a \Gamma + \alpha_n (1 - \Gamma), \tag{9.20}$$

where we let $\alpha_n = \alpha_p$ and $\Gamma_n = \Gamma_p$ and dropped the subscript from $\Gamma_a = \Gamma$ (the quantity $\Gamma$ is called the *beam confinement factor*). Finally, we can use $J = eNd/\tau$ to obtain the threshold current density

$$J_t = \left(\frac{ed}{\tau}\right)\left[N_{tr} + \frac{\gamma_t}{\Gamma\sigma}\right]. \tag{9.21}$$

We should note that for very small values of $d$, the confinement factor $\Gamma$ is proportional to $d^2$ (this can be obtained from optical waveguide theory). Thus, the threshold current is proportional to $d$ until the active region becomes very thin, at which point the second term in brackets dominates, and the threshold current will *increase* with further reductions in $d$. When the two bracketed terms are equal, half of the injected current is used merely to establish quasi-Fermi levels appropriate for laser gain. The minimum threshold current in GaAs occurs when $d \approx 0.1$ $\mu$m.

We can estimate the threshold current density using the following parameters: $\Gamma \approx 1$, $r_1^2 = 0.32$ (GaAs), $r_2 = 1$ and $l = 500$ $\mu$m. The mirror loss is 23 cm$^{-1}$ and if we let $\alpha_a = 10$ cm$^{-1}$, the threshold gain is $\gamma_t = 33$ cm$^{-1}$. The values of $\tau$, $N_{tr}$ and $\sigma$ are taken from Table 9.1. Assuming $d = 0.1$ $\mu$m, the threshold current density is about 800 A/cm$^2$, which is about a factor of 10 less than the threshold in a homojunction laser.

For the laser shown in Fig. 9.3, the confinement of the field along the plane of the junction is not nearly as good as the confinement perpendicular to the plane. This laser will therefore not, in general, operate in the lowest *transverse* mode. To improve the transverse mode behavior, the *stripe geometry* shown in Fig. 9.5 is used. Two variants are in the figure: a *gain-guided* laser and an *index-guided* laser. The

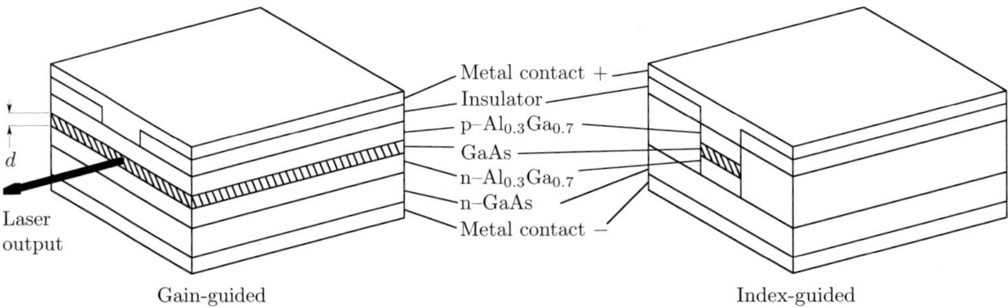


**Fig. 9.5** A *gain-guided* (left) and *index-guided* (right) double heterostructure stripe laser.

gain-guided configuration uses a very narrow electrode to inject the carriers into a narrow region of the active area of the laser, thereby ensuring that laser gain only occurs there. The index-guided approach uses in addition a very narrow stripe of active GaAs, surrounded on all sides by media with a lower index of refraction. The field is confined in a two-dimensional optical waveguide formed by the active medium and the lower index cladding. This is occasionally called a *buried heterostructure laser* and is much better at confining the field than the gain-guided approach.

## 9.4  Quantum well lasers

In the pursuit of lower thresholds by reducing the thickness of the active layer, we saw that the confinement factor suffers when $d$ is reduced to less than about 0.1 $\mu$m. However, there are significant advantages to be obtained when $d$ is on the order of the de Broglie wavelength of the carriers (about 100 Å for electrons). Quantum effects then become prominent and useful changes occur to the density of states and other parameters. Lasers with these very small active regions are called *quantum well lasers* and are the dominant type of semiconductor laser in use today (2009), due to their extremely low thresholds and high efficiency. The very thin active region provides little or no field confinement and these lasers therefore usually employ a separate field confinement structure which is similar to the one used in a double heterostructure laser. The size of the optical confinement region is on the order of the *optical wavelength*, and the size of the active gain region is on the order of the *electron de Broglie wavelength*. Such lasers are called *separate confinement heterostructure quantum well lasers*.

An AlGaAs-GaAs double heterostructure single quantum well laser with a separated confinement region is shown in Fig. 9.6 (with apologies to the reader for the jargon, which can get fairly dense at times; we will not burden the reader with the numerous and confusing acronyms which proliferate in this field). The active region is

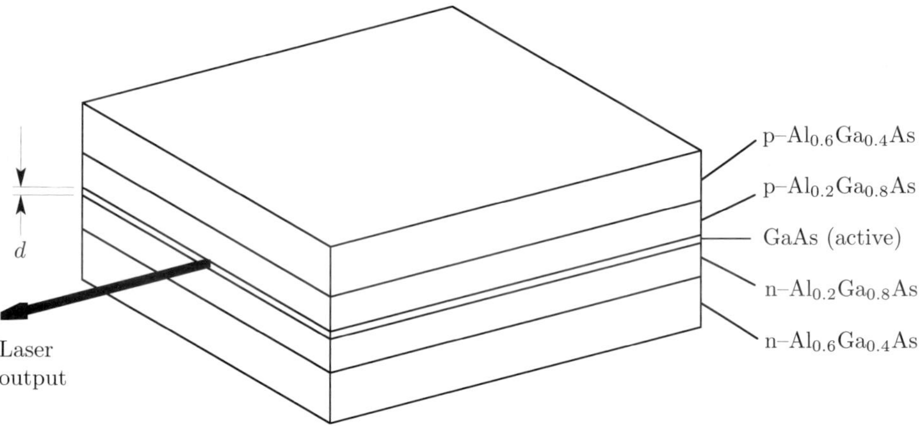


**Fig. 9.6**  A heterostructure quantum well laser with a separate *step index* confinement region.

a 100-Å wide layer of GaAs surrounded by 0.1 $\mu$m layers of Al$_{0.2}$Ga$_{0.8}$As. The cladding for the field confinement consists of layers of Al$_{0.6}$Ga$_{0.4}$As, whose refractive index is smaller than that of Al$_{0.2}$Ga$_{0.8}$As.

The band structure of a quantum well laser is shown in Fig. 9.7. Instead of having the abrupt *step index* shown in Fig. 9.6, the layer on either side of the active GaAs region has a *smooth transition* between Al$_{0.6}$Ga$_{0.4}$As and Al$_{0.2}$Ga$_{0.8}$As, resulting in the linear change in band energy shown in the figure. (These lasers are called GRINSCH lasers: GRaded INdex Separate Confinement Heterostructure). This arrangement does a better job than a step index in confining the carriers to the extremely thin active

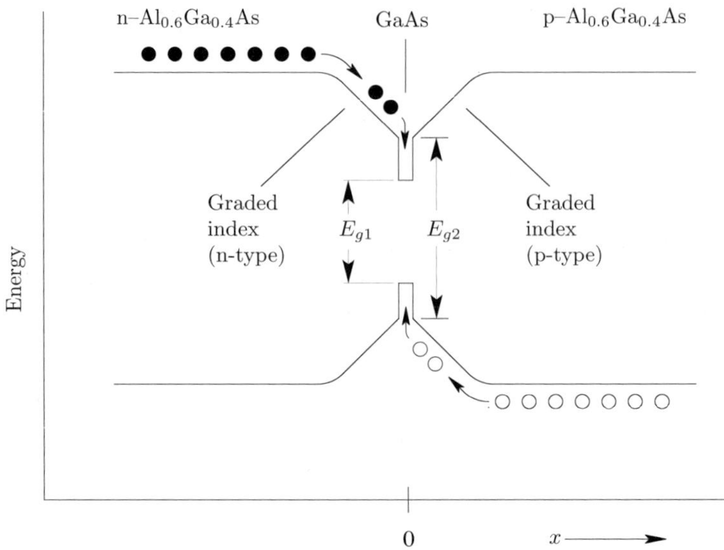


**Fig. 9.7** Band structure of a quantum well laser using a graded index.

GaAs region. The field confinement is due to the optical waveguide structure formed by the graded index. The thinness of the active region requires a different approach to the electron wavefunction and density of states.

When we discussed the electron wavefunction in *bulk semiconductors* (those with dimensions large compared to the carrier de Broglie wavelength), we assumed that the wavefunction was a modulated plane wave with a large number of possible values for the wave vector, $\boldsymbol{k}$. When one of the dimensions of the active material becomes comparable to the carrier de Broglie wavelength, one must treat the eigenfunction corresponding to that coordinate differently from the others. In the limit of an extremely thin layer, that wave vector component will have *a single value*. The Hamiltonian is written with the special coordinate separated from the others:

$$H = T + V = -\frac{\hbar^2}{2m}\nabla_\perp^2 - \frac{\hbar^2}{2m}\frac{\partial^2}{\partial x^2} + V, \tag{9.22}$$

where $\nabla_\perp$ is the transverse Laplacian, which operates on the coordinates perpendicular to the junction normal (i.e., parallel to the junction surface). Separating variables, the wavefunction is a product of a two-dimensional Bloch wavefunction (a function of the coordinates, $\boldsymbol{r}_\perp$) and a function of $x$, $\phi(x)$,

$$\psi(\boldsymbol{r}) = u_{k_\perp}(\boldsymbol{r}_\perp)e^{i\boldsymbol{k}_\perp \cdot \boldsymbol{r}_\perp}\phi(x). \tag{9.23}$$

It is very straightforward to solve Schrödinger's equation for an *infinite barrier* in the $x$-direction:

$$\phi(x) = \sin(l\pi x/L_x), \qquad l = 1, 2, \cdots \infty \tag{9.24}$$

$$E_l = \frac{l^2\hbar^2\pi^2}{2m^*L_x^2} = l^2 E_1 \qquad\qquad E_1 = \frac{\hbar^2\pi^2}{2m^*L_x^2}, \tag{9.25}$$

where, to be consistent with the literature, we call the dimensions of the active region $L_x, L_y$ and $L_z$ and the origin $(x = 0)$ is at the left edge of the active region. Thus, the wavefunctions and energies are

$$\psi(\boldsymbol{r}) = u_{k_\perp}(\boldsymbol{r}_\perp)e^{i\boldsymbol{k}_\perp \cdot \boldsymbol{r}_\perp}\sin(l\pi x/L_x) \tag{9.26}$$

$$E = \frac{\hbar^2 k_\perp^2}{2m^*} + l^2 E_1. \tag{9.27}$$

The energy expression is applicable to either electrons or holes, where one uses the effective mass $(m^*)$ appropriate to the type of carrier being considered. While an infinite barrier is not very realistic, it demonstrates all of the *qualitative* features of a quantum well without excessive complications (it overestimates some of the energies by a factor of about 2).

It is worth noting that a quantum well is essentially a *two-dimensional* physical system and all of its useful properties are a result of this reduction in the number of dimensions. One can reduce the dimensionality even further and construct *quantum wires* (one-dimensional) and *quantum dots* (zero-dimensional). Lasers have been made from both of these latter structures and they have some advantages over quantum wells. As the dimensionality is reduced, the states become more clustered around the laser oscillation energy, and the density of states and gain become larger (though the gain bandwidth will be reduced). The extreme is a zero-dimension quantum dot, which is essentially an artificial atom with only discrete energy levels. All of these structures are made possible by the recent advances in the ability to grow very thin layers of semiconductor materials using techniques such as molecular beam epitaxy. Remarkably, the transitions between the different layers occur on the scale of *a few atomic diameters*.

We will save some steps in the calculation of the gain by recognizing that the only quantities which are different from the bulk semiconductor case are the *density of states* and the *joint density of states*. Thus, we will use eqn 8.64 for the gain and replace $\rho_j(\nu)$ with the quantity appropriate for a quantum well. We start with the density of states.

The density of states in two dimensions is calculated in exactly the same way as it is in three dimensions. We use *periodic boundary conditions*, for which the allowed values of $\boldsymbol{k}_\perp$ are

$$\boldsymbol{k}_\perp = \left(\frac{2\pi m}{L_y}, \frac{2\pi n}{L_z}\right), \quad m, n \text{ integers.} \tag{9.28}$$

The area of each cell is $4\pi^2/(L_y L_z) = 4\pi^2/A$, where $A = L_y L_z$ is the perpendicular area of the active region. Thus, the number of states between $k_\perp$ and $k_\perp + dk_\perp$ is

$$\text{Number of states} = 2\left(\frac{A}{4\pi^2}\right)(2\pi k_\perp dk_\perp) = \frac{A k_\perp}{\pi}dk_\perp, \tag{9.29}$$

where the factor of 2 is due to the number of spin states. From the equation for the energy due to the $y$ and $z$ motion (eqn 9.27), the number of states per unit area for $l = 1$ is

$$\rho(E)dE = \frac{m^*}{\pi\hbar^2}dE. \tag{9.30}$$

When $l = 2$, the carrier can be found in either the $l = 1$ or $l = 2$ band and the density doubles. Generalizing to any value of $l$,

$$\rho(E) = \sum_{l=1}^{\infty} \frac{m^*}{\pi \hbar^2} H(E - E_l), \tag{9.31}$$

where

$$H(x) = \begin{cases} 0 & : \quad x < 0 \\ 1 & : \quad x \geq 0. \end{cases} \tag{9.32}$$

Graphs of $\rho(E)$ for a quantum well and a bulk semiconductor are shown in Fig. 9.8. Since the separation between the various values of $E_l$ is proportional to $1/L_x^2$, the

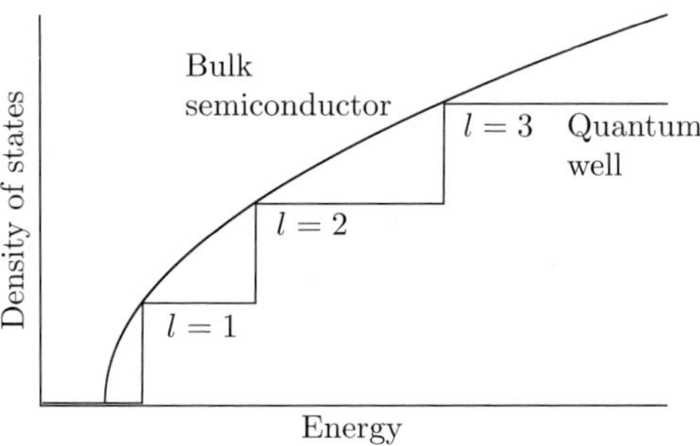


**Fig. 9.8** Density of states for quantum well and bulk semiconductor (for reference).

quantum well steps rapidly approach the continuous bulk semiconductor curve as the size of the well is increased.

In an earlier section, we derived selection rules for momentum, energy and spin in a bulk semiconductor. Quantum wells have an additional quantum number, $l$. The selection rule for $l$ can be determined by writing down the dipole matrix element for a transition between the conduction band and the valence band using the quantum well wavefunctions. For simplicity, we will assume that the wave is polarized in the $y$-direction, so the operator responsible for optical transitions can be written as

$$\text{Interaction operator } (y\text{-polarization}): \quad e\boldsymbol{E}(\boldsymbol{r}_\perp) \cdot \boldsymbol{r}_\perp e^{i\boldsymbol{k}_{opt} \cdot \boldsymbol{r}_\perp}, \tag{9.33}$$

where $\boldsymbol{k}_{opt}$ is the wave vector for the radiation. Although the field is $E\hat{\boldsymbol{y}}$, we write the dot product in terms of $\boldsymbol{r}_\perp$ to facilitate the interpretation of the first factor of the matrix element. The transition matrix element can be factored into the product of two integrals: one operating on the transverse coordinates ($y$ and $z$) and one acting on the $x$ coordinate. Due to the direction of polarization, the latter has no coordinate dependence. The matrix element is then

$$\langle \psi_c(\boldsymbol{r})|H_I|\psi_v(\boldsymbol{r})\rangle = \int u_{kc}^*(\boldsymbol{r}_\perp)e^{-i\boldsymbol{k}\cdot\boldsymbol{r}_\perp}(e\boldsymbol{E}\cdot\boldsymbol{r}_\perp e^{i\boldsymbol{k}_{opt}\cdot\boldsymbol{r}_\perp})u_{kv}(\boldsymbol{r}_\perp)e^{i\boldsymbol{k}\cdot\boldsymbol{r}_\perp}\,dy\,dz$$

$$\times \int \phi_c^*(x)\phi_v(x)\,dx. \tag{9.34}$$

The first integral yields the $\boldsymbol{k}$ selection rule already considered. From the orthogonality of the $\phi$ functions (sines) for different values of $l$, the selection rule for $l$ is

$$\text{Quantum well selection rule on } l: \quad \Delta l = 0. \tag{9.35}$$

Although we used a convenient polarization in deriving this rule, it can be shown that the rule is valid for any polarization.

One very useful feature of quantum wells is the fact that one can *change the effective band gap* by changing $L_x$. If we measure the energies from the bottom of the conduction and top of the valence band of the *bulk material*, the conduction band energy, $E'_c$, and the valence band energy, $E'_v$, are given by (for $l = 1$)

$$\text{Conduction band:} \quad E'_c = E_{1c} + \frac{\hbar^2 k_\perp^2}{2m_c^*} \tag{9.36}$$

$$\text{Valence band:} \quad E'_v = E_{1v} + \frac{\hbar^2 k_\perp^2}{2m_v^*}, \tag{9.37}$$

where $E_{1c}$ and $E_{1v}$ are the energies of the first discrete level ($E_1$) in the conduction and valence bands and the valence band energy increases *downward*. The transition energy, $E' = h\nu$, is $E_{g1} + E'_c + E'_v$. Thus, we have

$$h\nu = E' = E_{g1} + (E_{1c} + E_{1v}) + \frac{\hbar^2 k_\perp^2}{2m_r^*}. \tag{9.38}$$

We see from this that the minimum transition energy is not the band gap, $E_{g1}$, as it would be in a bulk semiconductor, but instead it is the band gap plus the quantity $E_{1c} + E_{1v}$. These terms are proportional to $1/L_x^2$ and we therefore have the ability to change the effective band gap by changing $L_x$.

The joint density of states is particularly simple to derive since $\rho(E)$ is independent of $E$ for a given $l$. We will begin by assuming that the transitions occur between states with $l = 1$, which is certainly the case at low injection current levels. We then need to express the density of states as a function of the transition frequency, $\nu$. This is done in two steps. First, we use the relationship

$$\rho(k)dk = \rho(E')dE', \tag{9.39}$$

where, again, $E'$ is the *transition energy*. We saw above that $\rho(k) = k_\perp/\pi$. From this, we obtain $\rho(E')$

$$\rho(E') = \rho(k)\frac{dk}{dE'} = \left(\frac{k_\perp}{\pi}\right)\frac{m_r^*}{\hbar^2 k_\perp} = \frac{m_r^*}{\pi\hbar^2}, \tag{9.40}$$

where the derivative is obtained by differentiating eqn 9.38. Note that this differs from the $\rho(E)$ derived earlier since it is a function of the *transition energy*, $E'$. The second step uses $E' = h\nu$ and $\rho(E')dE' = \rho_j(\nu)d\nu$,

$$\rho_j(\nu) = \rho(E')\frac{dE'}{d\nu} = \left(\frac{m_r^*}{\pi\hbar^2}\right)h = \frac{2m_r^*}{\hbar}. \tag{9.41}$$

When we substitute $\rho_j(\nu)$ into the gain equation, we need a *volumetric* density. To convert the $\rho_j(\nu)$ that we have derived into a volumetric density, we simply divide by $L_x$:

$$\rho_j(\nu) \longrightarrow \frac{\rho_j(\nu)}{L_x}. \tag{9.42}$$

Substituting the joint density into the gain equation, eqn (8.64), we obtain

$$\gamma(\nu) = A\frac{\lambda^2}{8\pi n^2}\left(\frac{2m_r^*}{\hbar L_x}\right)[f_v(E_1) - f_c(E_2)]. \tag{9.43}$$

The actual gain will need to take into account the fairly small overlap between the active region and the field, given by the quantity $\Gamma$. If the lateral extent of the field is $W$, the beam confinement factor is

$$\Gamma = \frac{\int_{-L_x/2}^{L_x/2} E^2 dx}{\int_{-\infty}^{\infty} E^2 dx} \approx \frac{L_x}{W}. \tag{9.44}$$

Note that this is a different result from that which was claimed in connection with the double heterostructure laser, where it was stated that $\Gamma$ is proportional to $d^2$ when $d$ is very small. The difference is due to the presence of a *separate confinement structure* in the quantum well laser so the field mode width is *independent* of $L_x$. Making the substitution $\gamma(\nu) \longrightarrow \Gamma\gamma(\nu)$, we obtain

$$\text{Quantum well gain:} \quad \gamma(\nu) = A\frac{\lambda^2}{8\pi n^2}\left(\frac{2m_r^*}{\hbar W}\right)[f_v(E_1) - f_c(E_2)]. \tag{9.45}$$

The gain can be estimated using $A = 3 \times 10^8$ s$^{-1}$ (3 ns carrier lifetime), $\lambda = 1$ $\mu$m, $m_r^* = .058 \times m_{electron}$, $n = 3.6$ and $W = 0.1$ $\mu$m (the same as the $d$ in the double heterostructure laser). Assuming that the second bracketed factor is unity, the result is

$$\text{Quantum well gain:} \quad \gamma \approx 102 \text{ cm}^{-1}. \tag{9.46}$$

This is fairly respectable (a total gain of $\approx 3$ in typical diodes) and is an upper bound achieved with a very large injection current, which will cause $f_v(E_1) - f_c(E_2)$ to approach unity. The gain can be increased even further by the use of *multiple quantum wells*. The improvement is less than proportional to the number of wells, particularly at lower injection currents.

It can be shown that the *transparency density* in a quantum well is roughly the same as that in a bulk semiconductor. The threshold current (eqn 9.21) relative to a double heterostructure laser is therefore reduced by the ratio of the active region widths ($d$). The quantum well laser is the final step in a sequence of lasers (homojunction$\rightarrow$double heterostructure$\rightarrow$quantum well) in which the threshold current is reduced by reducing the size of the active region.

We conclude this section by mentioning *strained quantum well lasers*, where lattice matching is deliberately violated to introduce strains in the interface between the layers. An example is a laser constructed from an InGaAs active region surrounded by AlGaAs. By varying the In/Ga molar ratio, one can change the band gap by a factor of 1.5, allowing great flexibility in the peak wavelength (0.84 $\mu$m to 1.3 $\mu$m). The lattice constants can differ by more than 3%, which causes considerable strain at the interface. This changes the effective mass of the holes (they are affected by the lattice parameters to a greater extent than the electrons). A factor of 2 reduction in the heavy hole mass will reduce the density of states and the transparency density by a factor of two and increase the differential gain ($\sigma = d\gamma/dN$) by the same factor (the explanations for these changes require a more involved treatment then we gave above).

In summary, the quantum well laser can provide useful laser gain with a threshold current density which is roughly 10 times smaller than that of a bulk semiconductor double heterostructure laser. The laser wavelength can to some extent be tailored by changing the well size. Further reductions in the threshold current and increases in the gain can be obtained using the strained quantum well configuration.

## 9.5  Distributed feedback lasers

In the section on heterostructure lasers, it was stated that *transverse modes* could be controlled by confining the field with a two-dimensional structure in a *stripe geometry laser*. The *index-guided* configuration is particularly well suited to ensuring oscillation in the *lowest transverse mode*. In pursuit of a *spectrally narrow* source of radiation (needed for both telecommunication and atomic physics experiments), it is also important to control the *longitudinal modes*, since semiconductor lasers are inhomogeneously broadened (mainly from spatial hole burning) and will tend to oscillate in a number of longitudinal modes, as shown in Fig. 9.9.

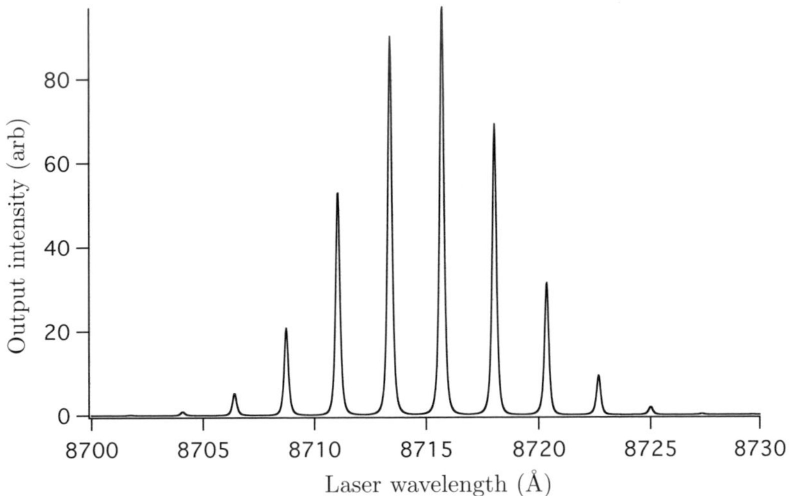

**Fig. 9.9** Computer-generated output spectrum of a 380-$\mu$m-long Fabry–Perot laser displaying multiple longitudinal modes.

Longitudinal mode control in a monolithic semiconductor laser requires an integral frequency-selective element. This is accomplished by either using frequency-selective end mirrors or by arranging that the entire active region provides the necessary feedback only at certain selected frequencies. The latter type of laser is called a *distributed feedback laser* (DFB laser) and the former is called a *distributed Bragg reflection laser* (DBR laser). The advantage of these techniques is that they can be employed in a monolithic device, with all of the attendant advantages of small size, simplicity, economy, etc. In a later section, we will describe *external cavity diode lasers*, which use an *extended cavity* to both force single-mode operation and reduce the Schawlow–Townes linewidth, which can be large due to the very small size (and consequently large linewidth) of the semiconductor laser cavities.

The *distributed feedback laser* dispenses with the reflecting facets for feedback and uses instead a spatially periodic medium for this purpose (the facets can still be reflecting; for simplicity we will assume that they are anti-reflection coated). The *index-coupled* variant employs a periodic *index of refraction* and the *gain-coupled* approach uses a periodic *gain*. We will discuss the index-coupled double heterostructure laser, which is displayed in Fig. 9.10.

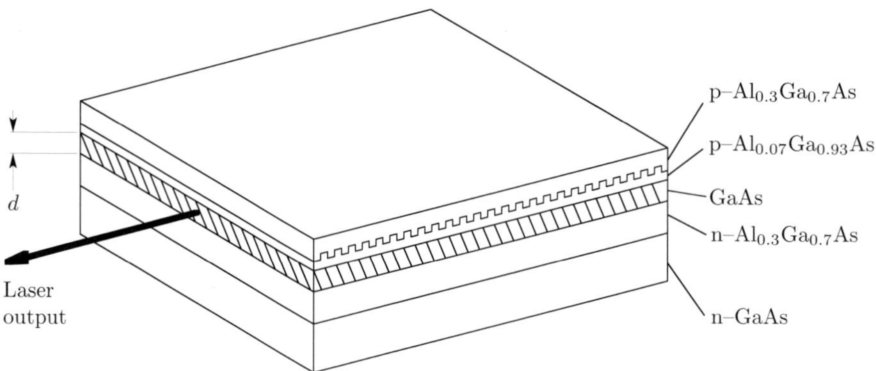


**Fig. 9.10** Distributed feedback double heterostructure index-coupled laser. The index is square-wave modulated since the Al$_{0.3}$Ga$_{0.7}$As layer and the Al$_{0.07}$Ga$_{0.93}$As layer have different indices of refraction and the mode extends into both layers.

The basic principle is that the reflected wave is generated by *Bragg reflection* from the periodic index of refraction, which is produced by the structure shown in the figure. The optical wave samples both sides of the corrugated interface between two different media and thus sees a spatially modulated index of refraction. The periodic index forms a one-dimensional *grating*, defined by the vector $\boldsymbol{K}$ of magnitude $K = 2\pi/\Lambda$, where $\Lambda$ is the spatial period. One can consider $K$ to be the *spatial frequency of the grating*. If the wave vectors of the forward and reflected optical waves are $\boldsymbol{k}_f$ and $\boldsymbol{k}_r$, the grating can be considered to apply a *spatial phase modulation* to the forward wave, whose spatial frequency is $k_f$. The modulation will generate *sidebands* ($\boldsymbol{k}_r$) at integral multiples of the spatial frequency of the grating:

$$\boldsymbol{k}_r = \boldsymbol{k}_f - q\boldsymbol{K}, \qquad q = 1, 2, 3, \cdots \infty, \tag{9.47}$$

where we ignore the "upper sidebands". If the magnitudes of $\boldsymbol{k}_f$ and $\boldsymbol{k}_r$ are the same, as is necessary for laser feedback, and the directions are *opposite*, one requires that

$$k_f = k_r = \frac{q\pi}{\Lambda} \implies \lambda = 2n\Lambda/q, \tag{9.48}$$

where $\lambda$ is the optical wavelength and $n$ is the *average* index of refraction seen by the wave. This is called the *Bragg condition*. The vector relations (for $q = 1$) are shown in Fig. 9.11. Detailed analysis will show that the Bragg condition is not exactly satisfied

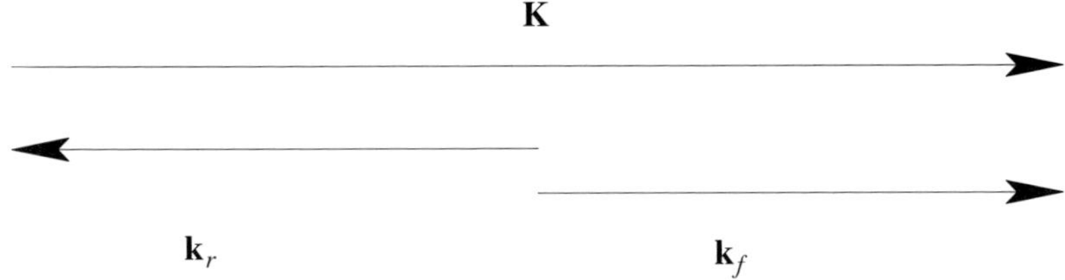


**Fig. 9.11** Vector relations illustrating Bragg condition $k_f = K/2$.

using the structure shown in Fig. 9.10 and to make a satisfactory distributed feedback laser, one needs to slightly modify the geometry.

The behavior of a distributed feedback laser can be obtained using the simple, one-dimensional model of Kogelnik and Shank (1972). We assume that the wave propagates in the $\pm z$ direction and the field satisfies the time-independent wave equation,

$$\frac{\partial^2 E}{\partial z^2} + k^2 E = 0, \tag{9.49}$$

where $E$ is the transverse electric field and $k$ is the wave vector of the wave in the medium. We further assume that the medium has a periodic gain, $\gamma(z)$, and a periodic index of refraction, $n(z)$, which have a simple sinusoidal dependence on $z$:

$$n(z) = n_0 + n_1 \cos 2\beta_0 z \tag{9.50}$$
$$\gamma(z) = \gamma_0 + \gamma_1 \cos 2\beta_0 z. \tag{9.51}$$

An index-coupled grating has $\gamma_1 = 0$ and a gain-coupled grating has $n_1 = 0$. The *average* values of the index of refraction and gain are $n_0$ and $\gamma_0$. For consistency with the above reference, the gain is defined for the *field* (not the intensity) and is *positive* for absorption. For this grating, the Bragg condition is

$$\text{Bragg condition:} \quad \lambda = \frac{2n\pi}{\beta_0} \quad \text{or} \quad k_f = \beta_0, \tag{9.52}$$

where $k_f$ is the wave vector of the radiation in the absence of the grating.

The modulations of $n(z)$ and $\gamma(z)$ are considered to be small ($n_1 \ll n_0$, $\gamma_1 \ll \gamma_0$) and the field variation due to the gain is fractionally small over one period of the grating ($\gamma_0 \ll \beta_0$). The square of the field's wave vector ($k^2$) in the presence of the grating is approximately

$$k^2 = k_f^2 + 2i\gamma_0 k_f + 4\kappa k_f \cos 2\beta_0 z, \tag{9.53}$$

where $\kappa$ is the *coupling constant* between the forward and reflected wave due to the grating and is given by

$$\text{Coupling constant:} \quad \kappa = \frac{\pi n_1}{\lambda} + \frac{i\gamma_1}{2} \tag{9.54}$$

and we have dropped terms of second order in small quantities (including $\gamma_0$). Note that the index coupling and gain coupling are in quadrature – this is a very important difference with significant consequences. The coupling, $\kappa$, is equal to the amount of feedback per unit length provided by the grating ($\kappa L$ is the total coupling, equivalent to the mirror reflectivity in a Fabry–Perot laser).

If the field, $E(z)$, *approximately* satisfies the Bragg condition, it can be expressed as the sum of a forward wave with amplitude, $A_f(z)$, and a reflected wave with amplitude, $A_r(z)$,

$$E(z) = A_f(z)e^{-i\beta_0 z} + A_r(z)e^{i\beta_0 z}, \tag{9.55}$$

where the amplitudes are slowly varying. The field expression can be substituted into the wave equation (including eqn 9.53), yielding two coupled first-order differential equations when the coefficients of different exponentials are equated to zero:

$$\begin{aligned}
i\kappa A_r &= -\frac{dA_f}{dz} + (\gamma_0 - i\delta)A_f \\
i\kappa A_f &= \frac{dA_r}{dz} + (\gamma_0 - i\delta)A_r.
\end{aligned} \tag{9.56}$$

The second derivatives of the amplitudes were ignored and $\delta$ is a measure of the difference between the laser frequency, $\omega$, and the Bragg frequency, $\omega_0 = \beta_0 c/n_0$,

$$\delta = \frac{k^2 - \beta_0^2}{2\beta_0} \approx \frac{n_0(\omega - \omega_0)}{c}. \tag{9.57}$$

The coupled differential equations can be solved using a trial solution of the form,

$$A_f = f_1 e^{\gamma' z} + f_2 e^{-\gamma' z} \tag{9.58}$$

$$A_r = r_1 e^{\gamma' z} + r_2 e^{-\gamma' z}, \tag{9.59}$$

where $\gamma'$ is

$$\gamma'^2 = \kappa^2 + (\gamma_0 - i\delta)^2. \tag{9.60}$$

If we take the origin of the coordinate system at the midpoint of the medium, the *boundary conditions* at each end are

$$\text{Boundary conditions:} \quad A_f(-L/2) = A_r(L/2) = 0, \tag{9.61}$$

since the forward field grows from zero as the wave propagates to the right and the reflected field behaves similarly starting from the right side of the medium. From the coupled differential equations (eqns 9.56), one observes that the equations are symmetric under $z \rightarrow -z$ and $A_f \leftrightarrow A_r$. We therefore expect *symmetric* $(E(z) = E(-z))$ and *antisymmetric* $(E(z) = -E(-z))$ solutions. Additional relations among the coefficients are obtained using the boundary conditions at each end. Thus, one has

$$f_1 = \pm r_2 \tag{9.62}$$

$$f_2 = \pm r_1 \tag{9.63}$$

$$\frac{f_1}{f_2} = \frac{r_2}{r_1} = -e^{\gamma' L}, \tag{9.64}$$

and the solutions are

$$A_f = \sinh \gamma'(z + L/2) \tag{9.65}$$

$$A_r = \pm \sinh \gamma'(z - L/2), \tag{9.66}$$

where an overall factor which is the same for $A_f$ and $A_r$ has been dropped.

The *eigenvalues* $(\gamma')$ can be obtained by substituting the above solutions into the coupled equations (eqns 9.56). After a bit of algebra, one obtains

$$
\begin{aligned}
\gamma' + (\gamma_0 - i\delta) &= \pm i\kappa e^{\gamma' L} \\
\gamma' - (\gamma_0 - i\delta) &= \mp i\kappa e^{-\gamma' L}.
\end{aligned}
\tag{9.67}
$$

In the limit of *weak coupling* $(\kappa \ll \gamma_0)$, one can use eqn 9.60 to obtain

$$\gamma' \approx \pm(\gamma_0 - i\delta). \tag{9.68}$$

Substituting this into eqns 9.67,

$$2(\gamma_0 - i\delta) \approx \pm i\kappa e^{(\gamma_0 - i\delta)L}, \tag{9.69}$$

whose square modulus gives us the threshold condition for the laser in the *high gain* limit:

$$\text{Threshold condition } (\kappa \ll \gamma_0): \qquad 4(\gamma_0^2 + \delta^2) \approx |\kappa|^2 e^{2\gamma_0 L}. \tag{9.70}$$

(Note that in the high gain limit, we still require that $\gamma_0 \ll \beta_0$). The spectral properties of the grating can be obtained from the left-hand side of this equation: when $\delta = \gamma_0$, the coupling squared $(|\kappa|^2)$ will need to be increased by a factor of 2 from its value at resonance $(\delta = 0)$ to maintain threshold. If we consider $\gamma_0$ to be the *spectral width* of the grating, the wavelength and frequency resolution is

$$\frac{\Delta \lambda}{\lambda} = -\frac{\Delta \nu}{\nu} \approx \frac{\gamma_0}{k_f}. \tag{9.71}$$

We thus obtain the familiar result that the width is inversely proportional to the *length of the grating* (or the number of grating lines) since the round trip gain is proportional to $\gamma_0 L$ and this is fixed for a laser with a fixed loss.

The location of the laser *grating* resonances can be obtained from eqn 9.69 by comparing the real and imaginary parts of both sides. We will use a *polar* representation of $\kappa$:

$$\kappa = |\kappa|e^{i\phi}. \tag{9.72}$$

Near the Bragg frequency ($\delta \ll \gamma_0$),

$$\delta L = (q + \tfrac{1}{2})\pi + \phi \qquad \text{(integral } q\text{)}. \tag{9.73}$$

Expressing this as a ratio of the deviation from the Bragg frequency ($c\delta/2\pi n_0$) to the Fabry–Perot free spectral range ($c/2n_0 L$):

$$\text{Number of free spectral ranges from Bragg frequency} = q + \tfrac{1}{2} + \frac{\phi}{\pi}. \tag{9.74}$$

This demonstrates that the modes have approximately the same separation as in a Fabry–Perot laser. A purely *gain-coupled* laser ($n_1 = 0$) will have $\phi = \pi/2$ and will therefore have a resonance at the exact Bragg frequency. An *index-coupled* laser ($\gamma_1 = 0$) has a *real* $\kappa$ and therefore *will not have any resonances at the Bragg frequency*. Furthermore, an index-coupled laser will have two low threshold modes, separated by approximately one free spectral range. Since these modes have an equal gain, the laser can oscillate on *either mode*, a not very desirable state of affairs. There are several remedies for this; the most commonly used cure is to place an abrupt $\pi/2$ *phase shift* in the grating at the exact center. This is shown in Fig. 9.12. A laser with this kind

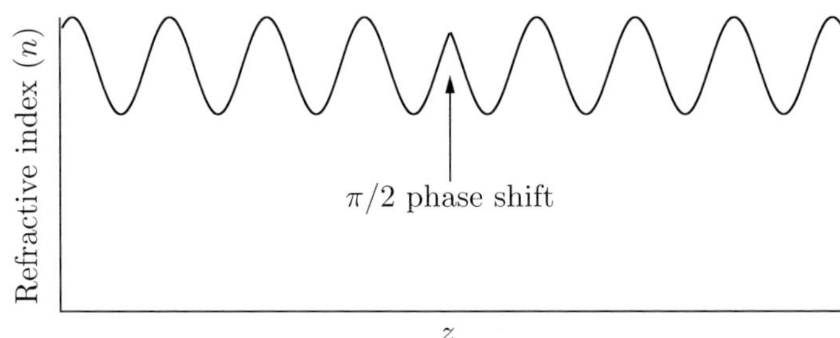


**Fig. 9.12** Sinusoidal index modulation with $\pi/2$ phase shift at the center.

of grating will have the lowest threshold mode at the Bragg frequency.

Another approach to longitudinal mode control is to use Bragg reflection gratings at each end of the laser in place of the reflecting facets which are typical in a Fabry–Perot laser. A laser of this kind is called a distributed Bragg reflection (DBR) laser and is shown schematically in Fig. 9.13. These lasers are slightly easier to fabricate since the gain region is physically separate from the gratings. They have the disadvantage of using shorter gratings which have poorer frequency selectivity than in the DFB laser. This is compensated a bit by the shorter active cavity, whose modes are further apart. DFB lasers seem to be the preferred single-longitudinal-mode monolithic semiconductor laser choice at present.

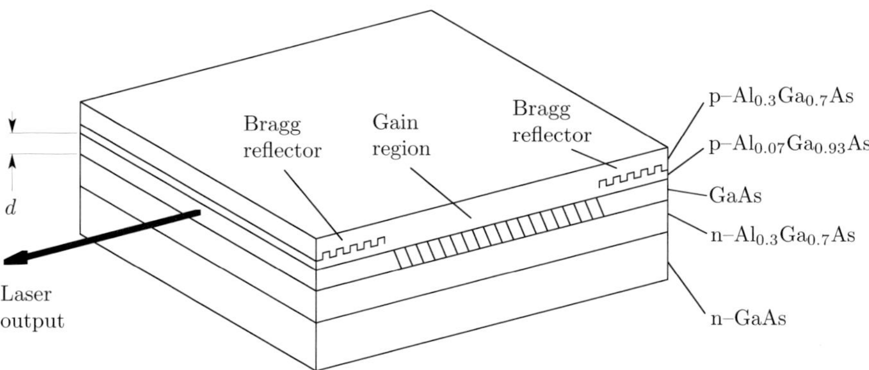


**Fig. 9.13** Distributed Bragg reflection laser.

## 9.6 The rate equations and relaxation oscillations

Although we derived the laser rate equations in Chapter 5, we will do so again using the approach of Petermann (1991). Our notation will be similar to his and will differ in some cases from our previous notation. We will derive a rate equation for both the photon number and the *phase* (of the electric field); the latter is needed to properly account for modulation and noise behaviors of semiconductor lasers.

First, we will obtain an expression for the gain as a function of the stimulated emission rate, $R$. We will proceed in the same way as in Chapter 5, but will anticipate dispersion in the gain medium and distinguish between the *group velocity* and *phase velocity* where previously we assumed that they were the same (no dispersion). The stimulated emission rate is defined as the probability per unit time that a photon incident on the gain medium will stimulate the emission of another photon. The rate $R$ therefore satisfies

$$\frac{dS}{dt} = RS, \tag{9.75}$$

where $S$ is the number density of photons. (Although we used $n$ in the earlier chapters, we use $S$ here to be consistent with the literature.) The velocity of *energy transport* is the *group velocity*, $v_g$, defined as

$$v_g = \frac{d\omega}{dk}, \tag{9.76}$$

where we assume that the medium is weakly dispersive and that the index of refraction increases with frequency (*normal dispersion*). The magnitude of the photon flux is $Sv_g$ and this quantity satisfies a *continuity equation* in one dimension

$$\frac{\partial(Sv_g)}{\partial z} + \frac{\partial S}{\partial t} = \langle\text{source rate}\rangle, \tag{9.77}$$

where the *source rate* is the rate of increase in the photon number density due to some external source, in this case the stimulated emission. In the steady state the time derivative is zero and we obtain

$$\frac{d(Sv_g)}{dz} = RS. \tag{9.78}$$

If we multiply both sides by $h\nu$, we obtain an equation for the growth of the *intensity*, $I = h\nu Sv_g$

$$\frac{dI}{dz} = \frac{R}{v_g} I. \tag{9.79}$$

From the definition of the *laser gain* due to stimulated emission, $g_{st}$,

$$g_{st} = \frac{R}{v_g}. \tag{9.80}$$

The actual gain, $g$, is obtained by multiplying $g_{st}$ by the *mode confinement factor*, $\Gamma$:

$$\text{Gain:} \quad g = \Gamma g_{st} = \Gamma R/v_g. \tag{9.81}$$

In a semiconductor gain medium of length, $L$, with absorption coefficient, $a$, the electric field of a wave propagating in the $z$-direction will be

$$E(z) = E_0 e^{-i\beta z + (g-a)z/2}, \tag{9.82}$$

where $\beta$ is the *propagation constant* of the wave (we use $\beta$ instead of $k$ since the wave is in a *guiding structure* rather than in free space). From this equation and the requirement that the field replicate itself in one round trip, the laser mode frequencies are given by

$$\nu = \frac{qc}{2L\mu}, \tag{9.83}$$

where $q$ is an integer and $\mu$ is the refractive index. We will now consider a laser slightly above threshold, where the threshold carrier density is $N_{th}$ and the threshold oscillation frequency is $\nu_{th}$. We will assume that the refractive index depends only upon the frequency (due to dispersion) and carrier density. Expanding the refractive index around the threshold values of frequency and carrier density,

$$\mu = \mu_0 + \frac{\partial \mu}{\partial \nu}(\nu - \nu_{th}) + \frac{\partial \mu}{\partial N}(N - N_{th}). \tag{9.84}$$

If we assume that the last two terms are much smaller than the first, we can substitute this equation into eqn 9.83 and obtain

$$\left\{ 1 + \frac{\nu_{th}}{\mu_0} \frac{\partial \mu}{\partial \nu} \right\} (\nu - \nu_{th}) = -\frac{\nu_{th}}{\mu_0} \frac{\partial \mu}{\partial N}(N - N_{th}), \tag{9.85}$$

where $\nu_{th} = qc/2L\mu_0$. It is easy to show that the factor in curly brackets is $\mu_g/\mu_0$, where $\mu_g$ is the *group index* defined by

$$\mu_g \equiv \frac{c}{v_g}. \tag{9.86}$$

(Group velocity is discussed in some detail in Chapter 10). From this, we obtain the equation governing semiconductor laser frequency modulation:

$$\nu - \nu_{th} = -\frac{\nu_{th}}{\mu_g}\frac{\partial \mu}{\partial N}(N - N_{th}).$$

(9.87)

After one round trip, one can easily see from eqn 9.82 that the field is multiplied by the *round-trip gain*, $G_r$

$$G_r = r_1 r_2 e^{-2i\beta L + (g-a)L},$$

(9.88)

where $r_{1,2}$ are the field reflectivities of the end facets. Since the conditions are slightly different from those at threshold, we use the expansion of the refractive index (eqn 9.84) to determine $\beta$

$$\beta = \frac{\omega}{c}\mu = \frac{\omega_{th}}{c}\left\{\mu_0 + \frac{\partial \mu}{\partial N}(N - N_{th}) + \frac{\mu_g - \mu_0}{\omega_{th}}(\omega - \omega_{th})\right\},$$

(9.89)

where $\omega_{th}$ and $\omega$ are angular frequencies at threshold and slightly above threshold and the third term in brackets is due to a well-know relation between $\mu_g$ and $\mu_0$ (eqn 10.31). The gain can be written as the product of a frequency independent factor, $G_1$, and a frequency dependent factor, $G_2$. From eqns 9.88 and 9.89, the two gain factors are

$$G_1 = r_1 r_2 e^{(g-a)L}e^{-i\phi_G} \qquad \text{where } \phi_G = \frac{2\omega_{th}L}{c}\frac{\partial \mu}{\partial N}(N - N_{th})$$

(9.90)

$$G_2 = e^{-it_{rt}(\omega - \omega_{th})}.$$

(9.91)

In the second equation, $t_{rt}$ is the round trip time and is equal to $2L\mu_g/c$. In deriving the expression for $G_2$, we used eqn 9.87 and the fact that $\omega_{th}L\mu_0/c$ and $\omega L\mu_g/c$ are integral multiples of $2\pi$.

In eqn 9.82, we gave the electric field as a function of $z$. Since we seek a *rate equation* for the electric field, we now consider the time dependence of the field, which we write as the product of a factor oscillating at $\omega_{th}$ and a slowly varying part, $E_s(t)$:

$$E(t) = E_s(t)e^{i\omega_{th}t}.$$

(9.92)

We will invoke the usual consistency argument and require that after one round trip the field replicates itself,

$$E(t) = G_1 G_2 E(t) = G_1 e^{it_{rt}\omega_{th}}\left\{e^{-i\omega t_{rt}}E(t)\right\}.$$

(9.93)

The bracketed quantity on the right can be shown to be equal to $E(t - t_{rt})$. This equality is demonstrated by expanding $E(t - t_{rt})$ in a Taylor series and evaluating the various time derivatives using $d/dt \to i\omega$ (since $E(t) \propto e^{i\omega t}$). An expansion of the factor in brackets will yield the same result. Thus, we obtain:

$$E_s(t) = G_1 E_s(t - t_{rt}).$$

(9.94)

Since $E_s$ is *slowly varying*, we can expand it to first order in $t_{rt}$,

$$E_s(t) = G_1 E_s(t - t_{rt}) = G_1\left(E_s(t) - t_{rt}\frac{dE_s(t)}{dt}\right).$$

(9.95)

Placing the derivative on the left-hand side yields a differential equation for the field:

$$\frac{dE_s}{dt} = \frac{1}{t_{rt}}(1 - 1/G_1)E_s.$$

(9.96)

Since $1/G_1$ is close to unity, one can expand it:

$$\frac{1}{G_1} = e^{-\ln(r_1 r_2) - (g-a)L + i\phi_G} \approx 1 - \ln(r_1 r_2) - (g-a)L + i\phi_G.$$

(9.97)

The final result is obtained by substituting this into the differential equation, eqn 9.96:

$$\text{Equation for field:} \qquad \frac{dE_s}{dt} = \left[i(\omega - \omega_{th}) + \tfrac{1}{2}(\Gamma R - 1/\tau_p)\right] E_s.$$

(9.98)

The term involving $R$ comes from eqn 9.81:

$$g\frac{c}{\mu_g} = \Gamma R,$$

(9.99)

the $\omega - \omega_{th}$ term is from eqns 9.87 and 9.90 and the *photon lifetime*, $\tau_p$, is

$$\frac{1}{\tau_p} = v_g \left(a - \frac{1}{L}\ln(r_1 r_2)\right).$$

(9.100)

The rate equation for the electric field (eqn 9.98) has a very simple interpretation. The equation states that $E_s(t)$ has an oscillatory component at frequency $\omega - \omega_{th}$ (and thus $E(t)$ oscillates at $\omega$, as expected) and grows exponentially at a rate equal to the difference between $\Gamma R/2$ and $1/(2\tau_p)$. Indeed, a non-rigorous derivation could have made from the these considerations.

The field is normalized so that the photon number density, $S$, is $|E_s|^2$ and the phase, $\phi$, is just the phase of $E_s$. With these assignments, it is easy to show that the rate equations for the photon number density and electric field phase can be obtained from:

$$\frac{dS}{dt} = 2\mathrm{Re}\left\{E_s^* \frac{dE_s}{dt}\right\}$$

(9.101)

$$\frac{d\phi}{dt} = \frac{1}{S}\mathrm{Im}\left\{E_s^* \frac{dE_s}{dt}\right\}.$$

(9.102)

The rate equations are:

$$\text{Number density:} \qquad \frac{dS}{dt} = (\Gamma R - 1/\tau_p)S$$

(9.103)

$$\text{Phase:} \qquad \frac{d\phi}{dt} = \omega - \omega_{th}.$$

(9.104)

We have neglected spontaneous emission in deriving these equations.

The phase rate equation can be reformulated to make it more useful for linewidth calculations, which will be carried out later in this chapter. From eqns 9.87 and 9.104,

$$\frac{d\phi}{dt} = \omega - \omega_{th} = -\frac{\omega_{th}}{\mu_g}\frac{\partial\mu}{\partial N}(N - N_{th}) = -\frac{\omega_{th}}{\mu_g}\frac{\partial\mu'}{\partial\mu''}\frac{\partial\mu''}{\partial N}(N - N_{th}). \tag{9.105}$$

The quantities $\mu'$ and $\mu''$ are respectively the real and imaginary parts of the refractive index, which is given by

$$\mu = \mu' - i\mu''. \tag{9.106}$$

The imaginary part is proportional to the gain, $g$, which is

$$g = -\frac{4\pi\mu''}{\lambda}, \tag{9.107}$$

while the real part is the usual index of refraction. (We did something similar using a complex susceptibility in Chapter 5.) The derivative of the real part with respect to the imaginary part is very important in semiconductor theory and is called the *linewidth enhancement factor*, $\alpha$:

$$\text{Linewidth enhancement factor:} \quad \alpha = \frac{\partial\mu'}{\partial\mu''} = \frac{\partial\mu'/\partial N}{\partial\mu''/\partial N}. \tag{9.108}$$

This quantity is responsible for a substantial increase in the Shawlow–Townes linewidth in diode lasers and has a numerical value typically between 3 and 10. It describes the coupling between the gain and index of refraction in semiconductors. Since $g_{st}$ increases with $N$ while the index of refraction, $\mu'$, decreases, $\alpha$ is always positive (due to the minus sign used for the imaginary part of $\mu$). As a result of the inverse relation between $\mu'$ and $N$, the index of refraction in the *wings* of the wave will be increased relative to the center and the confinement of the field in the laser diode waveguide structure will be harmed. The parameter $\alpha$ is a measure of this (adverse) tendency and it is therefore often called the *anti-guiding parameter*. The reason that $\alpha \neq 0$ in semiconductor lasers is due to the somewhat *unsymmetric* gain curve in semiconductors (Chapter 8). Many laser gain curves (e.g., Lorentzians and Gaussians) are symmetric about their peak and thus the dispersion relation of the index of refraction has a zero at the peak of the gain curve. A medium with such a dispersion relation will have a zero $\alpha$.

Using the $\alpha$ parameter and the relation between the gain and $\mu''$, we obtain

$$\frac{d\phi}{dt} = \tfrac{1}{2}\alpha v_g\frac{\partial g}{\partial N}(N - N_{th}). \tag{9.109}$$

A final transformation of this equation is based upon the relation between $g$ and $R$ and the fact that the value of $\Gamma R$ at threshold is just $1/\tau_p$ (from the photon number density rate equation)

$$\frac{d\phi}{dt} = \tfrac{1}{2}\alpha(\Gamma R - 1/\tau_p), \tag{9.110}$$

where the product of the derivative and $N - N_{th}$ is just $\Delta g$, the gain increment above threshold.

Spontaneous emission is accounted for by adding the term involving $R_{sp}$ to the right side of the photon number equation

$$\frac{dS}{dt} = (v_g g - 1/\tau_p)S + \frac{R_{sp}}{\tau_r} N, \tag{9.111}$$

where $\tau_r$ is the spontaneous emission lifetime (lifetime for electron–hole recombination). The added term, which is small compared to the other terms, is the rate of spontaneous photons which are emitted into the laser mode.

A third (and final) rate equation is that for the carriers and will be presented without a formal derivation, since the terms are fairly self-explanatory:

$$\text{Carriers:} \quad \frac{dN}{dt} = \frac{I}{eV} - \frac{N}{\tau_r} - \Gamma RS, \tag{9.112}$$

where $I$ is the injected current and $V$ is the volume of the active region. The first term on the right is the electron injection rate, the second is the recombination rate and the third is the rate of stimulated emission (which causes recombination). In summary, the three rate equations are

$$\text{Photon density:} \quad \frac{dS}{dt} = (v_g g - 1/\tau_p)S + \frac{R_{sp}}{\tau_r} N \tag{9.113}$$

$$\text{Electric field phase:} \quad \frac{d\phi}{dt} = \tfrac{1}{2}\alpha(v_g g - 1/\tau_p) \tag{9.114}$$

$$\text{Carrier density:} \quad \frac{dN}{dt} = \frac{I}{eV} - \frac{N}{\tau_r} - v_g g S. \tag{9.115}$$

We will derive the small signal behavior by considering *small, time-dependent displacements* of $I, N$ and $S$ from their steady state values $I_0, N_0$ and $S_0$

$$\begin{aligned}
I(t) &= I_0 + I_1(t) \\
N(t) &= N_0 + N_1(t) \\
S(t) &= S_0 + S_1(t).
\end{aligned} \tag{9.116}$$

Since the gain depends upon the carrier density and photon density, we will also expand it around the operating point defined by $I_0$, $N_0$, and $S_0$:

$$g = g_0 + g_N N_1 + g_S S_1, \tag{9.117}$$

where $g_N$ and $g_S$ are derivatives of the gain with respect to $N$ and $S$. The parameter $g_S$ is a measure of nonlinear *gain compression* due to saturation and it is less than zero. Substituting into the rate equations and dropping second-order quantities in the small displacements we obtain

$$\begin{aligned}
\frac{dN_1}{dt} &= \frac{I_1}{eV} - \left(\frac{1}{\tau_r} + v_g g_N S_0\right) N_1 - v_g(g_0 + g_S S_0)S_1 \\
\frac{dS_1}{dt} &= \left(v_g g_N S_0 + \frac{R_{sp}}{\tau_r}\right) N_1 - \left(\frac{R_{sp} N_0}{\tau_r S_0} - v_g g_S S_0\right) S_1,
\end{aligned} \tag{9.118}$$

where we also used the relations among the steady-state quantities obtained by setting the time derivatives in the original rate equations equal to zero. Differentiating the second equation with respect to time and using the first equation for $dN_1/dt$ and the second equation for $N_1$, we obtain

$$\frac{d^2 S_1}{dt^2} + A\frac{dS_1}{dt} + BS_1 = \frac{v_g g_N S_0 + R_{sp}/\tau_r}{eV}I_1, \tag{9.119}$$

where

$$A \approx \frac{1}{\tau_r} + v_g(g_N - g_S)S_0 \tag{9.120}$$

$$B \approx v_g\left(\frac{g_N}{\tau_p} - \frac{g_S}{\tau_r}\right)S_0. \tag{9.121}$$

In the above, we have ignored the spontaneous emission terms ($R_{sp} \to 0$) and have used the approximate expression $v_g g_0 \approx 1/\tau_p$ (gain equals loss), which is valid near threshold.

A formally identical differential equation will be encountered in a chapter on control systems and is solved for a step change in the driving term ($I_1$) using Laplace transforms. The driving term is

$$I_1(t) = I_{10}u(t), \tag{9.122}$$

where

$$u(t) = \begin{cases} 0 & : & t \leq 0 \\ 1 & : & t > 0. \end{cases} \tag{9.123}$$

We will write the equation in a more standard form:

$$\frac{d^2 S_1}{dt^2} + 2\zeta\omega_n\frac{dS_1}{dt} + \omega_n^2 S_1 = \frac{v_g g_N S_0}{eV}I_{10}u(t), \tag{9.124}$$

where we have ignored spontaneous emission in the right-hand driving term. In most cases, the gain saturation term, $g_S$, is much smaller than the gain carrier dependence, $g_N$. Ignoring $g_S$, the parameters $\omega_n$ and $\zeta$ are

$$\omega_n^2 = \frac{v_g g_N S_0}{\tau_p} \tag{9.125}$$

$$\zeta = \frac{1}{2\omega_n}\left(\frac{1}{\tau_r} + v_g g_N S_0\right) = \frac{1}{2}\sqrt{\frac{\tau_p}{v_g g_N S_0}}\left(\frac{1}{\tau_r} + v_g g_N S_0\right). \tag{9.126}$$

The quantities $v_g g_N S_0$ and $1/\tau_r$ are both approximately $10^9$ s$^{-1}$. The photon lifetime, $\tau_p$, is given by eqn 9.100 and, for a typical laser, $\tau_p \approx 10^{-12}$ s. Thus, $\omega_n \approx 3 \times 10^{10}$ s$^{-1}$ (about 5 GHz) and $\zeta \approx 0.03$ (extreme *underdamping*). The solution for $S_1(t)$ is

$$S_1(t) = \frac{v_g g_N S_0}{eV}\left(1 - \frac{1}{\beta}e^{-\zeta\omega_n t}\sin(\omega_n\beta t + \theta)\right) \tag{9.127}$$

$$\text{where} \quad \beta = \sqrt{1 - \zeta^2}$$

$$\theta = \tan^{-1}\beta/\zeta.$$

The damping time is $1/\zeta\omega_n \approx 10^{-9}$ s. A plot of the photon number (which is proportional to the laser power) as a function of time for a step change in injection current at $t = 0$ appears at the top of Fig. 9.14. The damped oscillation is called *relaxation*

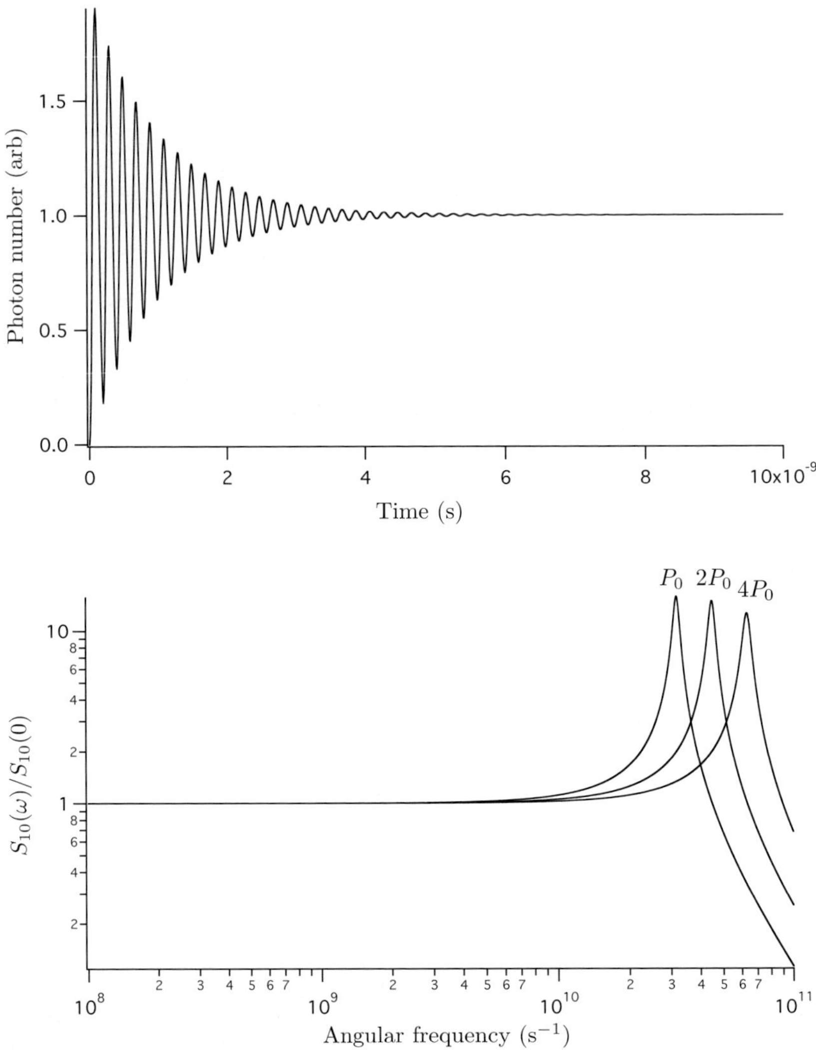

**Fig. 9.14** The transient response to a step in injection current (top) and the modulation response as a function of angular frequency (bottom). Three laser powers differing by factors of 2 are plotted, showing the power dependence of the modulation bandwidth ($P_0$ is an arbitrary starting point).

*oscillation* and occurs to some extent in all lasers. The oscillation frequency is very high (several GHz) in semiconductor lasers and is proportional to the square root of the laser power. A small amount of relaxation oscillation is always present and is driven by technical noise and by spontaneous emission, which continually generates very small power *steps* which ring as shown in the plot.

The response to a *sinusoidal* modulation of the injection current can be obtained by substituting

$$S_1(t) = S_{10}e^{st} \tag{9.128}$$

$$I_1(t) = I_{10}e^{st}, \tag{9.129}$$

where the *complex frequency* is $s = \sigma + i\omega$ and we ignore the real part in what follows. The size of the amplitude modulation is

$$S_{10} = \frac{1}{s^2 + 2\zeta\omega_n s + \omega_n^2} \frac{v_g g_N S_0}{eV} I_{10}. \tag{9.130}$$

A normalized plot of the modulation response is shown at the bottom of Fig. 9.14 for three values of the steady-state power. As can be seen, the bandwidth increases with the power, and the peak of the underdamped response is at approximately $\omega_n$. From the approximate expression for $\omega_n$, the DC response is

$$S_{10}(0) = \frac{\tau_p}{eV} I_{10} \tag{9.131}$$

and the response at $s = i\omega$ is

$$S_{10}(\omega) = \frac{\omega_n^2}{-\omega^2 + 2i\zeta\omega_n\omega + \omega_n^2} \frac{\tau_p}{eV} I_{10}(\omega). \tag{9.132}$$

## 9.7 Diode laser frequency control and linewidth

There are two methods for controlling the output frequency of a monolithic diode laser. We will discuss in some detail the use of the injection current for this purpose. The frequency of a monolithic diode laser can also be controlled using the temperature since the cavity geometry and refractive index depend upon the temperature.

The injection current controls the laser frequency through two mechanisms: the conversion of the amplitude changes to frequency changes via the linewidth broadening factor, $\alpha$, (*carrier effects*) and the changes in the cavity geometry and index of refraction via temperature changes which depend upon the diode current. The temperature effects usually dominate at temperatures below the thermal roll-off frequency of about 1–10 MHz. The $\alpha$-dependent carrier effect rolls off near the frequency of relaxation oscillations (several GHz). We will discuss the temperature-dependent effects first.

The frequency deviation, $\Delta\nu$, due to a change in temperature, $\Delta T$, is given by

$$\frac{\Delta\nu}{\nu} = -(\alpha_L + \alpha_\mu)\Delta T. \tag{9.133}$$

The quantity $\alpha_L$ is the thermal expansion coefficient and is about $0.6 \times 10^{-5}$/degree for GaAs, while $\alpha_\mu$ is the thermal refractive index coefficient $((1/\mu)d\mu/dT)$ and is about $10^{-4}$/degree for GaAs. Thus, for a 850 nm laser, the *static* (DC) temperature tuning sensitivity of a GaAs laser is about 37 GHz/degree. It is important to distinguish this frequency shift, which is due to change in the laser cavity resonance frequency, from the frequency shift due to the *dependence of the peak of the gain curve on temperature*. The latter value of $d\nu/dT$ is about 4 times as large as the former and is used for gross tuning of laser diodes. The shift of the gain peak with temperature will mainly result in *mode hops* as the temperature is changed. The fact that there are two temperature tuning mechanisms with different temperature sensitivities can cause there to be frequency *gaps* in the tuning of the lasers; these can only be removed with an eternal cavity or by injection locking.

The time dependence of the diode temperature can be obtained from the thermal equation in the active medium:

$$C_t\frac{\partial T}{\partial t} - k_t\nabla^2 T = W_i, \tag{9.134}$$

where $C_t$ is the heat capacity, $k_t$ is the thermal conductivity to the environment and $W_i$ is the heat source due to the injection current. The exact solution is rather complicated and is obtained by solving this equation in each layer of the diode. From simple considerations, one can show that the *transfer function* for the temperature is given by

$$\frac{T(\omega)}{I(\omega)} = \frac{C_{th}}{1 + i(\omega/\omega_{th})}, \tag{9.135}$$

where $I(\omega), T(\omega)$ are the Fourier transforms of the current and temperature and $\omega_{th}$ is the *thermal cut-off frequency*.

Treatment of the carrier effect begins with eqn 9.87, which we will reproduce here:

$$\nu - \nu_{th} = -\frac{\nu_{th}}{\mu_g}\frac{\partial\mu}{\partial N}(N - N_{th}).$$

Using the definition of $\alpha$ and $\mu''$, this equation is equivalent to

$$\nu - \nu_{th} = \frac{v_g\alpha}{4\pi}\frac{\partial g}{\partial N}(N - N_{th}). \tag{9.136}$$

One would think that obtaining the derivative of $N$ with respect to $I$ would complete the calculation. Unfortunately, this would significantly underestimate the size of the frequency shift. For frequencies below about 1 GHz, the main source of the frequency shift is actually the *gain compression*, parametrized with $g_S$. This behavior is similar to that of an anharmonic oscillator, whose resonant frequency depends upon the oscillation amplitude. To study this analytically, we will first rewrite the photon number rate equation as

$$\frac{dS}{dt} = v_g\left[\frac{\partial g}{\partial N}(N - N_{th}) + g_S S\right]S + \frac{R_{sp}N}{\tau_r}. \tag{9.137}$$

We have simply expanded the gain around the threshold carrier density ($N_{th}$) and used the facts that $g_N = \partial g/\partial N$ and $v_g g_0 = 1/\tau_p$. Eliminating the term involving $\partial g/\partial N$ in the above two equations, we obtain

$$\nu - \nu_{th} = \frac{\alpha}{4\pi} \left[ \frac{1}{S} \frac{dS}{dt} - v_g g_S S - \frac{R_{sp} N}{S \tau_r} \right].$$ (9.138)

Well above threshold, we can ignore the third term in brackets (the spontaneous emission term). If we consider a sinusoidal variation at $\omega_m$ of both $\nu$ and $S$ about their average values, $\bar{\nu}, \bar{S}$

$$\nu(t) = \bar{\nu} + \Delta\nu e^{i\omega_m t}$$
$$S(t) = \bar{S} + \Delta S e^{i\omega_m t},$$ (9.139)

then the ratio of $\Delta\nu$ to $\Delta S$ is

$$\frac{\Delta\nu}{\Delta S} = \frac{\alpha}{4\pi\bar{S}}(i\omega_m - \omega_g),$$ (9.140)

where

$$\omega_g = v_g g_S \bar{S}$$ (9.141)

and we assume that the changes in the parameters are much smaller than their average values. (Note that the $\omega_g$ contribution is positive since $g_S$ is *negative*.) The magnitude of $\omega_g$ is in the GHz region for diode lasers and therefore the frequency response is fairly flat between about 10 MHz and 1 GHz. The dependence of the laser frequency on the injection current is obtained by multiplying the derivatives

$$\frac{d\nu}{dI} = \frac{d\nu}{dS} \frac{dS}{dI}.$$ (9.142)

The second factor is eqn 9.132. Including the temperature effects (eqn 9.135),

$$\frac{d\nu}{dI} = \left( \frac{\alpha \tau_p}{4\pi e V \bar{S}} \right) \frac{\omega_n^2(i\omega_m - \omega_g)}{-\omega_m^2 + 2i\zeta\omega_n\omega_m + \omega_n^2} - \frac{\nu(\alpha_L + \alpha_\mu)C_{th}}{1 + i(\omega_m/\omega_{th})}.$$ (9.143)

This function is plotted in Fig. 9.15 for $\omega_n = 3 \times 10^{10}$ s$^{-1}$, $|\omega_g| = \omega_n$, $\zeta = 0.03$, a DC thermal sensitivity of 1.6 GHz/mA and a DC carrier sensitivity of about 0.1 GHz/mA.

We will conclude this section by justifying the designation of $\alpha$ as the linewidth enhancement factor: we will derive the expression for the linewidth of a bare diode laser, using the approach of Henry (1982). As in the preceding, we assume that the square modulus of the electric field, $E$, represents the number density of photons, $S$, and that the phase of the field is $\phi$:

$$E = \sqrt{S} e^{i\phi}.$$ (9.144)

We are interested in the changes in $\phi$ and $S$ due to a single spontaneous emission. A phasor plot of the situation appears in Fig. 9.16. The change in the field, $\Delta E$, is of unit magnitude (since it represents a single photon) and is given by

$$\Delta E = e^{i(\phi+\theta)},$$ (9.145)

where $\theta$ is random. There will be two contributions to the change in the photon phase: one due to geometry and indicated in the figure and the other due to the coupling of $\phi$

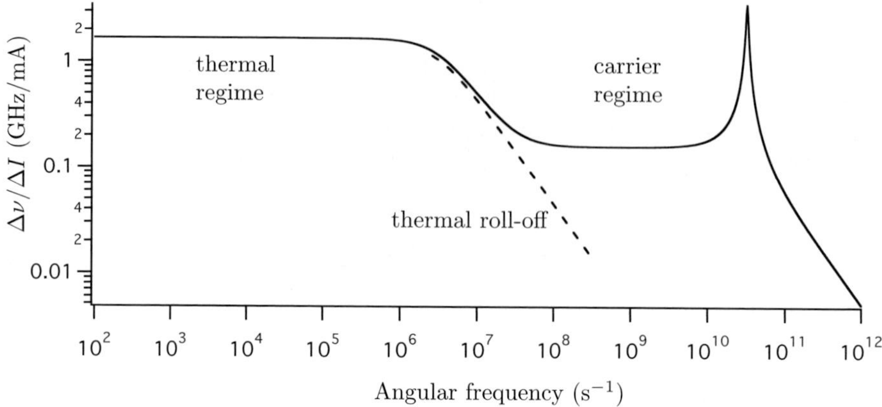

**Fig. 9.15** Theoretical plot of $\Delta\nu/\Delta I$.

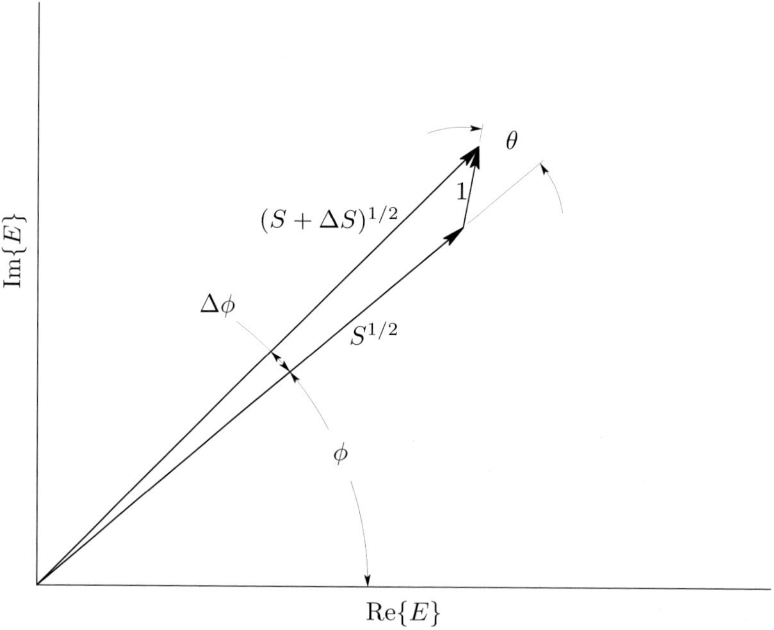

**Fig. 9.16** Phasor plot showing change in photon number density, $S$, and electric field phase, $\phi$, due to the spontaneous emission of one photon.

and $S$ via the rate equations. The geometrical change in phase, $\Delta\phi_1$, due to the single emission is

$$\Delta\phi_1 = \frac{\sin\theta}{\sqrt{S}} \qquad (\theta \text{ random}). \tag{9.146}$$

Using the law of cosines, the change in $S$ is

$$\Delta S = 1 + 2\sqrt{S}\cos\theta. \tag{9.147}$$

The average value of $\Delta S$ is one, as expected, since the spontaneous event should add one photon to the field.

To determine the second contribution to the phase change, we will use the rate equations for phase and photon number density, without the spontaneous emission term:

$$\frac{dS}{dt} = (v_g g - 1/\tau_p)S \tag{9.148}$$

$$\frac{d\phi}{dt} = \tfrac{1}{2}\alpha(v_g g - 1/\tau_p). \tag{9.149}$$

From these, we obtain

$$\frac{d\phi}{dt} = \frac{\alpha}{2S}\frac{dS}{dt}. \tag{9.150}$$

The rate equation contribution to the phase change, $\Delta\phi_2$, is therefore

$$\Delta\phi_2 = -\frac{\alpha}{2S}\Delta S = -\frac{\alpha}{2S}(1 + 2\sqrt{S}\cos\theta). \tag{9.151}$$

The minus sign is due the nature of the spontaneous emission process. When a spontaneous photon is emitted, there is a *step change* in the field intensity which causes damped relaxation oscillations, after which the intensity returns to its steady state. Thus, $S$ changes from $S + \Delta S$ (immediately after emission) to the steady state value $S$ and the change is negative. This latter process – the restoration of the steady state – is the source for this component of the line broadening, not the spontaneous emission (whose contribution is the $\Delta\phi_1$ term). The total phase change is

$$\Delta\phi = \Delta\phi_1 + \Delta\phi_2 = -\frac{\alpha}{2S} + \frac{1}{\sqrt{S}}[\sin\theta - \alpha\cos\theta]. \tag{9.152}$$

The first term is constant and it causes the phase to increase at rate

$$\Delta\omega = \frac{d}{dt}\Delta\phi = -\frac{\alpha R_{spont}}{2S}, \tag{9.153}$$

where $R_{spont}$ is the spontaneous emission rate. This is just a constant frequency shift ($\Delta\omega$) due to a reduction in the stimulated rate by spontaneous emission. The second term results in phase *fluctuations*, whose accumulated square value is

$$\langle\Delta\phi^2\rangle = \left\langle\left(\sum_{i=1}^{N} S^{-1/2}[\sin\theta_i - \alpha\cos\theta_i]\right)^2\right\rangle, \tag{9.154}$$

where $N$ is the (large) number of observed emissions and the $\langle\rangle$ denotes an *ensemble average*. This can easily be summed and averaged since the cross terms all average to zero ($\theta_i$ is random) and the $\sin^2$ and $\cos^2$ average to $1/2$. The sum can be performed by multiplying each (identical, after averaging) term by $R_{spont}t$, where $t$ is the observation time:

$$\langle \Delta \phi^2 \rangle = \frac{R_{spont}(1 + \alpha^2)t}{2S}. \tag{9.155}$$

The resultant electric field executes a *random walk* after each emission event. From the theory of random walks, the autocorrelation function of the field is

$$\langle E^*(t)E(0) \rangle = |E(0)|^2 e^{-|t|/t_{coh}}, \tag{9.156}$$

where the coherence time, $t_{coh}$, is given by

$$1/t_{coh} = \frac{\langle \Delta \phi^2 \rangle}{2t} \tag{9.157}$$

and the power spectrum is the Fourier transform of the autocorrelation function. The power spectrum is a Lorentzian whose full width at half maximum is

$$\Delta \nu = \frac{1}{\pi t_{coh}} = \frac{R_{spont}}{4\pi S}(1 + \alpha^2). \tag{9.158}$$

This last result is the important one. Carrying the calculation to completion will yield the standard Shawlow–Townes linewidth multiplied by $1 + \alpha^2$. Thus, diode lasers will have linewidths which are ten to a hundred times as large as expected from an *ideal* laser with the same power and cavity configuration ($\alpha$ varies from 3 to 10). When this is combined with the small cavity length, we obtain linewidths of tens of MHz or greater. The external cavity, described in the next section, allows one to greatly reduce this large linewidth.

## 9.8   External cavity diode lasers (ECDLs)

We have seen that monolithic distributed feedback lasers can be operated in both a single transverse and single longitudinal mode. However, the spectral breadth can be fairly large since the Shawlow–Townes line breadth is proportional to the square of the cavity width and this is very large in the extremely short feedback structures found in monolithic DFB lasers. Moreover, the linewidth enhancement factor further increases the spectral breadth by the factor $1 + \alpha^2$. A significant electronic tuning range is difficult to achieve in a DFB laser since, as was shown in the last section, the tuning range due to changes in the injection current is fairly small and one is forced to use the fairly slow mechanism of diode temperature changes to tune the diode over a large range. Both of these limitations can be overcome by using the diode as a gain medium in a much larger cavity composed of mirrors, gratings and other possible frequency-selection elements. Such a laser is called an external cavity diode laser (ECDL). A longer cavity reduces the Shawlow–Townes width by the inverse of the square of the cavity length and eliminates the effects of the $\alpha$ parameter since the active medium length is a very small fraction of the overall cavity length. With the use of a coarse frequency selecting element (usually a grating) together with cavity length changes, a large tuning range can be obtained without changing the diode current or temperature.

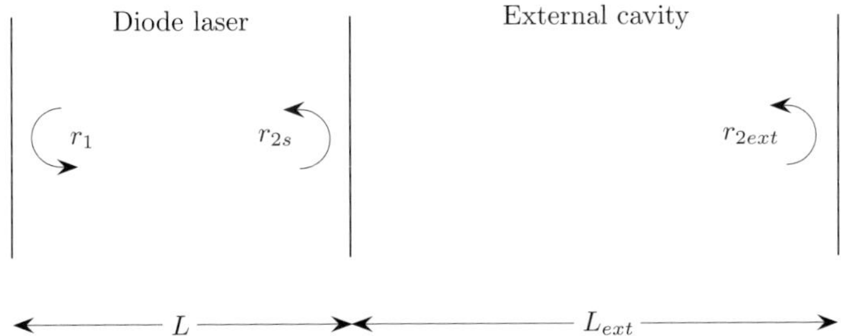


**Fig. 9.17** Schematic of laser cavity reflectors together with external reflector.

Although commercial ECDLs are available, many atomic physics experimentalists manufacture their own units using commercial monolithic diodes and readily available optical components. If one chooses this route, one is immediately confronted with the problem of the existing coatings (if any) on the diode. The best results are obtained if one of the diode facets has zero reflectivity and the other is highly reflecting. Then, the diode is truly just a gain element in a larger cavity. On the other hand, if the facets are not antireflection coated, one is left with a much more complicated system since the laser resonator will consist of two coupled cavities: one due to the monolithic laser diode and the other due to the external cavity. In the following we will analyze the coupled cavities in an external cavity laser and describe some geometries which facilitate tuning over a relatively wide frequency range.

Our analysis is due to Petermann (1991) and can be used to treat small to moderate amounts of feedback. Thus, we can describe an external cavity laser without anti-reflection coatings and analyze the adverse effects of small amounts of feedback on a solitary diode laser. We will assume that there is a reflective element with field reflectivity $r_{2ext}$ placed a distance $L_{ext}$ from the output facet of a laser diode whose facets have field reflectivities $r_1$ and $r_{2s}$ and whose length is $L$ (Fig. 9.17). This analysis assumes that the external cavity supports only a single reflection, which is a fairly realistic assumption since $r_{2s}$ due to an uncoated facet is about 0.57 (in GaAs). The effective reflectivity, $r_2$, of the output facet is now frequency-dependent and is given by

$$r_2(\omega) = r_{2s} + r_{2ext}(1 - |r_{2s}|^2)e^{-i\omega\tau_{ext}}, \tag{9.159}$$

where $\tau_{ext} = 2L_{ext}/c$ is the round trip time in the external cavity and the bracketed expression describes *two* transmissions through the output facet. The two laser oscillation conditions are

$$r_1|r_2|e^{(g'_{th}-a)L} = 1 \tag{9.160}$$

$$2\beta L + \phi_r = 2\pi m \qquad \text{(integral } m\text{),} \tag{9.161}$$

where $\phi_r$ is the phase of the complex number, $r_2$, $g'_{th} - a$ is the threshold gain minus loss of the *compound* cavity and $\beta$ is the propagation constant in the guiding structure of the diode. We can define a *coupling parameter*, $\kappa_{ext}$,

$$\kappa_{ext} = \frac{r_{2ext}}{r_{2s}}(1 - |r_{2s}|^2). \tag{9.162}$$

When $\kappa_{ext} \ll 1$, $|r_2|$ and $\phi_r$ are given by

$$|r_2| = r_{2s}(1 + \kappa_{ext}\cos\omega\tau_{ext}) \tag{9.163}$$

$$\phi_r = -\kappa_{ext}\sin\omega\tau_{ext}. \tag{9.164}$$

We seek the change in threshold gain and phase $(\phi_r)$ due to the presence of the feedback. From the gain condition (eqn 9.160) and eqn 9.163, the gain change is

$$g'_{th} - g_{th} = -\frac{\kappa_{ext}}{L}\cos\omega\tau_{ext}, \tag{9.165}$$

where $g_{th}$ is the threshold gain in the *absence* of the external cavity. The phase equations with and without feedback are

$$\text{No feedback:} \quad \frac{2\mu L\omega_{th}}{c} = 2\pi m \tag{9.166}$$

$$\text{With feedback:} \quad \frac{2L}{c}(\mu + \Delta\mu)(\omega_{th} + \Delta\omega) + \phi_r = 2\pi m + \Delta\phi, \tag{9.167}$$

where we used $\mu\omega/c$ for $\beta$ and $\Delta\omega = \omega - \omega_{th}$. The quantity $\Delta\phi$ is the difference between the round trip phase change and $2\pi m$; in the presence of feedback, we will require that $\Delta\phi$ be zero (or a multiple of $2\pi$) for laser oscillation to take place. Expanding the above equation and keeping only terms to first order in small quantities,

$$\Delta\phi = \frac{2L}{c}(\mu\Delta\omega + \omega_{th}\Delta\mu) + \phi_r, \tag{9.168}$$

where we used the first equation to eliminate zero-order quantities. We will make our by now familiar expansion of $\mu$ around the threshold values of the carrier density and frequency:

$$\Delta\mu = \frac{\partial\mu}{\partial N}(N - N_{th}) + \frac{\partial\mu}{\partial\omega}(\omega - \omega_{th}), \tag{9.169}$$

where $N_{th}$ is the carrier density without feedback. Substituting this into the equation for $\Delta\phi$ yields

$$\Delta\phi = \frac{2L}{c}\left\{\Delta\omega\left[\mu + \omega_{th}\frac{\partial\mu}{\partial\omega}\right] + \omega_{th}\frac{\partial\mu}{\partial N}(N - N_{th})\right\} + \phi_r. \tag{9.170}$$

The expression in the left-hand inner brackets is just, $\mu_g$, the group index of refraction. The term on the right can be expressed in terms of $\alpha$

$$\frac{\partial\mu}{\partial N}(N - N_{th}) = -\frac{\alpha c}{2\omega_{th}}\frac{\partial g}{\partial N}(N - N_{th}) = -\frac{\alpha c}{2\omega_{th}}(g'_{th} - g_{th}). \tag{9.171}$$

Including this and the expression for the change in gain (eqn 9.165) and $\phi_r$ in the $\Delta\phi$ equation results in

$$\Delta\phi = \frac{2\mu_g L}{c}(\omega - \omega_{th}) + \kappa_{ext}[-\sin\omega\tau_{ext} + \alpha\cos\omega\tau_{ext}]. \tag{9.172}$$

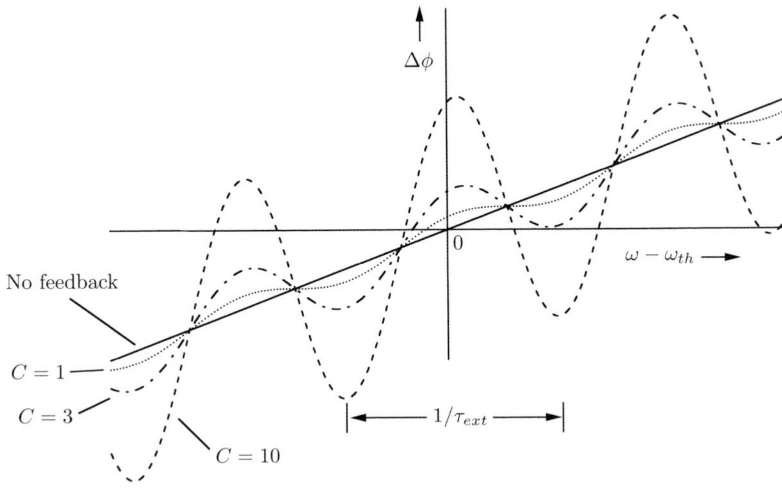



**Fig. 9.18** Plots of $\Delta\phi$ versus $\omega - \omega_{th}$ for a variety of values of the coupling parameter, $C$.

Writing this in terms of the round-trip time in the diode ($\tau_L = 2\mu_g L/c$) and combining the trigonometric functions

$$\Delta\phi = \tau_L(\omega - \omega_{th}) + \kappa_{ext}\sqrt{1+\alpha^2}\sin(\omega\tau_{ext} - \arctan\alpha). \tag{9.173}$$

A slightly more useful coupling parameter than $\kappa_{ext}$ is the parameter $C$ defined as

$$C \equiv \frac{\tau_{ext}}{\tau_L}\kappa_{ext}\sqrt{1+\alpha^2}. \tag{9.174}$$

The final version of our equation for $\Delta\phi$ is

$$\Delta\phi = \tau_L(\omega - \omega_{th}) + C\frac{\tau_L}{\tau_{ext}}\sin(\omega\tau_{ext} - \arctan\alpha). \tag{9.175}$$

Although the $\Delta\phi$ vs $\Delta\omega$ function might seem a bit obscure, it actually provides quite a bit of insight into the behavior of a diode laser under varying amounts of feedback. A family of plots of this function for a variety of values of $C$ appears in Fig. 9.18. The diode laser with external feedback can only oscillate when $\Delta\phi$ is zero or a multiple of $2\pi$. Without feedback, the function is a straight line through the origin, confirming that the laser operates only at $\omega_{th}$ ($\Delta\phi = 0$). The parameter $C$ is defined so that the function is strictly monotonic when $C < 1$. Thus, with weak feedback ($C < 1$), the laser still oscillates at a single frequency, though the frequency will be shifted. This frequency shift is a sensitive function of $\tau_{ext}$, so even a small change of the position of the external reflector (on the order of a wavelength of light) will have a large effect on the laser frequency. One can immediately see from eqn 9.173 that the maximum frequency shift when $C < 1$ is

$$\Delta\omega_{max} = \frac{\kappa_{ext}}{\tau_L}\sqrt{1+\alpha^2}. \tag{9.176}$$

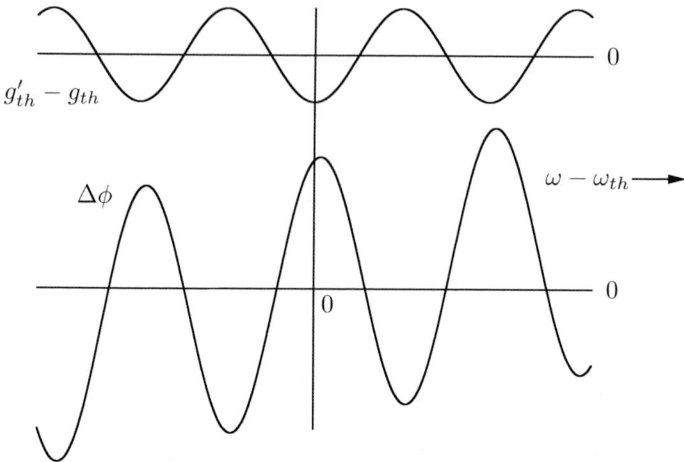

**Fig. 9.19** Plots of $\Delta\phi$ and $g'_{th} - g_{th}$ versus $\omega - \omega_{th}$ for a Littrow configuration external cavity diode laser $(C = 28)$.

For example, for a small external field reflectivity of $10^{-4}$ and a laser cavity length of 300 $\mu$m, the shift is about 18 MHz $(\alpha = 5)$ and will occur when $L_{ext}$ changes by $\lambda/4$. Thus optical isolators are necessary if frequency stability is important.

An external cavity diode laser uses for feedback either the first-order light back-reflected from a grating (Littrow configuration) or the light reflected from the combination of a grating and mirror (Littman configuration). In the Littrow configuration, about 25% of the light can be returned to the diode $(r_{2ext} = 0.5)$ and the coupling parameter, $\kappa_{ext}$, for an uncoated diode $(r_{2s} = 0.56)$ is about 0.6, which is on the edge of the range assumed in the above calculations. For a diode length of 300 $\mu$m and external cavity length of 1 cm, the parameter $C$ is about 28 $(\alpha = 5)$. Curves of $\Delta\phi$ and threshold gain versus frequency are shown in Fig. 9.19. There are a number of possible laser frequencies, separated by about one half the free spectral range of the external cavity. The likely laser frequencies will be those with the *smallest* threshold gain, which eliminates alternate zero crossings of $\Delta\phi$ and results in the expected mode separation.

According to the Shawlow–Townes linewidth formula, the diode laser spectral linewidth is proportional to the *square* of the laser cavity linewidth, which is proportional to $1/\tau_{eff}^2$, where $\tau_{eff}$ is the *effective round trip delay*. The latter is the round trip delay which yields the correct *photon lifetime* and can be given in general by

$$\tau_{eff} = \frac{d\Delta\phi}{d\omega} = \tau_L F, \tag{9.177}$$

where

$$F = 1 + C\cos(\omega\tau_{ext} - \arctan\alpha). \tag{9.178}$$

Thus, the modification to the linewidth due to the external cavity is

$$\Delta\nu = \frac{\Delta\nu_0}{F^2},\tag{9.179}$$

where $\Delta\nu_0$ is the linewidth without the external cavity. For a large $C$, the oscillation frequencies are those for which

$$\text{Oscillation frequencies:}\quad \sin(\omega\tau_{ext} - \arctan\alpha) \approx 0 \qquad (C \gg 1).\tag{9.180}$$

Thus, $F \approx 1 + C$ at the laser emission frequencies and

$$\Delta\nu = \frac{\Delta\nu_0}{(1+C)^2}.\tag{9.181}$$

For the above example, the external cavity reduces the linewidth by nearly a factor of 1000.

We will state without proof that the frequency modulation sensitivity will be *reduced* by the factor $F$, which approaches $1 + C$ in the strong feedback regime:

$$\left(\frac{\partial\nu}{\partial I}\right)_{feedback} = \frac{1}{F}\left(\frac{\partial\nu}{\partial I}\right)_{nofeedback} \implies \frac{1}{1+C}\left(\frac{\partial\nu}{\partial I}\right)_{nofeedback}.\tag{9.182}$$

One can make the following plausibility argument for this result. In an external cavity laser with an antireflection coated output facet, the ratio of the modulation sensitivity to that of a bare diode should be the inverse of the ratio of the cavity lengths. From eqn 9.177, this ratio is $F$, since it is the ratio of $\tau_{eff}$ to $\tau_L$ and $\tau_{eff}$ is the *round trip time* for the extended cavity laser when the output facet has no reflections.

We will complete our theoretical examination of external cavity diode lasers with a discussion of potential instabilities which can occur for intermediate levels of feedback. There are two regimes in which we expect unconditional stability: the low feedback regime, with $C < 1$, where there can be adverse effects on the laser frequency due to changes in the feedback phase but in which the laser will otherwise behave normally and the high feedback regime, where the output facet is antireflection coated and the feedback is very strong. In the latter case, the laser will have a short active region in a larger optical cavity and will behave like the lasers discussed in earlier chapters and therefore warrants little additional analysis.

Researchers Tkach and Chraplyvy (1986) analyzed the effect of feedback on DFB lasers and established five distinct regimes. Distributed feedback lasers behave similarly to Fabry–Perot lasers in this regard, though $\tau_L$ will be a little shorter than expected for a laser of a given length and $r_{2s}$ will depend upon the DFB coupling parameter $(\kappa)$ and is a bit smaller than the facet reflectivity in an uncoated Fabry–Perot laser. Their conclusions are graphically portrayed in Fig. 9.20. In region I, $C < 1$ and the laser is unconditionally stable as discussed above. In region II, the possibility of multimode emission exists and the laser will often jump between modes causing the linewidth to greatly increase. As the feedback is increased, one will enter another region of stability, region III. There the laser will oscillate on the line with the smallest linewidth rather than the line with the lowest threshold gain. Region IV is the regime of *coherence collapse*, where the linewidth becomes extremely great (up to 100 GHz)

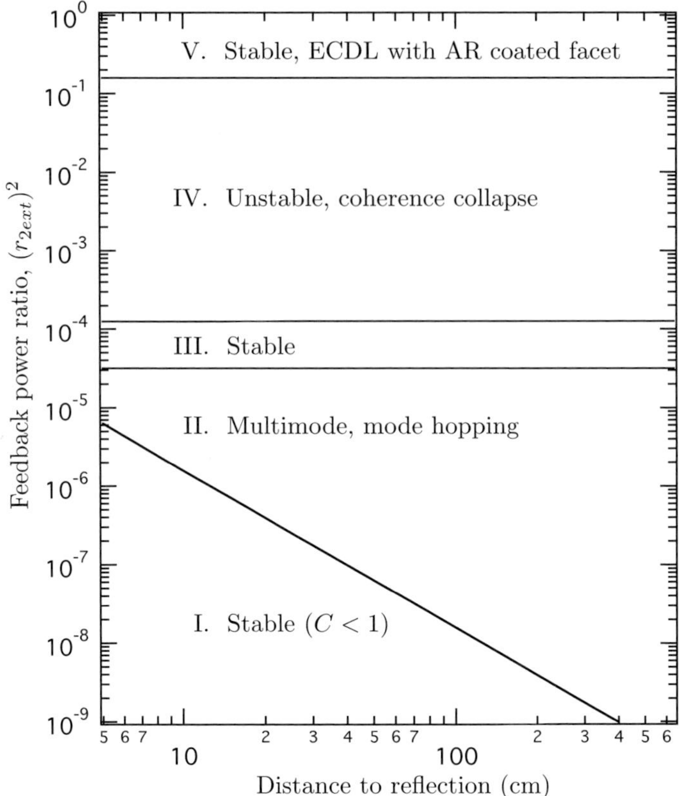



**Fig. 9.20** Stability regions for 1.5 $\mu$m DFB laser (from Tkach (1986)).

and the relaxation oscillations (at several GHz) are continuously excited. It is speculated (Henry (1986)) that this instability occurs when the coupling coefficient ($\kappa_{ext}/\tau_L$ in Henry's theory) becomes comparable to the relaxation frequency. This would occur when $\kappa_{ext} \approx 2 \times 10^{-2}$. The transition to coherence collapse should be independent of the length of the external cavity. The final region, Region V, is the desirable place to operate an external cavity diode laser and is in effect when the laser output facet is antireflection coated and the feedback ratio is fairly large. Unfortunately, an external cavity diode laser with uncoated facets can operate close to region IV and possibly experience the adverse effects of this regime.

We will complete this section with a discussion of some of the practical aspects of external cavity design and describe the *Littrow* and *Littman* feedback approaches. The Littrow feedback scheme is shown schematically in Fig. 9.21. The feedback element is a grating arranged in the *Littrow configuration*, where the diffracted light is retro-reflected from the grating back to the laser along the same path as the incident light. The laser emission is expanded with a collimating lens to illuminate as many lines of the grating as possible in order to increase its frequency selectivity. Due to the extreme feedback sensitivity of the laser diode, the lens should be very well antireflection coated;

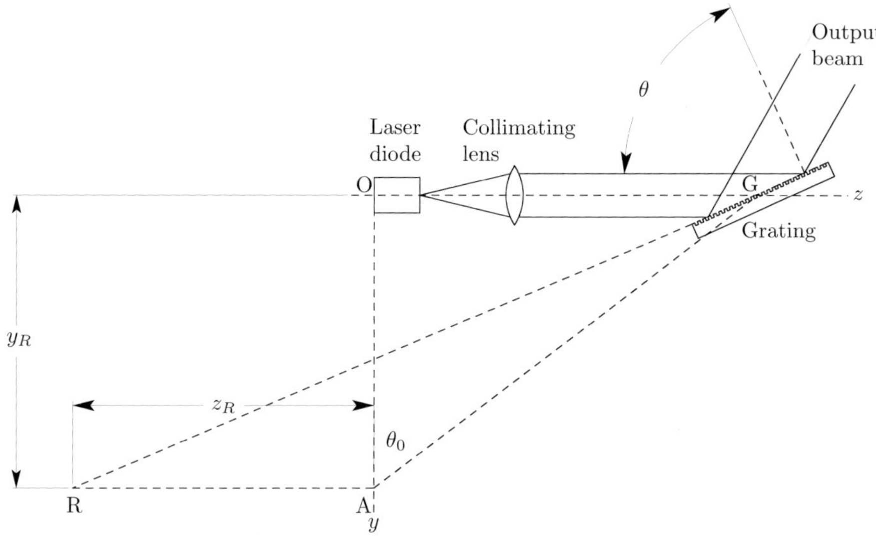


**Fig. 9.21** External cavity diode laser using the Littrow configuration.

in practice, it might still be necessary to slightly tilt the lens to eliminate feedback problems. Coarse tuning of the laser is accomplished by tilting the grating and fine tuning is done by translating the grating along the $z$-axis without changing its angle. The latter can be done electronically using a piezo-electric transducer. The laser output is the specular reflection from the grating.

For wide tunability, it is desirable to tilt the grating about an axis which maintains *tracking* between the grating center frequency and the cavity resonance. Otherwise, mode hops can occur as the laser is tuned with the grating. We will now determine the location of the pivot point which accomplishes this (following the treatment of Labachelerie (1993)). Referring to the figure, the grating wavelength, $\lambda_r$, is

$$\lambda_r = 2a \sin\theta, \tag{9.183}$$

where $a$ is the grating period. If $L$ is distance $OG$ in the figure (the cavity length), cavity resonances occur when

$$L = m\lambda_m/2, \tag{9.184}$$

where $m$ is an integer (we will ignore the diode refractive index at present) and $\lambda_m$ is the cavity resonance wavelength. We will first adjust the geometry so that, at $\theta_0$ and $L_0$, the two wavelengths are the same:

$$\lambda_m(L_0) = \lambda_r(\theta_0). \tag{9.185}$$

We seek a location of the pivot point (whose coordinates are $y_R$ and $z_R$) which preserves this equality over the largest possible range of wavelengths. The wavelength difference is $F(\theta)$:

$$F(\theta) = \lambda_m(\theta) - \lambda_r(\theta). \tag{9.186}$$

In attempting to minimize $F(\theta)$, we will find the location of the pivot point for which the first two derivatives with respect to $\theta$ are zero.

There is a fairly subtle point which, if overlooked, will give a less than optimum result. It turns out that there is a *phase shift* when the grating is translated in its plane by amount $t_0$; this shift advances the phase by $-2\pi$ when $t_0$ increases by the grating period, $a$. Thus the round-trip phase shift is

$$\Delta\phi = \frac{4\pi}{\lambda}L - \frac{2\pi t_0}{a}. \tag{9.187}$$

The function $F(\theta)$ is now

$$F(\theta) = \frac{2L(\theta)}{m + t_0(\theta)/a} - 2a\sin\theta. \tag{9.188}$$

If $t(\theta_0) = 0$, the expression for $F(\theta)$ is

$$F(\theta) = 2a\left[\frac{L(\theta)\sin\theta_0}{L_0 + t_0(\theta)\sin\theta_0} - \sin\theta\right]. \tag{9.189}$$

From simple geometry, $L(\theta)$ is given by

$$L(\theta) = -y_R\left[\frac{\sin\theta_0}{\cos\theta} - \tan\theta\right] - z_R\left[\frac{\cos\theta_0}{\cos\theta} - 1\right] + L_0\frac{\cos\theta_0}{\cos\theta}. \tag{9.190}$$

This is an equation for $L(\theta)$ in terms of $L_0$ and the corresponding angle $\theta_0$, which satisfies the equation $F(\theta_0) = 0$. The first derivative of $F(\theta)$ vanishes when

$$y_R = \frac{L_0}{\tan\theta_0}, \tag{9.191}$$

where we have used for $t_0(\theta)$, to first order

$$t_0(\theta) = \left[\frac{L_0 - z_R}{\cos\theta_0}\right](\theta - \theta_0). \tag{9.192}$$

The pivot point is thus constrained to a horizontal line which passes through point $A$ in the figure. We can refine the pivot point location by obtaining the second derivative of $F(\theta)$ along the above horizontal line. We find that $z_R = -L_0$: the pivot point is behind the rear facet of the diode by $L_0$.

The advantages of the Littrow approach are simplicity and efficiency (it introduces less loss than the Littman configuration). Its principal disadvantage is that the output beam angle changes with wavelength tuning. This angle problem can be eliminated by rigidly attaching to the grating assembly a mirror which is parallel to the grating (Fig. 9.22). This preserves the angle of the output beam, although the beam will be displaced parallel to itself by an amount whose upper bound is $\approx 2L\Delta\theta$ for an angular change of $\Delta\theta$, where $L$ is the distance between the grating and the mirror.

The Littman approach (Liu (1981)) uses a grating at near grazing incidence for gross frequency tuning and reflects the first-order diffracted beam back using a mirror

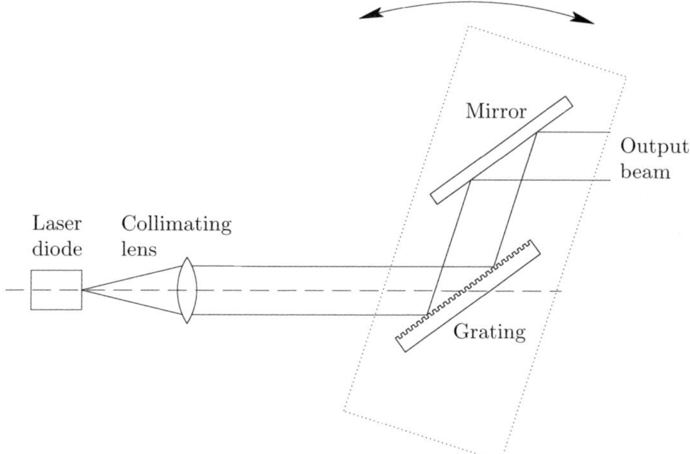



**Fig. 9.22** Arrangement which eliminates angle changes with wavelength tuning.

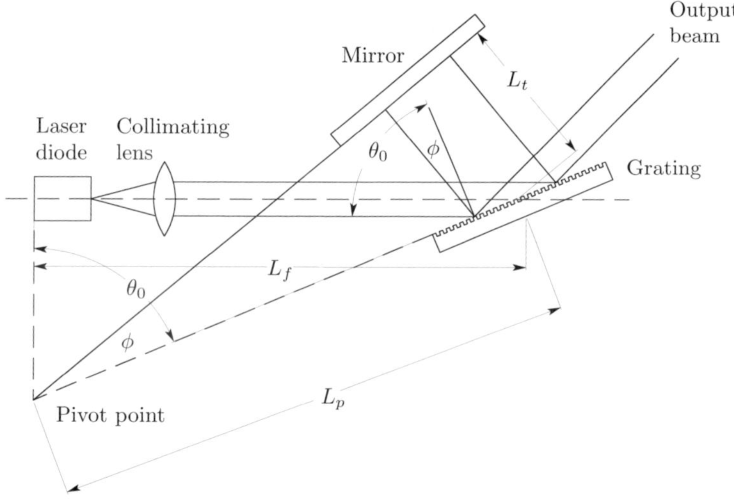



**Fig. 9.23** Geometry used in Littman approach with self-tracking.

which can be rotated. As with the Littrow approach, the output beam is the specular reflection from the grating, whose orientation is fixed. Tuning is accomplished by rotating the mirror (which also changes the cavity length). This scheme was originally used in dye lasers and the pivot point of the mirror was chosen to achieve tracking between cavity modes and the grating frequency. The tracking for the Littman configuration is *exact* over the entire tuning range of the grating. Figure 9.23 illustrates the geometry and defines the distances and angles used in the following analysis. As with the Littrow calculation, we will assume the optical length of the diode is equal to its physical length.

For tracking, the following two wavelengths need to be the same

$$\text{Cavity mode:} \quad \lambda_m = \frac{2}{m} L(\phi) = \frac{2}{m}[L_f + L_t(\phi)] \tag{9.193}$$

$$\text{Grating resonance:} \quad \lambda_r = \frac{a}{p}(\sin\theta_0 + \sin\phi), \tag{9.194}$$

where $a$ is the grating period, $p$ is the grating order, $m$ is the cavity axial order, $L(\phi)$ is the optical path length between the end facet of the diode and the mirror and $\phi$ is the angle between the grating and mirror normals. The first equation is the resonance condition for the cavity and the second equation is the standard grating master relationship. From the figure,

$$L(\phi) = L_f + L_p \sin\phi. \tag{9.195}$$

Thus, we have

$$\frac{2}{m}(L_f + L_p \sin\phi) = \frac{a}{p}(\sin\theta_0 + \sin\phi). \tag{9.196}$$

This single equation implies that tracking will occur if the following two equations are satsified:

$$\frac{2}{m}L_f = \frac{a}{p}\sin\theta_0 \tag{9.197}$$

$$\frac{2}{m}L_p = \frac{a}{p}. \tag{9.198}$$

Thus, $L_f/L_p = \sin\theta_0$, which implies, as the figure indicates, that the pivot point is at the intersection of the mirror and grating planes and directly below the rear facet of the diode.

## 9.9 Semiconductor laser amplifiers and injection locking

Despite the efforts of many semiconductor laser researchers, it has proven to be very difficult to obtain high power (1 W or more) from a diode laser together with a narrow frequency spectrum and good spatial mode. An output power of more than $\approx$100 mW results in optical damage to the output facet of the small area diode lasers described in this chapter. Increasing the width of the active medium from several $\mu$m to 100 $\mu$m or more to allow higher power operation severely degrades the transverse mode of the laser. These wide stripe lasers can produce more than 1 W and are useful for pumping other lasers but are very poor sources of narrowband radiation with a well-behaved transverse mode. The best solution at present is to use a semiconductor *traveling wave amplifier* to increase the power of a narrowband diode to useful levels.

Traveling wave amplifiers are ordinary diode lasers with antireflection coatings at each end. They suffer from the same problems as diode lasers: optical damage unless the cross-sectional area is made fairly large. While a broad-area amplifier will probably not have a problem with transverse mode quality, it is fairly inefficient and has relatively low gain. The best solution is to combine the efficiency of a narrow-area device with the robustness of a broad-area diode and make the active region *flared*. If

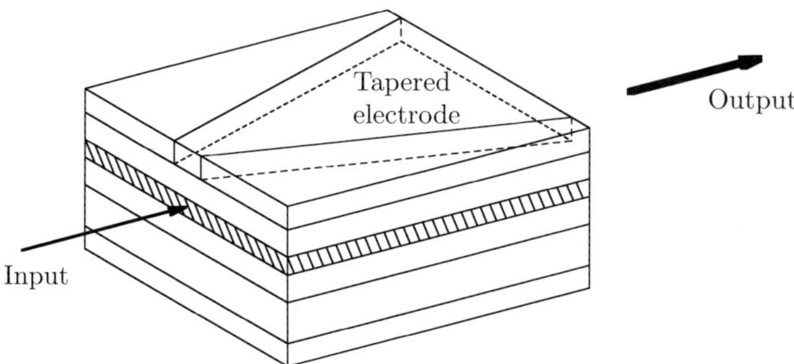



**Fig. 9.24** Gain-guided tapered amplifier. The tapered optical waveguide is defined by the tapered injection current distribution.

the flare is smooth and gradual, the mode will adiabatically fill the area as the wave progresses and the intensity (and therefore electric field) at the output facet will be below the damage threshold even for output powers in the 1 W range. A schematic of a gain-guided flared traveling wave amplifier appears in Fig. 9.24.

The tapered (flared) amplifier will also solve another problem: the power limitation due to saturation of the active medium in a diode with a narrow channel. To see why this is so, we will analyze a general traveling wave amplifier.

The behavior of the intensity as a function of position in a traveling wave amplifier with uniform cross section is described by the equation

$$\frac{dI}{dz} = \left( \frac{g_0}{1 + I/I_s} - a \right) I,$$ (9.199)

where $g_0$ is the unsaturated gain, $I_s$ is the saturation intensity and $a$ is the absorption coefficient due to loss in the active medium. The first term in brackets is just the saturated gain (eqn 5.148). This equation does not have a closed solution. We will examine certain limiting cases by first ignoring the absorption term, $a$. One can *formally* integrate the equation, obtaining

$$\ln \frac{I_{out}}{I_{in}} + \frac{I_{out} - I_{in}}{I_s} = g_0 L,$$ (9.200)

where $L$ is the length of the amplifier. The power gain of the amplifier, $G$, is then

$$G = \frac{I_{out}}{I_{in}} = G_0 \exp \left( -\frac{I_{out} - I_{in}}{I_s} \right) = G_0 \exp \left( -\frac{(G - 1)I_{out}}{GI_s} \right),$$ (9.201)

where $G_0$ is the small-signal gain (in the absence of saturation)

$$G_0 = e^{g_0 L}.$$ (9.202)

If $G_0 \gg 1$, the value of $I_{out}$ which reduces the gain to one-half its small-signal value is

$$I_{out} \approx (\ln 2) I_s.$$ (9.203)

Thus, we see that the gain rapidly diminishes when the intensity approaches the saturation intensity. When $I \gg I_s$ at the input, the gain is unity and the amplifier is essentially *transparent*. This interesting result can be obtained by integrating eqn 9.199 in the limit $I \gg I_s$ and $I \gg g_0 I_s L$. Typically, saturating *powers* in laser diodes are 100 mW or less.

We turn next to the efficiency of the amplifier. The extracted intensity is defined by $I_{out} - I_{in}$ and is given by

$$I_{extr} \equiv I_{out} - I_{in} = I_s \ln \frac{G_0}{G}. \tag{9.204}$$

The *maximum extracted intensity*, $I_{avail}$ (*available intensity*) is given by

$$I_{avail} = I_s \ln G_0, \tag{9.205}$$

since $G = 1$ at high power. If we define the *efficiency*, $\eta$, as $I_{extr}/I_{avail}$,

$$\eta = 1 - \frac{\ln G}{\ln G_0}. \tag{9.206}$$

The last term is just the ratio of the gains expressed in dB units. Thus, to extract half of the power, we lose half of the gain, expressed in dB. We therefore have the conflicting requirements that to obtain the maximum output intensity and efficiency, we need the output to be near the saturating level, but not too near, since the gain suffers when the output begins to saturate the gain medium.

If we include the absorption term in eqn 9.199, we can obtain another implicit formula for $I_{out}/I_{in}$ by integration

$$\ln \frac{I_{out}}{I_{in}} = (g_0 - a)L + \frac{g_0}{a} \ln \left( \frac{g_0 - a(1 + I_{out}/I_s)}{g_0 - a(1 + I_{in}/I_s)} \right). \tag{9.207}$$

From eqn 9.199, one can see that the intensity approaches an asymptote, $I_{max}$, when the expression in brackets approaches zero:

$$I_{max} = (g_0/a - 1)I_s. \tag{9.208}$$

The value of $z$ where the intensity is near maximum is given by $L_{max}$

$$L_{max} \approx \frac{1}{g_0 - a} \ln \left[ \frac{g_0 - a}{a} \frac{I_s}{I_{in}} \right]. \tag{9.209}$$

Thus, there is no point in making the amplifier longer than $L_{max}$. Beyond this length, all of the stimulated photons are absorbed and the traveling wave intensity does not increase.

The above expressions are all written in terms of the *intensity*, which is simply proportional to the power in an amplifier with uniform cross section. The saturation intensity is *fixed* and is given approximately by

$$I_s = \frac{h\nu}{\sigma\tau}, \tag{9.210}$$

for a transition whose cross section is $\sigma$ and whose decay time is $\tau$. The saturating *power* can, however, be changed by changing the cross-sectional area of the active

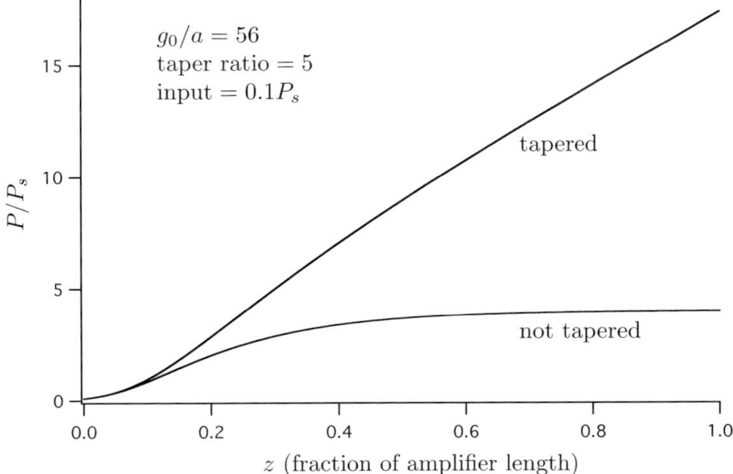


**Fig. 9.25** Plots of power as a function of position in tapered and uniform amplifier (the differential equations were easily integrated using the free computer program, *Octave*). $P_s$ is the saturating power at the amplifier *input*.

medium. If the width of the active medium is linearly tapered, the saturating power is given by

$$P_s = P_{s0}(1 + k_t z),\tag{9.211}$$

where $P_{s0}$ is the saturating power at the input end ($z = 0$) and the taper is described by the constant $k_t$. For example, if the width increases by a factor of 10, $k_t$ would be $9/L$. The equation for the traveling wave amplifier, written in terms of power, is now

$$\frac{dP}{dz} = \left(\frac{g_0}{1 + \frac{P}{P_{s0}(1+k_t z)}} - a\right)P.\tag{9.212}$$

It should come as no surprise that the taper removes many of the limitations on the amplifier gain due to saturation. Plots of the growth of the power of the traveling wave for both a tapered and a uniform amplifier appear in Fig. 9.25. The input power is $0.1 \times P_s$ and both amplifiers show the expected exponential growth until the power approaches $P_s$. Then, the uniform amplifier power ceases to grow while the tapered amplifier power grows in a linear fashion allowed by the linear taper in the amplifier width.

A traveling wave amplifier operates *below threshold* for laser oscillation, since the facets are antireflection coated. It is also possible to use an *above-threshold* laser oscillator to coherently increase the power of a low-power laser. The latter approach is called *injection locking*.

Injection locking works by sending a relatively weak laser beam, the *master beam*, into an already oscillating laser which is emitting the *slave beam*. The frequency difference between the two beams is normally fairly small and the master beam will be amplified by the gain medium in the slave laser. When the amplified master intensity is above the slave laser intensity, the master beam can *take over* the slave laser and the latter's emission will be at the exact master frequency. This phenomenon is called *injection locking* and the *locking bandwidth* is the maximum frequency difference between the two lasers which will result in injection locking; it depends upon the powers of the two lasers and the mode separation in the slave laser. We will derive the locking bandwidth using the rate equations for the laser field and phase.

We will first write down the time-dependence of the two fields:

$$\text{Slave laser:} \quad E_{slave}(t) = E_s(t)e^{i\omega_s t + \phi(t)} \tag{9.213}$$

$$\text{Master laser:} \quad E_{master}(t) = E_i e^{i\omega_m t}. \tag{9.214}$$

The rate equations, including the injected field, are derived from eqns 9.98 and 9.104:

$$\frac{dE_s}{dt} - \tfrac{1}{2}(\Gamma R - 1/\tau_p)E_s = \eta f_d E_i \cos(\Delta(t)) \tag{9.215}$$

$$\frac{d\phi}{dt} - (\omega_s - \omega_{th}) = \eta f_d (E_i/E_s)\sin(\Delta(t)). \tag{9.216}$$

Note that the first equation is lacking the $\omega_s - \omega_{th}$ term, which is present in the rate equation derived earlier in the chapter. This is due to the definition of $E_s(t)$ in eqn 9.213: the rate equations derived earlier defined $E_s$ by factoring out the time dependence $e^{i\omega_{th}t}$ from the electric field. The above definition factors out $e^{i\omega_s t}$; thus, the $\omega_s - \omega_{th}$ term is not needed.

The right-hand terms have the following rationale. The *geometrical* relations are illustrated in Fig. 9.26. First, the injected field adds *in phase* to $E_s$ and *in quadrature* to $\phi$. Thus, the injected field increases the internal *field* by $E_i \cos(\Delta(t))$ and the *phase* by $(E_i/E_s)\sin(\Delta(t))$, where $\Delta(t)$ is the instantaneous phase angle between $E_i$ and $E_s$ and is given by

$$\Delta(t) = \Delta\omega t - \phi(t). \tag{9.217}$$

The quantity $\Delta\omega = \omega_m - \omega_s$ is the difference between the two laser frequencies. The *coupling constant* is $\eta f_d$, where $f_d = 1/t_{rt}$ is the mode spacing ($t_{rt}$ is the *round trip time*) and $\eta$ is the *coupling efficiency* into the slave laser, including mode matching and transmission efficiencies. The internal field, $E_s$, bounces back and forth at the rate $f_d$ and therefore the *rate of change* of $E_s$ and $\phi$ due to the injected field is equal to the rate at which the internal field presents itself to the injected field ($f_d$) times the increment in the field and phase due to the injection.

If $\Delta N$ is the increment of the carrier density over threshold, then

$$\omega_s = \omega_{th} + \tfrac{1}{2}\alpha v_g g_N \Delta N \tag{9.218}$$

$$v_g g(N) = v_g(g_{th} + g_N \Delta N) = 1/\tau_p + v_g g_N \Delta N = \Gamma R, \tag{9.219}$$

where we used eqn 9.109, and the rate equations are now

$$\frac{dE_s}{dt} = \tfrac{1}{2}v_g g_N \Delta N E_s + \eta f_d E_i \cos \Delta(t) \tag{9.220}$$

$$\frac{d\phi}{dt} = \tfrac{1}{2}\alpha v_g g_N \Delta N + \eta f_d (E_i/E_s)\sin \Delta(t), \tag{9.221}$$

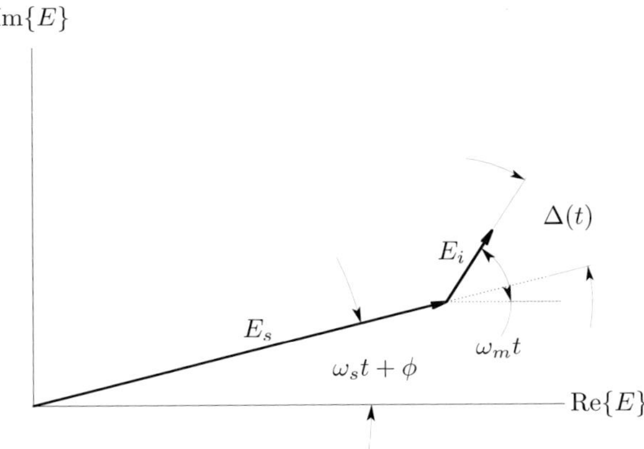


**Fig. 9.26** Phasors of slave laser field envelope $(E_s)$ and injected field amplitude $(E_i)$ justifying $E_i \cos \Delta$ for the field increment and $(E_i/E_s) \sin \Delta$ for the phase increment. These increments become *rates* after multiplication by $f_d$.

If $\phi_L$ is the (constant) phase difference between the lasers when the slave is *locked* to the master, then

$$\phi(t) = \Delta \omega t + \phi_L. \tag{9.222}$$

There is a subtle difference between this $\Delta \omega$ and the previous one. This one is the frequency change of the slave laser from its free-running (no injection) value to its locked frequency. Since the two $\Delta \omega$ values are *numerically* the same, from eqn 9.217 we obtain

$$\Delta(t) = -\phi_L. \tag{9.223}$$

The steady-state solution to the first rate equation is obtained by setting $dE_s/dt = 0$, from which we can obtain

$$\Delta N = -\frac{2\eta f_d}{v_g g_N} \frac{E_i}{E_s} \cos \phi_L. \tag{9.224}$$

From the second rate equation and eqns 9.222 and 9.224, we obtain

$$\Delta \omega = -\eta f_d \frac{E_i}{E_s} [\sin \phi_L + \alpha \cos \phi_L]. \tag{9.225}$$

Setting $\eta = 1$ and combining the trigonometric functions,

$$\Delta \omega = -f_d \frac{E_i}{E_s} \sqrt{1 + \alpha^2} \sin(\phi_L + \arctan \alpha). \tag{9.226}$$

Since the sine function is constrained to be in the interval $[-1, 1]$, the *locking half bandwidth* is

$$\Delta \omega_{locking} = \frac{c}{2L\mu} \frac{E_i}{E_s} \sqrt{1 + \alpha^2}. \tag{9.227}$$

We notice the characteristic $\sqrt{1 + \alpha^2}$ factor in the expression for the bandwidth. This factor would not have appeared if we had used the general result for the injection locking bandwidth found in a number of books and papers on the subject. On the other hand, part of this enhanced bandwidth might be unstable, as discussed in a number of papers (Mogensen (1985), for example).

## 9.10  Miscellaneous characteristics of semiconductor lasers

**Table 9.2** Wavelengths and materials for currently available diode lasers.

| Material | Wavelength range | Notes |
| --- | --- | --- |
| AlGaN | 350–400 nm | Developmental |
| GaInN | 375–440 nm | Mostly at $\approx 400$ nm |
| AlGaInP/GaAs | 620–680 nm | |
| $Ga_{0.5}In_{0.5}P$/GaAs | 670–680 nm | |
| GaAlAs/GaAs | 750–900 nm | |
| GaAs/GAs | 904 nm | |
| InGaAs/GaAs | 915–1050 nm | Strained layer |
| InGaAsP/InP | 1100–1650 nm | |

We will conclude this chapter with a discussion of some of the laser diode properties which have not been mentioned earlier in this and the preceding chapter. First, the laser emission wavelength is, of course, material dependent. We briefly mentioned some binary, ternary and quaternary compounds which are used in diode lasers. Table 9.2 lists the wavelengths and materials of some currently available semiconductor lasers.

We have discussed the spectral characteristics of semiconductor lasers, but have not mentioned the far field pattern, which is usually elliptical. If the elliptical spot at the output of an *index-guided* laser has semi-major axes of length $w_x$ and $w_y$, the angular divergences in the $x$ and $y$ planes are

$$\theta_{x,y} = \sin^{-1}\left(\frac{\lambda}{w_{x,y}\pi}\right).$$

(9.228)

We specified an index-guided laser since the output facet of such a laser can be considered as a source of a spherical wave whose center is the center of the facet but whose divergence angles are different in the two orthogonal planes. On the other hand, the source of the spherical wave for the $x$-plane is displaced from that for the $y$-plane in a *gain-guided* laser and therefore the beam is astigmatic.

In order to obtain a *circular beam* from a diode laser, one must use a corrective optical element which has a different magnification in the $x$- and $y$-planes: this is called an *anamorphic* optical system. There are two commonly used approaches. One can use one or more cylindrical lenses to individually adjust the magnification in the two planes. A system with two cylindrical lenses appears in Fig. 9.27. If the focal points of *both* lenses are placed at the output facet (for an index-guided laser) and

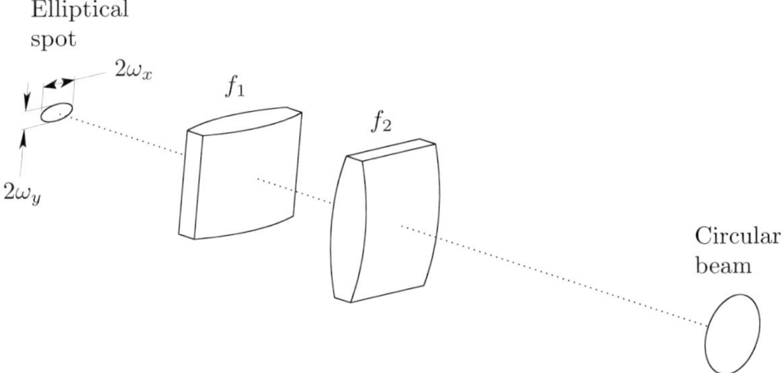


**Fig. 9.27** Use of a pair of cylindrical lenses to make a circular beam from an elliptical spot generated by a diode laser.

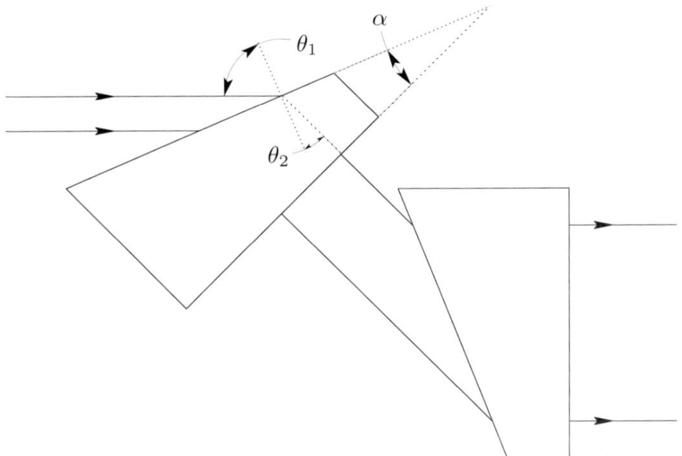


**Fig. 9.28** Use of a pair of anamorphic prisms to magnify a beam in one plane.

the focal lengths are $f_1$ and $f_2$, the different magnifications will generate a circular collimated beam when

$$\frac{f_1}{f_2} = \frac{w_y}{w_x}. \tag{9.229}$$

The advantage of this approach is that the beam axis is not changed and small changes can be made by changes in the separation between the lenses. A possible disadvantage is due to the greater difficulty in manufacturing cylindrical lenses with $\lambda/4$ surface figures.

One can also use a pair of prisms to expand the beam along one plane. The common approach is to use *anamorphic prisms* in the configuration shown in Fig. 9.28. The two prisms are identical and are cut so that the beam enters each prism at Brewster's angle and exits normal to the output surface. If $L$ is the distance along the prism surface between the two input rays, then it is easy to see that

$$w_1 = L\cos\theta_1 \tag{9.230}$$

$$w_2 = L\cos\theta_2, \tag{9.231}$$

where $w_1$ and $w_2$ are the incident and transmitted beam widths for each prism. Dividing these equations, the magnification $M$ is

$$M = \frac{w_2}{w_1} = \frac{\cos\theta_2}{\cos\theta_1}. \tag{9.232}$$

At Brewster's angle, $\theta_1 = \tan^{-1} n$, where $n$ is the index of refraction of the prism material. Using this expression in the above equation, one can easily show that

$$\text{One prism at Brewster's angle:} \quad M = n. \tag{9.233}$$

From the requirement that the beam exit normal to the prism surface, it is easy to see that $\alpha = \theta_2$ and

$$\alpha = \tan^{-1}\frac{1}{n}. \tag{9.234}$$

The magnification for two prisms is $M^2 = n^2$. One can *tune* the magnification to some extent by rotating both prisms, though one will sacrifice either the Brewster incident angle or the normal exit beam. This system displaces the beam and only works well for one polarization: the orthogonal polarization suffers great Fresnel loss at the near grazing incidence to the prism input surfaces.

## 9.11   Further reading

The three types of semiconductor laser in order of development (homojunction, double heterostructure and quantum well) are well described in the books by Yariv (1989) Svelto (2004), Milonni (1988) (except for the quantum well), Verdeyen (1995) and Suhara (2004). The distributed feedback laser is discussed in the classic paper by Kogelnik (1972) and in the books by Yariv, Verdeyen and Svelto. One of the best and most complete discussions of diode laser modulation and noise is in the book by Petermann (1991). Our discussion of the linewidth of semiconductor lasers is based upon the classic treatment by Henry (1982). Petermann's book also contains a good treatment of the effect of optical feedback and an early discussion of external cavity diode lasers. Our discussion of the various feedback regions is based upon the paper by Tkach and Chraplyvy (1986). The treatment of *tracking* using the Littrow configuration is in the paper by Labachelerie (1993) and the analogous treatment using the Littman configuration is in the paper by Lui (1981). A complete discussion of flared amplifiers is in the paper by Lang (1993) and our analysis of injection locked diode lasers is based upon the paper by Petitbon (1988). The book by Siegman (1986) also contains some useful material on both laser amplifiers and injection locking.

## 9.12   Problems

(9.1) A diode laser which uses GaAs as the active medium has a cavity length of 300 $\mu$m (uncoated facets). Calculate the mode spacing and the Schawlow–Townes linewidth assuming an anti-guiding factor equal to 5.

(9.2) A double heterostructure diode laser has an active layer "normalized" thickness, $D$, given by

$$D = 2\pi(n_{act}^2 - n_{clad}^2)^{1/2}\frac{d}{\lambda} \qquad \begin{aligned} n_{act} &= \text{active index} \\ n_{clad} &= \text{cladding index} \end{aligned} \qquad (9.235)$$

and a confinement parameter, $\Gamma$, which is approximately

$$\Gamma \approx \frac{D^2}{2 + D^2}, \qquad (9.236)$$

where $d$ is the *physical* thickness of the active layer. Using these expressions (and others) determine the approximate value of $d$ which minimizes the threshold in GaAs at 850 nm. (Assume that $D \ll 2$, the facets are uncoated, bulk absorption can be ignored and the cladding index is 3.4.)

(9.3) Calculate the output power of a diode laser using the following approach. If $\eta$ is the *quantum efficiency* (ratio of number of internal stimulated photons to number of injected electrons), first calculate the number of stimulated photons as a function of $\eta$, the injected current, $I$, and the *threshold* current, $I_{th}$. Next, convert this to an *internal* power (using the frequency, $\nu$) and finally determine the output power by multiplying by the ratio of the output coupling to the total internal plus external loss. Using this expression, determine the *slope efficiency* $(dP/dI)$.

(9.4) Assume that a laser amplifier is based upon the quantum well structure described in the text. If the bulk absorption is 10 cm$^{-1}$, what is the longest *practical* length of an amplifier before the overall gain is limited by saturation?

# 10

# Mode-locked lasers and frequency metrology

## 10.1   Introduction

All of the lasers discussed so far have emitted continuous wave (CW) radiation. Despite a promise in the Preface that we would consider CW lasers exclusively, there is one type of laser which has aspects of both a CW laser and a pulsed laser and has recently become so important to the frequency metrology community that we would be remiss in not describing it in some detail. This is the *mode-locked laser*, which is capable of generating such extremely short pulses that in some realizations it is called a *femtosecond* laser, implying that it emits pulses containing only *a few cycles of optical radiation*. We will present the theory of mode-locked lasers, describe a common implementation (the mode-locked Ti-sapphire laser) and discuss the application of these lasers in frequency metrology. The latter application led to the Nobel Prize in Physics for John L. Hall and Theodor W. Hänsch in 2005.

## 10.2   Theory of mode locking

Our treatment of mode locking will not be very rigorous; for more rigor, one can refer to a large number of excellent books and papers on the subject, some of which are mentioned at the end of this chapter. Initially, we will assume there is no *dispersion* in the laser cavity: the propagation speed of a wave packet (group velocity) is the same as the phase velocity. We begin by making the (somewhat unrealistic) simplifying assumption that the spectrum of an inhomogeneously broadened laser consists of a series of lines of constant amplitude spaced by the cavity free spectral range, $\nu_c$. If there are $N$ lines, the total width, $\Delta\nu_{inhom.}$, is

$$\Delta\nu_{inhom.} = (N - 1)\nu_c. \tag{10.1}$$

Figure 10.1 displays the rectangular spectrum of this laser. If the center frequency is $\omega_0$, the total electric field is

$$E_{total} = E_0 \sum_{-(N-1)/2}^{+(N-1)/2} \exp[i(\omega_0 + n\omega_c)t + i\phi_n(t)], \tag{10.2}$$

where $\phi_n(t)$ is a random function of time uniformly distributed between 0 and $2\pi$ and $E_0$ is the field for each line. It is easy to see that the time-averaged intensity is

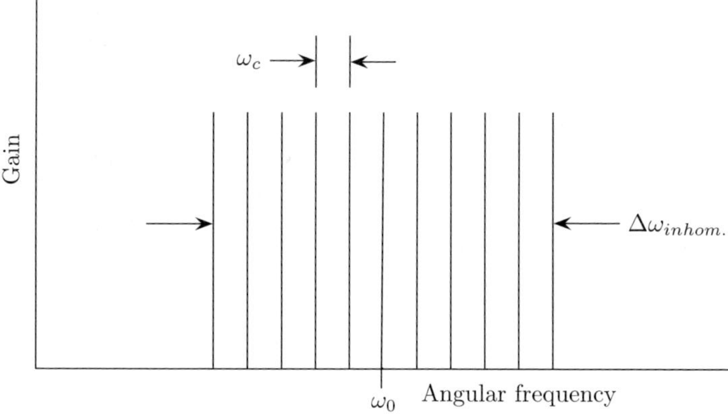


**Fig. 10.1** Simplified rectangular spectrum of inhomogeneously broadened laser.

$$\langle I \rangle = N I_0 \qquad I_0 = \text{intensity of each line,} \tag{10.3}$$

since the cross-terms average to zero when one takes the square modulus of the field. The behavior with time of the intensity would be extremely noisy due to the random phases (left side of Fig. 10.2), though any laboratory measurement of the intensity would average for a sufficient length of time to remove most of the noise. Note that the noise is entirely due to the *cross-terms* when one takes the square modulus of the field: it is a random *interference* among the laser modes. Now let us assume that we have a mechanism for *locking* the phases of all of the lines together, so that we may take $\phi_n(t)$ out of the sum. The total field is then

$$E_{total} = E_0 e^{i\omega_0 t + i\phi(t)} \sum_{-(N-1)/2}^{+(N-1)/2} e^{in\omega_c t} = E_0 e^{i\omega_0 t + i\phi(t)} \frac{\sin N\omega_c t/2}{\sin \omega_c t/2} \tag{10.4}$$

and the intensity is

$$I = I_0 \left[ \frac{\sin N\omega_c t/2}{\sin \omega_c t/2} \right]^2, \tag{10.5}$$

where $I_0$ is again the intensity in each line. Using l'Hôpital's rule, the peak intensity of this function is

$$I_{peak} = N^2 I_0. \tag{10.6}$$

The curve of intensity versus time when the modes are locked is shown on the right side of Fig. 10.2. The temporal width, $\Delta t_p$, of each pulse is roughly

$$\Delta t_p \approx \frac{1}{N\nu_c} \qquad \nu_c = \omega_c/2\pi. \tag{10.7}$$

We therefore conclude that this process increases the peak power for each pulse over the CW laser average power by a factor of $N$ and generates pulses with a duty cycle $\nu_c \Delta t_p = 1/N$. The average power is unchanged over that of the CW laser but the peak

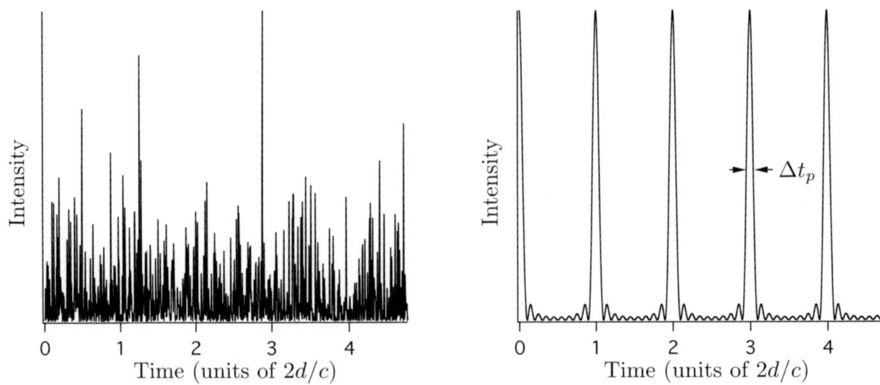

**Fig. 10.2** Intensity versus time for laser with rectangular frequency spectrum and 10 modes. On the left the modes have random phase and on the right the phase is constant. The time scale is in units of the round-trip time $(2d/c)$ which is about 10 ns. Thus, the noise on the left would be normally averaged out with most laboratory detectors.

power can be extremely high when $\Delta t_p$ is in the femtosecond ($10^{-15}$ $s$) range. Finally, the pulse width is very simply

$$\Delta t_p \approx \frac{1}{\Delta\nu_{inhom.}} \qquad \Delta\nu_{inhom.} = \Delta\omega_{inhom.}/2\pi. \qquad (10.8)$$

The reader should recognize eqn 10.5 as essentially the same function as that describing the interference pattern from $N$ slits uniformly illuminated by coherent light (except that the $N$-slit function depends upon a spatial coordinate). We have on several occasions used a temporal analogy to help illustrate a spatial behavior; we now reverse this and use a familiar spatial situation ($N$-slit interference) to help explain a temporal one. The important point is that the $N$ waves interfere with one another to produce noise if they have random phases (equivalent to illuminating the slits with *incoherent light*) and a regular pattern of sharp lines if the phases are locked together (equivalent to using *coherent* light to illuminate the slits).

To illustrate how the phases can be locked together, we use a conceptual diagram of a mode-locked laser (Fig. 10.3). All mode-locked lasers use some variation of this technique. The key element which distinguishes a mode-locked laser from a simple CW laser is an intracavity *shutter* which is arranged to cyclically open and close with a period exactly equal to the round-trip time of the intracavity wave. The only field distribution which can survive in such a laser is a regular series of pulses with a repetition rate equal to the shutter cycling frequency (which is also equal to the free spectral range of the cavity and therefore to its mode separation). One can work backwards (using Fourier theory) and show that the only field frequency dependence which can generate these regular pulses is a series of uniformly spaced modes whose separation is the cavity free spectral range. The amplitudes might be different from the ones we have used, but the phases of all the components are locked together. We should mention that the phases need not be all identical to obtain mode locking; they can also depend linearly on the summation index, $n$. Finally, judicious placement of

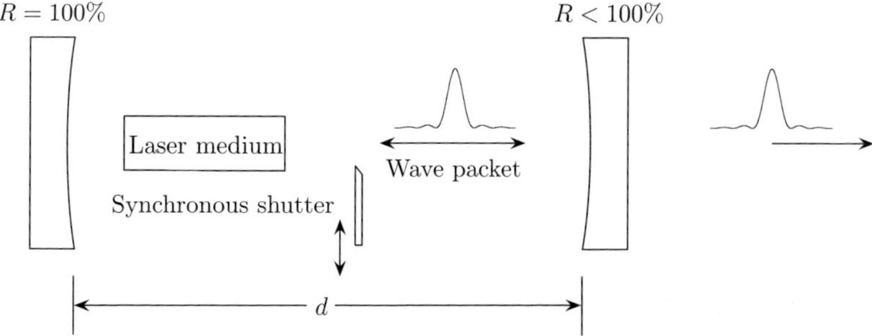



**Fig. 10.3** Conceptual diagram of mode-locked laser.

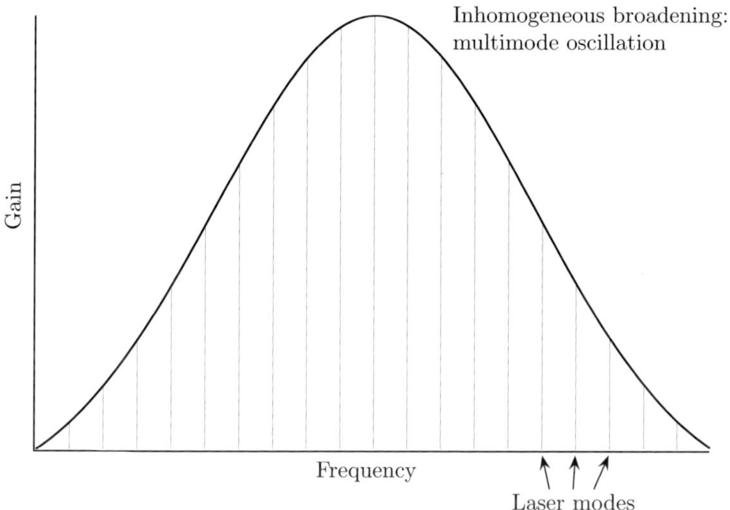



**Fig. 10.4** Inhomogeneously broadened laser modes (Gaussian envelope).

the shutter in the cavity allows mode locking at *harmonics* of the mode spacing $(c/2d)$; this is called *harmonic mode locking*.

A much more realistic mode distribution in an inhomogeneously broadened laser is shown in Fig. 10.4, where the envelope of modes is a Gaussian. The electric field at the $n^{th}$ mode is

$$E_n = E_0 \exp\left[-2\ln 2 \left(\frac{n\omega_c}{\Delta\omega}\right)^2\right], \tag{10.9}$$

where $E_0$ is the field at the peak of the Gaussian and $\Delta\omega$ is the inhomogeneous width in angular units. In the time domain, the field is

$$E(t) = \sum_{-(N-1)/2}^{+(N-1)/2} E_n e^{i\omega_0 t + in\omega_c t}$$

$$= E_0 e^{i\omega_0 t} \sum_{-(N-1)/2}^{+(N-1)/2} \left\{ \exp\left[ -2\ln 2 \left( \frac{n\omega_c}{\Delta\omega} \right)^2 \right] \right\} e^{in\omega_c t}. \tag{10.10}$$

This series can be summed when $N$ is very large by converting it into an integral using the following substitutions

$$x = n\omega_c \tag{10.11}$$

$$dx = (n+1/2)\omega_c - (n-1/2)\omega_c = \omega_c. \tag{10.12}$$

The field is then

$$E(t) = \frac{E_0 e^{i\omega_0 t}}{\omega_c} \int_{-\infty}^{\infty} \left\{ \exp\left[ -\frac{2(\ln 2)x^2}{(\Delta\omega)^2} \right] \right\} e^{ixt}\, dx. \tag{10.13}$$

(We introduced the $dx$ factor by multiplying the exponential by $dx/\omega_c = 1$.) The integral should be readily recognized as the Fourier transform of the expression in the curly brackets, which is a Gaussian. The transform is also a Gaussian,

$$E(t) = E_0 e^{i\omega_0 t} \left( \frac{\pi}{2\ln 2} \right)^{1/2} \left( \frac{\Delta\omega}{\omega_c} \right) \exp\left[ -\left( \frac{\Delta\omega t}{2\sqrt{\ln 2}} \right)^2 \right]. \tag{10.14}$$

The intensity is

$$I(t) = I_0 \exp\left[ -2\left( \frac{\Delta\omega t}{2\sqrt{\ln 2}} \right)^2 \right], \tag{10.15}$$

where $I_0$ is the peak intensity. This is a Gaussian-shaped pulse of width

$$\Delta t_p = \frac{2\ln 2}{\pi}\frac{1}{\Delta\nu} = \frac{0.44}{\Delta\nu} \qquad \Delta\nu = \Delta\omega/2\pi. \tag{10.16}$$

This equation only describes a *single* pulse since the periodic structure was lost when we converted the sum to an integral: the mode spacing approached zero in the frequency domain while the pulse separation became infinite in the time domain.

We began our theoretical analysis with an inhomogeneously broadened laser to avoid a possible paradox: since a homogeneously broadened laser only oscillates on a *single mode*, one might conclude (erroneously) that it cannot be mode locked. Here the frequency domain picture leads us astray: viewed from the time domain, it should be clear that the use of a shutter allows only a repetitive pulse emission regardless of the type of broadening in the amplifying medium. A Fourier analysis of the laser field would give the same result as for an inhomogeneously broadened laser. The resolution of the paradox lies in the behavior of the shutter. The shutter, interrupting the gain at the rate $c/2d$, actually *imposes the needed mode structure* on the homogeneously broadened laser, allowing it to behave like the inhomogeneously broadened laser described above.

Thus, the results for a homogeneously broadened laser are essentially the same as for an inhomogeneously broadended laser and the former laser type can be mode locked using the same techniques as are used in the latter. Since the modulation in the homogeneously broadened laser serves two purposes (imposing a mode structure and coupling the modes together), a somewhat larger modulation amplitude is needed to achieve mode locking than in an inhomogeneously broadened laser.

## 10.3  Mode locking techniques

Mode locking is implemented in two general ways. *Active mode locking* uses an externally operated shutter which opens and closes at the mode spacing frequency, as was suggested in the conceptual model introduced above. A more common set of techniques, called *passive mode locking*, uses the fields themselves to operate an internal "shutter" which is based on some nonlinear optical effect (such as saturation) in the intracavity medium.

The shutter in an actively mode locked laser is usually either an amplitude or phase modulator placed inside the cavity. A third type of shutter which is externally operated is the *pump laser*: a mode-locked pump laser with the same mode spacing (and cavity length) will amplitude modulate the gain of the pumped laser in just the right manner to achieve mode locking. This is occasionally called *synchronous pumping* for obvious reasons.

When a modulator is used in a standing wave cavity, it is advantageous to place it close to one of the laser mirrors. The reason for this is to prevent other pulse configurations from developing; these can occur if the pulse passes the shutter *twice* per round trip with similar delays between each pass. If the shutter is near the mirror, the two shutter passes will occur with very different delays. In a ring laser, the shutter placement is not critical. There is one situation where a shutter is deliberately placed halfway between the mirrors in a standing wave cavity. Such a configuration can support two counter-propagating pulses and is called the *colliding pulse* scheme. These lasers can actually generate very short pulses.

The theory of amplitude modulated mode locking is essentially the same as that presented in the conceptual model. The modulator provides a synchronism between adjacent modes and ensures that the only internal field distribution supported by the laser is a narrow pulse which passes through the modulator when it allows the maximum transmission. Amplitude modulated mode locking is usually accomplished with an intracavity acousto-optic modulator (AOM). The modulator is a bit different from that used outside of a laser: the acoustical termination on the far side of the crystal from the transducer is removed so that there are acoustical standing waves inside the modulator medium. The loss from the modulator is greatest when the acoustical standing wave amplitude is greatest and this happens at twice the modulation frequency. Thus, the modulator is driven at one half of the mode separation frequency. Figure 10.5 is a schematic of an AM mode-locked laser.

Frequency modulation (FM) mode locking works on a somewhat different principle from AM mode locking. A phase modulator (usually an electro-optic modulator or Pockel cell) is inserted into the cavity very near to one of the cavity mirrors. Phase modulation is equivalent to a modulation of the optical path length in the cavity. A

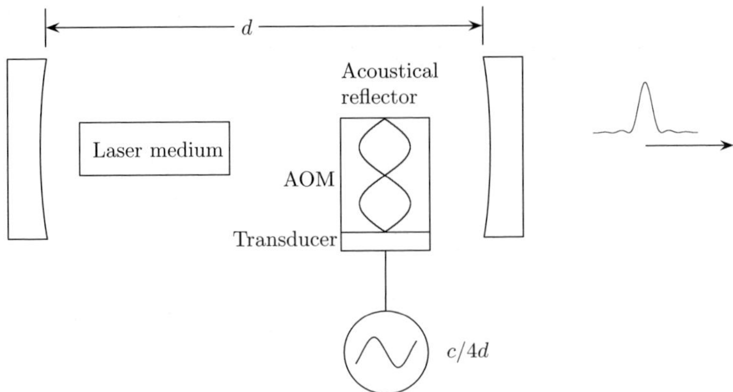



**Fig. 10.5** Mode locking using an acousto-optic modulator driven at one half the cavity mode spacing ($c/4d$). The acoustical standing waves in the AOM are shown.

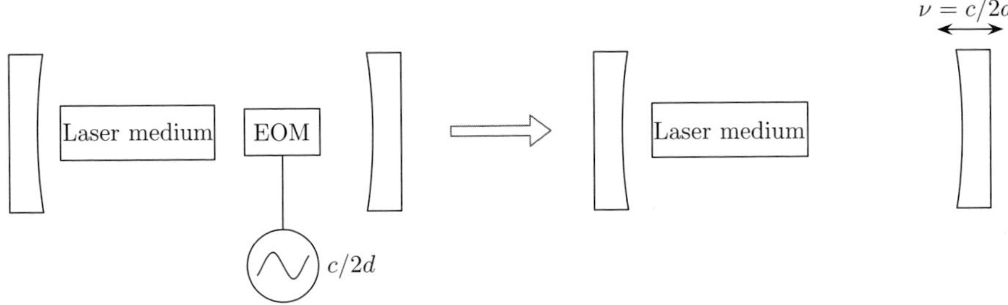



**Fig. 10.6** FM mode locking using an electro-optic modulator. The equivalent oscillating mirror model is shown on the right.

useful model for this behavior is to consider one of the cavity mirrors as moving back and forth at the modulation frequency. A wave packet incident on the moving mirror will be Doppler shifted after reflection by an amount which depends upon the speed of the mirror when the wave reaches it. Thus, such a wave will be subject to a succession of Doppler shifts which will cumulatively change the frequency of the intracavity wave by an amount which is sufficient to prevent laser oscillation. On the other hand, if the wave impinges on the mirror when it is reversing direction, there will be no Doppler shift and laser oscillation can take place. A schematic of an FM mode-locked laser together with the equivalent *moving mirror* model appears in Fig. 10.6.

FM mode locking is *bistable*: there are two possible times in the mirror motion (at the two turning points) which can lead to laser oscillation, and the laser can jump between these two possible modes leading to an irregular or noisy pulse train. The pulses are also slightly frequency chirped since the effective mirror is never completely at rest. These are two disadvantages of FM mode locking and it is employed somewhat less frequently than AM active mode locking.

The most common mode locking scheme exploits some nonlinear behavior of the intracavity medium and is called *passive mode locking*. We will discuss two passive

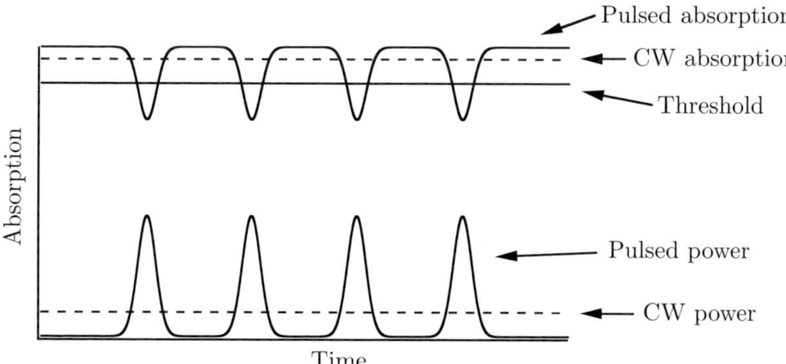



**Fig. 10.7** Behavior of saturating medium in laser cavity. The dotted lines are for CW operation with the same average power as pulsed. Only the pulsed operation dips below threshold absorption.

approaches: the use of a *saturable absorber* and the *Kerr-lens* technique. These schemes are simpler to implement than active schemes and are routinely generating very short (femtosecond) pulses.

A saturable absorber is one whose absorption, due to saturation, *decreases* with increasing intensity. We will present a simplified theory of mode locking using a saturable absorber (from Garmire (1967)). We will assume without proof that a saturable absorber results in mode-locked behavior. The motivation for this is suggested by Fig. 10.7, which plots the pulse power and absorption of the medium as a function of time and shows that a CW wave having the same average power (area under the curve) as the pulses will be below threshold and not oscillate but the pulsed wave will be above threshold near the peak of the pulses and the laser will therefore oscillate at those times.

Our calculation will determine the dependence of the fractional absorption of the saturating medium on the pulse width, $t_p$. We assume that each pulse has $E$ photons per unit area and therefore an average flux per pulse of $E/t_p$. For simplicity, the pulse time dependence is assumed to be *rectangular*, of width $t_p$. The saturating medium will be modeled as a two-level system with level population densities $N_2$ and $N_1$. The treatment is very similar to the derivation of laser gain in Chapter 5. The population densities satisfy a rate equation at each point in space

$$\frac{dN_1}{dt} = R(N_2 - N_1) + N_2/\tau, \tag{10.17}$$

where $R$ is the stimulated rate and $\tau$ is the lifetime of level 2. We will assume that the wave propagates along the $z$-axis. The loss in photon flux per unit length is equal to the number of absorptions per unit time per unit volume,

$$\frac{d}{dz}\left(\frac{E}{t_p}\right) = -R(N_1 - N_2). \tag{10.18}$$

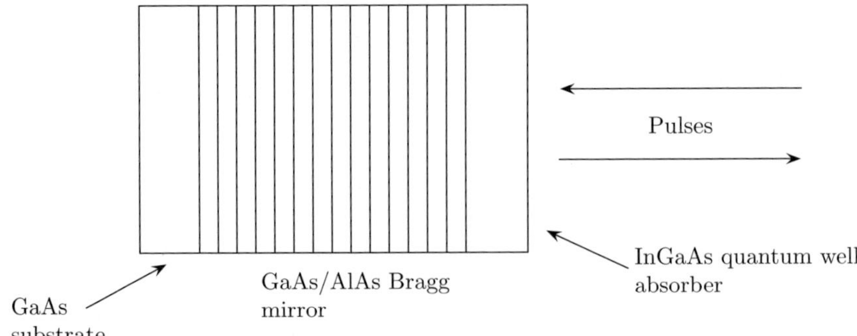


**Fig. 10.8** Schematic of semiconductor saturable absorber mirror (SESAM).

We assume that all of the molecules (the saturating medium is usually a molecular dye) are in the ground state at $t = 0$ and the initial population is $N_{10}$. The above equations can be readily solved with these initial conditions. The solution provides the fractional absorption per unit length, $\alpha$

$$\alpha = \frac{N_{10}\sigma s t_p}{2\sigma E + s t_p} + \frac{2N_{10}\sigma^2 E}{(2\sigma E + s t_p)^2}[1 - e^{-(2\sigma E + s t_p)}], \qquad (10.19)$$

where $s = 1/\tau$ and $\sigma = R t_p / E$. Two conclusions can be drawn from this result. First, $\alpha$ is a *monotonically decreasing* function of $1/t_p$: as the pulses get shorter the loss *always* decreases. This is the reason that passive mode locking favors very short pulses and is therefore the scheme invariably used in femtosecond lasers. Second, the loss is nearly independent of $t_p$ when $s t_p < 1$: there is little improvement in transmission of the molecules when the pulse width is smaller than the molecular lifetime. Thus, the pulse width will tend to approach the molecular lifetime, which is often some tens of femtoseconds. Of course, this assumes that the bandwidth, $\Delta\nu$, of the gain medium is sufficient to amplify all of the modes which make up this short pulse: $\Delta\nu \gg s$.

Although historically the saturable medium has often been a molecular dye, another effective saturable medium is a *semiconductor*. It is often combined with one of the cavity mirrors to form a *semiconductor saturable absorber mirror* (SESAM). A schematic of a SESAM is shown in Fig. 10.8. The mirror is constructed on a GaAs substrate. The reflection is due to Bragg reflection from a periodic structure consisting of alternating layers of GaAs and AlAs. The saturable region faces the intracavity radiation and consists of an InGaAs quantum well absorber whose band gap is tuned to absorb radiation at the desired wavelength (usually around 1 $\mu$m). The region between the Bragg mirror and the front surface forms an optical cavity which can adversely affect the functioning of the device; the device is usually designed to be *antiresonant* at the desired wavelength (since the absorber can be very thin – as little as 10 nm thick – the free spectral range of this cavity is very large). The time behavior of the absorber shows a very rapid saturation (in less than 1 ps) followed by a longer recovery (typically several picoseconds to hundreds of picoseconds). The advantages of these mirrors is their simplicity; their disadvantage is their possibly insufficient bandwidth for the shortest pulses.

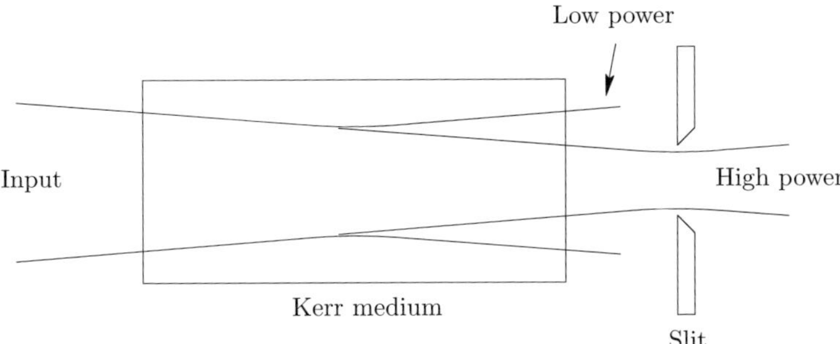


**Fig. 10.9** The Kerr effect combined with a slit can be used to favor high intensities and therefore pulsed behavior.

Kerr-lens mode locking exploits the *optical Kerr effect*, which is the second order electro-optic effect. Thus, the refractive index is proportional to the electric field squared (the intensity, $I$):

$$\text{Optical Kerr effect:} \quad n = n_0 + n_2 I. \tag{10.20}$$

Since this effect is *even* in the electric field, most media display a Kerr effect (unlike the first-order Pockel effect, which requires a non-centrosymmetric medium). The radial intensity distribution of a Gaussian beam is

$$I = I_0 e^{-2r^2/\omega_0^2}. \tag{10.21}$$

For a medium of length $L$, the overall phase shift, $\phi$, is

$$\phi = \frac{2\pi n L}{\lambda}. \tag{10.22}$$

Thus, the change, $d\phi$, in the phase shift due to the intensity distribution is

$$d\phi = \frac{2\pi L}{\lambda} dn = \frac{2\pi L n_2 I_0}{\lambda} e^{-2r^2/\omega_0^2} \approx \frac{2\pi L n_2 I_0}{\lambda}(1 - 2r^2/\omega_0^2) \tag{10.23}$$

near the center of the beam ($r \ll \omega_0$). This *quadratic* dependence of the phase shift with radial distance is characteristic of a *positive lens*. Since the strength of the lens is proportional to the intensity, intense beams will be focused and weak beams will not. By placing a slit downstream from the Kerr medium, one can cause the loss for weak (and therefore non-focused) beams to be much greater than that for strong (focused) beams. This has the same effect as a saturable medium in favoring a pulsed time-dependence instead of a continuous one. Since the Kerr effect is often not sufficiently discriminatory, the optical cavity is frequently designed to be *close to unstability* to increase the effectiveness of the Kerr effect. Figure 10.9 illustrates the use of the Kerr effect for mode locking.

## 10.4  Dispersion and its compensation

In optics, *dispersion* refers to the dependence of the index of refraction, $n(\omega)$, on the frequency, $\omega$. The term is usually only employed when there is a non-trivial dependence:

$n \neq$ constant. Thus, in a dispersive system, the wave vector, $k$, will no longer have the simple linear dependence on $\omega$ ($k = n\omega/c$) that a non-dispersive system has. Until now, we have only been tangentially interested in dispersion and have only mentioned it when discussing laser frequency pulling and in semiconductor diode lasers. The presence of dispersion in the laser medium or cavity optics can, however, make it very difficult to obtain very narrow pulses in a mode-locked laser since it can cause pulse broadening. The pulse is composed of a spread of frequencies and each frequency travels with a slightly different phase velocity causing the pulse shape to change as the pulse propagates. We will discuss this phenomenon and several techniques for eliminating it in the following.

When describing *wave packets* (i.e., pulses), it is useful to distinguish between two velocities: the *group velocity*, which is the velocity of the packet (and usually energy transport) and the *phase velocity*, which is the velocity of the constant phase surfaces of the wave which we refer to as the *wavefronts*. The usual procedure for describing a wave packet is to consider it to be a continuous sum of components whose frequencies differ only slightly from each other and which are arranged to cancel outside of some finite time interval, $\Delta t_p$. One can sum over $\omega$ (temporal frequency) or $k$ (spatial frequency); we will use the former. The sum is the Fourier integral,

$$E(z,t) = \int_{-\infty}^{\infty} A_\omega(\omega - \omega_0) e^{ik(\omega)z - i\omega t} \, d\omega, \tag{10.24}$$

where we assume the wave depends upon only one spatial coordinate ($z$) and $\omega_0$ is the center frequency of the wave.

We will first expand the wave vector around $\omega_0$

$$k(\omega) = k_0 + \left(\frac{\partial k}{\partial \omega}\right)_{\omega_0} (\omega - \omega_0) + \cdots . \tag{10.25}$$

Substituting this into the Fourier integral,

$$E(z,t) = e^{i(k_0 z - \omega_0 t)} \int_{-\infty}^{\infty} A_\omega(\Delta\omega) \exp\left\{ -i\Delta\omega \left[ t - \left(\frac{\partial k}{\partial \omega}\right)_{\omega_0} z \right] \right\} d\Delta\omega, \tag{10.26}$$

where $\Delta\omega = \omega - \omega_0$. The integral, which we will call $A()$, depends only upon $t - (\partial k/\partial \omega)z$ and the field is

$$E(z,t) = A\left[ t - \left(\frac{\partial k}{\partial \omega}\right)_{\omega_0} z \right] e^{i(k_0 z - \omega_0 t)}. \tag{10.27}$$

The field is therefore the product of a plane wave propagating to the right with speed $\omega_0/k_0$ and an *envelope function*, $A(t - (\partial k/\partial \omega)z)$, propagating to the right with (possibly different) speed $\partial \omega/\partial k$. When the dispersion function is *linear*, these two speeds are the same. As mentioned above, the speed of the envelope is called the *group velocity*, $v_g$, defined as

$$\text{Group velocity:} \quad v_g \equiv \left(\frac{\partial \omega}{\partial k}\right)_{\omega_0}. \tag{10.28}$$

In all that follows, we assume that the wave vector depends *weakly* on the frequency and that the index of refraction is a monotonically increasing function of the frequency (normal dispersion). With these assumptions, the group velocity has a simple interpretation as the velocity of energy transport and we avoid certain non-physical behaviors, such as the case of a group velocity that is greater than $c$. In a medium with refractive index $n(k)$, the surfaces of constant phase travel at the *phase velocity*, $v_\phi$

$$\text{Phase velocity:} \quad v_\phi = \frac{\omega(k)}{k} = \frac{c}{n(k)}. \tag{10.29}$$

Differentiating the second equality with respect to $k$ yields the following relation for $v_g$:

$$v_g = \frac{c}{n(k) + \omega\frac{dn(\omega)}{d\omega}}. \tag{10.30}$$

One often defines the *group index of refraction* as $n_g = c/v_g$; it is related to the ordinary index $(n)$ by

$$n_g = n + \omega\frac{dn(\omega)}{d\omega}. \tag{10.31}$$

Except in the vicinity of an absorption, most materials show *normal dispersion*, defined as $dn/d\omega > 0$. In these cases, the group velocity is less than the phase velocity. Near an absorption, one can have *anomalous dispersion* $(dn/d\omega < 0)$ which can result in the group velocity being greater than the phase velocity. While a group velocity which differs greatly from the phase velocity is possible (even one greater than $c$), we exclude such curiosities in this book as being somewhat non-physical and not consistent with our assumption of *weak* dispersion.

In a dispersive medium ($n$ not constant), a pulse whose temporal width at $z = 0$ is $t_p$ will spread an additional $\Delta t_p$ at $z = l$ by approximately

$$\Delta t_p \approx \frac{l}{(v_g)_{min}} - \frac{l}{(v_g)_{max}} \approx l\frac{\Delta\left(\frac{1}{v_g}\right)}{\Delta\omega}\Delta\omega = l\frac{\Delta\left(\frac{\partial k}{\partial\omega}\right)}{\Delta\omega}\Delta\omega = l\left(\frac{\partial^2 k}{\partial\omega^2}\right)_{\omega_0}\Delta\omega, \tag{10.32}$$

where $\Delta\omega$ is the spread of frequencies in the wave.

In the absence of dispersion, $k$ depends linearly on $\omega$ and the group and phase velocities are the same. The simplest description of dispersion is to assume an additional quadratic dependence on $\omega$:

$$k(\omega) = k_0 + \left(\frac{\partial k}{\partial\omega}\right)_{\omega_0}(\omega - \omega_0) + \frac{1}{2}\left(\frac{\partial^2 k}{\partial\omega^2}\right)_{\omega_0}(\omega - \omega_0)^2. \tag{10.33}$$

Equation 10.32 shows that a pulse in a medium described by this simple dispersion relation will spread (if the second derivative $> 0$) as it propagates, and the spread will be proportional to the distance traveled. The second derivative of $k$ with respect to $\omega$ is called the *group velocity dispersion* (GVD):

$$\text{Group velocity dispersion} = \left(\frac{\partial^2 k}{\partial\omega^2}\right)_{\omega_0} = \left(\frac{\partial\frac{1}{v_g}}{\partial\omega}\right)_{\omega_0}. \tag{10.34}$$

If we multiply the GVD by the distance traveled, we obtain the *group delay dispersion* (GDD). Since the *phase*, $\phi$, accumulated by a wave in distance $l$ is $kl$, the GDD is

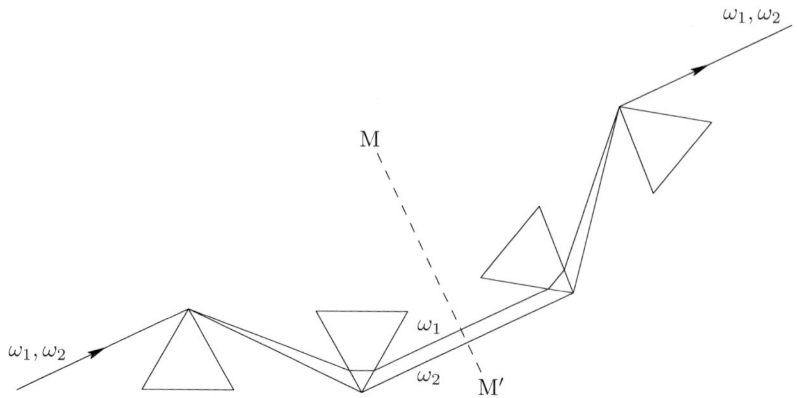


**Fig. 10.10** Dispersion compensation using four mirrors (or two in a standing wave laser).

occasionally called $\phi''$ (where the two primes signify double differentiation with respect to $\omega$). Most media have a positive $\phi''$ and thus lead to pulse spreading. Dispersion compensation is the addition of an element having a negative $\phi''$ into the laser cavity.

A common approach to dispersion compensation is to use the symmetrical four-prism sequence shown in Fig. 10.10. The figure shows two waves, one at $\omega_1$ and one at $\omega_2$. The prisms are cut and arranged so that the light enters each prism at or near Brewster's angle to reduce the loss and at the minimum deviation angle. Adjacent surfaces in each prism pair are parallel to each other. Analysis by Fork, et al. (1984) demonstrated that one can produce a negative group velocity dispersion using this configuration and the size of the dispersion can be changed by changing the separation between the prisms within each pair. In a standing-wave laser, the symmetry plane (MM′) can be replaced with a mirror enabling one to use only two prisms.

A more recent approach to dispersion compensation is to replace at least one cavity mirror with a *chirped mirror*. These are a special kind of *dielectric mirror*, similar to the mirrors used in virtually all laser cavities. Highly reflective coatings using transparent dielectric materials are usually of the *Bragg* variety, which consist of a number (up to 100) of alternating layers of materials with two different refractive indices whose optical thicknesses are one quarter of the central wavelength. If one includes the 180° phase shift when reflecting from a higher index to a lower index, it is easy to see that reflections from alternate layers are all in phase yielding a cumulative high reflectivity. Using a larger number of coatings increases both the bandwidth and the reflectivity.

The layers of dielectric material in a chirped mirror are *chirped in thickness* (i.e., uniformly varying with depth in the mirror). This causes *different* wavelengths to be reflected from *different* quarter-wave stacks. The wavelength-dependent penetration depth results in a wavelength-dependent phase shift which can be tailored to compensate the group delay dispersion of the laser. Figure 10.11 illustrates how the longer wavelengths penetrate the mirror to a greater extent than the shorter wavelengths. The original chirped mirrors based upon these simple ideas unfortunately didn't work very well. The group delay had large amplitude oscillations which prevented the mirrors from being used to generate femtosecond pulses. This problem was resolved with

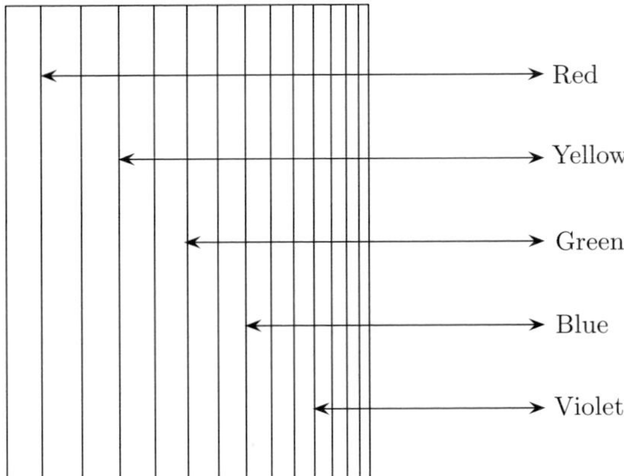


**Fig. 10.11** Operating principle of chirped mirror, providing negative group delay dispersion.

the *double chirped* configuration. In addition to the *chirp* of the Bragg wavelength, the thickness ratio of the high index to low index layers was adjusted to smooth the transition from the highly reflective region to the low-reflectance region. In addition, a broadband antireflective coating was placed over the front surface of the mirror. The double chirped mirrors provide a fairly wavelength-independent negative group-delay correction and have a larger bandwidth than prisms, which have some wavelength dependence of their group delay due to dispersion in the prism material. These mirrors are therefore better suited for correcting dispersion in femtosecond lasers.

## 10.5   The mode-locked Ti-sapphire laser

The Ti-sapphire laser is the pre-eminent source of femtosecond pulses since its gain bandwidth is larger than that of any other laser in the visible or near infrared. The emission spectrum of a Ti-sapphire mode-locked laser is shown in Fig. 10.12. There are a number of commercial mode-locked Ti-sapphire lasers that deliver respectably narrow pulses at a range of repetition rates from a bit less than 100 MHz to above 1 GHz. The field is rapidly growing and the state-of-the-art is in the non-commercial research laboratories. We will discuss some of the characteristics of a representative sample of the latter group of lasers.

Before discussing a specific system, we need to slightly refine our discussion of mode-locked theory and introduce the *soliton-type* mode-locked laser, which can generate the narrowest pulses. The soliton laser was developed in the 1980s (see, for example, Mollenauer (1984)) and initially used an optical fiber to produce the solitons. A soliton is a short pulse which is a solution to the nonlinear Schrödinger equation and possesses the very useful characteristic of propagating either with unchanging shape (lowest order) or with a shape which is a periodic function of distance (higher order). In a mode-locked laser, solitons can appear when the medium has a *negative* group velocity dispersion combined with a Kerr nonlinearity, in which the index of refraction is a linear function of the intensity. The former condition can be obtained by overcompensating the dispersion using one of the techniques described in the previous

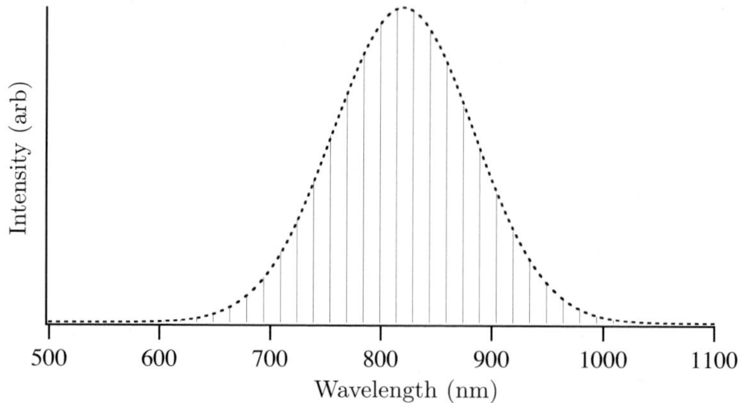

**Fig. 10.12** Emission spectrum of mode-locked Ti-sapphire laser.

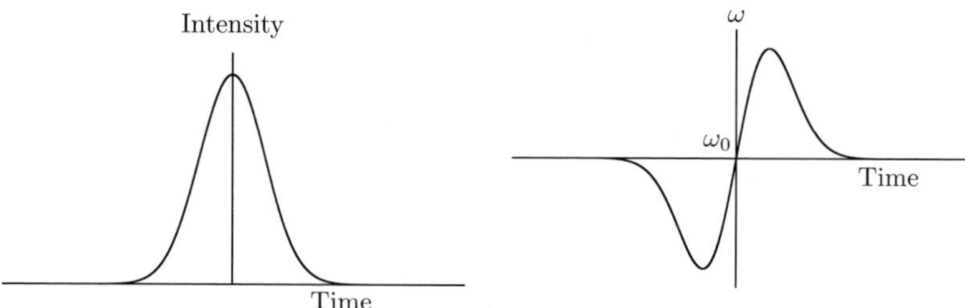

**Fig. 10.13** Time dependence of the pulse frequency (right) for Gaussian pulse (left) due to the Kerr effect.

section. The Kerr nonlinearity occurs in the Ti-sapphire gain crystal and will produce *self phase modulation* (SPM).

Self phase modulation is simply a *phase modulation* of the pulse which is due to the Kerr effect and the rapid temporal variation of the pulse intensity. When the Kerr effect is present, the total phase, $\phi(t)$, of a pulse whose carrier frequency is $\omega_0$ is given by

$$\phi(t) = \omega_0 t - k(z)z = \omega_0 t - \frac{n(z)\omega_0}{c}z = \omega_0 t - \frac{(n_0 + n_2 I)\omega_0}{c}z. \tag{10.35}$$

The instantaneous frequency of the pulse is

$$\omega = \frac{d\phi}{dt} = \omega_0 - \frac{\omega_0 n_2 z}{c}\frac{\partial I}{\partial t}. \tag{10.36}$$

The behavior of the pulse frequency as a function of time is shown on the right side of Fig. 10.13. Notice that the pulse acquires a positive-going frequency *chirp* near the peak of the pulse. We will now show that the negative group velocity dispersion due to other elements in the laser cavity will produce a negative-going frequency chirp which will counter the one due to the Kerr effect in the gain medium.

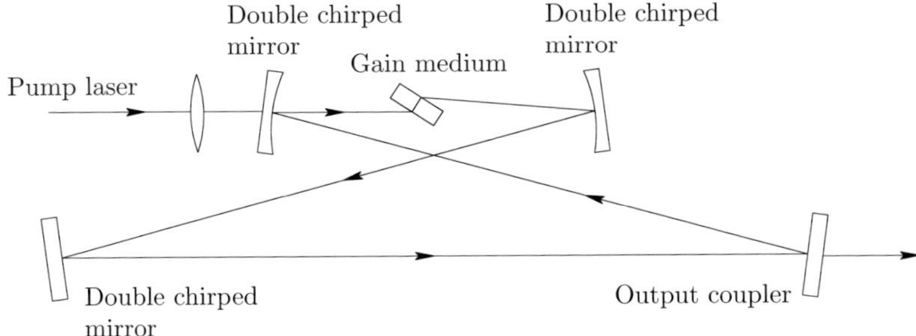


**Fig. 10.14** Realization of GHz mode locked Ti-Sapphire ring laser used at the National Institute of Standards and Technology (NIST).

In the presence of a small amount of dispersion, the expression for a Gaussian pulse can be shown to be (Svelto, Appendix G)

$$E(z, t') = \frac{E_0 \tau_p}{(\tau_p^2 + i D_2)^{1/2}} \exp\left(-\frac{t'^2}{2(\tau_p^2 + i D_2 z)}\right) e^{i\omega_0 t}, \qquad (10.37)$$

where $\tau_p$ is the pulse width and $D_2$ is the group velocity dispersion,

$$D_2 = \frac{\partial^2 k}{\partial \omega^2} \qquad (10.38)$$

and $t'$ and $z$ are related by the fact that the pulse envelope propagates at the group velocity

$$t' = t - \frac{z}{v_g}. \qquad (10.39)$$

If we calculate the *phase* of this expression (including the $e^{i\omega_0 t}$ factor) and differentiate to obtain the instantaneous frequency, we obtain

$$\omega = \omega_0 + \frac{D_2}{|D_2|} \frac{z/L_D}{1 + (z/L_D)^2} \frac{2t'}{\tau_p^2}, \qquad (10.40)$$

where $L_D = \tau_p^2/|D_2|$ is the *dispersion length* of the medium. We thus see that a negative group velocity dispersion will produce a negative-going frequency chirp. If we adjust the laser parameters so that the two chirps cancel, we can obtain a soliton-type pulse behavior, where the pulse propagates with constant shape. From soliton theory, it can be shown that the laser *power*, $P$, and pulse width, $\tau_p$, are now related as

$$\tau_p^2 \propto \frac{|D_2|}{P}. \qquad (10.41)$$

A realization of a GHz mode-locked Ti-sapphire ring laser used by workers at the National Institute of Standards and technology (NIST) in Boulder, Colorado (Ye (2005)) is shown in Fig. 10.14. The cavity is the familiar astigmatically compensated ring discussed in some detail in Chapters 2 and 3. The net negative group velocity dispersion is provided by the three double chirped cavity mirrors. The Ti-sapphire crystal is fairly short in order to ensure a negative group velocity dispersion since the crystal itself contributes a positive GVD proportional to its length. The repetition rate is 2 GHz for a fairly short cavity. The output coupler has a transmission of 2%. This laser produced pulses as short as 14 fs, while a variant with two additional folded chirped mirrors and a longer Ti-sapphire crystal produced $\approx$ 40 fs pulses.

## 10.6 Frequency metrology using a femtosecond laser

While mode-locked lasers are excellent tools for making time-resolved studies of extremely short-lived phenomena, their principal use in the atomic physics laboratory is in optical frequency metrology. Formerly, optical frequencies were accurately measured by coherently linking them to a number of microwave standards using an elaborate chain of lasers and microwave oscillators. Some ingenious techniques were suggested for making a coherent link in one step, such as phase locking a relativistic electron in a Penning trap to a laser and detecting the electron's radio-frequency synchrotron radiation. The most practical technique for accomplishing this single-step link, however, is to exploit the extraordinary uniformity of the *comb* of frequencies emitted by a mode-locked laser. The already very broad comb from a femtosecond laser is further broadened using the recently developed *microstructure fiber*, allowing the comb to span more than an octave and enabling direct measurement of optical frequencies. This section will only touch on the high points of this research, since it is being very rapidly developed at present and there is already a large literature on the subject (including some books, e.g. Ye (2005)).

The utility of a femtosecond comb lies in the extraordinary uniformity of the frequency spacing of the lines generated by a mode-locked laser. Originally, by comparison of two combs, the error due to possible non-uniformities was considered to be about $4 \times 10^{-17}$; more recently (Zimmermann (2004)) the error bound was reduced to $6.6 \times 10^{-21}$. Thus, the comb introduces essentially no error in frequency measurements. The frequency, $f_n$, of the $n^{th}$ comb element is given by

$$f_n = nf_r + f_{ceo} \qquad (10.42)$$

where $n$ is an integer, $f_r$ is the repetition rate of the laser and $f_{ceo}$ is an offset called the *carrier envelope offset frequency* and is due to dispersion. We will discuss this term in detail in the next section; for now, we assume that it is a stable and constant quantity (although this is not strictly true).

Femtosecond combs were originally used to *compare* two optical frequencies with great accuracy using an apparatus similar to that shown in Fig. 10.15. At the top of the figure is a spectrum of the laser including the two optical frequencies being compared. The quantities $\delta_1$ and $\delta_2$ are the differences between the laser frequencies and the nearest comb line. Since $f_r$ is $\approx$ 100 MHz and $\delta_1, \delta_2 < f_r$, these two frequencies are easily manipulated using standard radiofrequency practices. The difference frequencies are measured by making the three beams colinear and directing them to a grating which separates the composite beam into two beams centered around $f_1$ and $f_2$. The difference frequencies appear in each diode photocurrent since the latter is proportional

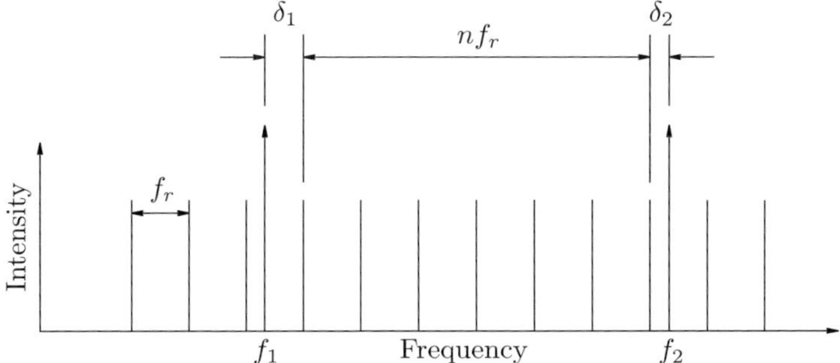

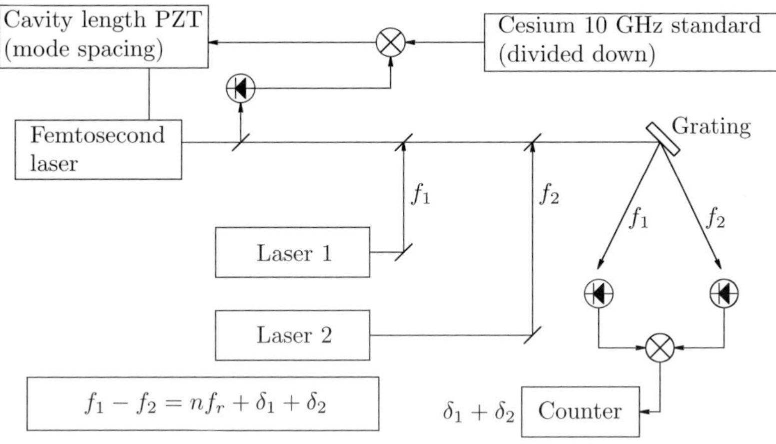

**Fig. 10.15** Femtosecond comb apparatus used to very accurately compare two optical frequencies.

to the *square* of the incident electric field. The difference frequencies, $\delta_1$ and $\delta_2$, can be mixed together to obtain $\delta_1 + \delta_2$. The difference between $f_1$ and $f_2$ is then

$$f_1 - f_2 = nf_r + \delta_1 + \delta_2, \tag{10.43}$$

where the integer $n$ can be obtained if the frequencies are known to the accuracy of a common digital wavemeter (about 30 MHz). The repetition rate, $f_r$, is locked to some subharmonic of a microwave frequency standard using a piezo-electric driven mirror in the laser cavity to control the mode spacing and a photodiode to sample the laser output and detect the mode spacing. A simple feedback system will make $f_r$ an exact submultiple of the frequency of the microwave standard. The ultimate fractional error can be equal to the error in the microwave standard since the error in the difference frequencies is negligible compared to $n$ times the error in the subharmonic of the microwave standard ($n$ can be $10^6$ or more).

Although many of the earlier Ti-sapphire mode-locked lasers had a repetition rate of about 100 MHz, there has recently been a drive to increase this to 1 GHz or greater.

A larger $f_r$ has two advantages. First, since there are fewer teeth in the comb, each will have more power, improving the signal-to-noise of the various beat signals used in a frequency measurement. Second, ambiguities in the determination of the mode number, $n$, are eased when the repetition rate is very large. A 100 MHz $f_r$ pushes current digital frequency meters to their accuracy limit in obtaining the mode number.

A breakthrough occurred when the *microstructure* fiber was developed and the comb was made to span more than an octave. Assuming that the fiber did not degrade the uniformity of the mode spacing, an absolute frequency measurement could be made by doubling the unknown frequency, $f$, and comparing $f$ to $2f$. In this case, $f_1 - f_2 = f$, providing the unknown frequency directly. This scheme is called a *self-referenced* frequency measurement or a "$\nu$-to-$2\nu$ measurement". Difference frequency measurements and $\nu$-to-$2\nu$ measurements solve the problem of the unknown $f_{ceo}$ by measuring the *difference* between two frequencies, which cancels the inhomogeneous term, $f_{ceo}$. Another approach is to *control* $f_{ceo}$, usually forcing it be some fixed frequency. The latter scheme will be discussed in the next section.

The key to the microstructure fiber is its very small core diameter: as little as 1.7 $\mu$m, compared to the $\approx 5$ $\mu$m diameter in a conventional single-mode fiber. This greatly increases the intensity of the guided wave and thereby enhances the nonlinearities which broaden the comb width by generating numerous intermodulation products of the comb frequencies. From fiber theory, a reduction in the core diameter necessitates an increase in the difference between the refractive indices of the core and the cladding (surrounding layer). The lowest cladding index is obtained when the cladding is *air*. The microstructure fiber surrounds the core by a number of hollow tubes, making the effective cladding index close to unity (Fig. 10.16). In addition to a small core diameter for the highest nonlinearity, it is also necessary to have very small group velocity dispersion so that the pulse does not broaden and lose intensity as it travels down the fiber. By careful design, the microstructure fiber shown in the figure can possess both of these useful properties.

Although the ability to obtain an octave spanning comb using a microstructure fiber is a remarkable achievement, it has some drawbacks. Because of the small core diameter and consequent high intensity of the guided light, optical damage is a problem with these fibers and their lifetime can be as short as a few hours. There has been some success in spanning an octave directly with a properly designed Ti-sapphire laser, using the nonlinearity of the laser active medium.

Finally, it should be pointed out that the definition of an octave span can be operationally relaxed quite a bit, since it is possible to obtain useful beats when the intensities of components at the edge of the gain curve are four orders of magnitude below the peak intensity. Thus, a comb whose full width at half maximum is well under an octave can still be considered to be "octave spanning" and used in the *self-referenced* frequency measurements described above.

## 10.7   The carrier envelope offset

Most of the frequency measurement schemes which utilize femtosecond combs are dominated by the need to cope with the inhomogeneous offset, $f_{ceo}$. In this section, we will discuss the origin of this quantity and a technique for controlling it.

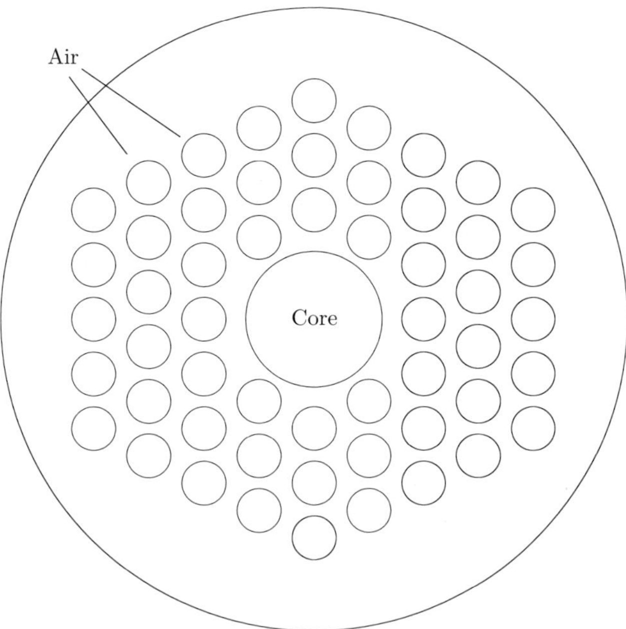



**Fig. 10.16** Schematic cross section of the microstructure with very small core diameter and low group velocity dispersion due to air holes surrounding the core.

An explanation of the origin of the carrier envelope offset will exploit the analogy with multiple slit interference which was used earlier in this chapter. The usual technique for analyzing multislit interference is to consider the accumulated *phase difference* of the field between adjacent slits as the field propagates to a screen where the interference pattern can be observed. Constructive interference occurs when the phase difference is an integral multiple of $2\pi$. From this requirement, we obtain the possible values of the wave number, $\boldsymbol{k}$ (a *spatial frequency*), which give constructive interference. The treatment of the mode-locked laser is exactly the same, except that it is in the *time* domain rather than the *spatial* domain of multislit interference. Fig. 10.17 illustrates the concepts used in our treatment. In order to determine the *frequencies* which result in constructive interference, we determine the amount by which the phase has evolved between successive pulses and require that it be an integral multiple of $2\pi$. In the absence of dispersion, constructive interference occurs when

$$\omega\tau = 2\pi n \qquad \tau = 1/f_r, \ \ \text{integral } n$$
$$\implies f_n = n f_r, \tag{10.44}$$

where $\tau = 1/f_r$ is the temporal separation between pulses. When dispersion is present, the group velocity will differ from the phase velocity. Thus, the pulse envelope, which propagates at the group velocity will travel at a different speed from the underlying sinusoidal wave, which moves at the phase velocity. If the *additional* phase difference between successive pulses is $\Delta\phi$, then the condition for constructive interference will

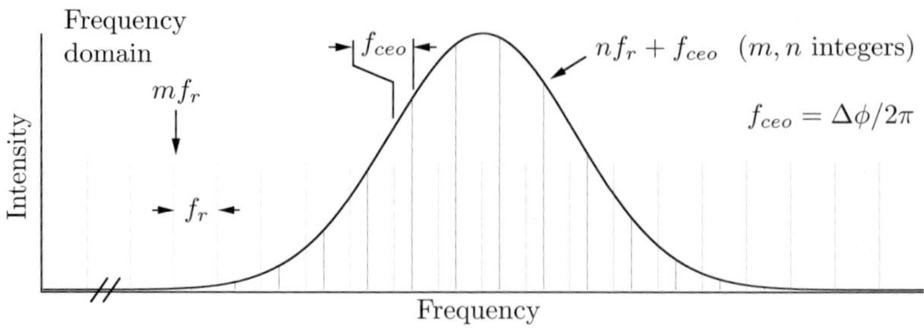


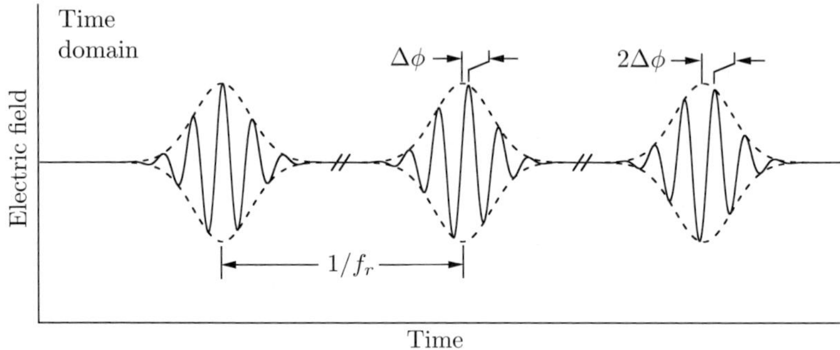


**Fig. 10.17** Illustration of "carrier envelope slippage" which results in non-zero $f_{ceo}$.

be

$$\omega\tau = 2\pi n + \Delta\phi$$
$$\Longrightarrow f_n = n f_r + \Delta\phi f_r / 2\pi. \tag{10.45}$$

Thus,

$$f_{ceo} = \Delta\phi f_r / 2\pi. \tag{10.46}$$

As shown in the figure, there is a *slippage* between the wave ("carrier") and its envelope.

When dispersion is present, the mode spacing of a mode-locked laser cavity is *different* from the pulse separation, since the mode separation is determined by the phase velocity while the pulse separation is determined by the group velocity and the two are not exactly the same. The very fact that the pulses are extremely narrow is a *proof* of the uniformity of the frequency modes: any slight non-uniformity would make it impossible to obtain pulses containing only a few optical cycles since the different Fourier components of a non-uniform spectrum would cause the pulses to broaden. The narrow pulses are, of course, also an indication of the success in eliminating the intracavity dispersion. The remaining dispersion is responsible for a non-zero $f_{ceo}$.

A standard technique for controlling $f_{ceo}$ is shown schematically in Fig. 10.18. It is a variant of the self-referenced $\nu$-to-$2\nu$ approach but applied to the comb rather than the unknown frequency. The laser output is broadened to about an octave using a microstructure fiber and separated into low frequencies and high frequencies using

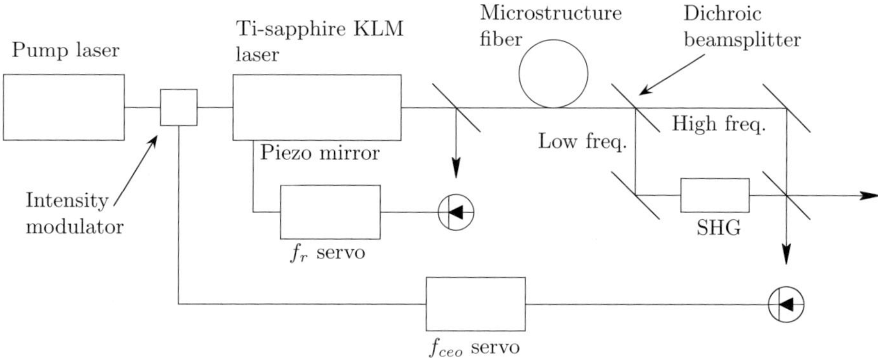

**Fig. 10.18** Technique for controlling $f_{ceo}$ in a Kerr-lens mode-locked (KLM) laser.

a dichroic beamsplitter. The low frequency part is frequency doubled and beat with the high frequencies, yielding $f_{ceo}$ directly when $2n = m$ ($n$, $m$ are integers):

$$\text{beat frequency} = 2(nf_r + f_{ceo}) - (mf_r + f_{ceo}) = (2n - m)f_r + f_{ceo}. \tag{10.47}$$

The Mach–Zehnder interferometer shown in the figure ensures that the high frequency pulses and the frequency-doubled low-frequency pulses overlap. The carrier offset frequency is locked to a fixed value using a phase-locked loop with a loop bandwidth of about 10 kHz since the *noise* in $f_{ceo}$ is in the kHz region (it might be higher if an argon laser is used for a pump). The value of $f_{ceo}$ can be controlled either through the pump power or through the tilt angle of one of the cavity mirrors; the figure shows the former method.

## 10.8 Further reading

The books on lasers and quantum electronics mentioned in previous chapters all have good descriptions of mode-locked lasers. None, however, discuss frequency metrology using a comb. A good reference for femtosecond combs and frequency metrology is the recent book by Ye and Cundiff (2005). The use of a saturable absorber for passive mode locking is described in a paper by Garmire and Yariv (1967). Dispersion compensation using prisms is discussed in the paper by Fork et al. (1984) and double-chirped mirrors for the same purpose are described in a paper by Paschotta et al. (1999). Soliton lasers are described in the paper by Mollenauer and Stolen (1984). Numerous papers discuss the use of femtoseconds lasers in frequency metrology; a recent example is the paper by Zimmermann et al. (2004).

## 10.9 Problems

(10.1) Given two sinusoidal waves (representing the electric field, for example) propagating to the right with equal amplitudes and with frequencies $\omega_1$ and $\omega_2$ and wave vectors $k_1$ and $k_2$, there will be an envelope of "beats" which propagates at the group velocity, $v_g$, which will, in general, be different from the phase velocity, $v_\phi$. For this example, write an expression for both the phase and group velocities. (Hint: the group velocity is usually given by a derivative with respect to $k$. With only two waves, this doesn't make much sense. You will find that the derivative is replaced by something which applies to two discrete values of $k$).

(10.2) Consider a 623.8 nm HeNe laser (lasing atom is Ne, which is Doppler broadened at room temperature).

   (a) Estimate the shortest pulse that can be produced by such a laser.

   (b) What is the separation between pulses if the mirror separation is 40 cm?

   (c) Why do liquid-dye and many solid-state lasers produce shorter pulses than typical gas lasers (such as a HeNe laser)?

(10.3) Show that mode locking can occur if the *phase* of each mode is a *linear* function of the mode number, $n$. What happens if the phase depends *quadratically* on the mode number?

(10.4) What is the minimum FWHM of the envelope of modes in a mode-locked laser employed in the measurement of an unknown optical frequency using the *self-referential* method if a useable signal can be obtained when the modes which are compared to the unknown frequency are 40 db lower than the peak?

# 11

# Laser frequency stabilization and control systems

## 11.1 Introduction

Lasers have a well-deserved reputation as sources of extremely spectrally narrow optical radiation. For example, a *free-running* (unstabilized) dye or Ti-sapphire ring laser has a short-term linewidth of less than one MHz, which corresponds to a *relative* instability of less than $10^{-8}$. External cavity diode lasers can have linewidths of 50 kHz and the YAG non-planar-ring-oscillator has a *measured* linewidth of about 3 kHz, which corresponds to a relative instability of $10^{-11}$. The dominant source of frequency noise in all of these systems is *technical noise*: noise which is not fundamental in nature and which is due to fluctuations in the properties of the laser medium or support structures. We have seen that monolithic diode lasers are usually dominated by enhanced Shawlow–Townes noise and have linewidths of a bit more than 1 MHz. There are a number of applications which demand even smaller spectral linewidths than are available from a free-running laser and therefore require some form of active *frequency stabilization*.

We will discuss the techniques of frequency stabilization using negative feedback and a stable frequency reference, which is usually a passive optical cavity. These techniques can also be applied, with some modification, to the stabilization of a laser to an atomic or molecular reference. Discriminants from atomic and molecular resonances will be discussed in a later chapter. In this chapter, it will be assumed that the frequency discriminant will be generated using the Pound–Drever–Hall approach discussed in detail in Chapter 4. We will first derive the basic results using fairly self-evident principles. Next, we will review some simple linear system theory using Laplace transforms. We will then use this theory to determine the dynamical behavior of laser stabilization systems. Since the theory can be used for any control system, we will also briefly describe other servo systems often found in the atomic physics laboratory, such as temperature stabilization systems.

## 11.2 Laser frequency stabilization – a first look

Before resorting to the machinery of linear network theory to analyze a frequency stabilization system, we will use some simple considerations to obtain the important features of such a system. The basic idea is that we use the frequency discriminant to *correct* deviations of the instantaneous laser frequency from the resonance frequency of the cavity. Since we are using *negative* (degenerative) feedback, the correction signal must have a sign which is *opposite* to that of the error so that it *corrects* the error by reducing it. A block diagram of the stabilization system appears in Fig. 11.1 The

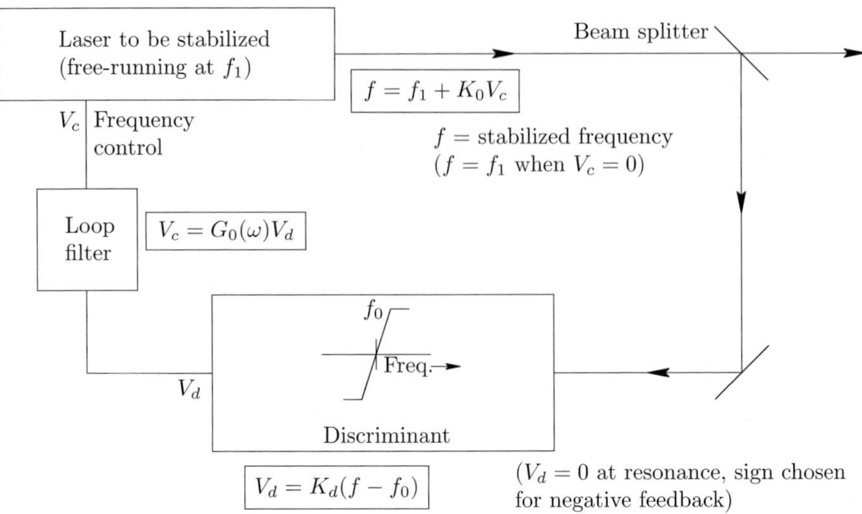


**Fig. 11.1**  Block diagram of frequency stabilization system.

laser frequency, $f$, is *linearly* controlled by an external voltage, $V_c$, and the voltage to frequency characteristic is

$$\text{Laser:} \quad f = f_1 + K_0 V_c, \tag{11.1}$$

where the free running frequency ($V_c = 0$) is $f_1$. The frequency discriminator (cavity and Pound–Drever–Hall apparatus) has a response voltage, $V_d$, given by

$$\text{Cavity:} \quad V_d = K_d(f - f_0), \tag{11.2}$$

where $f_0$ is the resonance frequency of the cavity. Between the discriminator output and laser frequency control input, there is a *loop filter*, described by the *transfer function*, $G_0(\omega)$, which depends on the frequency, $\omega$, of the discriminator output. The behavior of the loop filter is described by

$$\text{Loop filter:} \quad V_c = G_0(\omega)V_d. \tag{11.3}$$

Any frequency dependence of the laser frequency control or discriminant can be included in the loop filter response, $G_0(\omega)$.

The *loop equation* can be written down by inspection

$$f = f_1 + K_0 V_c = f_1 + K_0 G_0(\omega)V_d = f_1 + K_0 K_d G_0(\omega)(f_0 - f), \tag{11.4}$$

where we have chosen a sign for $K_d$ which results in negative feedback. Solving for $f$, one obtains

$$f = \frac{f_1}{1 + G(\omega)} + \frac{G(\omega)f_0}{1 + G(\omega)}, \quad \text{where} \quad G(\omega) \equiv K_0 K_d G_0(\omega). \tag{11.5}$$

The quantity $G(\omega)$ is called the *open-loop gain*: it is the gain for a trip around the loop if one breaks the loop at *any* point. The *closed-loop gain* is the ratio of $f$ to $f_0$ if one ignores $f_1$ (by setting it equal to zero): it is a measure of how well the laser frequency tracks changes in the *reference cavity frequency*:

$$\text{Closed loop gain} = \frac{G(\omega)}{1 + G(\omega)}. \tag{11.6}$$

The *closed-loop error* is defined as $f - f_0$ with the loop in effect (closed) and is given by

$$\text{Closed loop error} \equiv f - f_0 = \frac{f_1 - f_0}{1 + G(\omega)}. \tag{11.7}$$

Assume that we *independently* change $f_0$ by $\Delta f_0$ and $f_1$ by $\Delta f_1$. Then the change in the laser frequency, $\Delta f$ is

$$\Delta f = \frac{\Delta f_1}{1 + G(\omega)} + \frac{G(\omega)\Delta f_0}{1 + G(\omega)}. \tag{11.8}$$

From this we see that, when $|G(\omega)| \gg 1$, the *free running jitter* $(\Delta f_1)$ is reduced by $G(\omega)$ while the laser accurately tracks changes in the reference frequency $(\Delta f_0)$ since the closed loop gain is nearly unity. Thus, making the open-loop gain as large as possible *reduces noise due to the laser* and improves the *tracking of the reference frequency*. We further see that, at frequencies outside the bandwidth of $G(\omega)$ (i.e., where $G(\omega) \leq 1$), the laser is dominated by its own frequency noise and poorly tracks the reference cavity. Thus, one can consider the frequency, $\omega$, for which $G(\omega) \approx 1$ as the point below which the laser's frequency characteristics will be dominated by the reference and above which it will be dominated by its intrinsic behavior. These characteristics are common to all control systems which use negative feedback.

## 11.3   The effect of the loop filter

We will assume that the loop filter has a frequency behavior which is characterized by extremely high gain at very low frequencies and a gain which decreases with increasing frequency. We will call it an *integrator*, since this is the behavior of a single integrator (we will discuss this in detail in our section on linear system theory). We will then perform a simple experiment: we will inject a small signal, $V_i$, at one of two places: either before the integrator or after the integrator. The signals will be small sinusoidal voltages at a frequency for which $G(\omega) \gg 1$. This procedure is shown in Fig. 11.2. One can determine the loop behavior due to each of the two injected signals by observing the response at the output of the discriminator (with the loop closed). If one injects the signal *before the integrator*, the laser frequency will change from its former value, $f$, to

$$\text{Before integrator:} \quad f \longrightarrow f + \frac{V_i}{K_d} \quad (G(\omega) \gg 1). \tag{11.9}$$

This is equivalent to a *modulation of the cavity resonance frequency*, and the laser will track it very well. On the other hand, if one adds the same signal *after the integrator*, the change in the laser frequency will be

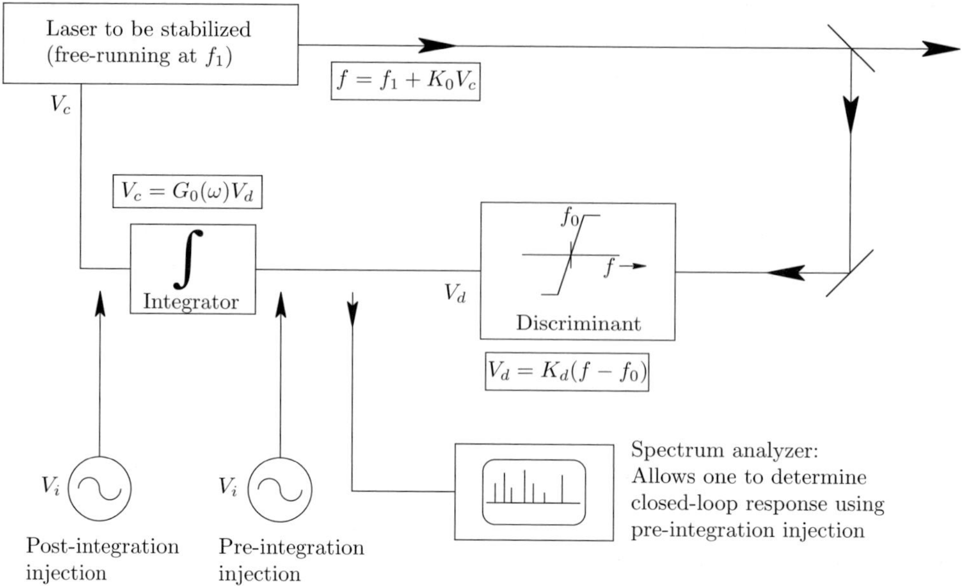


**Fig. 11.2** Injection of small signals before and after the loop filter (integrator).

$$\text{After integrator:} \quad f \longrightarrow f + \frac{K_0 V_i}{1 + G(\omega)} \longrightarrow f \quad (G(\omega) \gg 1). \tag{11.10}$$

Unlike the previous case, the signal will be *suppressed by the loop*, since it will be treated as an addition to the laser frequency jitter. These two very different behaviors are due to the extremely high gain of an integrator at low frequencies. One can use this procedure to determine the frequency response of the closed loop if one uses a sweeping source after the integrator and observes the output of the discriminator as the frequency is swept (one can inject the signal before the integrator if care is taken to isolate the injected signal from the observed response). The bandwidth of the loop will be the frequency of $V_i$ at which a signal just appears at the discriminator output. Under normal functioning of the closed loop, the signal *after the integrator* can be interpreted as the *correction signal needed to eliminate the free-running laser jitter* and the signal *before the integrator* can be interpreted as a *direct measure of the residual frequency jitter not removed by the loop* (assuming a perfectly stable cavity).

## 11.4 Elementary noise considerations

The ultimate performance of any frequency stabilization system is limited by the shot noise of the signal at the output of the discriminator. A practical system will, of course, have other noise contributions, but these can usually be reduced or eliminated by good design practices. The shot noise is fundamental and cannot be eliminated. We can make some crude estimates of the noise-limited performance with the aid of Fig. 11.3. The noise source will be modeled as a noise voltage, $V_n$, injected into the

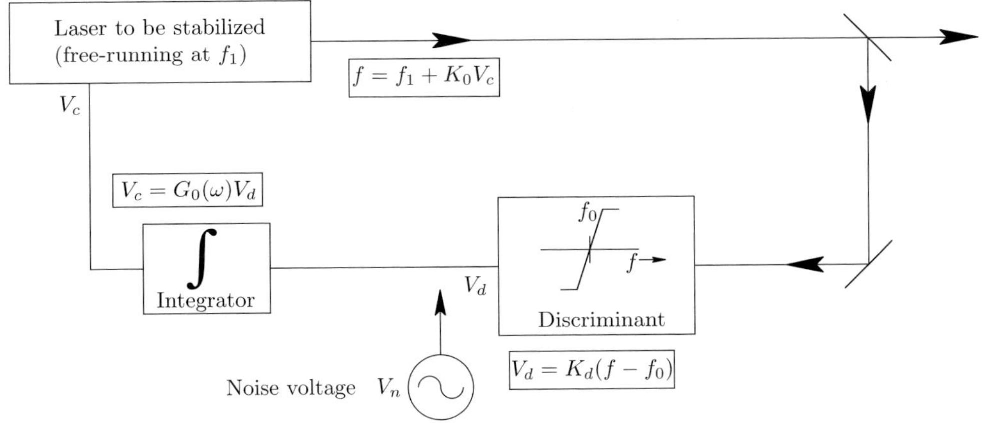


**Fig. 11.3** Block diagram of stabilization system for noise analysis.

loop immediately after the discriminator. The frequency jitter, $\Delta f_{laser}$, within the bandwidth of the loop due to this noise is

$$\Delta f_{laser} = \frac{V_n}{K_d}. \tag{11.11}$$

The *discriminator slope*, $K_d$, is obtained from eqn 4.52:

$$\text{Slope} = 8GP_0 J_0(\beta) J_1(\beta) \frac{1}{\delta\omega}, \tag{11.12}$$

where $G$ is the responsivity of the detector in volt/watt and $P_0$ is the incident power. Evaluating the Bessel functions at the optimum $\beta$ (1.08), we obtain

$$\text{Slope} = \frac{2.71 GP_0}{\delta\omega}. \tag{11.13}$$

Ignoring the factor $2.71/2\pi$ ($\delta\omega = 2\pi \Delta f_{cavity}$), the value of $K_d$ can be written approximately as

$$K_d \approx \frac{\text{signal}}{\text{cavity width}}, \tag{11.14}$$

since the cavity reflectivity is close to unity and the signal voltage is $\approx GP_0$. This is a general result which holds for all types of discriminants (with slightly different coefficients). Thus,

$$\Delta f_{laser} \approx \frac{\text{noise}}{\text{signal/width}} = \frac{\Delta f_{cavity}}{\langle \text{SNR} \rangle}, \tag{11.15}$$

where SNR is the *signal-to-noise ratio*. From this, we obtain the general rule that the ultimate stabilized frequency jitter is equal to the *cavity width divided by the signal-to-noise ratio*. We occasionally say we can *split* the cavity width by the SNR. The shot noise limited SNR is

$$\text{SNR} = \frac{I_s}{I_n} = \frac{I_s}{\sqrt{2eI_s \Delta\nu}}, \tag{11.16}$$

where $\Delta\nu = 1/2\pi\tau$ is the noise bandwidth for an integration time of $\tau$ and $I_s$ is the signal current, which is given by

$$I_s = \frac{eP\eta}{\hbar\omega} \tag{11.17}$$

for a laser power into the cavity of $P$ and detector quantum efficiency (number of photo-electrons per photon) of $\eta$. The shot-noise-limited laser jitter is therefore

$$\text{Laser frequency jitter:} \quad \Delta f_{laser} = \Delta f_{cavity} \left(\frac{\hbar\omega}{\pi P\eta}\right)^{1/2} \left(\frac{1}{\tau}\right)^{1/2}. \tag{11.18}$$

All of these results are *rough estimates* and will be refined later in the chapter. It should be emphasized that $\Delta f_{laser}$ is a *lower bound* on the laser jitter obtained by a very well designed servo system. In practice, it is almost impossible to achieve this level of performance over a large range of noise frequencies. It will be useful to estimate the stabilized laser linewidth if we could actually realize shot-noise-limited behavior. Ultra-high finesse cavities are available with $\Delta f_{cavity} = 2$ kHz. If the power is 0.1 mW and $\eta = 1$, the shot-noise-limited laser jitter at $\lambda = 1$ $\mu$m is $\Delta f_{laser} = 5 \times 10^{-5}$ Hz for an integration time $(\tau)$ of 1 s, which is very narrow indeed. The relative instability in 1 second is $2.5 \times 10^{-19}$.

Before leaving this topic, it would be instructive to investigate the actual linewidth of a frequency-locked laser. The jitter calculated above is just the *measured* frequency noise over time $\tau$. If one had an extremely sharp optical spectrum analyzer, what would the laser line look like? It turns out that a laser whose frequency noise is *spectrally uniform* (such as the shot-noise-limited laser being discussed) will have a *Lorentzian* lineshape whose width, $\Delta f$, is given by

$$\text{Linewidth (white noise):} \quad \Delta f = \pi S_{\Delta f}, \tag{11.19}$$

where $S_{\Delta f}$ is the *power* spectral density of the laser frequency noise (in $\text{Hz}^2/\text{Hz}$). This can be justified with a calculation of the autocorrelation function of the frequency noise and an application of the Wiener-Khintchine theorem (the calculation is carried out in Elliot et al. (1982)). Using this result, the laser linewidth due to shot noise is

$$\Delta f = \pi \frac{2eI_s}{(K_d)^2} = \frac{2\pi(\Delta f_{cavity})^2 h\nu}{\eta P}, \tag{11.20}$$

where we used eqn 11.17 for the current and estimated $K_d$ (in A/Hz) as

$$K_d \approx \frac{e\eta P}{h\nu\Delta f_{cavity}} \quad \text{(in A/Hz).} \tag{11.21}$$

(The A/Hz units for $K_d$ are appropriate since the shot noise is represented as a *current*.) If one compares $\Delta f$ to the Schawlow–Townes expression for the quantum limit on the laser linewidth (eqn 6.84), one observes that the linewidth of a frequency-locked laser is described by essentially the same formula, replacing the laser cavity width with the external cavity width. Since $\Delta f$ is proportional to the *square* of the cavity linewidth, the improvement can be many orders of magnitude – a factor of $10^{12}$ for diode lasers! Again, this is a theoretical limit which is never realized in practice.

## 11.5   Some linear system theory

From the preceding section, we concluded that the best results are obtained when one constructs a laser stabilization system with the largest open loop gain possible over the largest possible range of frequencies. In order to realize this, one needs to examine the dynamical properties of servo systems, since the maximum gain and bandwidth will be limited by the requirement that the loop be *stable*. This section will provide a background in linear system theory as a prelude to a discussion of loop stability and other matters.

A *linear system* is one which satisfies the following *linearity conditions*:

$$f_1(t) \longrightarrow r_1(t) \tag{11.22}$$

$$f_2(t) \longrightarrow r_2(t) \tag{11.23}$$

$$\Rightarrow \alpha f_1(t) + \beta f_2(t) \longrightarrow \alpha r_1(t) + \beta r_2(t), \tag{11.24}$$

where $\alpha, \beta$ are complex constants, $r_1$ and $r_2$ are the *responses* to *driving functions* $f_1, f_2$, and the arrows connect the responses to the driving function. A general response, $r(t)$, is connected to a driving function, $f(t)$, by the differential equation

$$a_n \frac{d^n r}{dt^n} + a_{n-1} \frac{d^{n-1} r}{dt^{n-1}} + \cdots + a_1 \frac{dr}{dt} + a_0 r(t) \tag{11.25}$$
$$= b_m \frac{d^m f}{dt^m} + b_{m-1} \frac{d^{m-1}}{dt^{m-1}} + \cdots + b_1 \frac{df}{dt} + b_0 f(t),$$

where the coefficients are independent of time (a system described by a differential equation with constant coefficients is called a *time-invariant* system: its properties don't change with time). We will first assume a simple exponential dependence for both $r(t)$ and $f(t)$

$$f(t) = f_0 e^{st} \tag{11.26}$$

$$r(t) = r_0 e^{st}, \tag{11.27}$$

where $s$ is a *complex number* and is called the *complex frequency*. By including a real part in $s$, we allow for exponentially damped and exponentially increasing signals. Substituting these functions into the differential equation and solving for $r_0/f_0$, we obtain

$$H(s) = \frac{r_0}{f_0} = \frac{b_m s^m + b_{m-1} s^{m-1} + \cdots + b_1 s + b_0}{a_n s^n + a_{n-1} s^{n-1} + \cdots + a_1 s + a_0}. \tag{11.28}$$

The quantity $H(s)$ is called the *transfer function* for the system and it is a complex-valued rational fraction. It is simply the ratio of the response to the driving function. It should be fairly obvious that the overall transfer function of *cascaded systems* (systems placed one after the other) is simply the *product* of the transfer functions of the individual systems. We see that an exponential driving function will give rise to an exponential response with the *same* complex frequency.

Most driving and response functions are not exponentials. In order to describe a *general* function, we decompose it into a *continuous sum of damped, eternal sinusoidal functions* using the *Laplace transform*. For a function of time, $f(t)$, the Laplace transform is $F(s)$; by convention, the time function is in lower case and the Laplace transform is in upper case using the same letter. The Laplace transform and its inverse are defined by

$$F(s) = \mathcal{L}[f(t)] = \int_{0^-}^{\infty} f(t)e^{-st}dt \tag{11.29}$$

$$f(t) = \mathcal{L}^{-1}[F(s)] = \frac{1}{2\pi i}\int_{\sigma-i\infty}^{\sigma+i\infty} F(s)e^{st}d\omega, \tag{11.30}$$

where $s = \sigma + i\omega$, and the inverse transform integration takes place in the complex plane along a vertical line whose real intercept is $\sigma$; any line which is to the right of the singularities of $F(s)$ can be used. The lower limit $(t = 0^-)$ in the first integral refers to a time which is infinitesimally before $t = 0$ to include in the integration a possible *impulse* at $t = 0$. The Laplace transform is a generalization of the *Fourier transform* to include damped exponentials, which makes this transform easier to apply to certain problems. Due to the presence of a damped factor, the Laplace transform is more likely to exist than the Fourier transform. The Laplace transform of $f(t)$ exists for $s = \sigma + i\omega$ if the integral

$$\int_{-\infty}^{\infty} |f(t)|e^{-\sigma t}dt \tag{11.31}$$

is finite. One can show that the latter condition is satisfied for all $\sigma > \alpha$ if there exists a real positive number $\alpha$ satisfying

$$\lim_{t\to\infty} [e^{-\alpha t}f(t)] = 0. \tag{11.32}$$

The following is a list of some of the properties of the Laplace transform:

**Linearity**: If $f_1(t) \leftrightarrow F_1(s)$ and $f_2(t) \leftrightarrow F_2(s)$, then:

$$a_1 f_1(t) + a_2 f_2(t) \leftrightarrow a_1 F_1(s) + a_2 F_2(s)$$

**Time-differentiation**: If $f(t) \leftrightarrow F(s)$, then

$$\frac{df}{dt} \leftrightarrow sF(s) - f(0^-)$$

**Time-integration**: If $f(t) \leftrightarrow F(s)$, then

$$\int_0^t f(t)dt \leftrightarrow \frac{F(s)}{s}$$

**Time-shifting**: For $t_0 > 0$,

$$f(t - t_0) \leftrightarrow F(s)e^{-st_0}$$

**Time convolution**: If $f_1(t) \leftrightarrow F_1(s)$ and $f_2(t) \leftrightarrow F_2(s)$, then:

$$\int_{-\infty}^{\infty} f_1(\tau)f_2(t - \tau)d\tau \leftrightarrow F_1(s)F_2(s)$$

**Initial-value theorem:** If $f(t)$ and $df/dt$ have Laplace transforms, then the *initial value* of $f(t)$ is given by:

$$f(0^+) = \lim_{s \to \infty} [sF(s)]$$

**Final-value theorem:** If $f(t)$ and $df/dt$ have Laplace transforms, then the *final value* of $f(t)$ is given by:

$$f(\infty) = \lim_{s \to 0} [sF(s)]$$

The double arrow ($\leftrightarrow$) identifies a *Laplace transform pair* and the symbols $0^+$ and $0^-$ indicate a time infinitesimally after and infinitesimally before $t = 0$. A list of the most important Laplace transform pairs is given in Table 11.1.

**Table 11.1**  Table of Laplace transforms.

| $f(t) = \mathcal{L}^{-1}[F(s)]$ | $F(s) = \mathcal{L}[f(t)]$ |
| --- | --- |
| $\delta(t)$ (delta function) | $1$ |
| $u(t)$ (unit step at $t = 0$) | $\dfrac{1}{s}$ |
| $tu(t)$ (unit ramp) | $\dfrac{1}{s^2}$ |
| $u(t)\sin(\omega_0 t)$ | $\dfrac{\omega_0}{s^2 + \omega_0^2}$ |
| $u(t)\cos(\omega_0 t)$ | $\dfrac{s}{s^2 + \omega_0^2}$ |
| $u(t)e^{at}$ | $\dfrac{1}{s - a}$ |

Functions of time, $f(t)$, which have the property that $f(t) = 0$ when $t < 0$ are called *causal functions*. Since, by causality, the response cannot anticipate the driving function, the response must also be zero for $t < 0$. Most functions we will encounter are causal.

If we assume that the driving function is causal and take the Laplace transform of both sides of eqn 11.25, we obtain the same result for the transfer function as was obtained for an exponential driving function:

$$H(s) = \frac{R(s)}{F(s)} = \frac{b_m s^m + b_{m-1} s^{m-1} + \cdots + b_1 s + b_0}{a_n s^n + a_{n-1} s^{n-1} + \cdots + a_1 s + a_0}, \tag{11.33}$$

where we assume the system is initially de-energised and set the initial values equal to zero. Note that, from the initial value theorem, the transfer function at $t = 0$ will be infinite unless $m < n$; such a rational fraction is called a *proper fraction*.

The response function, $r(t)$, is obtained in three steps: first, we determine the transfer function, then take the Laplace transform of the driving function, $f(t)$, and finally invert the product:

$$r(t) = \mathcal{L}^{-1}[H(s)F(s)]. \tag{11.34}$$

The beauty of this approach is that the driving function transform can usually be obtained from a table and the transfer function can be obtained using the *standard rules for combining impedances*, where some elementary impedances are

$$\text{Capacitor:} \ \ Z_C = \frac{1}{sC} \tag{11.35}$$

$$\text{Inductor:} \ \ Z_L = sL \tag{11.36}$$

$$\text{Resistor:} \ \ Z_R = R. \tag{11.37}$$

In order to use the table of Laplace transforms, it is usually necessary to simplify $H(s)$ by factoring the denominator and expanding $H(s)$ into *partial fractions*. If we divide out $a_n$ and factor the denominator,

$$H(s) = \frac{b_m s^m + b_{m-1} s^{m-1} + \cdots + b_1 s + b_0}{(s - s_1)(s - s_2) \cdots (s - s_l)(s - s_0)^p}, \tag{11.38}$$

where there are $l$ *simple roots* and one root with multiplicity $p$ $(n = l + p)$. The partial fraction expansion yields

$$H(s) = \frac{c_1}{s - s_1} + \frac{c_2}{s - s_2} + \cdots + \frac{c_l}{s - s_l} \qquad \text{(simple roots)} \tag{11.39}$$

$$+ \frac{k_0}{(s - s_0)^p} + \frac{k_1}{(s - s_0)^{p-1}} + \cdots + \frac{k_{p-1}}{s - s_0} \qquad \text{(multiple root).} \tag{11.40}$$

There are numerous procedures for obtaining partial fraction expansions; we will briefly describe one for simple roots and one for multiple roots. First assume that the above partial fraction expansion contains only simple roots. Then, the coefficient $c_r$ is

$$\text{Simple roots:} \ \ c_r = H(s)(s - s_r)|_{s=s_r}, \tag{11.41}$$

where $s_r$ is one of the roots. The explanation for this should be self-evident. For a multiple root with *multiplicity* (maximum power) equal to $p$, the expansion is

$$H(s) = \frac{k_0}{(s - s_0)^p} + \frac{k_1}{(s - s_0)^{p-1}} + \cdots + \frac{k_j}{(s - s_0)^{p-j}} + \cdots + \frac{k_{p-1}}{s - s_0} \tag{11.42}$$

and the $j$th coefficient is

$$k_j = \frac{1}{j!} \frac{d^j}{ds^j} \left[ H(s)(s - s_0)^p \right] \Bigg|_{s=s_0}. \tag{11.43}$$

An example should illustrate the power and simplicity of the use of Laplace transforms to analyze an electrical network. The circuit appears in Fig. 11.4. We seek an expression for the current as a function of time if the switch is closed at $t = 0$. The

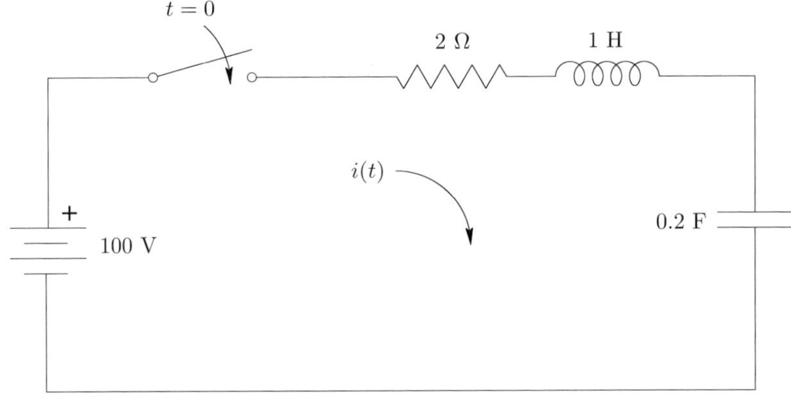



**Fig. 11.4** RLC circuit analyzed using Laplace transforms.

driving function is a step of magnitude equal to 100 V and which occurs at $t = 0$. From the table, the transform of the driving function is

$$F(s) = \mathcal{L}[100u(t)] = \frac{100}{s}. \tag{11.44}$$

The transfer function is the reciprocal of the total impedance

$$H(s) = \frac{1}{Z(s)} = \frac{1}{2 + s + 1/0.2s}, \tag{11.45}$$

where we used the expressions for the individual impedances of the elements and combined them using the well-known rule for series impedances. The current is

$$
\begin{aligned}
i(t) &= \mathcal{L}^{-1}[H(s)F(s)] \\
&= \mathcal{L}^{-1}\left[\frac{100}{s^2 + 2s + 5}\right] = 25i\mathcal{L}^{-1}\left(\frac{1}{s+1+2i} - \frac{1}{s+1-2i}\right) \\
&= 25i(e^{(-1-2i)t} - e^{(-1+2i)t}) = 50e^{-t}\sin 2t.
\end{aligned}
\tag{11.46}
$$

## 11.6 The stability of a linear system

The most important and severe constraint on a control system is the need to prevent undesired sustained *oscillation*. In order to examine the conditions for oscillation (instability), we need to understand the behavior of the *natural modes* in a system; these are the responses which occur in the *absence of a driving function*. We begin by discussing the *impulse response* of a linear system.

A *unit impulse* which occurs at time $t'$ can be represented by a *delta function*, $\delta(t - t')$, defined by

$$\delta(t - t') = 0, \quad \text{when } t \neq t' \tag{11.47}$$

$$\int_{-\infty}^{\infty} \delta(t)dt = 1. \tag{11.48}$$

This is a *pulse* of infinite height and infinitesimal width which occurs at $t = t'$. The Laplace transform of a delta function, $\delta(t)$, is 1 (from the table). Therefore, the *response* to a *unit impulse* is $h(t)$ and is given by

$$h(t) = \mathcal{L}^{-1}[H(s)], \tag{11.49}$$

for an impulse which occurs at $t' = 0$. Since the system is time-invariant, this can also be written as $h(t - t')$, which is interpreted as the response at observation time $t$ due to an impulse which occurs at time $t'$. The quantity $h(t - t')$ is called the *impulse response* of the system and, for a *time-invariant* system, depends only on $t - t'$. One can consider the transfer function, $H(s)$, to be the *Laplace transform of the impulse response*.

From the definition of the delta function, any function can be expanded in unit impulses:

$$f(t) = \int_{-\infty}^{\infty} f(t')\delta(t - t')dt'. \tag{11.50}$$

Since $h(t - t')$ is the response at $t$ to a single impulse at $t'$, by *superposition* (a consequence of linearity), the response to a general driving function can be written as a *continuous sum* of the impulse response, $h(t - t')$, multiplied by the driving function evaluated at time $t'$:

$$r(t) = \int_{-\infty}^{\infty} h(t - t')f(t')dt'. \tag{11.51}$$

This is a *convolution integral* and the use of the *time convolution property* yields the familiar relation, $R(s) = H(s)F(s)$.

An impulse driving function will excite the *natural modes* of a linear system. To see why this is so, we expand the transfer function in partial fractions:

$$h(t) = \mathcal{L}^{-1}[H(s)] = \mathcal{L}^{-1}\left[\frac{c_1}{s - s_1} + \frac{c_2}{s - s_2} + \cdots + \frac{c_n}{s - s_n}\right] \tag{11.52}$$

$$= c_1 e^{s_1 t} + c_2 e^{s_2 t} + \cdots + c_m e^{s_n t}, \tag{11.53}$$

where we assumed that all of the roots of the denominator are simple. Thus, the response to an impulse will be a series of exponentials at frequencies which are the *roots of the denominator of the transfer function*. Since the delta function is active only for an infinitesimal duration at $t = 0$, these responses occur in the *absence* of a sustained drive and the frequencies are called the *natural frequencies of a system*. (The sound of a *bell* which is briefly struck is a familiar example of this.) This result can also be obtained by solving the *homogeneous* equation which is obtained by setting the right-hand side of eqn 11.25 equal to zero (i.e., removing the driving function). The roots of the denominator of the transfer function are called the *poles* of $H(s)$ and their locations in the complex plane determine the stability of a linear system. The roots of the *numerator* of the transfer function are called the *zeros* of $H(s)$.

Figure 11.5 displays the natural responses of a system for a number of pole locations in the complex $s$-plane. We observe that the poles in the left-hand plane will have damped behaviors and those in the right-hand plane will have exponentially increasing

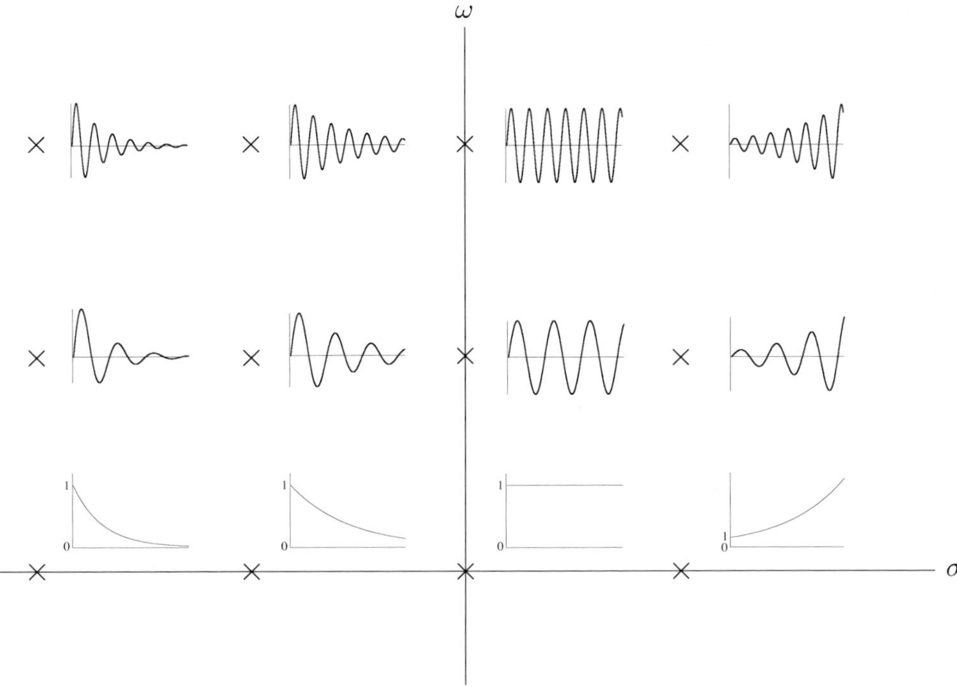

**Fig. 11.5** Natural responses of a system for various pole locations (the "×"s).

behaviors. Clearly, the latter are undesirable. The *simple* poles on the imaginary axis will correspond to a stable sinusoid; while this is not usually desirable, it is not considered an unstable behavior. Finally, the existence of poles with multiplicity greater than one on the imaginary axis results in at least a linearly increasing sinusoid and therefore is unstable behavior. To see why this is so, we evaluate the inverse transform of $1/(s-i\omega)^2$

$$\mathcal{L}^{-1}[1/(s-i\omega)^2] = te^{i\omega t}u(t). \tag{11.54}$$

Poles with higher multiplicity will be associated with a response that increases even more rapidly with time. We thus arrive at the following rule. *A system is stable if and only if the poles of the transfer function are either all in the left-hand plane or are on the imaginary axis and are simple.*

There are a number of techniques for determining the stability of a system using the behavior of the poles. We will not discuss these but will instead use a more easily understood approach which is uniquely applicable to feedback systems. It is called the *Bode plot* and will be presented after a general discussion of feedback.

## 11.7   Negative feedback

A block diagram of a generic system employing negative feedback appears in Fig. 11.6. In this configuration, the output is made to *track* the input by comparing the output to the input, taking the difference, and applying the difference (the error signal, $E(s)$)

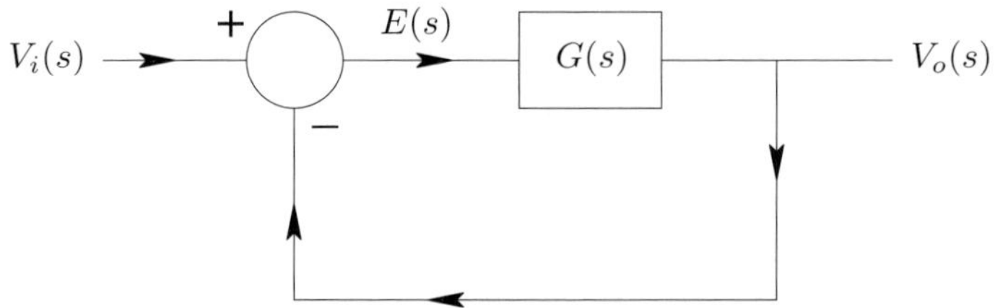

**Fig. 11.6**  A generic negative feedback system.

to the element whose transfer function is $G(s)$. After some simple algebra, the error
signal is

$$\text{Error:}\quad \frac{E(s)}{V_i(s)} = \frac{1}{1 + G(s)} \tag{11.55}$$

and the output signal is

$$\text{Output:}\quad \frac{V_0(s)}{V_i(s)} = H(s) = \frac{G(s)}{1 + G(s)}. \tag{11.56}$$

The quantity $1 + G(s)$ is called the *return difference*: its *zeros* (roots) are the natural
frequencies of the system (and determine its stability). The quantity $G(s)$ is the *open-
loop transfer function*: the gain around the loop if it is broken at any point.

Some treatments of feedback use other network topologies. These are easily transformed into the geometry depicted in Fig. 11.6 using *block diagram equivalences*, some
of which appear in Fig. 11.7. For example, the top-left diagram is used in some other
treatments of feedback. With the aid of the indicated equivalence, one can immediately
write down the transfer function for this geometry:

$$\text{(Alternative geometry)}\quad H(s) = \left(\frac{1}{F(s)}\right)\frac{G(s)F(s)}{1 + G(s)F(s)} = \frac{G(s)}{1 + G(s)F(s)}. \tag{11.57}$$

This geometry is more general than the original one and can describe a system where
the output does *not* track the input. We will use the original geometry in all that
follows, since laser stabilization systems track the reference.

An important characteristic of a tracking feedback system is its *dynamic behavior*:
how well the tracking performs with a *changing* input. We usually consider inputs
which vary as powers of $t$: a step input (constant), a ramp input ($\propto t$) and an ac-
celerated input ($\propto t^2$). We would like the *error to approach zero after long times*. To
analyze the behavior as $t \to \infty$ we use the final value theorem. Thus, we take the limit

$$e_s(\infty) = \lim_{s \to 0}[sV_i(s)H_e(s)], \tag{11.58}$$

where $V_i(s)$ is the transform of the input and $H_e(s)$ is the transfer function for the
*error signal, $e_s(t)$:*

$$H_e(s) = \frac{1}{1 + G(s)}. \tag{11.59}$$

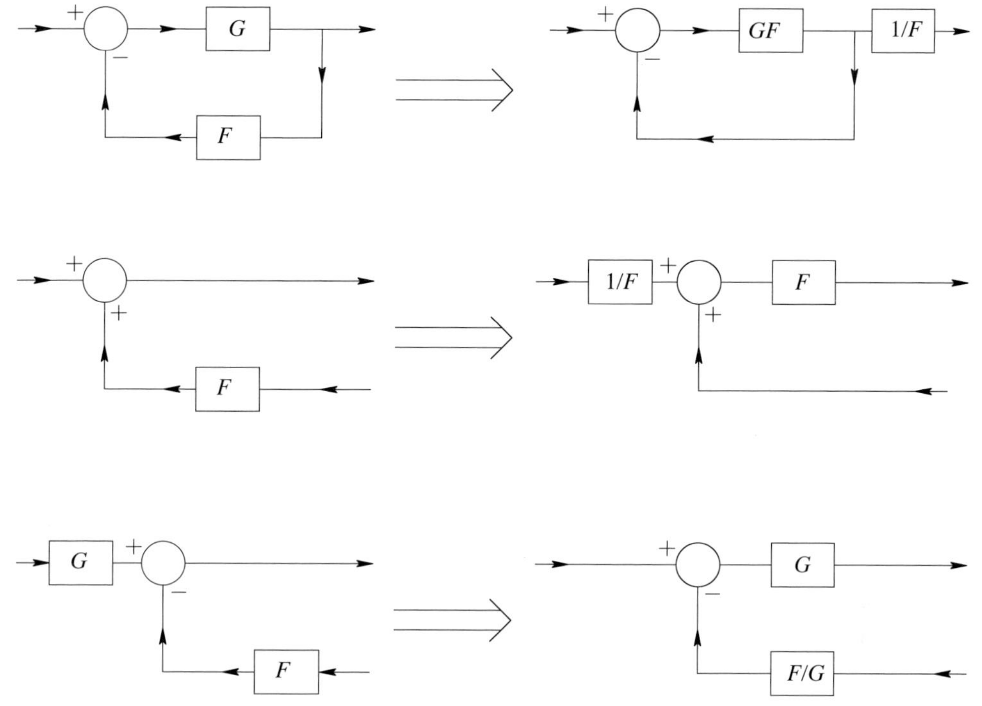

**Fig. 11.7** A few block diagram equivalences.

We first model the function $G(s)$ (the open loop transfer function) as one or more cascaded integrators. For a single integrator, $G(s)$ is

$$\text{Single integrator:}\quad g(t) = \int_0^t [\,]dt' \implies G(s) = \frac{1}{s} \tag{11.60}$$

from the integration rule for Laplace transforms (the integrand is not shown since the integral is treated as an *operator*). Clearly, for $n$ cascaded integrators, the transform is

$$n \text{ integrators:}\quad G(s) = \frac{1}{s^n}. \tag{11.61}$$

For a *step input*, the transform of the input is $V_i(s) = 1/s$. The error is

$$\text{Step input:}\quad e_s(\infty) = \lim_{s\to 0}\left( s\frac{1}{s}\frac{1}{1+G(s)} \right) = \frac{1}{1+G(0)}. \tag{11.62}$$

In order to force $e_s(\infty) \to 0$, it is necessary for $G(0) \to \infty$. This requires at least one integrator.

The transform of a *ramp input* is $V_i(s) = 1/s^2$. The error is

$$\text{Ramp input:}\quad e_s(\infty) = \lim_{s\to 0}\left( s\frac{1}{s^2}\frac{1}{1+G(s)} \right) = \lim_{s\to 0}\frac{1}{sG(s)}. \tag{11.63}$$

To reduce the error (occasionally called the *velocity* error) to zero, we clearly need at least two integrators.

Finally, for an *acceleration input*, $V_i(s) = 1/s^3$ and

$$\text{Acceleration input:}\quad e_s(\infty) = \lim_{s \to 0}\left(s\,\frac{1}{s^3}\,\frac{1}{1+G(s)}\right) = \lim_{s \to 0}\frac{1}{s^2 G(s)}. \tag{11.64}$$

It should come as no surprise that we need at least three integrators to eliminate the acceleration error at $t = \infty$. In the interest of stability, it is desirable to minimize the number of integrators, so it is possible that some of the tracking capabilities might be lost in an actual system.

Complementary to the above discussion of steady-state errors at $t = \infty$ is the *transient response*: the short-term behavior of the output of a closed loop with a step input at $t = 0$. To investigate this, we will describe a *canonical* second-order system, which is shown schematically in Fig. 11.8. The transfer function for the output is

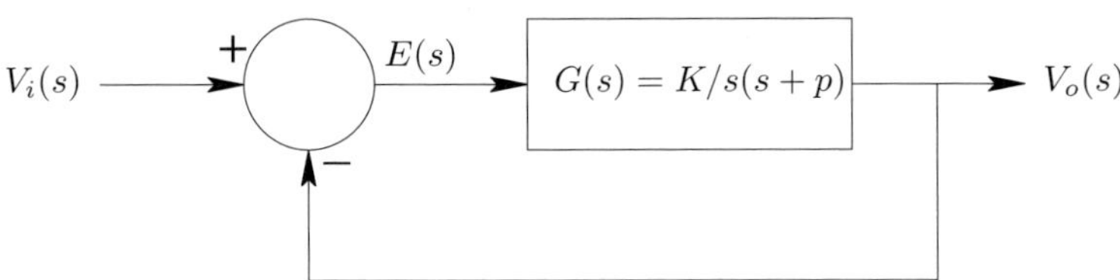


**Fig. 11.8**  A canonical second-order loop.

$$H(s) = \frac{G(s)}{1+G(s)} = \frac{K}{s^2 + ps + K} = \frac{\omega_n^2}{s^2 + 2\zeta\omega_n s + \omega_n^2}. \tag{11.65}$$

The constants $p$ and $K$ are positive and real. We have introduced two new parameters: the *natural frequency*, $\omega_n$, and the *damping constant*, $\zeta$. These are given by

$$\omega_n = \sqrt{K} = \text{ natural frequency} \tag{11.66}$$

$$\zeta = \frac{p}{2\sqrt{K}} = \text{ damping constant}\quad \begin{cases} \zeta > 1 & : \quad \text{overdamping} \\[2mm] \zeta = 1 & : \quad \text{critical damping} \\[2mm] \zeta < 1 & : \quad \text{underdamping}. \end{cases} \tag{11.67}$$

The poles of $H(s)$ are

$$\text{Poles:}\quad s_1, s_2 = -\zeta\omega_n \pm \omega_n\sqrt{\zeta^2 - 1}. \tag{11.68}$$

Note that the poles are real when $\zeta > 1$ (overdamping) and there is therefore no oscillatory component to the step response in this case. Furthermore, the system is *unconditionally stable*, since the poles are in the left-hand plane for all finite $\zeta$ and $\omega_n$.

The transient response is obtained by taking the inverse transform of $F(s)H(s)$. The result, for $\zeta \leq 1$, is

$$r(t) = \mathcal{L}^{-1}\left[\frac{\omega_n^2}{s(s^2 + 2\zeta\omega_n s + \omega_n^2)}\right] = 1 - \frac{1}{\beta}e^{-\zeta\omega_n t}\sin(\omega_n\beta t + \theta) \quad (11.69)$$

$$\text{where} \quad \beta = \sqrt{1 - \zeta^2} \quad (11.70)$$

$$\theta = \tan^{-1}\beta/\zeta. \quad (11.71)$$

For $\zeta > 1$, the result is

$$r(t) = 1 - \frac{1}{\beta}e^{-\zeta\omega_n t}\sinh(\omega_n\beta t + \theta) \quad (11.72)$$

$$\text{where} \quad \beta = \sqrt{\zeta^2 - 1} \quad (11.73)$$

$$\theta = \tanh^{-1}\beta/\zeta. \quad (11.74)$$

For $\zeta \ll 1$, the response of $H(s)$ to a *sinusoidal input* is strongly peaked at resonant frequency $\omega_n$. The response can be described by a $Q$ value:

$$Q = \frac{1}{2\zeta}. \quad (11.75)$$

The response to a step input for a number of damping factors $(\zeta)$ appears on the left of Fig. 11.9. The pole locations for the same values of $\zeta$ are shown in the graph on the

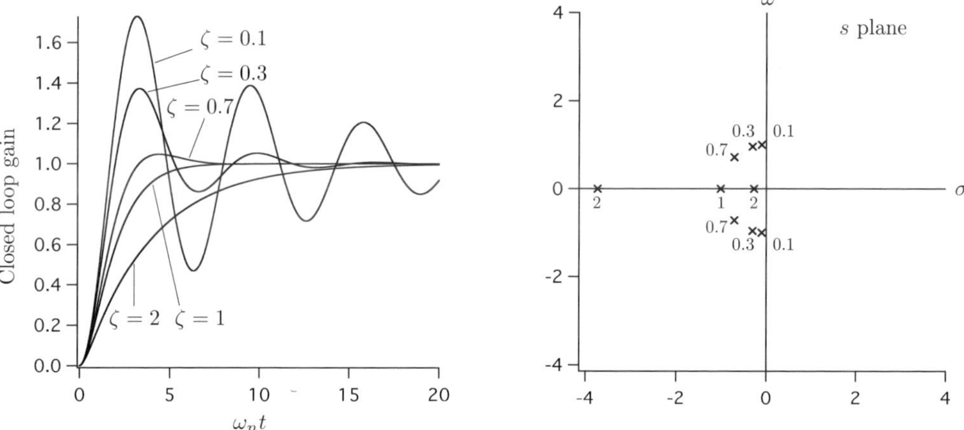

**Fig. 11.9** Transient response of a second-order loop. The curve on the left is the response curve for a number of damping factors and the pole locations for the same values of $\zeta$ are shown in the right-hand graph.

right. This sequence of poles plotted as a function of $\zeta$ is an example of the *root locus* technique of stability analysis, where the *locus* of the poles of the closed-loop transfer function is plotted as a function of some parameter and it is determined whether or not the poles are confined to the left hand of the complex $s$-plane. The second-order loop is simple enough that the pole locations can be easily determined from the algebraic expression (eqn 11.68) for the poles.

The *Bode plot* method of stability analysis is perhaps the most familiar and easiest to apply. It begins with a fairly obvious observation about the transfer function, $H(s)$,

$$H(s) = \frac{G(s)}{1 + G(s)}. \tag{11.76}$$

For the loop to be stable, the return difference, $1 + G(s)$, must be *non-zero* for *all frequencies*. This means that the *open-loop gain*, $G(s)$, must never have a magnitude of unity and a phase of $-180°$ *at the same time*. We usually design a system with stability in mind by starting at the *unity gain frequency* of $G(s)$ and making sure that the phase shift is less than $\pi$ radians at this frequency. The utility of this method is due to the fact that the unity gain frequency is often limited by inescapable time delays and, for best servo performance, one designs the loop filter to allow the maximum unity gain frequency. Furthermore, $|G(s)|$ is almost always a monotonically decreasing function of frequency and therefore, if one establishes stability when $|G(s)| = 1$, the possibility of instability will *never* be revisited at lower frequencies. We can then include a sufficient number of filter stages to provide adequate gain at *lower* frequencies in order to satisfy the error and tracking requirements without adding destabilizing phase shifts at the unity gain frequency. The unity gain frequency of the open-loop gain has another important characteristic: it is the frequency where the *closed-loop gain* begins to fall off (it is only the $-3$ dB point of the closed-loop gain). One is cautioned to use the minimum number of stages to avoid excessive time delays and to be careful with large phase shifts at gain $> 1$ since a disturbance can momentarily saturate the amplifiers and trigger oscillation.

We can see at a glance whether a system is stable by plotting both the *magnitude* of the open-loop gain and its *phase* versus frequency. The frequency is always plotted on a logarithmic scale and the open-loop gain magnitude is also plotted on a logarithmic scale. With a logarithmic scale, it is especially easy to draw the plot using a few simple rules. For cascaded systems, the total response is simply the *sum of the individual responses*. For the magnitude plot,

1. Each pole increases the slope by 20 db/decade (6 db/octave).
2. Each zero of $G(s)$ decreases the slope by 20 db/decade.
3. The actual pole/zero frequency occurs at $\pm 3$ db point.

For the phase plots (plotted on a *linear* scale for the phase and logarithmic scale for the frequency),

1. Each pole provides a $-90°$ phase shift (*lag*), asymptotically.
2. Each zero provides a $+90°$ phase shift (*lead*), asymptotically.
3. The lag (lead) at the exact pole (zero) frequency is $45°$.

It is interesting to note that each 20 dB/decade increase or decrease in *slope* is accompanied by a $90°$ increase or decrease in phase shift. This characteristic is a consequence of a general *dispersion relation* which applies to all physically realizable linear systems and has its origin in the requirement of *causality*: every response must *follow* its cause.

Typical Bode magnitude and phase plots are shown in Fig. 11.10. One can easily

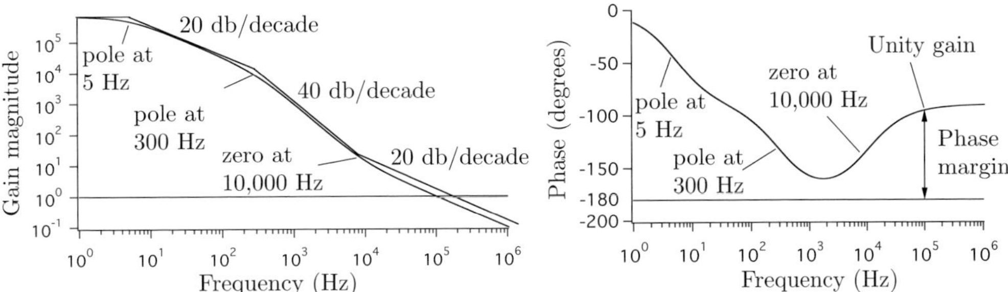


**Fig. 11.10** Bode magnitude and phase plots.

draw the plots by approximating them using straight lines with the slopes and values given by the above rules. Although Fig. 11.10 was for an arbitrary system, it does illustrate a technique for obtaining increased low-frequency gain with the judicious use of one or more *zeros* (provided by a *circuit element which differentiates*). While a system with two poles will not oscillate, it can have a *transient response* characterized by significant *ringing*, which can be thought of as a condition of *near instability*. By introducing a zero slightly below the unity gain frequency, one can greatly improve the transient response and still have adequate gain at lower frequencies.

The nearness to instability can be gauged by the *phase margin*: the difference between the actual phase shift and $-180°$ at the unity gain frequency The smaller the phase margin, the closer the loop is to being unstable. In order to better illustrate the significance of phase margin, we will display bode plots for a second-order loop with two different values of the damping constant, $\zeta$. The Bode plots on the left of Fig. 11.11 are for a damping constant of 0.707 and the plots on the right are for a damping constant of 0.1. The larger damping constant provides a comfortable 65° phase margin while the smaller damping constant provides a fairly small 11° phase margin. If one refers back to Fig. 11.9, it should be evident that the small phase margin is accompanied by a very poor transient behavior: the response to a step input takes several cycles before it settles down. The larger phase margin, on the other hand, quickly settles after a slight *overshoot*. For these reasons, the usual practice is to require a phase margin of 30° $(\pi/6)$, which provides a good protection against instability while ensuring reasonable transient behavior.

Before continuing, we should caution the reader about a source of possible mistakes for the beginner. We have been discussing two different transfer functions: the *open-loop gain* and the *closed-loop transfer function*. The Bode plot analysis of stability *only applies to the open-loop gain*. The closed-loop gain is useful for determining stability from the *pole positions* (must be in the left half-plane) and one can determine the transient response to a unit step by inverting the quantity $H(s)/s$, but the Bode plot approach to stability analysis can be used only with the open-loop gain. Finally, the poles of the open-loop transfer function are used in making the Bode plot according to

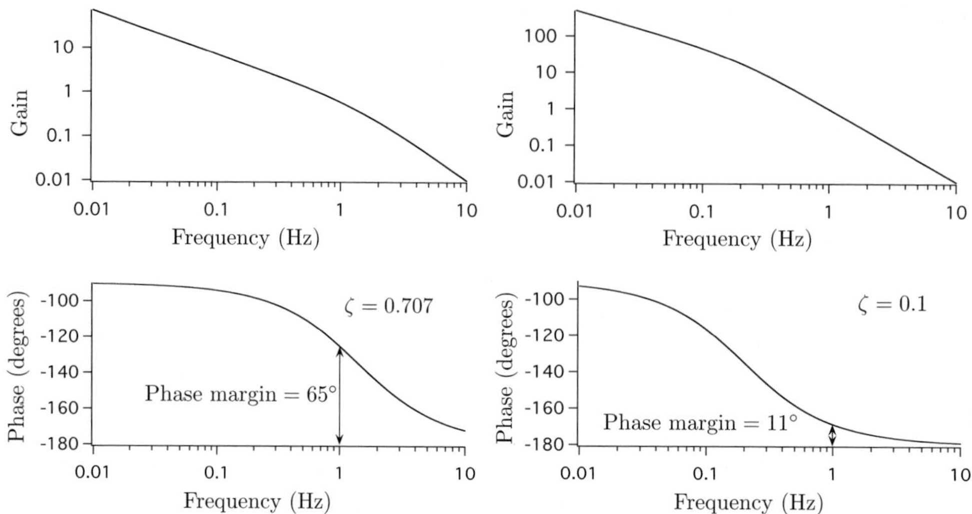

Fig. 11.11 Bode magnitude and phase plots for the second-order loop for $\zeta = 0.707$ (left) and $\zeta = 0.1$ (right).

the rules discussed earlier and the poles of the closed-loop transfer function are used for an analysis of stability and the transient response.

As a final illustration of Bode plots, we will provide an explanation for the well-known unity-gain instability of uncompensated operational amplifiers, familiar to anyone who has worked with them. An operational amplifier with feedback in the non-inverting configuration is shown schematically in Fig. 11.12. The *feedback factor, $T(s)$,*

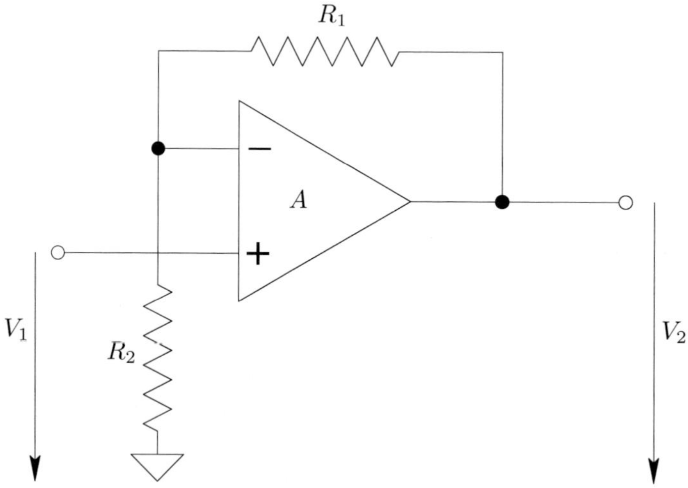

Fig. 11.12 Circuit of operational amplifier with feedback.

is equal to the voltage divider ratio,

$$T(s) = \frac{R_2}{R_1 + R_2}. \tag{11.77}$$

It is easy to show that the transfer function for the circuit is

$$H(s) = \frac{A(s)}{1 + A(s)T(s)}, \tag{11.78}$$

where $A(s)$ is the gain of the amplifier without feedback (the open-loop gain is $A(s)T(s)$). The closed-loop gain at frequencies for which $A(s)T(s) \gg 1$ is simply $1/T(s)$: the reciprocal of the feedback factor. This is one of the chief advantages of negative feedback: one can force desired properties of the circuit (the gain in this case) to be dependent upon *passive* elements which are much more stable and less prone to nonlinearities (which cause distortion) than active ones.

Bode plots for the bare gain of an operational amplifier without feedback (left) and the open-loop gain with two different levels of feedback (right) appear in Fig. 11.13. The bare operational amplifier is modeled as having a DC gain of $10^6$ and having

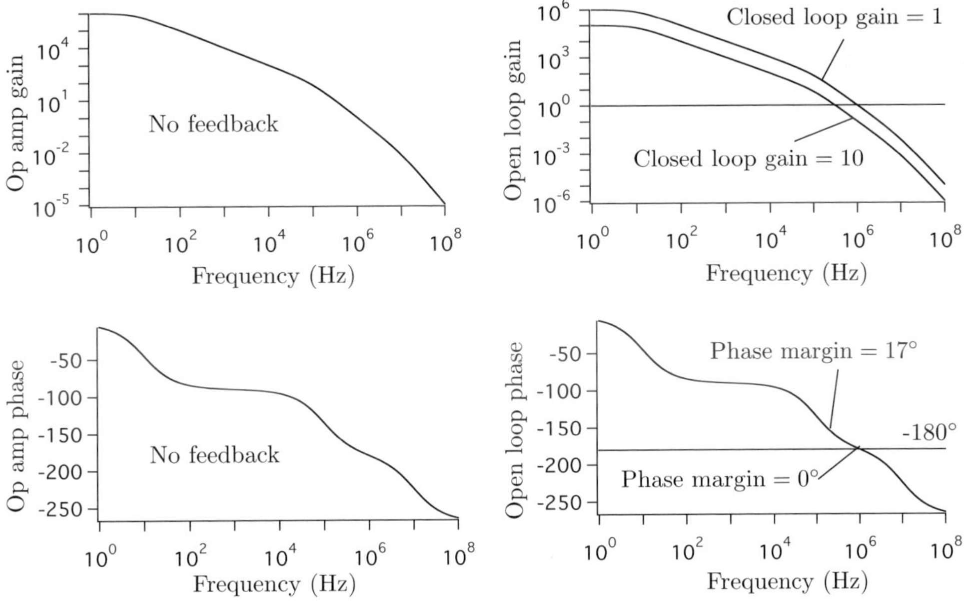

**Fig. 11.13** Bode plots for bare operational amplifier (left) and operational amplifier with a gain of 1 and gain of 10 (right). Note that the phase plot is the same for the two gains.

poles at 10 Hz, 100 kHz, and 10 MHz. The open-loop gain with a feedback factor of 1 ($R_2 \gg R_1$) is the same as the gain ($A(s)$) of the bare amplifier and with a feedback factor of 0.1 (closed-loop gain=10) the open-loop gain is $0.1A(s)$. When one considers the phase margin at both unity gain frequencies, the reason for the instability should be clear: the phase margin at a closed-loop gain of 1 is $0°$ and at a closed-loop gain of 10 is $17°$. One can use similar arguments to explain the instability of an amplifier with a capacitive load (it adds another pole due to the output impedance of the amplifier and the load capacitance).

We conclude this section with a discussion of the adverse effect on stability of *time delays*. From the time shifting property of Laplace transforms, a time shift of $\tau$ changes the transform by a factor of $e^{-s\tau}$:

$$f(t - \tau) \longrightarrow F(s)e^{-s\tau}. \tag{11.79}$$

Thus, there is a phase shift at angular frequency $\omega$ of $-\omega\tau$ radians. From this, we arrive at the effect of time delays on stability: *a time delay of $\tau$ seconds reduces the phase margin at the unity gain frequency, $\omega_0$, by $\omega_0\tau$ radians*. As an example, we will consider a loop filter consisting of four operational amplifiers plus 10 feet of coaxial cable. Assume the unity gain frequency of the loop is 1 MHz. A fast operational amplifier will have a time delay of about 10 ns. The delay in typical coaxial cables is about 1.5 ns per foot. The total delay is 55 ns which corresponds to a phase change of $20°$ at the unity gain frequency. This is quite significant and can cause an otherwise stable system to oscillate. The only cure is to use shorter cables and circuit topologies with fewer and shorter amplifier delays. Ultimately, time delays can force a lowering of the unity gain frequency, with likely adverse effects on the performance of the feedback system.

## 11.8   Some actual control systems

In this section, we will briefly describe several actual control systems but will defer a detailed analysis of two of them (temperature control and frequency lock) until later in the chapter. We will also describe some of the filters commonly used in control systems. We will adopt a convention which is common in the literature and factor the open loop transfer function into the product of a *controller* and a *plant*. Thus,

$$G(s) = G_c(s)G_p(s). \tag{11.80}$$

The plant, described by $G_p(s)$, refers to the process being controlled, including sensors and actuators. The controller, described by $G_c(s)$, is the additional circuitry (essentially a *filter*) which is added by the designer to obtain the desired servo performance. A block diagram which displays this factorization appears in Fig. 11.14.

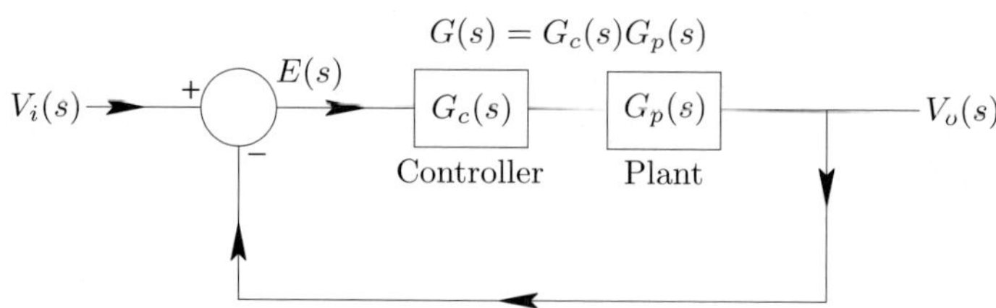


**Fig. 11.14** Block diagram of feedback system showing the controller and plant.

**Phase-locked loop:** The block diagram for a phase-locked loop appears in Fig. 11.15. The variable being controlled in a phase-locked loop is the *phase* of the output of a

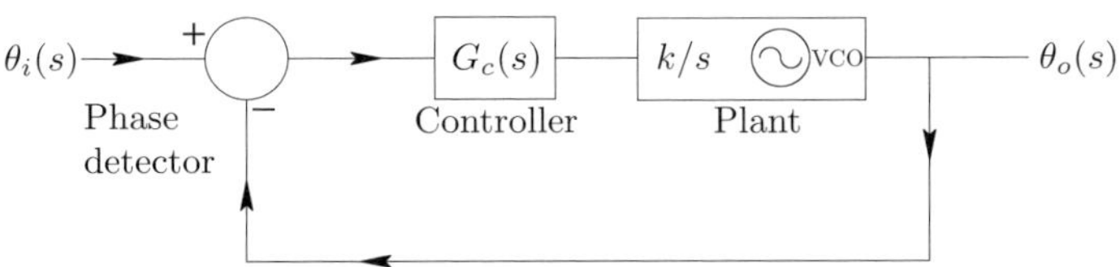


**Fig. 11.15** Block diagram of phase-locked loop.

voltage-controlled oscillator (VCO). Since the *frequency* of the VCO is controlled by its input, the plant contains an *implicit* integrator (phase is the time integral of frequency):

$$G_p(s) = \frac{k}{s}. \tag{11.81}$$

The phase of the VCO output is compared to that of an input *reference* signal in a *phase detector*, whose output is proportional to the phase difference between the two inputs.

Phase-locked loops are described by *type* and by *order*. The *type* is the number of poles at the *origin* of $G(s)$ (the open-loop gain) and essentially characterizes the loop's tracking behavior. The minimum type is 1, since there is at least one integrator (in the plant). Using the results of our discussion of the steady-state errors, we list the errors by loop type in Table 11.2.

**Table 11.2** Steady-state errors for various types of phase-locked loops.

|  | Type 1 | Type 2 | Type 3 |
| --- | --- | --- | --- |
| Step position | zero | zero | zero |
| Step velocity | constant | zero | zero |
| Step acceleration | unbounded | constant | zero |

The *order* of a loop is the degree of the numerator of the return difference $(1 + G(s))$. It is equal to the number of poles in the closed-loop transfer function. A common configuration has a *controller* whose gain is

$$G_c(s) = \frac{K}{s+p}. \tag{11.82}$$

When combined with the plant gain, this is just the open-loop gain of the second-order loop which we recently studied. Since it is only type 1, it might not be suitable for changing signals. Another common design uses a *lead-lag* controller, whose gain is

$$G_c(s) = K\frac{s+a}{s},\qquad(11.83)$$

where $a$ is a positive constant. The constant $a$ is all-important: it provides a zero in the left-hand plane. This is a type 2 second-order loop which has better tracking behavior than the type 1 loop. It is also unconditionally stable and its transient behavior can be analyzed in a similar manner to the type 1 second-order loop.

**Temperature control:** The block diagram for a temperature controller appears in Fig. 11.16. The plant consists of a sensor and heat generator. The sensor is usually

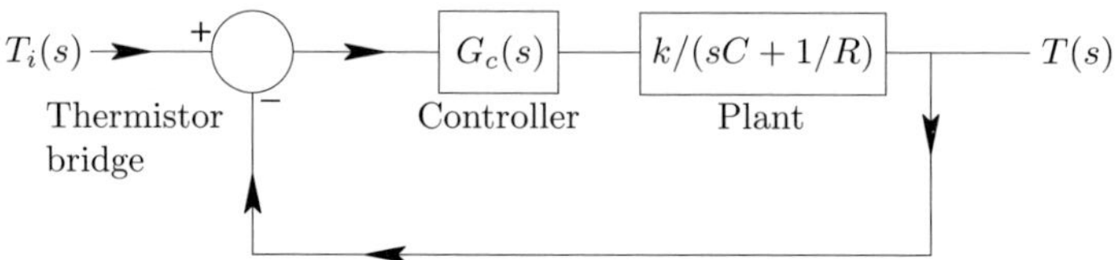


**Fig. 11.16** Block diagram of temperature controller.

a *thermistor bridge*, which will be described in detail later. It basically compares a temperature-sensitive resistor (thermistor) to a temperature-insensitive resistor and generates an output voltage proportional to the difference in resistance. The heat generator (often a simple resistive heater) accepts a power, $P$, and changes the temperature, $T$, according to the differential equation

$$P = C\frac{dT}{dt} + \frac{T-T_0}{R},\qquad(11.84)$$

where $C$ is the heat capacity of the system being controlled, $R$ is its thermal resistance to the environment and $T_0$ is the ambient temperature. If $T_0$ is constant, we can consider the controlled variable to be $\Delta T = T - T_0$. Taking the Laplace transform of this equation yields

$$\Delta T(s) = \frac{P(s)}{sC + 1/R}.\qquad(11.85)$$

If the heating element is a *thermoelectric cooler* (which can also heat), the power, $P(s)$, would be replaced with a current, since the output of this type of heater is proportional to the current, not the power. This behavior has a number of advantages, which will be discussed later. Thus, the plant is described by

$$G_p = \frac{k}{sC + 1/R},\qquad(11.86)$$

which is strictly valid only for a heater whose heat output is proportional to the drive current or voltage (e.g., a thermoelectric heater).

**Laser frequency stabilization:** A block diagram of a laser stabilization system appears in Fig. 11.17. The plant consists of the laser electronic frequency control

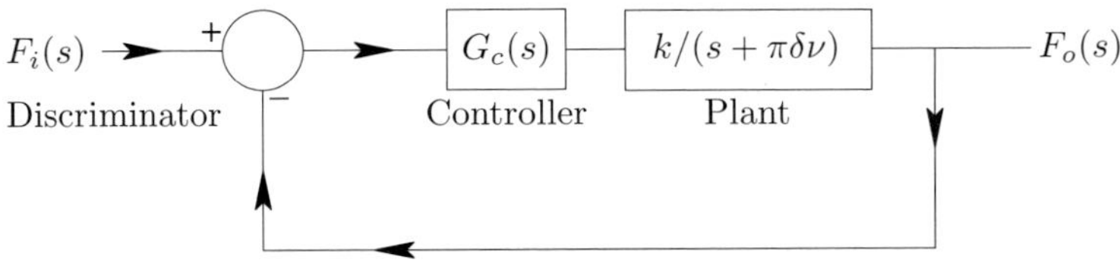

**Fig. 11.17** Block diagram of laser stabilization system.

and the discriminator, which was discussed in great detail in Chapter 4. It was shown in that chapter that the transfer function for an optical cavity of width $\delta\nu$ employing the Pound–Drever–Hall method for generating a discriminant is

$$G(\omega) = \frac{1}{1 + i\omega\tau}, \qquad \text{where} \quad \tau = \frac{2}{\delta\omega} = \frac{1}{\pi\delta\nu}. \tag{11.87}$$

Extending this to the whole $s$-plane,

$$G(s) \propto \frac{1}{s + \pi\delta\nu}. \tag{11.88}$$

There are a number of possible response functions for the electronic control of the laser frequency. For example, directly controlling the laser diode current has essentially unlimited bandwidth (from our point of view), but controlling the frequency in an external cavity diode laser using a PZT has a very limited bandwidth.

We will conclude this section with a description of some of the actual circuits used in loop filters, including the increasingly popular *proportional-integral-differential* (PID) controller.

**Integrator:** Perhaps the most important loop filter function is integration. An operational amplifier integrator together with its Bode plots is shown in Fig. 11.18. The transfer function, $G(s)$, of this circuit is

$$G(s) = k\frac{1}{1 + s\tau} \qquad \text{where} \quad k = \frac{R_2}{R_1}, \ \tau = R_2 C. \tag{11.89}$$

An ideal integrator would have $R_2 = \infty$; we have included a finite value to account for capacitor leakage and a finite (but very large) impedance across the operational amplifier. For an ideal integrator

$$G(s) = \frac{k}{s} \qquad \text{where} \quad k = \frac{1}{R_1 C}. \tag{11.90}$$

**Differentiator:** The inverse function to integration is differentiation and its schematic and Bode plots appear in Fig. 11.19. The transfer function is

$$G(s) = k(1 + s\tau) \qquad \text{where} \quad k = \frac{R_2}{R_1}, \ \tau = R_1 C. \tag{11.91}$$

An integrator circuit is the basic method for inserting *one pole* at $s = -1/\tau$ into the filter and a differentiator inserts *one zero* at $s = -1/\tau$.

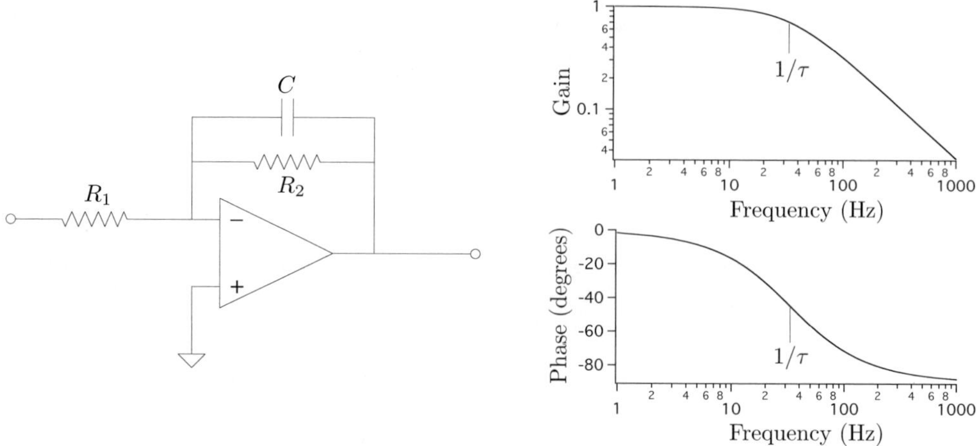

**Fig. 11.18** Integrator and its Bode plots.

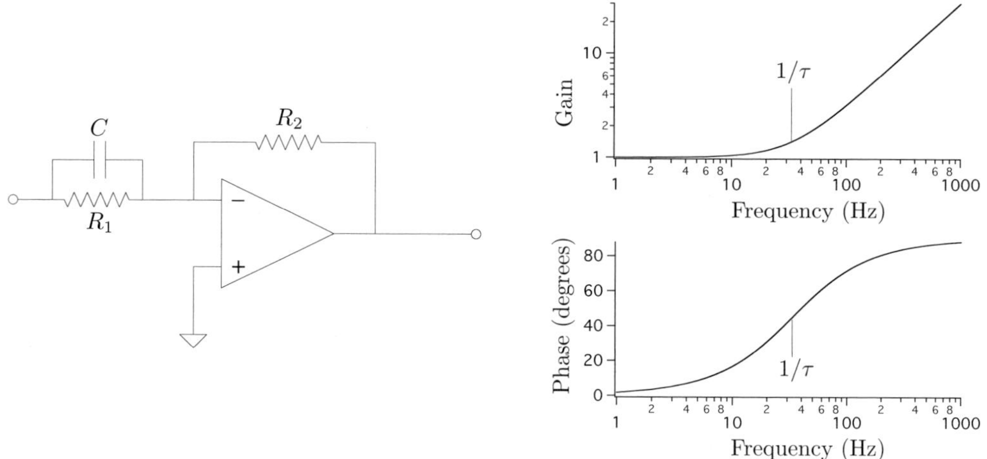

**Fig. 11.19** Differentiator and its Bode plots.

**Lead-lag:** A *lead-lag* circuit combines an integrator and a differentiator by inserting both a pole and a zero into the filter. Its transfer function is

$$G(s) = k\frac{1 + s\tau_1}{1 + s\tau_2}.$$
(11.92)

There are three circuit realizations of a lead-lag circuit, shown in Fig. 11.20 together with their Bode plots. As with the integrator, an *ideal* lead-lag filter would not have the 1 in the denominator; its transfer function would be

$$G(s) \propto \frac{1 + s\tau_1}{s}.$$
(11.93)

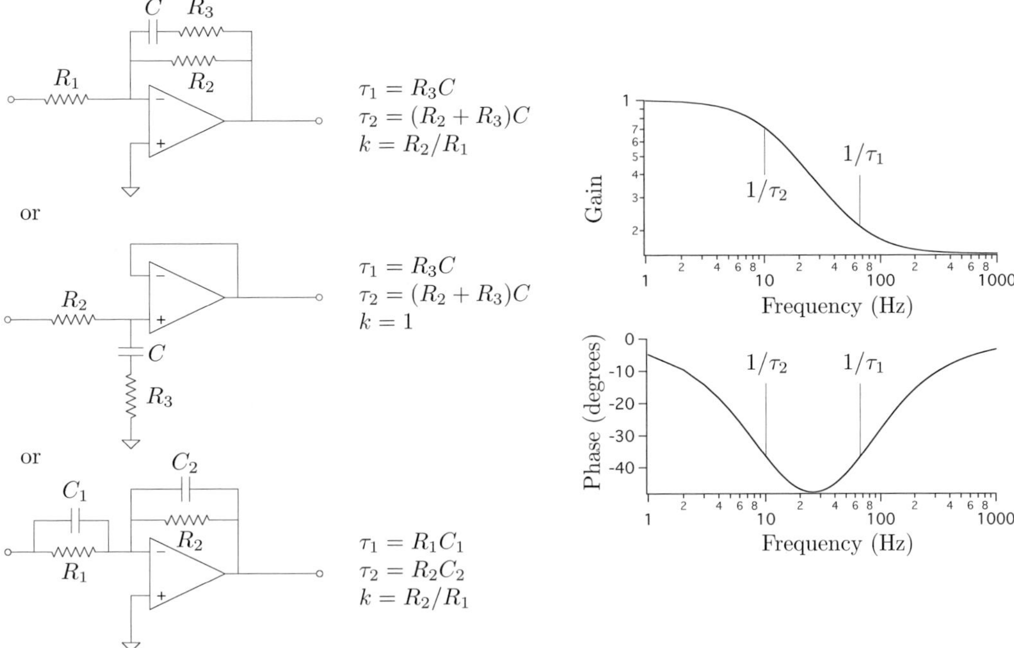

**Fig. 11.20**  Lead-lag circuits and their Bode plots.

**Proportional-integral-differential:**  The *proportional-integral-differential* controller has a single pole at the origin and two zeros. Its circuit and Bode plots are shown in Fig. 11.21. The transfer function is

$$G(s) = K_p + \frac{K_i}{s} + K_d s = \frac{K_d s^2 + K_p s + K_i}{s}. \tag{11.94}$$

The constants are

$$K_p = \frac{R_{p2} R_f}{R_{p1} R_p} \tag{11.95}$$

$$K_i = \frac{R_f}{R_{i1} R_i C_i} \tag{11.96}$$

$$K_d = \frac{R_{d2} C_d R_f}{R_d}. \tag{11.97}$$

The pole at $1/(R_{d1} C_d)$ in the differentiator is assumed to be well above the highest zero and is ignored. This controller circuit is occasionally used without the differentiator stage, in which case it is called a *proportional-integral* (PI) controller and is merely another realization of a lead-lag circuit. This circuit allows the three response functions to be controlled *independently*, in some cases with three "knobs" on the front panel of the controller box. The behaviors of the three components are:

- **Proportional ($K_p$):**

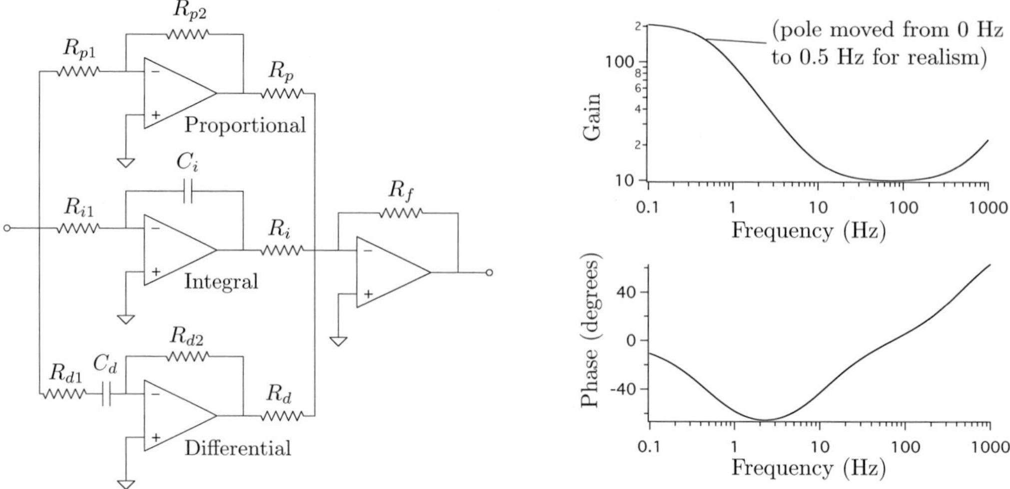



**Fig. 11.21** PID circuit and its Bode plots.

- * Reduces (but *does not* eliminate) steady-state error
- * Reduces rise time and increases overshoot
- **Integral ($K_i$):**
  - * Eliminates steady-state error and reduces rise time
  - * Worsens transient response (overshoot, settling time)
- **Differential ($K_d$):**
  - * Improves transient response (overshoot, settling time)
  - * Little effect on steady-state error or rise time

In summary, one would use the proportional and integral components to reduce the steady state error as much as possible and correct the resulting poor transient response with the differential component.

## 11.9  Temperature stabilization

We reproduce the block diagram for a temperature control system below in Fig. 11.22. Substituting the product, $G_c(s)G_p(s)$, into the expression for the closed-loop gain, we

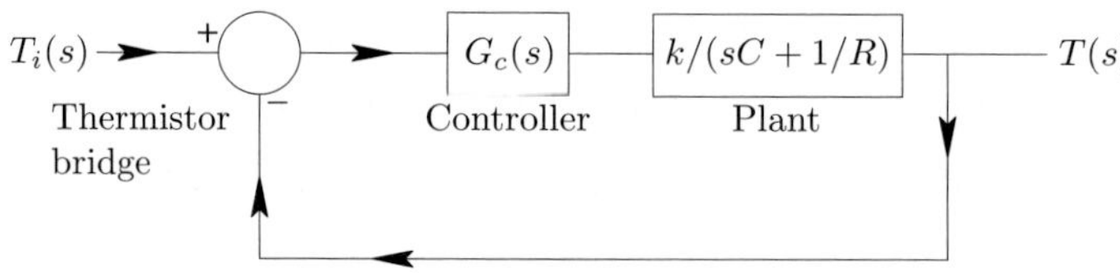



**Fig. 11.22** Block diagram of temperature controller.

obtain an expression for the final system temperature as a function of the reference, $T_i$:

$$T = \frac{G_c(s)\frac{k}{sC+1/R}}{1 + G_c(s)\frac{k}{sC+1/R}}T_i = \frac{kG_c(s)}{sC + 1/R + kG_c(s)}T_i, \tag{11.98}$$

where the expression for the controller function, $G_c(s)$, is yet to be determined.

An examination of the steady-state behavior of the system to a *step increase in ambient temperature* will help assess the minimum requirements for $G_c(s)$. We determine the response with the aid of Fig. 11.23. The figure requires some explanation.

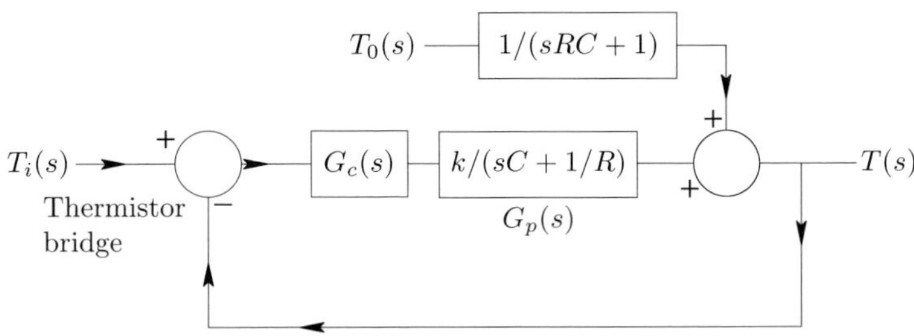


**Fig. 11.23** Block diagram of temperature controller with step change in ambient temperature.

Formerly, we assumed a *constant* $T_0$ and derived the plant response accordingly. If $T_0$ can vary with time, the differential equation and Laplace transform are

$$P = C\frac{d}{dt}T + \frac{T - T_0}{R} \tag{11.99}$$

$$T(s) = \frac{P(s)}{sC + 1/R} + \frac{T_0(s)/R}{sC + 1/R}. \tag{11.100}$$

Since the input to the plant is a *voltage* (we assume a thermoelectric cooler and driver), the *heating power* is related to the voltage via a constant of proportionality, $k$, and the plant transfer function is

$$G_p(s) = \frac{k}{sC + 1/R}. \tag{11.101}$$

We include the environmental influence by placing a summing junction *after* the plant, as shown in the figure, first multiplying $T_0(s)$ by the low-pass function, $1/(sRC + 1)$. Then, the equation for the controlled temperature, $T$, is

$$T(s) = \frac{G_c(s)G_p(s)}{1 + G_c(s)G_p(s)}T_i(s) + \frac{G_p(s)/(Rk)}{1 + G_c(s)G_p(s)}T_0(s). \tag{11.102}$$

The steady-state response, $\Delta T$, to a step in ambient temperature, $\Delta T_0$, is (with the aid of the final value theorem)

$$\Delta T = \frac{1}{1 + RkG_c(0)}\Delta T_0, \tag{11.103}$$

where we assume that $T_i(s)$ is constant. Thus, we see that we must have at least one integrator to eliminate errors due to changes in the ambient temperature.

The usual practice is to use a PID controller or lead-lag controller (sometimes called a "PI controller") for $G_c(s)$. The analysis of the loop behavior using a lead-lag controller begins with the open-loop transfer function

$$G(s) = G_c(s)G_p(S) = \left(\frac{s\tau_c + 1}{s}\right)\left(\frac{Rk}{s\tau_p + 1}\right), \tag{11.104}$$

where $\tau_p = RC$ is the thermal time constant of the system and $\tau_c$ is the time constant of the lead-lag filter. From this, we can write down the closed-loop transfer function, $H(s)$

$$H(s) = \frac{(Rk/\tau_p)(s\tau_c + 1)}{s^2 + [(Rk\tau_c + 1)/\tau_p]s + Rk/\tau_p}. \tag{11.105}$$

This can be written in a form similar to that of the second-order loop already studied,

$$H(s) = \frac{(s\tau_c + 1)\omega_n^2}{s^2 + 2\zeta\omega_n s + \omega_n^2}, \tag{11.106}$$

where

$$\omega_n = \sqrt{\frac{Rk}{\tau_p}} \tag{11.107}$$

$$\zeta = \frac{1 + Rk\tau_c}{2\sqrt{Rk\tau_p}}. \tag{11.108}$$

The design procedure is to use sufficient gain $(k)$ to obtain acceptable error and response time $(\propto 1/\omega_n)$ and adjust the controller parameter $(\tau_c)$ for a damping factor $\approx 1$ (since the *transient response* is determined mainly by the denominator of $H(s)$, the transient behavior of this loop would be very similar to that of the second-order loop discussed earlier). One can measure $\tau_p$ by manually turning on the heater until the system temperature is several degrees above ambient after which the heater is turned off and the temperature is plotted versus time. The time constant in the exponential decay of the temperature is $\tau_p$.

If a PID controller is used, there are a number of *algorithms* for setting the three knobs. The *Ziegler–Nichols* procedureis as follows:

- First set the I and D gains to zero.
- Increase the P gain until the system just begins to oscillate. This *critical gain* is called $K_c$ and the oscillation period is $T$.
- Set the P gain to $K_p = 0.6K_c$.
- Set the I gain to $K_i = 2K_p/T$.
- Set the D gain to $K_d = K_pT/8$.

At the beginning of this section, we alluded to a problem which occurs when one uses a resistive heating element. The problem is the *nonlinearity* of the system which is due to the fact that the temperature sensor produces a voltage proportional to the temperature error and the heater power is proportional to the voltage input squared.

There are three solutions. One is to linearize the heater power relation by expanding it around the operating point. The heater power is

$$P = \alpha^2 V_{in}^2 / R_h, \tag{11.109}$$

where $R_h$ is the heater resistance, $\alpha$ is the voltage gain of the driver and $V_{in}$ is the input voltage. If the desired temperature is $T_i$, the steady state power $(dT/dt = 0)$ is

$$P = (T_i - T_0)/R, \tag{11.110}$$

where $R$ is the *thermal resistance* to the environment and should not be confused with the heater resistance, $R_h$. For *small changes*, $\Delta T$, around the *output temperature, T,*

$$\Delta P = \Delta T / R. \tag{11.111}$$

From the heater power relation, this $\Delta P$ is

$$\Delta P = 2P \Delta V_{in} / V_{in}. \tag{11.112}$$

We thus obtain a linear relation

$$\Delta T = \left( \frac{2PR}{V_{in}} \right) \Delta V_{in}. \tag{11.113}$$

The problem with this is that the gain depends upon the operating point. It is much more desirable to use a *linear heater*. One can either use a thermoelectric cooler or a *pulse-width modulation* approach to the heater. In the latter system, a series of pulses whose width is proportional to the input voltage is delivered to the resistive heater; the heater averages the pulses and therefore the heating is proportional to the input voltage.

Finally, we will say a few words about the temperature sensor. It is standard practice to use a thermistor bridge, shown schematically in Fig. 11.24. The bridge is

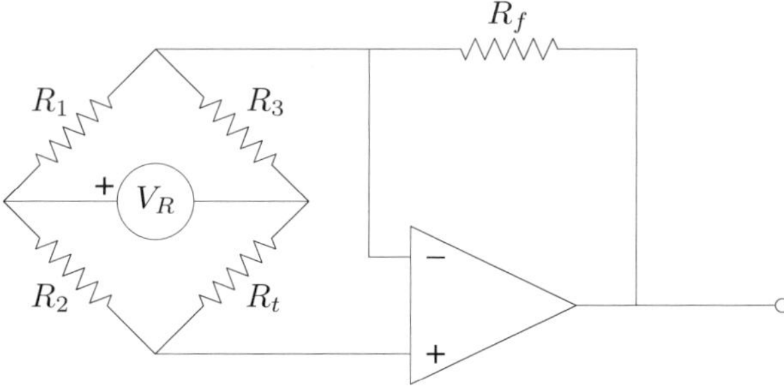

**Fig. 11.24** Thermistor bridge used as a temperature sensor.

*balanced* and the output is zero when opposite pairs of resistors have the same ratio. Three of the resistors ($R_1, R_2$ and $R_3$) have very small temperature coefficients, and

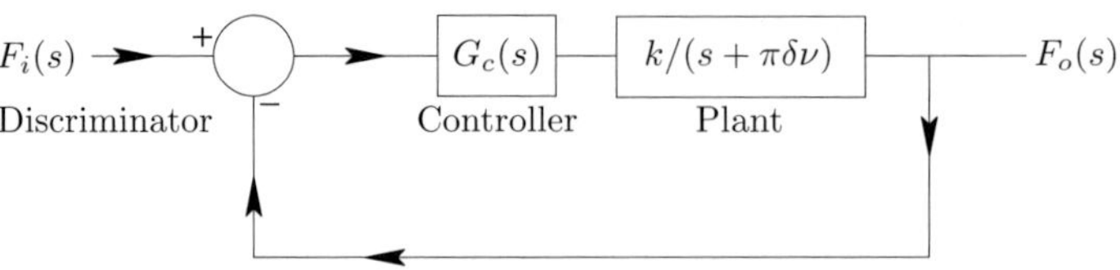



**Fig. 11.25** Block diagram of laser stabilization system.

$R_t$ (a *thermistor*) has a large temperature coefficient. The voltage, $V_R$, is derived from a very stable source. The output voltage is

$$V_{out} = \frac{V_R R_f}{R_2 + R_t} \left( \frac{R_t}{R_3} - \frac{R_2}{R_1} \right). \tag{11.114}$$

If we assume that $R_t/R_3 \approx R_2/R_1$ and consider only *small changes* in $R_t$, the output is approximately

$$\Delta V_{out} \approx \frac{V_R R_f}{R_t + R_2} \left( \frac{R_2}{R_1} \right) \frac{\Delta R_t}{R_t}. \tag{11.115}$$

This can be converted into a temperature to voltage function if the temperature to resistance characteristic of the thermistor is known. The errors due to the small temperature coefficients of $R_{1-3}$ can be eliminated by mounting these resistors in good thermal contact with the thermistor or by placing them on a heat sink and controlling its temperature using a simple commercial monolithic temperature controller.

## 11.10   Laser frequency stabilization

We finally return to the subject which was touched on at the beginning of the chapter: the frequency stabilization of a laser. We begin by summarizing the earlier results. The block diagram is reproduced (for the third time) in Fig. 11.25. Recalling the definitions of $K_0$ (laser frequency-to-voltage ratio) and $K_d$ (discriminator voltage-to-frequency ratio), the plant function is

$$G_p(s) = K_0 K_d \pi \delta \nu_c \frac{1}{s + \pi \delta \nu_c}. \tag{11.116}$$

The factor of $\pi\delta\nu_c$ in the numerator ensures that $G_p(0) = K_0 K_d$. The Bode plots of the plant are shown in Fig. 11.26. The transform of the output frequency is

$$F_o(s) = \frac{F_1(s)}{1 + G_c(s)G_p(s)} + \frac{G_c(s)G_p(s)F_0(s)}{1 + G_c(s)G_p(s)}, \tag{11.117}$$

where $F_1(s)$ is the free-running laser frequency and $F_0(s)$ is the cavity resonance frequency. The two components of the laser jitter are

$$\text{Laser fluctuations:} \qquad \Delta F_{laser}(s) = \frac{\Delta F_1(s)}{1 + G_c(s)G_p(s)} \tag{11.118}$$

$$\text{Cavity fluctuations:} \qquad \Delta F_{cavity}(s) = \frac{G_c(s)G_p(s)\Delta F_0(s)}{1 + G_c(s)G_p(s)}, \tag{11.119}$$

where $\Delta F_{laser}$ is a change in the laser frequency *due to* the laser and $\Delta F_{cavity}$ is a change in the laser frequency *due to* the cavity.

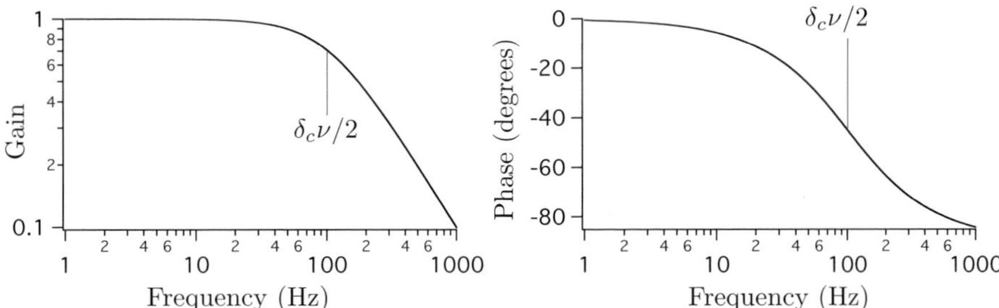

**Fig. 11.26** Bode plots for optical cavity used as a discriminator. The cavity linewidth is $\delta\nu_c$.

The choice of $G_c(s)$ (controller) is strongly dependent on the type of *frequency actuator*, which of course depends upon the type of laser being stabilized. The following is a list of possible actuators:

- **Laser diode current:**
  The laser diode current has an essentially unlimited bandwidth from the point of view of laser stabilization; the bandwidths are in the GHz range, which is far greater than any likely unity gain frequency. Controlling the current has the disadvantage of introducing amplitude modulation, which can increase the residual jitter of the stabilized laser.

- **Laser diode temperature:**
  The diode temperature allows a very large change of laser frequency at the expense of extremely low bandwidth. Even a fast temperature controller will still be limited to settling times of a second or so. It is often useful to use a separate loop to control the temperature; this keeps the laser frequency *close to the cavity resonance* and reduces the dynamic range requirements of the fast loop.

- **Piezo-electric positioner for ECDL mirror or grating:**
  The frequency of an extended cavity diode laser can be controlled either via the diode current (fast, but introduces AM) or the piezo-electrically actuated grating or mirror. The latter provides a very large sweep range but is only moderately fast. It is limited by acoustical resonances in the piezo crystal, which can be as low as a few kHz and as high as 100 kHz. The resonances introduce significant phase shifts at frequencies within a factor of 2 of the resonance and these can destabilize the loop. The resonances can be reduced by using a *notch filter*, allowing unity gain frequencies well above the resonance frequencies. It requires considerable expertise in the art of laser stabilization to achieve success in this endeavor.

- **Piezo-electric element for monolithic lasers:**
  Monolithic, optically pumped lasers are, almost by definition, difficult to electronically tune. Non-planar ring oscillators (NPROs) are normally tuned with a small piezo-electric element bonded to the YAG crystal – this stresses the crystal when a voltage is applied, changing the geometry and laser frequency. Since the piezo-electric actuator is very small, the resonant frequencies are quite high, in the 100 kHz range. The NPRO laser is intrinsically quiet and the piezo-electric

device will often be sufficient to obtain good stabilized performance without consideration of its resonances.

- **Galvonometrically driven plate:**
  Dye and Ti-sapphire lasers often use a small galvonometrically driven plate of glass at Brewster's angle in the internal beam to change the effective cavity length and therefore the laser frequency. This method of frequency control has a modest bandwidth – several hundred Hz. It can also cause small laser *pointing fluctuations*. Brewster plates are often used in conjunction with a faster frequency control (often a piezo-driven mirror) – the Brewster plate stabilizes against large frequency deviations occurring at low frequencies and the piezo-driven mirror eliminates small deviation higher frequency fluctuations.

- **Electro-optic modulators (EOMs):**
  The technique with the largest bandwidth for controlling the frequency of a dye or Ti-sapphire (or any laser with a large open beam path) is an electro-optic modulator. Although these devices occasionally have piezo-like resonances, they allow small deviation frequency control at frequencies of several MHz. They can require substantial voltages, but this is usually not a problem in most lasers since the high frequency noise usually has fairly small deviation and requires only a modest voltage applied to the EOM for its correction.

A number of the devices mentioned above require that modest voltages be applied to capacitive loads at a very high *slew rate*. Since $I = CdV/dt$, a significant current is needed to meet the slew rate requirement. Conventional high-voltage operational amplifiers usually don't have large current drive capacities together with fast response. There are a number of commercial parts, called *power boosters*, which do combine these two useful properties and these are recommended for driving piezos or EOMs.

We seek a controller which provides shot-noise-limited performance over as wide a frequency range as possible while maintaining stability (a phase margin of $30°$). We will begin our analysis by summarizing the results of Mor and Arie (1997) for a lead-lag (PI) filter whose transfer function is

$$G_c(s) = \frac{k_p s + k_i}{s}. \tag{11.120}$$

The laser noise *spectral density*, $S_{laser}$, (in $\text{Hz}^2/\text{Hz}$) is the sum of three components: white noise, *flicker* (1/f) noise and *random walk* noise, which is proportional to $1/f^2$, where $f$ is the frequency of the noise.

$$S_{laser} = \frac{\Delta\nu}{\pi} + \frac{k_f}{f} + \frac{k_r}{f^2}. \tag{11.121}$$

The laser lineshape due to white noise only is a Lorentzian with width $\Delta\nu$.

Three types of lasers will be considered:

- **Single mode diode**
  This laser is characterized by white noise only with a width of 50 MHz.

- **Distributed feedback diode**
  DFB lasers are also characterized by a white noise spectrum; we will assume the width is 1 MHz.

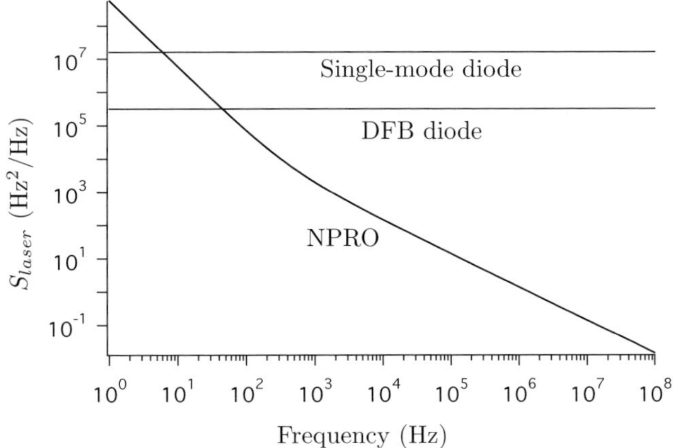



**Fig. 11.27** The frequency noise spectra for a single-mode diode, DFB diode and NPRO.

- **NPRO**
  Non-planar ring oscillators are dominated by flicker and random walk noise; the white noise can be ignored. We will assume that $k_f = 1.4 \times 10^6$ Hz$^2$ and $k_r = 5.7 \times 10^8$ Hz$^3$.

The noise spectra for the three lasers are plotted in Fig. 11.27.

The total closed-loop laser-noise spectral density $(S)$ will be the quadratic sum of the residual intrinsic laser jitter $(S_{laser})$ and the shot noise from the discriminator $(S_{disc})$ *referred to the laser output*, since the two noise sources are *uncorrelated*. From eqn 11.118, $S$ is

$$S = \frac{S_{laser} + |K_0 G_c|^2 S_{disc}}{|1 + G|^2}, \tag{11.122}$$

where $G = G_c G_p$, the discriminator noise is related to $\Delta F_0$ by the factor $K_0 G_c$ (proceeding from the discriminator output to the laser output) and we have dropped the $s$-dependence of all of the terms to reduce clutter. In order for the locked laser to be shot noise limited, we require that

$$\text{Shot noise limited:} \quad |K_0 G_c(s)|^2 > \frac{S_{laser}}{S_{disc}}. \tag{11.123}$$

The open-loop gain is

$$G(s) = \left( \frac{k_p s + k_i}{s} \right) \left( \frac{K_0 K_d \pi \delta \nu_c}{s + \pi \delta \nu_c} \right) = \frac{K(s + s_0)}{s(s + s_1)}, \tag{11.124}$$

where we have absorbed the multiplicative constants into $K = K_0 K_d k_p \pi \delta \nu_c$ and have a zero $s_0 = -k_i/k_p$ and a pole $s_1 = -\pi \delta \nu_c$. We note that this is formally very similar to the open-loop gain of a temperature stabilization loop with a PI (or lead-lag) filter. The only two parameters available for optimization are the overall gain $(K)$ and the zero $(-s_0)$ in the lead-lag filter.

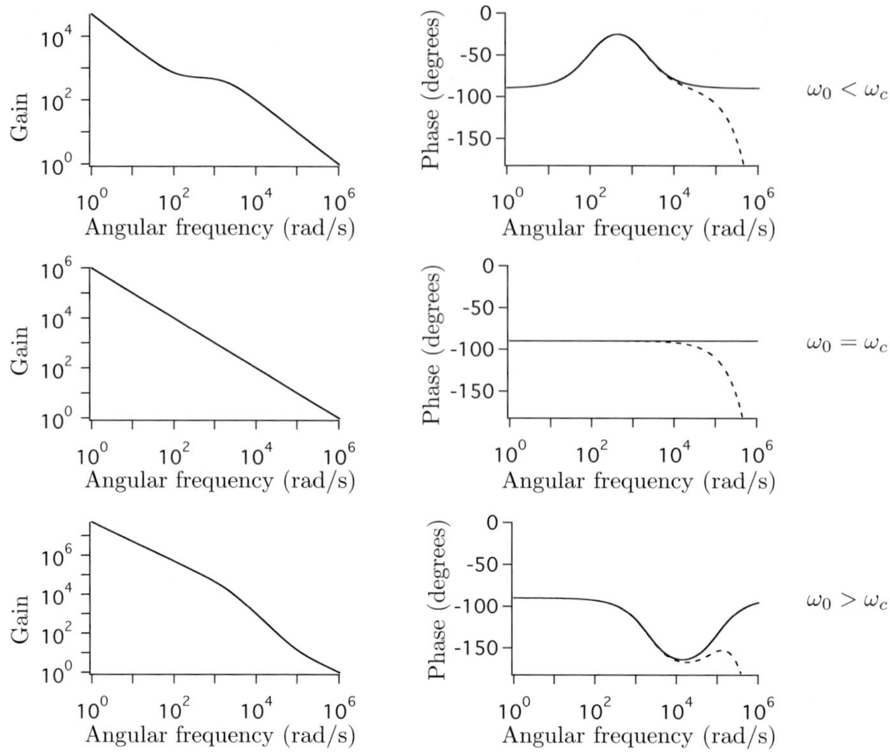

**Fig. 11.28** Bode plots for three locations of the zero ($\omega_0$) relative to the pole due to the cavity ($\omega_c$). The dashed curve includes 200 $\mu$s of time delay, which is greatly exaggerated to illustrate its effect on the phase. The pole frequency is $2 \times 10^3$ rad/s and the zero frequencies are 100, 2000 and 100,000 rad/s.

In order to obtain the largest gain at lower frequencies, it is desirable to have the largest possible unity gain frequency for the open-loop gain. For the single-mode diode and DFB diode, which have noise components into the MHz range, we can demonstrate that time delays set a limit on the unity gain frequency. The NPRO has less noise at higher frequencies and its stabilization system unity gain frequency is likely to be limited by transducer resonances rather than time delays.

We first substitute $s = i\omega$, $s_0 = \omega_0$ and $s_1 = \omega_c$ ($\omega_0$ is the zero location and $\omega_c$ is the pole due to the cavity) into eqn 11.124

$$G(\omega) = \frac{K(\omega - i\omega_0)}{i\omega(\omega - i\omega_c)}. \tag{11.125}$$

Figure 11.28 contains Bode plots for three possible locations of $\omega_0$ with respect to $\omega_c$: $\omega_0 < \omega_c$, $\omega_0 = \omega_c$ and $\omega_0 > \omega_c$. As can be seen from the figure, increasing the frequency of the zero increases the gain at lower frequencies. The risk is the increased possibility of instability, particularly if time delays are present. We will estimate the frequency range for which one can expect shot noise limited performance assuming that there is time delay, $\tau$, and that the zero frequency is somewhat above the cavity pole and about a factor of 5 below the unity gain frequency.

The magnitude and phase of the open-loop gain are

$$|G(\omega)| = \frac{K}{\omega}\sqrt{\frac{\omega^2 + \omega_0^2}{\omega^2 + \omega_c^2}} \tag{11.126}$$

$$\angle G(\omega) = \tan^{-1}\left(\frac{\omega_c\omega_0 + \omega^2}{\omega(\omega_0 - \omega_c)}\right). \tag{11.127}$$

If we include the effects of time delay ($\tau$) and insist on a phase margin of $30°$, the following equation must be solved for $\omega$:

$$\tan^{-1}\left(\frac{\omega_c\omega_0 + \omega^2}{\omega(\omega_0 - \omega_c)}\right) - \omega\tau = -\tfrac{5}{6}\pi. \tag{11.128}$$

The procedure used by Mor and Arie (1997) was to first neglect time delays and determine the frequency which produced a phase margin of $45°$; this is equivalent to setting the argument of the $\tan^{-1}$ equal to one. The resulting quadratic equation has two roots whose location can be seen in the lower right phase plot in Fig. 11.28: one is a little above the second pole and the other is a little above the zero (they are at the intersection of the line $\phi = -135°$ with the curve, and are on either side of the bowl-shaped region of rapid phase change). The authors of this work found the value of $\omega_0$ that provided the desired phase margin (the larger root), and placed the unity gain frequency approximately at $\omega_0$ for the highest possible gain at lower frequencies. While this procedure would probably work, it suffers from the presence of a large and rapidly varying phase very near the unity gain frequency, which can cause instability.

We will use a simpler and more conservative approach: we will place $\omega_0$ at a factor of 5 below the unity gain frequency, $\omega_{UG}$. Then the phase is varying very slowly at the unity gain frequency and it is very nearly $-90°$ (from the asymptotic behavior of two poles and one zero). The factor of 5 is a bit arbitrary; a smaller factor could be used, though it should be larger than 2 to avoid rapid phase changes near the unity gain frequency. The unity gain frequency will be determined from time delay considerations and $\omega_0$ will be set using $\omega_0 = \omega_{UG}/5$. Finally, the parameter $K$ will be determined from the requirement that $|G(\omega_{UG})| = 1$.

At $\omega_{UG}$, we have $-90°$ of phase shift; we are allowed an additional $-60°$ (due to time delays) to satisfy the phase margin requirement. The unity gain frequency is therefore

$$\omega_{UG} = \frac{\pi}{3\tau}. \tag{11.129}$$

From our assumption that $\omega_0$ is five times smaller than the unity gain frequency,

$$\omega_0 = \frac{\pi}{15\tau}. \tag{11.130}$$

Finally, $K$ is chosen from $|G(\omega_{UG})| = 1$

$$K = \omega_{UG}\sqrt{\frac{\omega_{UG}^2 + \omega_c^2}{\omega_{UG}^2 + \omega_0^2}}, \tag{11.131}$$

where $\omega_0$ and $\omega_c$ are determined from the previous equations.

We will estimate the behavior of a loop used to stabilize a DFB diode laser with the presence of 55 ns of delay. The unity gain frequency is 3 MHz ($19 \times 10^6$ rad/s). The zero ($\omega_0$) is placed at 600 kHz ($3.8 \times 10^6$ rad/s). We will assume a cavity linewidth of 2 kHz (typical of the highest finesse cavities); thus $\omega_c = 6.2 \times 10^3$ rad/s, which is much lower than $\omega_0$ as required. Since $\omega_c \ll \omega_{UG}$ and $\omega_0 = \omega_{UG}/5$, one can estimate $K$ as

$$K \approx \omega_{UG} \qquad (\omega_c \ll \omega_{UG},\ \omega_0 = \omega_{UG}/5). \tag{11.132}$$

From eqn 11.123, shot-noise-limited performance will occur at and below the frequency which satisfies

$$\left[\frac{\omega_{UG}}{K_{disc}\omega_c}\right]^2 \frac{\omega^2 + \omega_0^2}{\omega^2} \geq \frac{S_{laser}}{S_{disc}} \implies \frac{\omega^2 + \omega_0^2}{\omega^2} \geq \left[\frac{\omega_c K_{disc}}{\omega_{UG}}\right]^2 \frac{S_{laser}}{S_{disc}}, \tag{11.133}$$

where we divided $G(\omega)$ by $|G_p(\omega)/K_0|$ to obtain $|K_0 G_c(\omega)|$ on the left side (note that $\omega_c = \pi\delta\nu_c$). Anticipating that the right-hand side will be much greater than 1, we can simplify this to obtain

$$\omega_{snlp} = \omega_0 \left[\frac{\omega_{UG}}{\omega_c K_{disc}}\right] \sqrt{\frac{S_{disc}}{S_{laser}}} \tag{11.134}$$

where $\omega_{snlp}$ is a frequency below which the loop is shot noise limited. For the DFB laser, $S_{laser} = 3 \times 10^5$ Hz$^2$/Hz. The shot noise current per $\sqrt{\text{Hz}}$ at the output of the discriminant is

$$i_{noise} = \sqrt{2ei}. \tag{11.135}$$

At resonance, the carrier will be totally absorbed and the reflected power from the two sidebands will be $2J_1^2(\beta)P_i$ where $P_i$ is the incident power. The photocurrent will have an additional factor of $\sqrt{2}$ since the mixer will respond to signals and noise both above and below the reference frequency. The result is

$$\sqrt{S_{disc}} = i_{noise} = \sqrt{8eJ_1^2(\beta)\mathcal{R}P_i} \qquad (\text{in A}/\sqrt{\text{Hz}}), \tag{11.136}$$

where $\mathcal{R}$ is the *responsivity* (in A/watt) of the photodiode. In units of A/Hz, the expression for $K_{disc}$ is

$$K_{disc} = \frac{2.71\mathcal{R}P_i}{\delta\omega} = \frac{1.35\mathcal{R}P_i}{\omega_c}, \tag{11.137}$$

where $\delta\omega = 2\omega_c$ is the cavity width. If we assume that $P_i = 0.1$ mW and $\mathcal{R} \approx 1$, then $i_{noise} = 5.3 \times 10^{-12}$ A/$\sqrt{\text{Hz}}$ and $K_{disc} = 10^{-8}$ A/Hz. Putting all of these numbers into eqn 11.134, we have $\omega_{snlp} = 10^4$ rad/sec (1600 Hz). The fact that $\omega_{snlp} \ll \omega_0$ justifies our recent simplification. Note that the shot-noise-limited frequency range is proportional (up to a point) to $\omega_0$ and one would therefore like to make $\omega_0$ as large as possible without destabilizing the loop. The bare diode has 50× as much noise as the DFB laser; its stabilization procedure would be similar to the above and the shot-noise-limited performance would be somewhat worse than that of the DFB laser.

We mentioned above that the stabilization system for an NPRO would be subject to different design constraints from the DFB laser discussed above. The main difference is the actuator: the piezo-electric actuator has mechanical resonances which would

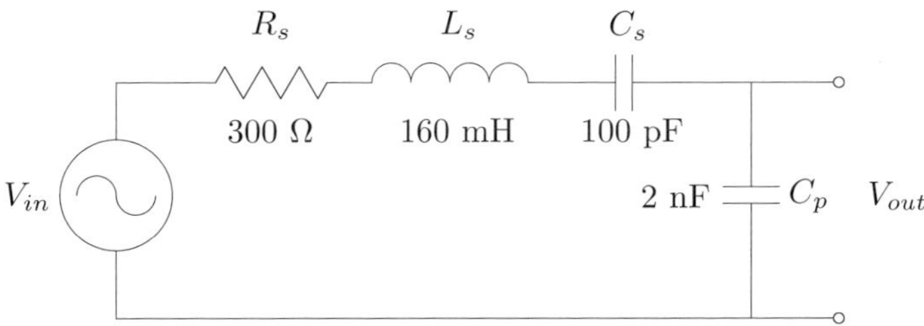

**Fig. 11.29** Equivalent circuit of a piezo-electric transducer. Typical values are indicated for a resonant frequency of 40 kHz.

produce destabilizing phase shifts and would be a much greater constraint on the unity gain frequency than time delays. This would be the case with any laser system which depends upon a piezo-electric actuator for fast frequency control.

A piezo-electric transducer is a *resonant, two-pole* electromechanical system which can be modeled with the equivalent circuit shown in Fig. 11.29. The mechanical mass is represented by the inductor, the stiffness by the series capacitor and the damping by a resistor (a quartz crystal for frequency control has the same equivalent circuit). The mechanical response to a drive voltage $V_{in}$ is proportional to the voltage, $V_{out}$, across $C_p$. The response can be written in the familiar form used in the analysis of a second-order loop:

$$\frac{V_{out}}{V_{in}} = \left(\frac{C_s}{C_p}\right)\frac{\omega_n^2}{s^2 + 2\zeta\omega_n s + \omega_n^2},\tag{11.138}$$

where

$$\omega_n = \frac{1}{\sqrt{L_s C_s}}\tag{11.139}$$

$$\zeta = \frac{R}{2\omega_n L}\tag{11.140}$$

$$Q = \frac{1}{2\zeta}.\tag{11.141}$$

The bode plots for a variety of damping factors appear in Fig. 11.30. Although the $Q$ of the transducer shown in the diagram is $> 100$, many piezo-electric devices have lower $Q$ values; tubular actuators can have a $Q$ of about 10 with a response corresponding approximately to the $\zeta = 0.1$ ($Q = 5$) curve. We note from the plot that the phase change is not that large until one is fairly close to the resonance. However, instability can still unexpectedly occur due to inadequate *gain margin*. The gain margin is the amount by which the gain is below unity (usually expressed in decibels) when the phase shift is 180°. The highly peaked response of the high-$Q$ transducers can destabilize the loop near the resonance by increasing the gain until it is equal to or greater than unity.

The usual conservative approach with piezo-electric actuators is to roll off the servo gain sufficiently rapidly so that the unity gain frequency is well below the mechanical resonance of the transducer, which can be as high as 100 kHz but is often much lower.

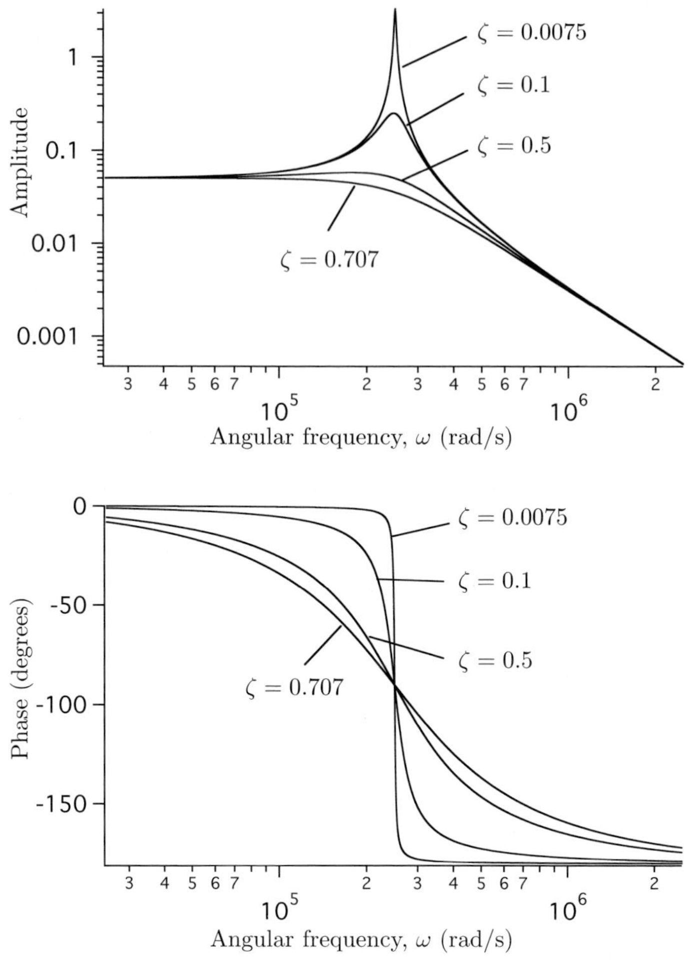


**Fig. 11.30** Bode plots for second-order system. The highly peaked piezo response is shown for $\zeta = 0.0075$.

Unfortunately, this sets a severe limit on the performance of the stabilization system. Some very expert workers in the field have sought to find a strategy for transcending this limitation and have reported on their successful efforts in the Handbook of Optics, Volume IV (Hall, et al. (2000)). We will summarize their findings.

The first attempt to surmount the limit due to the piezo-electric resonance was to use a PID circuit: the derivative adds a *phase lead* which can reduce the adverse effect of the phase shift due to the actuator. While this allowed a unity gain frequency somewhat above the piezo-electric resonance (by a factor of 2), the lead circuit reduces the low frequency gain and little is accomplished by using it.

The next step was to use a PI controller and a *notch filter* to suppress the resonance. The notch approximates the inverse of the resonance peak. This eliminated the

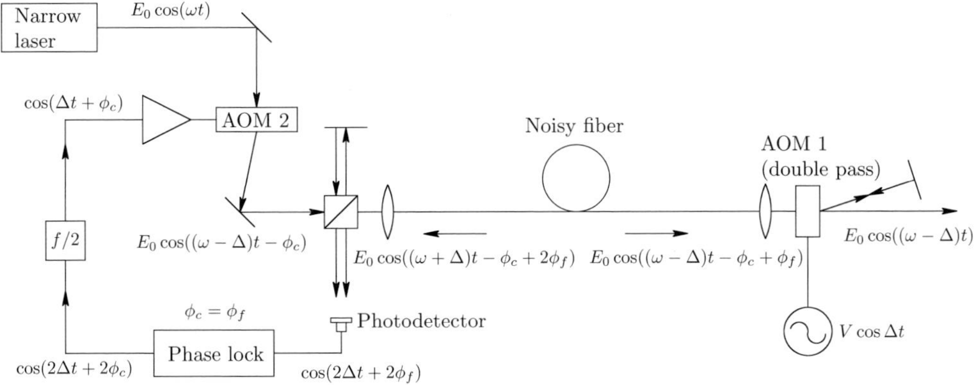


**Fig. 11.31** Setup for canceling phase noise introduced by a fiber.

undesired effects of the resonance and provided increased low-frequency gain compared to the PID circuit. The unity gain was slightly lower with the notch then with the PID (and no notch) but still above the resonant frequency. The best results were obtained using a *second* PI stage together with the notch. This provided good transient response and a gain improvement of more than 20 dB below 1 kHz (the resonance was at 25 kHz). This would be the best strategy for stabilizing an NPRO or other intrinsically stable monolithic lasers whose frequency is controlled by a piezo-electric actuator.

## 11.11  Optical fiber phase noise and its cancellation

Single-mode optical fibers provide a very convenient means of transporting laser beams from one place to another and can easily be made to preserve the beam properties and polarization. However, ambient acoustical noise can impose several radians of phase noise onto a beam transported by a fiber, broadening the laser by several kHz. For many applications, this is not a serious problem, but it is certainly undesirable for a laser whose linewidth has been reduced to 1 Hz or less. A number of methods for canceling this phase noise have been devised: they are all based upon the useful fact that the phase noise in a fiber is *independent* of the propagation direction. Thus, if one can retransmit the beam in a direction opposite to the original direction, the total phase noise will be exactly doubled. The cancellation will only work if the *coherence time* of the fiber noise ($\geq 0.1$ ms) is greater than the round trip travel time in the fiber: this limits the application of this technique to fiber lengths of about 10 km or less.

We will describe in detail one technique for fiber phase noise cancellation employed by workers at the Joint Institute for Laboratory Astrophysics (Ma (1994)). The scheme is shown schematically in Fig. 11.31. Although it is a conceptually simple idea, it appears rather complicated in the figure. In the following, we will trace the path of the beam through each step.

  1. The beam emitted by the laser is

$$\text{Initial laser beam:} \quad E_0 \cos \omega t \tag{11.142}$$

2. After passing though the first acousto-optic modulator, AOM 2, the beam will acquire a *correction phase*, $-\phi_c$ and a frequency shift by $-\Delta$, whose origin will be described shortly.

$$\text{After AOM 2:} \quad E_0 \cos((\omega - \Delta)t - \phi_c) \tag{11.143}$$

3. The beam crosses the beam splitter and enters the fiber, where it acquires an additional phase, $\phi_f$, due to the phase noise from the fiber.

$$\text{Upstream, after fiber:} \quad E_0 \cos((\omega - \Delta)t - \phi_c + \phi_f) \tag{11.144}$$

4. The effect of the second acousto-optic modulator, AOM 1, is to send a small portion of the beam back through the fiber shifted in frequency by $+2\Delta$, since the modulator is operated in the *double pass* configuration and it is driven at frequency $\Delta$. The quantities $\phi_c$ and $\phi_f$ are not affected by this AOM.

$$\text{Downstream, before fiber:} \quad E_0 \cos((\omega + \Delta)t - \phi_c + \phi_f) \tag{11.145}$$

5. After passing through the fiber for a second time, the fiber contribution to the phase noise will be doubled.

$$\text{Downstream, after fiber:} \quad E_0 \cos((\omega + \Delta)t - \phi_c + 2\phi_f) \tag{11.146}$$

6. The photodetector will receive two beams: one is from the fiber (downstream) and the other is a sample of the output of AOM 2, obtained from the beamsplitter and mirror on the left side of the diagram. These beams are:

$$\text{Beam 1:} \quad E_0 \cos((\omega + \Delta)t - \phi_c + 2\phi_f) \tag{11.147}$$
$$\text{Beam 2:} \quad E_0 \cos((\omega - \Delta)t - \phi_c) \tag{11.148}$$

The signal (photocurrent) at the output of the detector will be proportional to the square of the total electric field incident on the detector. The operation of squaring will generate sum and difference frequencies. The *difference frequency* signal is

$$\text{Output of photodetector:} \quad \cos(2\Delta t + 2\phi_f) \tag{11.149}$$

7. The phase-locked loop acts essentially like a filter, generating the low-frequency correction phase, $\phi_c$, which is equal to $\phi_f$. This operation eliminates external noise from various electronic sources and preserves the high frequency $\Delta$.

$$\text{Output of phase locked filter:} \quad \cos(2\Delta t + 2\phi_c) \tag{11.150}$$

8. The *divide by two* circuit, digitally divides the frequency (and phase) by two.

$$\text{Output of f/2:} \quad \cos(\Delta t + \phi_c) \tag{11.151}$$

9. Finally, the signal is amplified and inverted and applied to AOM 2. We observe the source of the upstream $\Delta$ modulation and the fact that the beam downstream of the fiber is shifted down by $\Delta$ and lacks the fiber-induced phase noise (since $\phi_c = \phi_f$).

$$\text{Fiber output (downstream):} \quad E_0 \cos((\omega - \Delta)t) \tag{11.152}$$

The scheme has been successful in reducing the fiber-induced phase changes to less than 0.3 radians, essentially eliminating the fiber as a source of phase noise.

## 11.12   Characterization of laser frequency stability

The techniques described in this chapter allow one to considerably reduce the frequency fluctuations of a laser by frequency locking it to a stable reference. In assessing the success of one's frequency stabilization efforts, it is worthwhile to establish some "yardstick" with which to measure frequency fluctuations. In this section, we will provide a summary of the standard measures of laser frequency stability. Before doing so, it is important to distinguish *accuracy* from *stability*. Accuracy is a measure of how close the measurement is to a conceptual (but fictitious) *exact value of a quantity*. It is usually expressed as a maximum fractional frequency error with no associated averaging time. The errors which contribute to an oscillator's inaccuracy are called *systematic errors*. Stability is a measure of how well the frequency has been determined, without reference to its "true value". An oscillator's stability is usually limited by *random fluctuations*, which can be due to the intrinsic measurement process or due to environmental effects such as temperature changes or vibration. Stability includes *drift*, though this can often be removed by comparing the oscillator to a more stable one. Stability is usually measured by comparing the number of oscillator cycles in adjacent measurement periods and therefore depends strongly on the measurement time. A stable oscillator is not necessarily accurate: there are oscillators with excellent stability but, due to systematic errors, are insufficiently accurate to be candidates for a primary frequency standard. On the other hand, common sense dictates that an accurate source should have an equivalent stability, since it would be very difficult to assess the accuracy of an unstable oscillator. We will be interested only in oscillator stability in the following.

As with all time-varying quantities, there is both a *time domain* and *frequency domain* approach to the measurement of frequency stability. Stability can be measured in the time domain by making repeated measurements of the number of cycles of the oscillator over fixed time intervals and obtaining some kind of variance of the deviations from a mean. The frequency domain approach might use a discriminator to convert frequency fluctuations to amplitude fluctuations and obtain the power spectrum of the latter. Since we are interested in the frequency noise in a *laser*, it is of course currently impossible to directly count the cycles of its radiation and we therefore usually make a frequency domain measurement of the laser's frequency noise. It is fortunate that it is fairly easy to convert the frequency domain measurement to one in the time domain.

We begin with the usual description of the output of an oscillator with phase noise

$$E(t) = E_0 \sin(2\pi\nu_0 t + \phi(t)), \tag{11.153}$$

where $\phi(t)$ is a random variable which depends upon time and $E(t)$ is the electric field. Since we are not interested in *amplitude noise*, we assume that $E_0$ is constant. The *instantaneous frequency*, $\nu(t)$, is defined as

$$\nu(t) \equiv \frac{1}{2\pi} \frac{d}{dt} (2\pi\nu_0 t + \phi(t)) = \nu_0 + \frac{1}{2\pi} \frac{d\phi}{dt}. \tag{11.154}$$

We are interested in the *fluctuations* of the frequency about the *fixed* $\nu_0$; we call this quantity $\Delta\nu(t)$:

$$\Delta\nu(t) = \nu(t) - \nu_0 = \frac{1}{2\pi} \frac{d\phi}{dt}. \tag{11.155}$$

We usually work with the *fractional frequency and phase fluctuations*, so we divide by $\nu_0$ to obtain $y(t)$ and $x(t)$:

$$\text{Frequency:} \quad y(t) \equiv \frac{\Delta\nu(t)}{\nu_0} \tag{11.156}$$

$$\text{Phase:} \quad x(t) \equiv \int_0^t y(t')dt' = \frac{\phi(t)}{2\pi\nu_0}. \tag{11.157}$$

Finally, the actual *measured* quantity is the *phase* (proportional to the number of oscillator cycles or the number of ticks of a clock) averaged over time $\tau$. The $i^{th}$ measurement is called $\bar{y}_i$ and is defined as

$$\bar{y}_i \equiv \frac{x(t_i + \tau) - x(t_i)}{\tau}. \tag{11.158}$$

In the time domain, the usual variance of the quantities $\{\bar{y}_i\}$ is given by

$$\text{Common variance:} \quad \sigma^2(N,\tau) = \frac{1}{N-1} \sum_{i=0}^{N} (\bar{y}_i - \langle \bar{y} \rangle)^2, \tag{11.159}$$

where $N$ measurements are made and $\langle \bar{y} \rangle$ is the *mean* of these measurements. This variance is fairly well behaved when the $\phi(t)$ is spectrally uniform "white noise". It is, however, divergent for many of the noise sources encountered by lasers and it is therefore an inadequate measure of laser stability. Instead, it is customary to use the *Allan variance*, defined by

$$\text{Allan variance:} \quad \sigma_y^2(\tau) = \left\langle \tfrac{1}{2}(\bar{y}(t+\tau) - \bar{y}(t))^2 \right\rangle, \tag{11.160}$$

where the angle brackets denote an infinite time average. One obtains the error by taking the difference between *adjacent samples* rather than the difference of a sample from a mean; this emphasizes random fluctuations and is insensitive, to first order, to slowly varying frequency changes. Of course, an infinite time average is not possible with experimental data; the Allan variance is very well approximated by taking $N$ samples and using the following formula:

$$\text{Allan variance:} \quad \sigma_y^2(\tau) \approx \frac{1}{2(N-1)} \sum_{i=0}^{N-1} (\bar{y}_{i+1} - \bar{y}_i)^2. \tag{11.161}$$

This variance is actually quite well behaved with most noise sources encountered in the laboratory. In all of the above definitions, we assume there is no *dead time*: the measurements are of duration $\tau$ and are separated by an interval equal to $\tau$.

We are often interested in the variation of $\sigma_y^2(\tau)$ with $\tau$. It turns out that this is easily obtained without retaking the data. In order to obtain the variance for $2\tau$, one simply adds adjacent measurements in pairs and applies the formula. This "trick" only works for the Allan variance of $y$ (with no dead time) and allows one to obtain $\sigma_y^2(n\tau)$ for integral $n$ from a single data set. The Allan variance obtained in this way is only reliable when $n\tau < T/m$, where $T$ is the total duration of the data and $m$ is an integer between 2 and 8.

In addition to providing a universal measure of oscillator performance, the Allan variance allows one to obtain very useful information about the character of the noise. Before discussing this, we will introduce the frequency domain measure of oscillator noise: the *spectral density of fractional frequency fluctuations*: $S_y(f)$. We begin with the spectral density of the phase noise, $S_\phi(f)$. This is simply the power spectrum of $\phi(t)$ and can be obtained using the Wiener-Khintchine theorem, which states that the power spectrum of some function of time is the Fourier transform of its auto-correlation function. If the autocorrelation function of the phase is $R_\phi(\tau)$,

$$R_\phi(\tau) = \lim_{T\to\infty} \frac{1}{T} \int_{-T/2}^{T/2} \phi(t+\tau)\phi(t)dt \tag{11.162}$$

$$S_\phi(\omega) = \frac{1}{2\pi} \int_{-\infty}^{\infty} R_\phi(\tau)e^{i\omega\tau}d\tau. \tag{11.163}$$

If the quantity being analyzed were a voltage or current (or electric field), $S_\phi(\omega)$ would be a true spectral density of the *power* whose units would be watt/Hz. It is still customary to call $S_\phi(\omega)$ a spectral density even though its units are radian$^2$/Hz. The spectral density of *frequency fluctuations*, $S_{\dot\phi}(\omega)$, is then

$$S_{\dot\phi}(\omega) = \omega^2 S_\phi(\omega). \tag{11.164}$$

This is simply a consequence of the fact that the Fourier transform of a time derivative is $i\omega$ times the original transform (applied twice, since $S_\phi$ is a *spectral density*, which is quadratic in the transform of $\phi$). From the autocorrelation relations at $\tau = 0$ and the inverse Fourier transform, we obtain

$$\langle \dot\phi^2 \rangle = R_{\dot\phi}(0) = \int_{-\infty}^{\infty} S_{\dot\phi}(\omega)d\omega, \tag{11.165}$$

which justifies our characterization of $S_{\dot\phi}(\omega)$ as a spectral density. Finally, transforming to frequencies in cycles per second, the spectral density of $y$ is easily shown to be

$$S_y(f) = \left(\frac{f}{\nu_0}\right)^2 S_\phi(f). \tag{11.166}$$

The spectral density of oscillator noise is conventionally modeled as a *power series*, where each power has a different physical significance:

$$S_y(f) = \sum_{\alpha=-2}^{+2} h_\alpha f^\alpha. \tag{11.167}$$

The five types of noise are

$$
\begin{aligned}
\text{White phase:} \quad & \alpha = 2 \\
\text{Flicker phase:} \quad & \alpha = 1 \\
\text{White frequency:} \quad & \alpha = 0 \\
\text{Flicker frequency:} \quad & \alpha = -1 \\
\text{Random-walk frequency:} \quad & \alpha = -2.
\end{aligned}
\tag{11.168}
$$

The familiar noises are white frequency and flicker frequency (also called $1/f$ noise). A remarkable property of the Allan variance is that, for the last four noise types, the Allan variance satisfies the following simple power law:

$$
\sigma_y^2(\tau) \propto \tau^{-\alpha-1} \qquad -2 \le \alpha \le 1.
\tag{11.169}
$$

The *Allan deviation* ($\sigma_y(\tau)$) is the square root of the Allan variance and provides the fractional frequency stability of the oscillator. From the above considerations, we obtain the well-known result that an oscillator which is limited by shot noise (white frequency) has a fractional frequency instability which is proportional to $\tau^{-1/2}$.

The effect of drift on the fractional frequency offset, $y(t)$, is to add a term,

$$
\text{Drift:} \quad x(t) = \tfrac{1}{2}Dt^2 \iff y(t) = Dt,
\tag{11.170}
$$

where $D$ is a constant (the drift rate). For an oscillator subject to only drift, $\bar{y}_{i+1} - \bar{y}_i = D\tau$ for all $i$ and the Allan deviation is obtained immediately from eqn 11.161:

$$
\text{Drift:} \quad \sigma_y(\tau) = \frac{D\tau}{\sqrt{2}}.
\tag{11.171}
$$

If the power series expansion of the noise is valid, the curve of $\log(\sigma_y(\tau))$ versus $\log(\tau)$ should have a concave shape with each successively *smaller* value of $\alpha$ becoming dominant as $\tau$ increases. Such a curve for a fictitious oscillator is shown in Fig. 11.32. One can use a curve of this kind together with a little detective work to determine the nature of the sources of oscillator frequency noise and drift.

The frequency domain measures can be used to directly determine the oscillator *linewidth* and can also be converted into an Allan variance. Experimentally, the spectral density of frequency fluctuations, $S_{\Delta f}(f)$, can be measured with a discriminator, which converts the frequency noise to amplitude noise after which the power density can be obtained using a radio-frequency spectrum analyzer. One can use the output of the frequency-locking discriminator for this purpose. Alternatively, one can use an external interferometer. The assumption is that the reference cavity or interferometer has very little noise over the frequencies of interest; if this assumption fails, the measurement becomes one which is relative to the cavity. The quantity $S_{\Delta f}(f)$ is defined via an auto-correlation function in the same way as are other similar quantities which are discussed above. The *mean square frequency deviation*, $\langle (\Delta f)^2 \rangle$, is

$$
\langle (\Delta f)^2 \rangle = \int_0^\infty S_{\Delta f}(f)df,
\tag{11.172}
$$

where we use a *one-sided spectral density* which is defined only for positive frequencies. Taking the square root of this quantity provides one with the root-mean-square

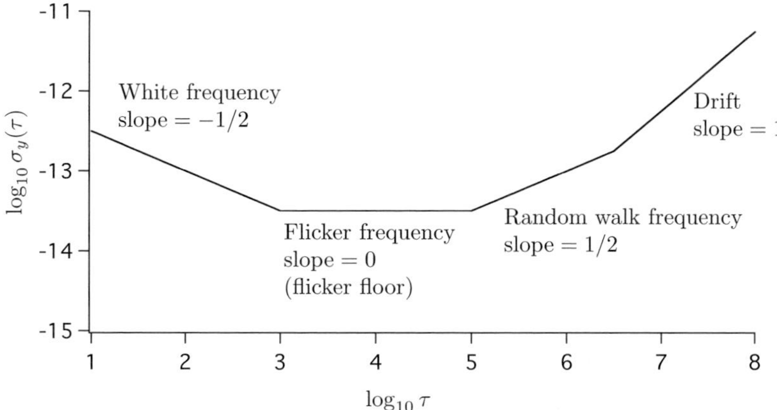

**Fig. 11.32** Typical Allan deviation curve showing how three different noise types and drift become dominant as $\tau$ increases.

linewidth of the oscillator. For white frequency noise, this integral diverges and one uses instead eqn 11.19:

$$\text{Linewidth with white frequency noise:} \quad \Delta f = \pi S_{\Delta f}. \tag{11.173}$$

It is more common to characterize the laser stability using the Allan variance, which can be obtained from $S_y(f)$ by evaluating an integral which is similar to eqn 11.172 but with a different weighting function:

$$\sigma_y^2(\tau) = 2 \int_0^\infty S_y(f) \frac{\sin^4(\pi f \tau)}{(\pi f \tau)^2} df. \tag{11.174}$$

Again, the Allan deviation is just the square root of this quantity and, together with its dependence on $\tau$, provides a fairly complete description of the stability of an oscillator. Note that the linewidth ($\Delta f$) is independent of the averaging time since it is a property of the laser beam and not of the measurement process. The linewidth can be measured, in principle, by analyzing the beam with an extremely narrow optical spectrum analyzer (Fabry-Perot). The Allan variance, on the other hand, depends upon the averaging time since it is an estimate of the error in determining the laser frequency after a measurement time equal to $\tau$.

We suggested that the linewidth of a laser can be measured by obtaining the power spectrum of the *error signal* from the discriminator used for frequency locking and calculating the RMS width from eqn 11.172. There is a frequency-domain alternative which yields the linewidth more directly and is probably the most reliable approach when the laser is extremely narrow (particularly when it is locked to an atomic resonance). This is to build a second *identical* oscillator and direct both sources to a fast photodetector, observing the spectrum of the beat note at the output of the detector. The beat note will reflect the *phase noise* spectrum of the laser if $|\phi(t)| \ll 1$, which is usually true for narrow sources. To see why this is so, we expand the electric field when $\phi(t)$ is small:

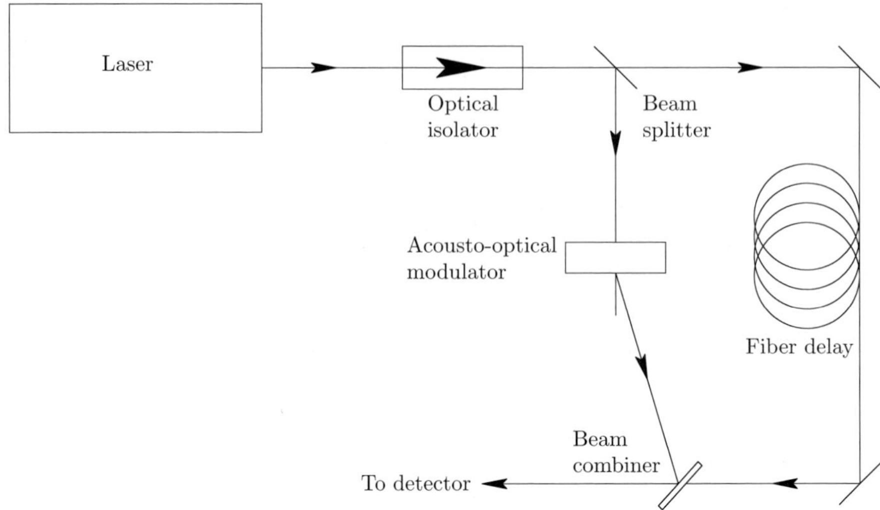


**Fig. 11.33** Block diagram of delayed heterodyne apparatus used to measure laser linewidths.

$$E(t) = E_0 \sin(2\pi\nu_0 t + \phi(t))$$
$$= E_0(\sin(2\pi\nu_0 t)\cos\phi(t) + \cos(2\pi\nu_0 t)\sin\phi(t)) \qquad (11.175)$$
$$\approx E_0(\sin(2\pi\nu_0 t) + \phi(t)\cos(2\pi\nu_0 t)).$$

The beat note for two uncorrelated sources, each of which is described by this expression, will have a low-frequency cross term that contains $\phi(t), \phi'(t)$ and is proportional to

$$(\phi(t) - \phi'(t))\sin(2\pi(\nu_0 - \nu_0')t), \qquad (11.176)$$

where the primed quantities refer to the second source and we ignore the $\phi(t)\phi'(t)$ term. The magnitude of the total phase fluctuations will be the square root of the quadratic sum of the two phase terms since the two sources are uncorrelated. If the sources are "identical", the width of the spectrum will be $\sqrt{2}$ greater than the width of *either* oscillator. This approach is often impractical, since the stabilized laser might be extremely complicated; a similar approach which is nearly as good is to use the *delayed homodyne* or *delayed heterodyne* technique, where we mix the laser beam with a *time-delayed* replica of itself. This is shown schematically in Fig. 11.33. If the delay time is greater than the coherence time of the laser, the beat note will have the same characteristics as one generated from two identical uncorrelated oscillators and the linewidth can be obtained as described above. It is usually more practical to shift the frequency of one of the beams: then the technique is called a time-delayed self-heterodyne measurement and the beat note will be at the shift frequency. A drawback of this approach is apparent when it is applied to the narrowest (sub-Hz) lasers: the length of the optical fiber can be impractically large. For example, some workers (Richter (1986)) have concluded that an accurate measurement requires a delay time which is six times the laser's coherence time. An external cavity diode laser with an expected 10 kHz linewidth would require 36 km of fiber to make an accurate linewidth measurement!

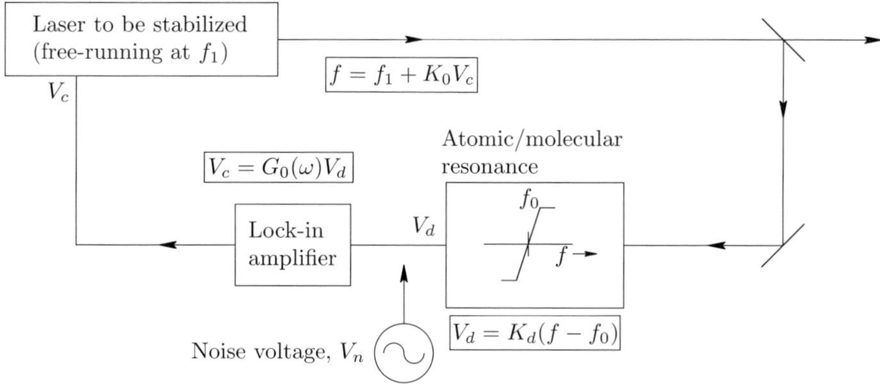


**Fig. 11.34** Block diagram of apparatus used to frequency lock to an atomic or molecular resonance using a lock-in amplifier.

The linewidth measurement of a sub-Hz laser would, of course, be impossible using this technique. These workers have shown how the measurement can be made using a variety of fibers whose delay times are less than the coherence time. Essentially, one is directly obtaining the autocorrelation function at several values of delay using this modified technique. Another approach is the use a *multipass* fiber, where the effective delay is increased by making the fiber into a *ring resonator* (see, for example, Yue, et al. (1988)). For the reasons mentioned above, the delayed homodyne and heterodyne approachs are most usefully employed on lasers with linewidths of $\approx 1$ MHz or greater.

## 11.13  Frequency locking to a noisy resonance

It is a common practice in the laboratory to frequency lock a laser to an atomic or molecular resonance by modulating some aspect of the resonance and obtaining the discriminant using a lock-in amplifier. The latter includes a simple low-pass filter with one or more poles after the phase-sensitive detector and has front panel controls which allow the operator to conveniently adjust the gain, time constant, number of poles, phase and modulation frequency. Thus, the lock-in very conveniently contains nearly all of the electronics needed to lock the laser to the resonance. We seek a strategy which optimizes the quality of the frequency lock. A block diagram of the apparatus appears in Fig. 11.34. The discriminant signal (possibly generated in the lock-in) is represented by voltage $V_d$ and the lock-in filter (including the gain circuitry) is represented by $G_0(\omega)$. These assignments allow us to use the previous theory to determine the noise characteristics of the loop.

One approach to adjusting frequency locking systems is to acquire a lock with a relatively low loop gain and then turn up the gain until something adverse occurs. There are two possible scenarios: either the loop will begin to oscillate or the locked laser will become very noisy due to the application of amplified source or electronic noise to the laser frequency control input. The former is the most common scenario when locking to a passive cavity and the latter when locking to an atomic or molecular resonance.

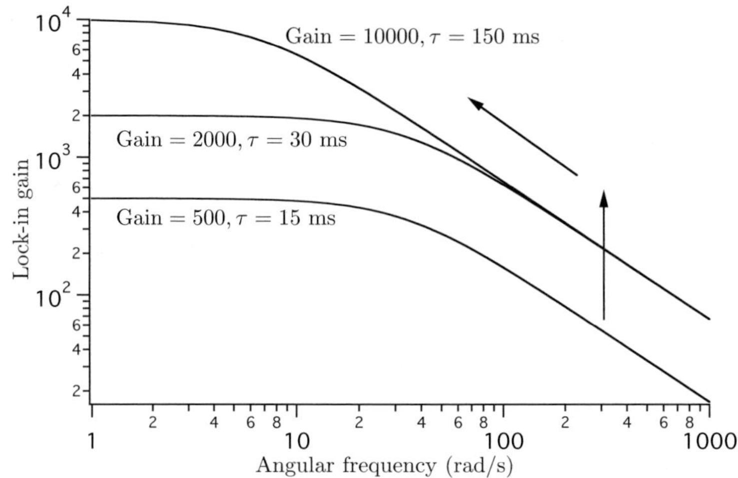


**Fig. 11.35** Gain versus frequency of lock-in as the gain is first increased without changing the time constant (vertical arrow) and then increased in proportion to the time constant (slanted arrow) preserving the high frequency gain characteristic.

If $V_n$ is the noise carried by the resonance, one can work backwards in the loop and observe that this will introduce a laser jitter equal to $V_n/K_d$ at frequencies where the loop gain is somewhat greater than unity. At low frequencies, the free-running intrinsic laser noise, $\Delta f_1$, will be reduced to

$$\text{Laser frequency noise reduction:} \quad \Delta f = \frac{\Delta f_1}{1 + K_0 K_d G_0(\omega)} \longrightarrow \frac{\Delta f_1}{K_0 K_d G_0(\omega)}.$$
$$(11.177)$$

The criterion for the source noise being the dominant contribution to the laser noise is then

$$\text{Source noise dominance:} \quad \frac{V_n}{K_d} \geq \frac{\Delta f_1}{K_0 K_d G(\omega)}. \qquad (11.178)$$

When $V_n$ is quite large, as in an atomic or molecular resonance, this inequality will be satisfied only with a fairly small value of $G(\omega)$ and the source noise will probably dominate well before the loop oscillates.

The gain is then set to a value which is just below the point where the source noise increases the laser jitter. This can be determined by either observing the lock-in output and looking for a sudden increase in the amplitude of the "correction" signal or by observing the laser's spectrum with an optical spectrum analyzer and noticing a sudden increase in the jitter of the resonance. The final step is to improve the low-frequency lock by increasing the lock-in time constant, $\tau$, and increasing the gain, $G_0$, by the same factor until the maximum time constant is reached. Plots of the lock-in gain versus frequency for three settings are shown in Fig. 11.35. As shown in the figure, first the gain is increased until the source noise is observed on the laser output (vertical arrow) and then both the gain and time constant are increased in the same ratio (slanted arrow). The latter operation preserves the medium to high frequency gain of the lock-in and should have little or no effect on the laser spectrum, since most of the source noise is in this frequency range. The very large DC gain will reduce the low-frequency loop error.

## 11.14  Further reading

There are a large number of books which cover servo and control systems with varying degrees of formalism. A classic text favored by the author is "Modern Control Systems" by Dorf (1967). An excellent short summary of control systems is in the book on phase-locked loops by Gardner (2004). There is likewise an enormous literature on linear system analysis and synthesis. The book used by the author is an undergraduate engineering text by Lathi (1965). Many of the references for frequency stabilization of lasers were given in the chapter on frequency discriminants. Three important papers which deserve being mentioned again are Drever, et al. (1983), Helmcke, et al. (1982) and Day, et al. (1992). A more recent paper in the Handbook of Optics, Volume IV (Hall (2000)) contains much useful information on more recent laser stabilization techniques. There is a large and extremely useful literature on oscillator frequency stability freely available online (currently) from NIST and indexed by TN 1337. Included are many classical papers on the subject, including some by D. W. Allan.

## 11.15  Problems

(11.1)  Obtain the transfer function (ratio of $v_2$ to $v_1$) for the circuit shown below.

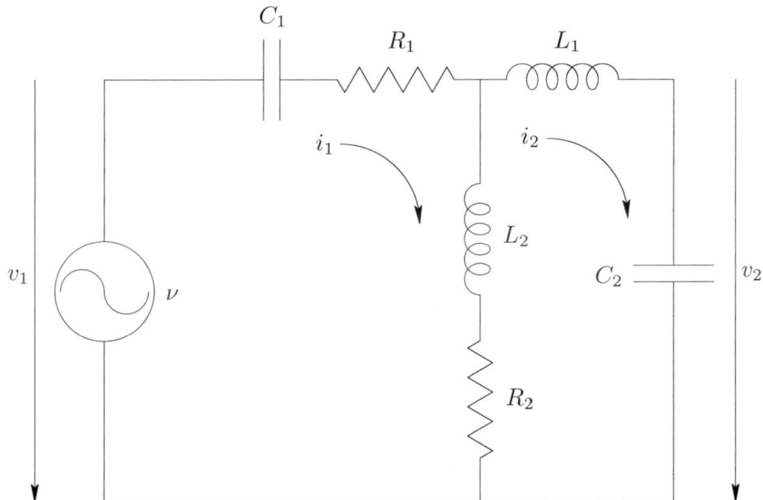

(11.2)  Determine the current, $i_2$, as a function of time if the voltage, $v_1$, is given by $v_1(t) = u(t)e^{-t/\tau}$ using the circuit immediately above. Assume that the initial currents and voltages are all zero.

(11.3)  Using the equation for $T(s)$ (eqn 11.102) for a temperature stabilization loop, find an expression for the temperature as a function of *time* when there is a step increase of $\Delta T_0$ in the *ambient* temperature. Assume that the controller is a lead-lag circuit.

(11.4) Using the technique described in the section on laser stabilization, determine the maximum frequency for shot-noise-limited behavior for the *single-mode diode*, whose noise spectrum is characterized in the text. Use a lead-lag filter and assume that there are 25 ns of time delay and that the zero of the filter is a factor of 2 below the unity gain frequency.

# 12
# Atomic and molecular discriminants

## 12.1 Introduction

Chapter 4 described in some detail the use of a *passive cavity* to generate a discriminant for laser stabilization. Passive cavities are fairly simple to use and can be very stable: a cavity constructed from Zerodur or ULE glass can have a drift rate as low as 0.1 Hz/second. However, it is often desirable to use an atomic or molecular resonance as a frequency reference. The long-term stability of such a reference can often be better than that of a passive cavity and, in some favorable cases, a rich molecular spectrum can also be used as a *poor man's frequency meter* to find and set the laser frequency much more accurately than is possible with a digital wavemeter. An atomic or molecular resonance will usually have a much smaller signal-to-noise ratio than a cavity resonance and will therefore usually be used only for long-term stabilization as the final stage in a hierarchy of increasingly stable references. The most extreme example of this is in the area of optical atomic frequency and time standards, where a very narrow resonance in a single ion or in a group of atoms trapped in an optical lattice is used to lock the laser frequency in the long term.

In this chapter, we will discuss some techniques used to obtain an atomic or molecular discriminant. The focus will be on the molecular spectra of $Te_2$ and $I_2$ obtained using sub-Doppler saturation spectroscopy. These molecules are very useful since they have rich vibrational-rotational spectra in the ranges 385–523 nm and 500–900 nm respectively and it is relatively easy to find a molecular resonance which is fairly close to any desired frequency in their ranges. For atomic resonances, the allied techniques of *polarization spectroscopy* and *dichroic atomic vapor laser lock* (DAVLL) will also be discussed. Finally, an example of the side-of-line approach with an atomic ion will be given.

## 12.2 Sub-Doppler saturation spectroscopy

In saturation spectroscopy, a laser beam is split into two (unequally intense) beams which are delivered to the atoms or molecules of interest using mirrors arranged so that the beams are counter-propagating and overlapping inside the sample. The stronger beam (*saturating beam*) burns a *hole* in the Doppler profile and the weaker beam (*probe beam*) is sent into a detector which measures the absorption as a function of frequency. Since the beams have opposite $k$-vectors, they interact with *Doppler subsets of atoms* which are displaced symmetrically about the center of the profile (the Doppler shift is $\boldsymbol{k} \cdot \boldsymbol{v}$). They only interact with the *same Doppler subset* when the detuning from resonance is less than the homogeneous width. The absorption

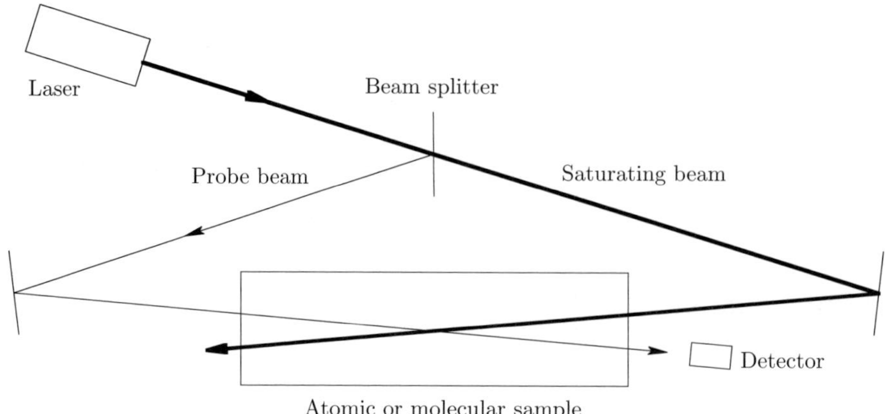


**Fig. 12.1** Schematic of apparatus for saturation spectroscopy.

of the probe beam versus laser frequency will show a broad Doppler profile with a pronounced *dip* when the two beams are interacting with the same atoms (those with approximately zero velocity along $\boldsymbol{k}$). The dip is Doppler-free and has a width equal to the homogeneous width of the sample species. The apparatus is shown in Fig. 12.1 and the resonant and off-resonant absorption profiles appear in Fig. 12.2. Not shown in the apparatus figure are lenses used to focus the saturating and probe beams into the sample; they facilitate obtaining the onset of saturation with modest laser power. A confocal parameter (eqn 1.20) of roughly the sample length is close to ideal. Finally, in addition to the resonances that occur at the various transition frequencies, there will be resonances halfway between pairs of transitions which share the same ground state and whose separations are less than the Doppler shift. These are called *cross-over* resonances and an explanation of their origin is left as an exercise for the reader.

One can obtain an analytic expression for the expected signal from eqn 5.159 in Chapter 5. That equation describes the gain experienced by a weak, tunable probe beam at frequency $\omega$ in the presence of a fixed frequency, strong saturating beam at $\omega'$. The result (called *hole burning*) will be repeated here for reference:

$$\text{Hole burning: } \gamma(\omega) \propto \int_{-\infty}^{\infty} N(v) g(\omega - \omega_0 - kv) \frac{1}{1 + \frac{I}{I_s}\bar{g}(\omega' - \omega_0 - k'v)} dv,$$

where $I$ is the intensity of the saturating laser and $I_s$ is the value of $I$ which produces saturation. This equation can be used to analyze saturation spectroscopy by setting $\omega = \omega'$ (probe and saturating frequencies are the same) and setting $k' = -k$ (beams are counter-propagating). The result is

$$\text{Saturation spectroscopy: } \gamma(\omega) \propto \int_{-\infty}^{\infty} N(v) g(\omega - \omega_0 - kv) \frac{1}{1 + \frac{I}{I_s}\bar{g}(\omega - \omega_0 + kv)} dv, \quad (12.1)$$

where $\bar{g}(\omega)$ is a Lorentzian normalized to $g(0) = 1$ and $N(v)dv$ is the number of atoms or molecules whose speed is between $v$ and $v + dv$. The quantity $\gamma(\omega)$ will describe

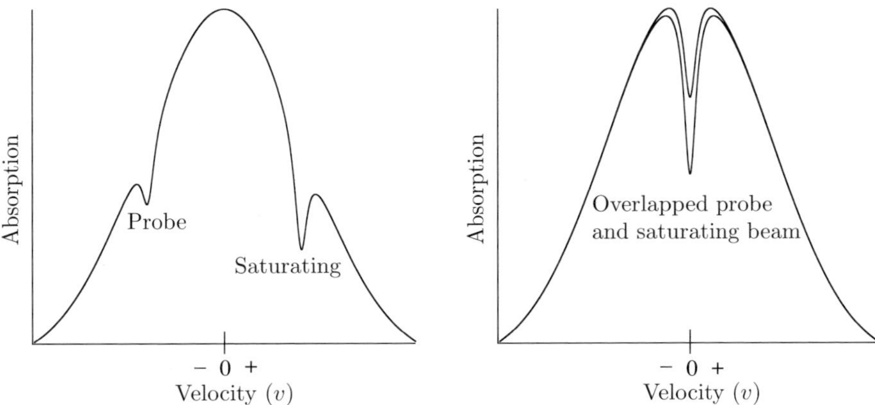



**Fig. 12.2** Doppler profiles in the presence of saturating and probe beams as a function of atomic or molecular velocity. The velocity, $v$, is converted to a frequency, $\omega$, using $\omega = kv + \omega_0$, where $\omega_0$ is the laser frequency. On the left, the lasers are away from rest-frame resonance; on the right they are on resonance.

*absorption* when the inversion ($\Delta N$) is positive (larger population in the ground state), which will always be the case in the apparatus under discussion. For simplicity, we assume that the homogeneous linewidth is much smaller than the inhomogeneous (Doppler) linewidth. In the limit of $I \ll I_s$, we can expand the fraction

$$\gamma(\omega) \propto \int_{-\infty}^{\infty} N(v)g(\omega - \omega_0 - kv)\left\{1 - \frac{I}{I_s}\bar{g}(\omega - \omega_0 + kv)\right\} dv. \qquad (12.2)$$

When we multiply the three factors and perform the integration, the first term is just a Doppler lineshape function, where we treat the $g(\omega - \omega_0 - kv)$ as a delta function. In evaluating the second term, we take the $N(v)$ factor out of the integration, since it is slowly varying, and the integral then becomes the convolution of two Lorentzians, each of which has width $\Delta\omega$. This convolution can be evaluated with the aid of the following formula, which can be verified with a contour integration:

$$\int_{-\infty}^{\infty} \frac{1}{a^2 + (y + x)^2} \frac{1}{b^2 + (y - x)^2} dx = \left(\frac{a + b}{ab}\right) \frac{\pi}{(2y)^2 + (a + b)^2}. \qquad (12.3)$$

The result is

$$\gamma(\omega) \propto N((\omega - \omega_0)/k)\left\{1 - \frac{I}{2I_s}\bar{g}(\omega - \omega_0)\right\}. \qquad (12.4)$$

The Lorentzian feature in the second term is thus a good replica of the original homogeneously broadened line multiplied by the slowly varying Gaussian function. Saturation spectroscopy is usually performed with $I \approx I_s$ for the best signals. A somewhat more involved analysis will yield the same qualitative result at higher powers: the observed line will have a Lorentzian lineshape with the homogeneous (power broadened) breadth.

Normally, one is not interested in the Doppler-broadened baseline and would remove it using one of a number of possible techniques. One way is to use a *thick* beam splitter which generates two parallel probe beams. If it is arranged that one of the probe beams does not overlap with the saturating beam, this probe beam can be used as a *reference* whose absorption signal would be subtracted from that of the other beam (which does overlap with the saturating beam). Since the reference signal contains *only* the Doppler-broadened baseline, the latter will be removed from the signal. Another method for eliminating the Doppler baseline is to *chop* the saturating beam (usually at some low audio frequency) and extract the Doppler-free signal using phase-sensitive detection. The validity of this approach can be seen from eqn 12.4: modulating the saturating beam will modulate the second term in the equation (which contains $I$) and the phase sensitive-detector will therefore yield the $\bar{g}(\omega - \omega_0)$ factor, which is multiplied by the modulated $I$. Either a mechanical chopper or an acousto-optic modulator can be used.

So far, we have explored the use of saturation spectroscopy to obtain Doppler-free atomic or molecular resonances. If we use the techniques described above, these resonances would be simple Lorentzian signals (eqn 12.4) and would not be very useful in that form for laser frequency stabilization. We will now discuss a technique which will generate a dispersion-shaped signal, which provides a good *discriminant* for laser stabilization. This scheme has some additional refinements which will be discussed below. One can, of course, generate a derivative signal by modulating the frequency of the laser at a low frequency (smaller than the resonance linewidth) and using phase-sensitive detection; the technique to be described has a number of advantages over simple frequency modulation despite its greater complexity.

The apparatus is shown in Fig. 12.3; the several modulation frequencies are specified merely for concreteness and would probably be different in an actual apparatus. The saturating beam is chopped at 10 kHz using an acousto-optic modulator (AOM) driven by a 100 MHz source shown in the upper right of the drawing. The upper mixer is used as a modulator to chop the AOM drive at 10 kHz. The probe beam is frequency modulated at 4 MHz by an electro-optic modulator and the detected signal is demodulated using a mixer. This is just a low frequency application of the Pound–Drever–Hall technique, described in detail in Chapter 4. Since the modulation frequency (4 MHz) is on the order of the natural linewidth of tellurium or iodine molecular transitions, the output will be a simple dispersion-shaped signal instead of the slightly more complicated signal which is seen when the modulation frequency is much larger than the resonance linewidth. The Pound–Drever–Hall signal is finally demodulated a second time (since it has a 10 kHz amplitude modulation from the chopped saturating beam) using the mixer near the bottom of the drawing; the demodulated signal is the desired discriminant.

In addition to providing a discriminant, this approach has the following additional features:

- The use of an AOM prevents interference fringes from stray beams. The probe and saturating beams are 100 MHz apart and any stray light from both beams can produce interference fringes which are *amplitude modulated* at 100 MHz and will likely be outside the passband of the detector and amplifier.

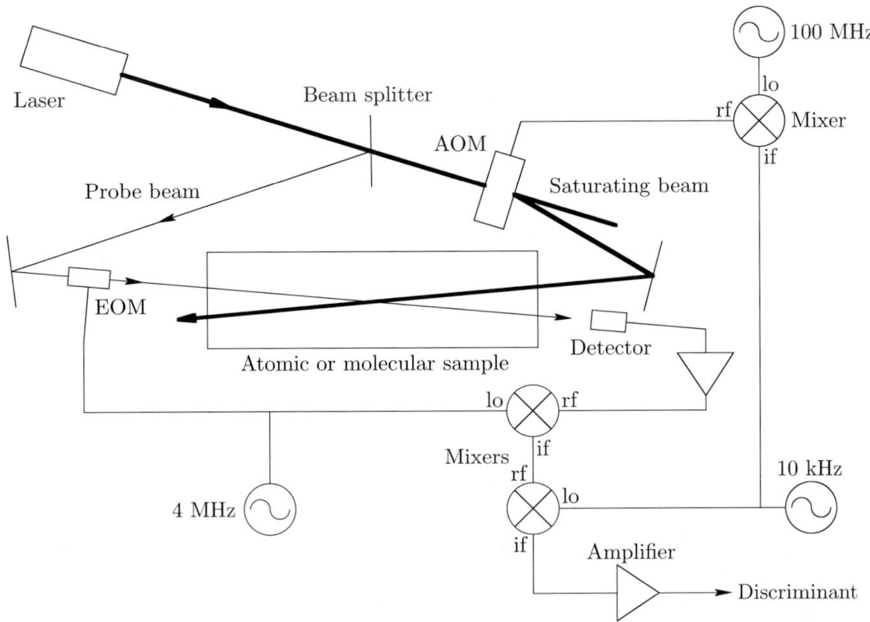


**Fig. 12.3** Schematic of improved saturation set-up.

- The first demodulation is at 4 MHz, which is usually above any amplitude noise in the laser beams and amplifiers. In practice, the author has observed amplitude noise from an external cavity diode laser modulated at 4 MHz; raising the modulation frequency to 20 MHz eliminated most of this noise.

- The amplitude modulation generated by the acousto-optic modulator can be at any frequency below the maximum allowed by the modulator (several hundred kHz for typical AOMs). Normally, one would not use frequencies below 1 kHz, since these would require filters with inconvenient time constants for observation of resonances while the laser is being swept.

- A little thought should convince the reader that the Doppler subset which will be simultaneously excited by both beams will be the one with a Doppler shift equal to one half of the AOM frequency shift (100 MHz in the figure). This allows the frequency of the resonance (and locked laser) to be precisely *tuned* by changing the AOM drive frequency. We will shortly discuss a technique which cancels the angular shift from the AOM so that a relatively wide tuning range is possible.

Since the ultimate signal is the product of a number of factors, including the saturating and probe intensities, one can also obtain the same results by placing the Pound–Drever–Hall phase modulator in the *saturating beam*. This approach is called *modulation transfer spectroscopy*.

We conclude this discussion with the description of a slightly more refined version of the apparatus just discussed. A schematic of the apparatus appears in Fig. 12.4, which shows an actual system used to lock an external cavity diode laser to resonances

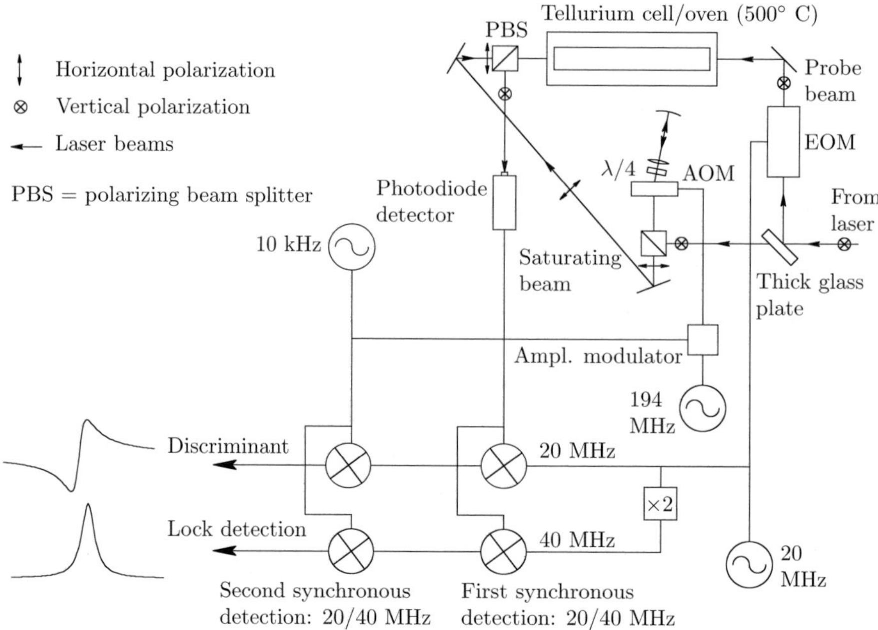


**Fig. 12.4** Schematic of improved saturation setup using a double-pass AOM and generating an additional lock-detection signal.

in tellurium molecules (the laser was frequency doubled). The setup is identical in function to that which was previously discussed: it uses a chopped saturating beam, Pound–Drever–Hall modulation of the probe beam and double demodulation. The following additional features are present:

- By exploiting the polarization manipulations shown in the figure, the saturating and probe beams can be made co-linear with each other with no loss of intensity since dissipative beam splitters are not used. The only conventional beam splitter is the thick glass plate which reflects $\approx 9\%$ of the light into the probe beam with the loss of 9% at the second surface.

- The acousto-optic modulator uses the *double-pass with cat's-eye* approach (discussed in Chapter 14) to cancel angular shifts and double the frequency shift. This scheme rotates the polarization by 90°, facilitating the use of the polarizing beam splitters as discussed in the previous point. Since there are no angular shifts with frequency, the resonance can be *tuned* in frequency by an amount roughly equal to the AOM tuning range (the factor of one half in the frequency shift cancels the factor of two due to the double pass AOM).

- By detecting the second harmonic of the Pound–Drever–Hall signal, an additional *symmetrical* resonance is obtained; this can be used as a *lock detector* to determine when the lock is broken. Additional electronics can be used to automatically sweep the laser over the resonance until the lock is reacquired. The reason for the symmetrical signal is a generalization of eqn 4.2: the signal observed with small

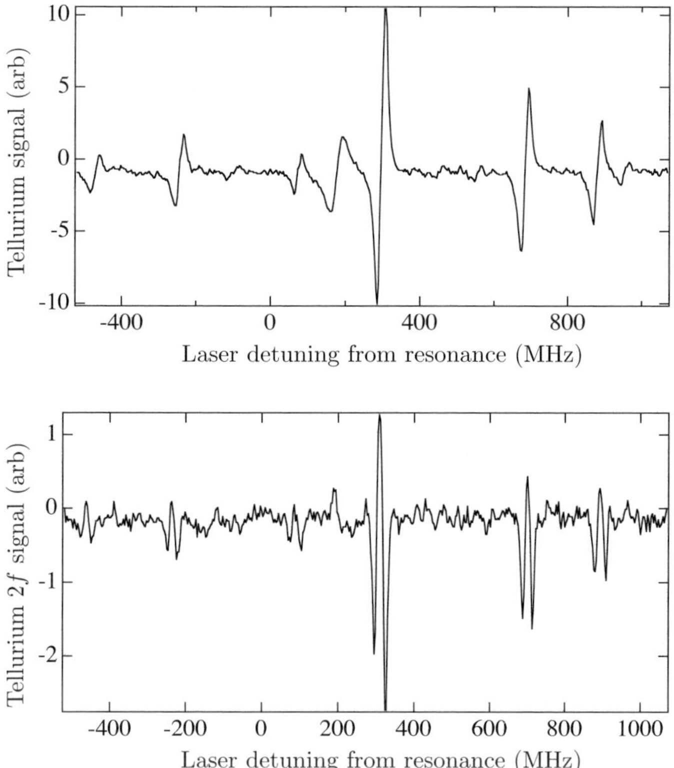

**Fig. 12.5** Saturated resonances in tellurium at 462 nm. The upper plot is the discriminant signal and the lower plot is the lock-detection signal.

deviation FM near a resonance is the *nth* derivative with respect to frequency of the resonance when the signal is detected at the *nth* harmonic of the modulation frequency. Thus, we are observing the *symmetric* second derivative.

Plots of actual discriminant and lock-detection resonances in tellurium are shown in Fig 12.5. A very short time constant was used, allowing a good signal-to-noise to be obtained in an oscilloscope display while sweeping the laser at a rate of about 10 Hz. The characteristic spectrum is readily memorized and serves as a *road map* which allows the operator to use the tellurium spectrometer as a frequency meter to reproducibly set the laser frequency with a precision of about 1 MHz.

## 12.3   Sub-Doppler dichroic atomic vapour laser locking (sub-Doppler DAVLL) and polarization spectroscopy

In the previous section, we discussed the use of the Pound–Drever–Hall technique to generate a discriminant from atomic or molecular resonances. When we discussed techniques for obtaining a discriminant from a passive cavity, we studied two approaches:

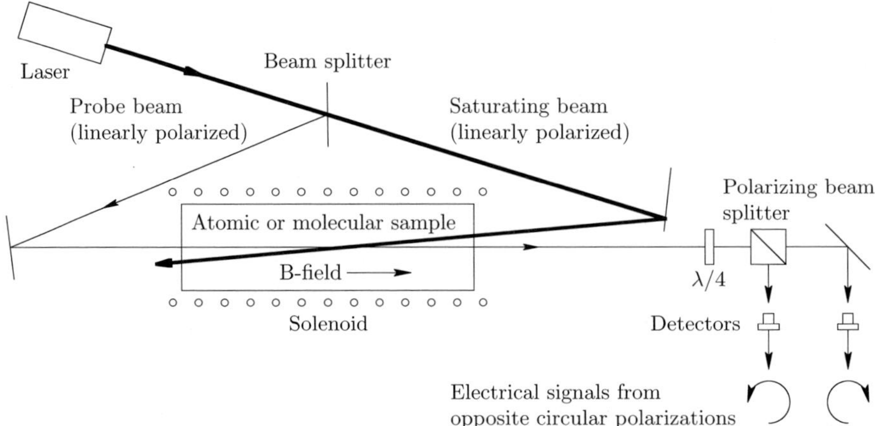


**Fig. 12.6** Apparatus for sub-Doppler DAVLL. The discriminant is the *difference* between the signals from the two detectors.

one which used frequency modulation (the Pound–Drever–Hall approach) and a second which dispensed with the modulation and used a polarization-dependent medium in the cavity (the polarization approach). We will discuss in this section the atomic and molecular analogue of the latter: the use of a *dichroic* medium to generate a discriminant. A dichroic medium is one in which the absorption depends upon the polarization of the light. When a counter-propagating saturating beam is used to remove the Doppler shifts, the technique is called sub-Doppler dichroic atomic vapour laser locking (sub-Doppler DAVLL). The closely allied technique of *polarization spectroscopy* will also be described.

The apparatus is shown in Fig. 12.6. The set-up is very similar to that used for ordinary saturation spectroscopy with the addition of a solenoid to generate a magnetic field parallel to the wave vector of the probe beam and a circular polarization analyzer for the probe beam. The circular analyzer detects both left and right circular polarization using a technique which was described in Chapter 4. The discriminant signal is the *difference* between the two polarization signals.

We will begin our explanation by assuming that the transition is between a $J = 0$ ground state and a $J = 1$ excited state in an atomic vapor. The quantization axis is along the magnetic field. One sense of circular polarization (which we will call $\sigma^+$) will cause a $\Delta m_J = +1$ transition and the other sense ($\sigma^-$) will cause a $\Delta m_J = -1$ transition. We will first consider plain DAVLL, which will not use a saturating beam and will be subject to Doppler broadening. The magnetic field will break the degeneracy between the $m_J = \pm 1$ levels and is large enough to produce Zeeman shifts which are on the order of the Doppler width (this requires a field of $> 100$ G). The probe beam polarization can be decomposed into equal parts of right and left circular polarization. The absorption lines for the two polarizations will be identical in shape but shifted in frequency by the Zeeman shift between the $m_J = \pm 1$ sublevels (there will be no $\Delta m_J = 0$ transition for this polarization). At the zero-field resonant frequency, the absorptions will be equal and their difference will be zero. On either side of resonance, the absorptions will be different and one will therefore obtain a signal which is an odd function of the laser frequency displacement from the zero-field resonance (i.e., a discriminant).

The analysis of saturated, sub-Doppler DAVLL is somewhat more complicated than that of simple saturation spectroscopy and will not be given here (a good reference for this is Maguire (2006)). The principle is, however, the same: the probe and saturating beams only interact with the same atoms when the laser frequency detuning from resonance is less than the homogeneous linewidth of the atoms. The dichroism of the saturated absorption causes a dispersion-shaped signal to be generated when the absorptions from the two senses of circular polarization are subtracted.

It is possible to discuss the optimum parameters without an analytic approach to the sub-Doppler DAVLL signal. The magnetic field need be only large enough to produce a Zeeman shift which is comparable to the homogeneous linewidth; for alkali atoms this is about 10 G. Unfortunately, there will be a Doppler-broadened background, which can be eliminated by chopping the saturating beam, but here we wish to avoid all modulations for the sake of simplicity. This background will make the location of the zero-crossing depend slightly on the magnetic field, which must be considered when locking a laser. A modest stabilization of the field should reduce this effect to a tolerable level. One should have a means of independently adjusting the powers in the saturating and probe beams. A simple method for accomplishing this is to use a half-wave plate and polarizing beam splitter to split the laser beam into the probe and saturating components. If the wave plate (which rotates the polarization) is between the laser and the beam splitter, rotating the wave plate will adjust the relative amounts of power in each beam (the orthogonal linear polarizations of the two beams will not be a problem). The saturating power should be near the onset of saturation (about 0.5 mW in a 1 mm diameter beam for alkali atoms): too much will broaden the transition, reducing the slope of the discriminant and too little will reduce the amplitude of the saturated resonance. The probe beam should be about ten times weaker than the saturating beam to avoid power broadening.

The *polarization spectroscopy* technique uses a circularly polarized saturating beam to polarize the atoms and detects the polarization using a linearly polarized probe beam. Historically, the technique used crossed polarizers in the probe beam on either side of the sample: near resonance, the polarization of the probe would be rotated (by the influence of the saturating beam on the atoms), producing a signal. Unfortunately, the signal does not have the required shape for a discriminant (it is symmetrical about the resonance frequency). A proper discriminant can be obtained by rotating the polarization of the probe (with a half-wave plate) by 45° before it encounters the sample and analyzing the polarization of the light after the sample in both the horizontal and vertical planes. The difference signal will clearly be zero away from resonance (the polarization is not rotated) and will be an odd function of the frequency displacement from resonance, as required for a discriminant. The apparatus is shown in Fig. 12.7.

Perhaps the most important distinction between polarization spectroscopy and DAVLL is that the latter detects the *dichroism* produced by a magnetic field and the former detects the *birefringence* (polarization-dependent refractive index) produced by

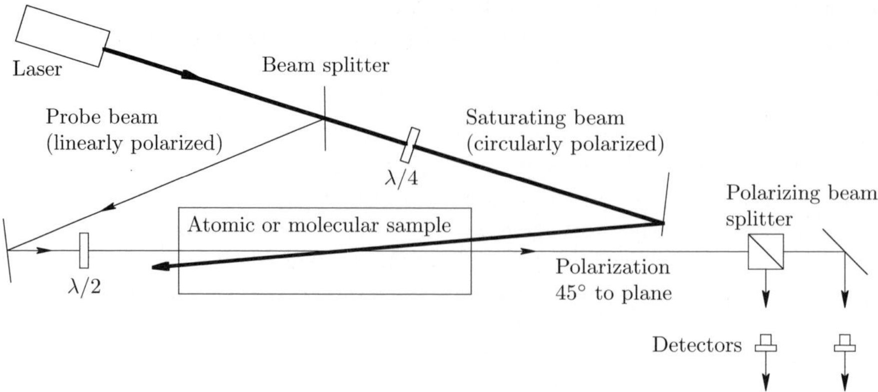



**Fig. 12.7** Apparatus for generating a discriminant using polarization spectroscopy. Again, the discriminant is the *difference* between the signals from the two detectors.

a circularly polarized saturating beam (it is easy to show that a small birefringence will rotate the polarization of linearly polarized light). One can consider the two techniques to be methods for analyzing the imaginary and real parts of the dielectric susceptibility.

We will derive an analytical expression for the polarization signal in two steps. First, we will determine the signal due to a birefringent medium. Then, we will specify in more detail the influence of the saturating beam on the medium, including the effects due the motion of the atoms. Using the Jones calculus (Chapter 4), the polarization of a wave incident on the sample and propagating in the $z$-direction is

$$
\boldsymbol{E} = \begin{pmatrix} E_x \\ E_y \end{pmatrix} = E_0 \begin{pmatrix} \cos\phi \\ \sin\phi \end{pmatrix} = \frac{E_0}{2} \left\{ e^{-i\phi} \begin{pmatrix} 1 \\ i \end{pmatrix} + e^{i\phi} \begin{pmatrix} 1 \\ -i \end{pmatrix} \right\},
\tag{12.5}
$$

where we have decomposed a beam whose linear polarization makes an angle $\phi$ with the $x$-axis into two opposite circular polarizations. After passing through a sample of length $L$, the field is

$$
\boldsymbol{E} = \frac{E_0}{2} \left\{ e^{-i\phi} \begin{pmatrix} 1 \\ i \end{pmatrix} e^{(-ik_+ - \alpha_+/2)L} + e^{i\phi} \begin{pmatrix} 1 \\ -i \end{pmatrix} e^{(-ik_- - \alpha_-/2)L} \right\},
\tag{12.6}
$$

where $k_\pm = \frac{\omega}{c} n_\pm$ and $\alpha_\pm$ are the real parts of the $k$-vectors and absorption coefficients for $\sigma^+$ and $\sigma^-$ circular polarization. One can factor out the average refractive index and absorption coefficient

$$\boldsymbol{E} = \frac{E_0}{2} e^{(-i\frac{\omega}{c}n - \frac{\alpha}{2})L} \left\{ e^{-i\phi} \begin{pmatrix} 1 \\ i \end{pmatrix} e^{-i\Omega} + e^{i\phi} \begin{pmatrix} 1 \\ -i \end{pmatrix} e^{+i\Omega} \right\}, \tag{12.7}$$

where

$$\Omega = \frac{\omega}{2c}\Delta n L - i\frac{L}{4}\Delta\alpha \tag{12.8}$$

$$n = \tfrac{1}{2}(n_+ + n_-) \tag{12.9}$$

$$\alpha = \tfrac{1}{2}(\alpha_+ + \alpha_-) \tag{12.10}$$

$$\Delta n = n_+ - n_- \tag{12.11}$$

$$\Delta\alpha = \alpha_+ - \alpha_-. \tag{12.12}$$

The signal intensity is proportional to $|E_x|^2 - |E_y|^2$ which is

$$I_{signal} = I_0 e^{-\alpha L} \cos\left(2\phi + L\Delta n\frac{\omega}{c}\right), \tag{12.13}$$

where $I_0$ is the signal without the sample. The first term in the cosine is twice the angle between the polarizers; we take $\phi$ to be $\pi/4$. The second term is due to the atoms and we assume that it is much less than one. Expanding the cosine about $\pi/2$, we obtain

$$I_{signal} = I_0 e^{-\alpha L} L\Delta n\frac{\omega}{c}. \tag{12.14}$$

The absorption of the probe has a Lorentzian frequency dependence

$$\Delta\alpha = \Delta\alpha_s \bar{g}(\omega - \omega_0), \tag{12.15}$$

where $\bar{g}(\omega - \omega_0)$ is a Lorentzian of width $\gamma$ normalized to unity at resonance and $\Delta\alpha_s$ is the absorption difference due to the saturating beam. One can obtain a *dispersion relation* from eqns 5.92 and 5.93 together with the relations between the real and imaginary components of the susceptibility and the index of refraction and absorption (eqns 5.83 and 5.84). The result is

$$\Delta n = \frac{c}{\omega_0} \frac{\omega_0 - \omega}{\gamma} \Delta\alpha. \tag{12.16}$$

Using this, the signal is:

$$I_{signal} = I_0 e^{-\alpha L} L\frac{\Delta\alpha_s}{\gamma}(\omega_0 - \omega)\bar{g}(\omega - \omega_0), \tag{12.17}$$

which is the expected dispersion-shaped signal obtained when $\phi = \pi/4$.

To complete the calculation, we need to know the effect of the saturating laser on the absorption difference, $\Delta\alpha_s$. We will assume that the dependence of $\Delta\alpha_s$ on the *saturating beam frequency* is also Lorentzian:

$$\Delta\alpha_s = \Delta\alpha_0 \bar{g}(\omega - \omega_0 - kv), \tag{12.18}$$

where $\Delta\alpha_0$ is the absorption difference when the saturating beam is on resonance; it depends upon details of the atomic structure together with the intensity and polarization of the saturating beam. We have included the Doppler shift in the above expression. To simplify the calculation, assume that $I \ll I_s$; the width is therefore the same as that due to the probe beam. Finally, we multiply the whole signal expression by the velocity distribution $N(v)$, include Doppler shifts for the probe beam, and integrate over all velocities:

$$I_{signal} \propto \int_{-\infty}^{\infty} N(v)(\omega - \omega_0 + kv)\bar{g}(\omega - \omega_0 + kv)\bar{g}(\omega - \omega_0 - kv)dv, \qquad (12.19)$$

where we have left out all of the multiplicative constants such as the intensities, $\Delta\alpha_0$, etc. One should note the different sign of the $kv$ term in the arguments of the probe and saturating lineshape expressions. As in the case of conventional saturation spectroscopy, we have a convolution, this time between Lorentzian and dispersion-shaped functions (we factor the Gaussian out of the integral since the inhomogeneous linewidth is much greater than the homogeneous linewidth). A straightforward (but slightly messy) contour integration yields the formula

$$\int_{-\infty}^{\infty} \frac{y + x}{a^2 + (y + x)^2} \frac{1}{b^2 + (y - x)^2} dx = \frac{2\pi y}{b[(2y)^2 + (a + b)^2]}, \qquad (12.20)$$

from which we can obtain the result

$$I_{signal} \propto (\omega - \omega_0)\bar{g}(\omega - \omega_0), \qquad (12.21)$$

which is the desired dispersion-shaped discriminant.

## 12.4  An example of a side-of-line atomic discriminant

We will conclude this chapter with a description of a side-of-line discriminant obtained using opto-galvanic detection.

The side-of-line locking technique was discussed in Chapter 4. We will briefly mention a variant of this scheme applied to locking a laser to a Doppler broadened *opto-galvanic* line. It was used in the author's laboratory to lock a frequency-doubled external cavity diode laser to a resonance in a barium discharge lamp; the laser was used to cool a single barium ion and was successfully locked, without interruption, for up to several weeks. The opto-galvanic technique exploits the change in the discharge current–voltage characteristic when a laser beam is resonant with one of the resonance lines of the medium in the discharge (in this case singly ionized barium). A schematic of the apparatus appears in Fig. 12.8.

The acousto-optic modulator (AOM) chops the laser beam (at 10 kHz in the figure). The discharge current is capacitively coupled and *summed* with a sample of the chopped laser intensity using the *undeflected* beam from the AOM. This latter beam is detected with a photodiode to generate a *reference*, which is subtracted from the opto-galvanic signal; it is summed in the operational amplifier since the two outputs of the AOM are modulated 180° out of phase. The output of the operational amplifier is the signal input to the lock-in, and a sample of the modulation voltage is the reference. Subtracting a sample of the laser intensity makes the lock point fairly robust against changes in the laser power. One would normally lock the laser several hundred MHz

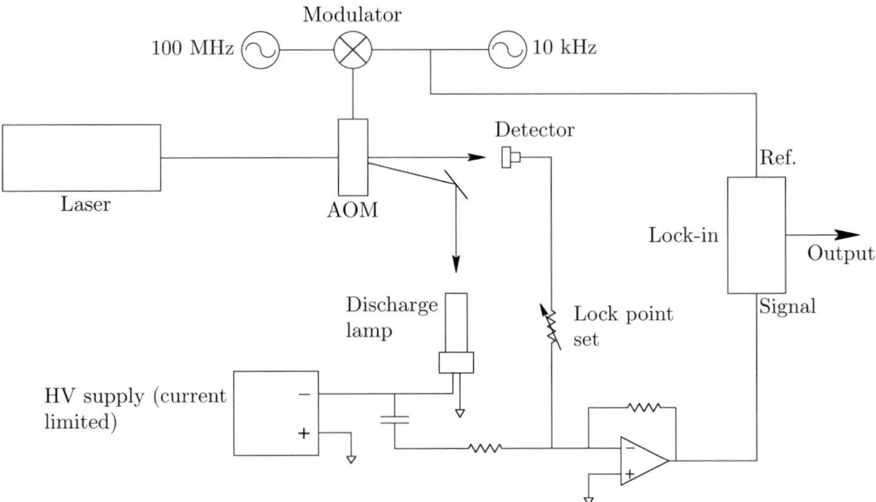

**Fig. 12.8** Apparatus for side-of-line locking to Doppler broadened opto-galvanic line in discharge tube.

below the opto-galvanic resonance (since it is *side-of-line* locking); the AOM frequency shift can be chosen to return the laser frequency to the exact resonance or any desired frequency (limited by the AOM tuning range). The lock point (and laser frequency) can be changed by changing the gain of the reference beam; this moves the lock point to a different place on the resonance line.

## 12.5 Further reading

Saturation spectroscopy is very well described in the book *Atomic Physics* by Foot (2005). The use of the Pound–Drever–Hall technique in saturation spectroscopy appears in a paper by Hall (1981). Sub-Doppler DAVLL is treated qualitatively in Harris, et al. (2008) and an analytical treatment of polarization spectroscopy and its application to laser locking appears in Pearman, et al. (2002).

## 12.6 Problems

(12.1) As mentioned in the text, so-called *cross-over resonances* occur in saturated spectroscopy when two allowed transitions share the same ground state and have transition frequencies whose difference is less than the Doppler width of either transition. They occur halfway between the two resonances. Explain the mechanism which allows these *spurious* resonances to occur.

(12.2) The Pound–Drever–Hall saturation technique produces a *symmetric* signal when the first mixer uses the *second harmonic* of the modulation frequency for a reference. Explain why the *nth* derivative of the resonance is obtained when the detection occurs at the *nth* harmonic. (This is strictly true only when the FM deviation and modulation frequency are somewhat smaller than the linewidth of the resonance.)

(12.3) Compare the discriminant slope obtained with the saturated Pound–Drever–Hall technique to that obtained using a side-of-line approach assuming that the linewidths are the same (the side-of-line would use a saturation approach). Assume that the laser in the side-of-line approach is tuned exactly halfway up the side of the line.

(12.4) The polarization scheme will generate a pure dispersion signal only when the polarizer axes are at $45°$ to each other. A polarization error, expressed as a deviation from $45°$, will cause the *lock point* to change. Determine the change in the lock point as a fraction of the atomic linewidth for a small angular change in one of the polarizers from the optimum orientation.

(12.5) Show that a small birefringence will rotate the polarization of linearly polarized light. Assume that the beam propagates in the $z$-direction and that, for polarization along the $x$-axis, the index is $n_x$, while for polarization along the $y$-axis, the index is $n_y$, where $|n_x - n_y| \ll n_x$. For simplicity, assume that the laser is polarized at $45°$ to the $x$-axis.

# 13
# Nonlinear optics

## 13.1   Introduction

In all of our discussions of electromagnetic wave propagation, we have assumed that the polarization is a linear function of the electric field. This is an excellent approximation for most of the phenomena observed in nature. With the advent of lasers, however, it is possible to generate extremely strong electric fields by tightly focusing the laser beam, and these strong fields can produce a polarization which depends on higher powers of the electric field. These phenomena are the subject matter of *nonlinear optics*. Aside from their intrinsic value as interesting physics, nonlinear interactions have a very practical value in the atomic physics laboratory. They enable one to generate useful amounts of coherent radiation at wavelengths not normally served by commonly available lasers. Thus, via second harmonic generation and sum-frequency mixing, one can generate radiation in the deep blue and ultraviolet regions of the spectrum. In addition, parametric processes can provide a useful supply of *entangled photons*, enabling experiments in measurement and quantum information theory.

This chapter will discuss the most common nonlinear interactions, beginning with second harmonic generation and proceeding to sum-frequency mixing and parametric phenomena. The principal techniques of *phase matching* will be discussed in detail. To facilitate this, we will begin with a discussion of *anisotropic crystals*, since crystal birefringence is exploited in the most common technique of phase matching. This introduction will also be useful in understanding the modulation techniques discussed in a later chapter.

## 13.2   Anisotropic crystals

An anisotropic crystal is one in which the behavior of an electromagnetic wave is dependent upon the direction of the electric field of the wave (i.e., the *polarization*). We will begin our analysis by writing down the time-dependent Maxwell equations and material relations for isotropic media and then modify the material relations to describe anisotropic media. We assume that there are no free charges or currents and that there is no magnetic material. Ampere's law and Faraday's law are

$$\nabla \times \boldsymbol{H} = \frac{\partial \boldsymbol{D}}{\partial t} \tag{13.1}$$

$$\nabla \times \boldsymbol{E} = -\frac{\partial \boldsymbol{B}}{\partial t}. \tag{13.2}$$

The material relationships are

$$D = \epsilon_0 E + P = \epsilon E \tag{13.3}$$

$$B = \mu_0(H + M) = \mu H. \tag{13.4}$$

Energy relations are described with the aid of the *Poynting vector, $S$* and the *energy density, $u$*:

$$S = E \times H \tag{13.5}$$

$$u = \tfrac{1}{2}(E \cdot D + B \cdot H). \tag{13.6}$$

These quantities satisfy an *energy continuity equation,*

$$\nabla \cdot S + \frac{\partial u}{\partial t} = 0, \tag{13.7}$$

in the absence of sources.

In an *anisotropic* medium, the polarization can point in a different direction from the field and $E$ and $D$ are not necessarily collinear. The dielectric constant, $\epsilon$, is now a *linear operator* represented by the *dielectric tensor*, $\epsilon_{ij}$,

$$D_i = \sum_j \epsilon_{ij} E_j \iff \begin{pmatrix} D_1 \\ D_2 \\ D_3 \end{pmatrix} = \begin{pmatrix} \epsilon_{11} & \epsilon_{12} & \epsilon_{13} \\ \epsilon_{21} & \epsilon_{22} & \epsilon_{23} \\ \epsilon_{31} & \epsilon_{32} & \epsilon_{33} \end{pmatrix} \begin{pmatrix} E_1 \\ E_2 \\ E_3 \end{pmatrix}. \tag{13.8}$$

The energy density of the *electric field, $u_e$*, and its time derivative, $\dot{u}_e$, are

$$u_e = \frac{1}{2} E \cdot D = \frac{1}{2} \sum_{ij} E_i \epsilon_{ij} E_j \implies \dot{u}_e = \frac{1}{2} \sum_{ij} \epsilon_{ij}(\dot{E}_i E_j + E_i \dot{E}_j). \tag{13.9}$$

From the *Poynting theorem,*

$$\nabla \cdot S + E \cdot \dot{D} + H \cdot \dot{B} = 0. \tag{13.10}$$

This implies that

$$\nabla \cdot S + \sum_{ij} E_i \epsilon_{ij} \dot{E}_j + H \cdot \dot{B} = 0 \implies \dot{u}_e = \sum_{ij} E_i \epsilon_{ij} \dot{E}_j. \tag{13.11}$$

Equating the two expressions for $\dot{u}_e$, we obtain the important result

$$\epsilon_{ij} = \epsilon_{ji}. \tag{13.12}$$

The dielectric tensor is therefore represented by a *symmetric matrix*, which can be *diagonalized* yielding three *orthogonal* eigenvectors (principal axes) and *real* eigenvalues (proportional to the refractive indices squared). Henceforth, we will always work in the coordinate system of the principal axes and will use the symbol $\epsilon_i$ for the diagonal elements of $\epsilon_{ij}$:

$$\epsilon_{ij} \longrightarrow \epsilon_i. \tag{13.13}$$

The angular relations among the vectors $\boldsymbol{D}, \boldsymbol{E}, \boldsymbol{H}, \boldsymbol{k}$ and $\boldsymbol{S}$ can be easily obtained for a *plane wave* propagating as $e^{i\boldsymbol{k}\cdot\boldsymbol{r}-i\omega t}$, where the following operator equivalences will be used:

$$\nabla\times \Longrightarrow i\boldsymbol{k} \times \tag{13.14}$$

$$\frac{\partial}{\partial t} \Longrightarrow -i\omega. \tag{13.15}$$

Maxwell's time-dependent equations are

$$\begin{aligned} \boldsymbol{k} \times \boldsymbol{H} &= -\omega\boldsymbol{D} \\ \boldsymbol{k} \times \boldsymbol{E} &= \mu\omega\boldsymbol{H}. \end{aligned} \tag{13.16}$$

From these two equations, one can draw the following conclusions:

- $\boldsymbol{D}$ and $\boldsymbol{H}$ are both perpendicular to $\boldsymbol{k}$.
- $\boldsymbol{D}$ is perpendicular to $\boldsymbol{H}$.
- $\boldsymbol{E}$ and $\boldsymbol{H}$ are perpendicular to each other and to $\boldsymbol{S} = \boldsymbol{E} \times \boldsymbol{H}$.

Thus, there are two sets of three mutually perpendicular vectors:

- $\boldsymbol{E}, \boldsymbol{H}$ and $\boldsymbol{S}$: *energy propagation direction* $= \hat{\boldsymbol{S}}$
- $\boldsymbol{D}, \boldsymbol{H}$ and $\boldsymbol{k}$: *phase propagation direction* $= \hat{\boldsymbol{k}}$

Note that $\boldsymbol{D}, \boldsymbol{E}, \boldsymbol{S}$ and $\boldsymbol{k}$ *are coplanar* (they are all perpendicular to $\boldsymbol{H}$). These relationships are illustrated in Fig. 13.1.

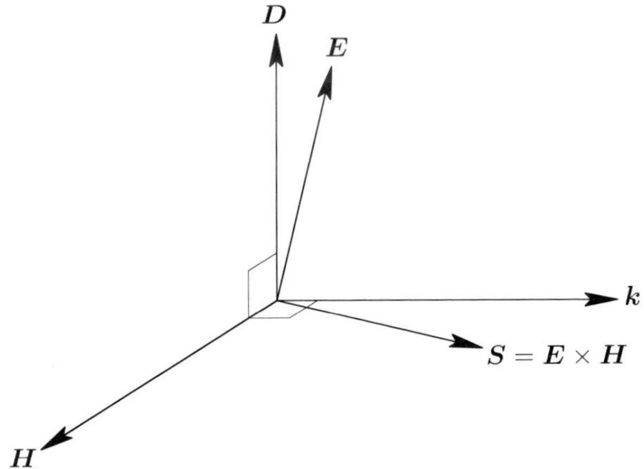


**Fig. 13.1** Angular relations among $\boldsymbol{D}, E, S, H$ and $\boldsymbol{k}$ in an anisotropic crystal.

The allowed values of the refractive index, $n$, can be obtained from the Maxwell equations (eqns 13.16). First, we will write $\boldsymbol{k}$ as

$$\boldsymbol{k} = k\hat{\boldsymbol{k}} = \frac{n\omega}{c}\hat{\boldsymbol{k}}. \tag{13.17}$$

Then, $\boldsymbol{D}$ and $\boldsymbol{H}$ are

$$\boldsymbol{D} = -\frac{n}{c}\hat{\boldsymbol{k}} \times \boldsymbol{H} \tag{13.18}$$

$$\boldsymbol{H} = \frac{n}{\mu c}\hat{\boldsymbol{k}} \times \boldsymbol{E}. \tag{13.19}$$

Substituting the second equation into the first, one obtains

$$\boldsymbol{D} = -\frac{n^2}{c^2\mu}\hat{\boldsymbol{k}} \times \hat{\boldsymbol{k}} \times \boldsymbol{E} = \frac{n^2}{c^2\mu}[\boldsymbol{E} - \hat{\boldsymbol{k}}(\hat{\boldsymbol{k}} \cdot \boldsymbol{E})] = \frac{n^2}{c^2\mu}\boldsymbol{E}_{transverse}, \tag{13.20}$$

where $\boldsymbol{E}_{transverse}$ is the component of $\boldsymbol{E}$ which is perpendicular to $\boldsymbol{k}$. Since we are in the coordinate system of the principal axes of $\epsilon_{ij}$, the dielectric matrix is diagonal and one can solve the above equation for each component of $\boldsymbol{E}$:

$$E_i = \frac{n^2\hat{k}_i(\hat{\boldsymbol{k}} \cdot \boldsymbol{E})}{n^2 - \epsilon_i/\epsilon_0}, \qquad i = x, y, z, \tag{13.21}$$

where we have used $c = 1/\sqrt{\mu_0\epsilon_0}$, $D_i = \epsilon_i E_i$ and $\mu = \mu_0$. Multiplying this equation by $\hat{k}_i$ and summing over $i$ yields

$$\hat{\boldsymbol{k}} \cdot \boldsymbol{E} = \hat{\boldsymbol{k}} \cdot \boldsymbol{E} \sum_{i=x,y,z} \frac{n^2\hat{k}_i^2}{n^2 - \epsilon_i/\epsilon_0}. \tag{13.22}$$

Cancelling the scalar product and dividing by $n^2$, one obtains the *Fresnel equation*:

$$\frac{\hat{k}_x^2}{n^2 - \epsilon_x/\epsilon_0} + \frac{\hat{k}_y^2}{n^2 - \epsilon_y/\epsilon_0} + \frac{\hat{k}_z^2}{n^2 - \epsilon_z/\epsilon_0} = \frac{1}{n^2}. \tag{13.23}$$

This is a quadratic equation in $n^2$ and therefore has *two positive solutions* for the allowed values of the refractive index.

One can pursue this approach further to determine the possible indices of refraction as a function of the propagation direction, but there is a much more useful *graphical* approach to the problem: the *index ellipsoid*. In a coordinate system where $\epsilon_{ij}$ is diagonal, the stored energy in the *electric field* is

$$u_e = \frac{1}{2}(\epsilon_x E_x^2 + \epsilon_y E_y^2 + \epsilon_z E_z^2) = \frac{1}{2}\left(\frac{D_x^2}{\epsilon_x} + \frac{D_y^2}{\epsilon_y} + \frac{D_z^2}{\epsilon_z}\right). \tag{13.24}$$

The index of refraction is given by $n = c/v_\phi$, where $v_\phi$ is the *phase velocity* in the medium. In a nonmagnetic medium, one can replace the three components of the dielectric tensor by three refractive indices which satisfy $\epsilon_i = n_i^2 \epsilon_0$. The energy density is

$$u_e = \frac{1}{2\epsilon_0} \left( \frac{D_x^2}{n_x^2} + \frac{D_y^2}{n_y^2} + \frac{D_z^2}{n_z^2} \right). \tag{13.25}$$

If we use a Cartesian coordinate system whose axes are the components of $\boldsymbol{r}$ and let

$$\boldsymbol{r} = \frac{\boldsymbol{D}}{\sqrt{2u_e\epsilon_0}}, \tag{13.26}$$

the surface of *constant energy density* is an ellipsoid, called the *index ellipsoid*:

$$\text{Index ellipsoid:} \quad \frac{x^2}{n_x^2} + \frac{y^2}{n_y^2} + \frac{z^2}{n_z^2} = 1, \qquad \left( \frac{\boldsymbol{D}}{\sqrt{2u_e\epsilon_0}} \to \boldsymbol{r} \right). \tag{13.27}$$

An index ellipsoid is portrayed in Fig. 13.2. The three *semi-major axes* are equal to the three indices of refraction. The following three steps describe the use of the index

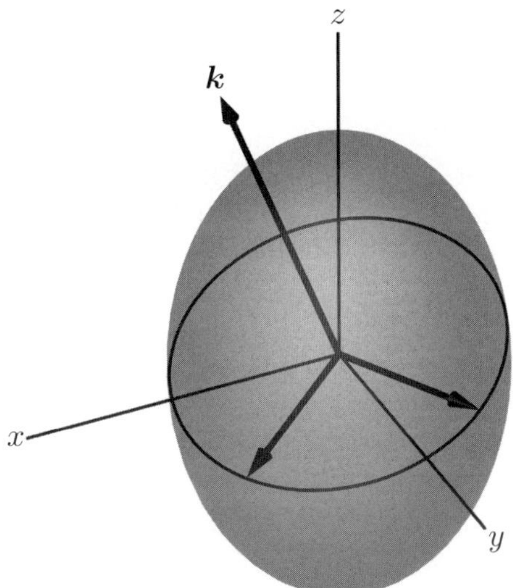


**Fig. 13.2** Index ellipsoid showing the $\boldsymbol{k}$-vector (nearly vertical) and the two possible polarizations of the $\boldsymbol{D}$ vector.

ellipsoid to determine the possible polarizations of $\boldsymbol{D}$ and indices of refraction when the $\boldsymbol{k}$-vector is known.

- First, plot an ellipsoid whose three semi-major axes are equal to three indices of refraction and whose center is at the origin of the coordinate system.
- Then draw a *plane* through the origin perpendicular to $\boldsymbol{k}$ (its equation is $\boldsymbol{r} \cdot \boldsymbol{k} = xk_x + yk_y + zk_z = 0$).
- The *ellipse* at the intersection of the plane and the ellipsoid has *two semi-axes*. *The allowed directions of $\boldsymbol{D}$ are along these semi-axes and the allowed values of n are equal to the lengths of the semi-axes.*

The validity of the above can be verified using the Fresnel equation. Implicit in the utility of this graphical approach is the fact that the index ellipsoid is a fixed object for a given crystal and for a given wavelength of light.

Some important conclusions about the behavior of wave propagation in anisotropic media can be immediately obtained from the index ellipsoid. First, it should be obvious that, for any ellipsoid, there is at least one propagation direction for which the ellipse of intersection is a *circle*. For these propagation directions, the index of refraction is *independent* of the polarization. These directions of $\hat{k}$ are called the *optical axes* of the medium. In general, an optical axis is defined as a *propagation direction in which the index of refraction is independent of the polarization.* The number and nature of the optical axes depend upon whether or not any of the refractive indices are *repeated.* If the medium is anisotropic, at least two indices are different. If only two are different, the crystal is called *uniaxial.* Uniaxial crystals for which $n_z > n_x = n_y$ are called *positive uniaxial,* and uniaxial crystals for which $n_z < n_x = n_y$ are called *negative uniaxial.* The ellipsoids for the three types of crystal are shown in Fig. 13.3. In drawings of unaxial index ellipsoids, it is customary for the *different* index to be along the $z$-axis (the two *equal* indices are along the $x$ and $y$ axes). It should be clear that the optical axis for uniaxial crystals is also along the $z$-axis and that there is only one optical axis.

If all three indices are different, one can convince oneself after a little reflection that there are *two* circles of intersection and that these crystals are therefore *biaxial.* The two optical axes of biaxial crystals are tilted symmetrically away from the $z$-axis at angles which depend upon the indices of refraction. The tilt angle, $\theta$, can easily be shown to be

$$\theta = \cos^{-1}\sqrt{\frac{1/n_y^2 - 1/n_z^2}{1/n_x^2 - 1/n_z^2}}, \tag{13.28}$$

where we have used the convention for biaxial crystals that $n_x < n_y < n_z$.

The two allowed polarizations discussed above can be considered to be *eigenpolarizations*: polarizations which propagate without change in the crystal. Of course, a general linearly polarized plane wave can also propagate in the crystal, but it will separate into two eigenpolarizations which propagate with *different phase velocities,* producing various degrees of *elliptical polarization* as it progresses. There is an important difference between the two eigenpolarizations in a uniaxial crystal (which can be seen in the top two ellipsoids in Fig. 13.3). As one changes the angle that $k$ makes with the $z$-axis, one of the indices will be constant and the other will change. The one which remains constant is called the *ordinary polarization* and the one which changes is called the *extraordinary polarization* (sometimes the associated waves are called ordinary and extraordinary waves, with the polarization being the salient difference). The reason for these terms will become clearer when we discuss *ray propagation.*

From the above geometrical considerations, one can calculate the indices of refraction for an arbitrary propagation direction in a uniaxial crystal (also in a biaxial crystal, but the algebra becomes fairly complicated). The calculation makes use of a construction called *normal surfaces,* which are shown in Fig. 13.4 for a positive and negative uniaxial crystal. Normal surfaces are ellipsoids of revolution, which can be reduced to simple ellipses for a uniaxial crystal, and are obtained by plotting, *in a*

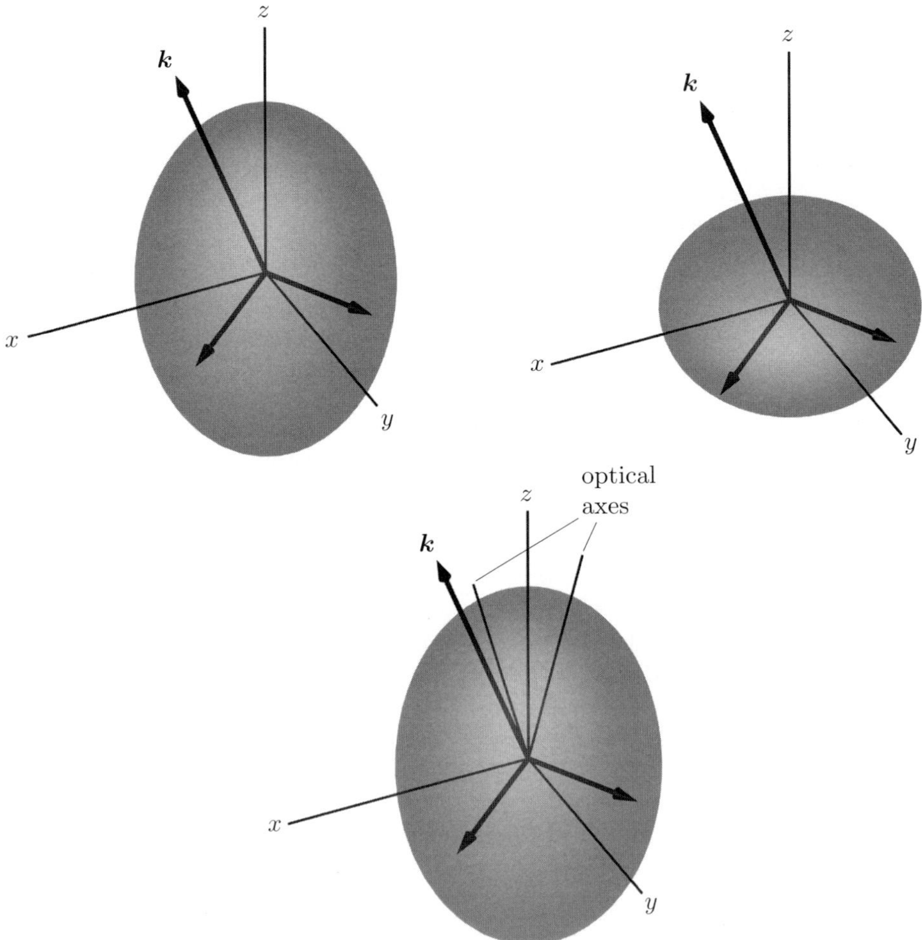



**Fig. 13.3** Index ellipsoid for positive uniaxial crystal (top left), negative uniaxial crystal (top right) and biaxial crystal (bottom). The two optical axes in the bottom figure are the lines without arrowheads.

*polar coordinate $(\rho, \theta)$ system*, the two indices $(\rho)$ as a function of the propagation direction $(\theta)$. Normal surfaces should be carefully distinguished from the index ellipsoid, which is a completely different object. The constructions are very simple, since the semi-major and semi-minor axes are just the indices of refraction. Notice that the curve for the ordinary index is a circle and the ellipse is inscribed in the circle for a negative crystal and the circle is inscribed in the ellipse for a positive crystal. Using these constructions, one can immediately write down equations for the indices

$$n_o(\theta) = n_0 \quad \text{(constant)} \tag{13.29}$$

$$\frac{1}{n_e^2(\theta)} = \frac{\cos^2 \theta}{n_o^2} + \frac{\sin^2 \theta}{n_e^2}, \tag{13.30}$$

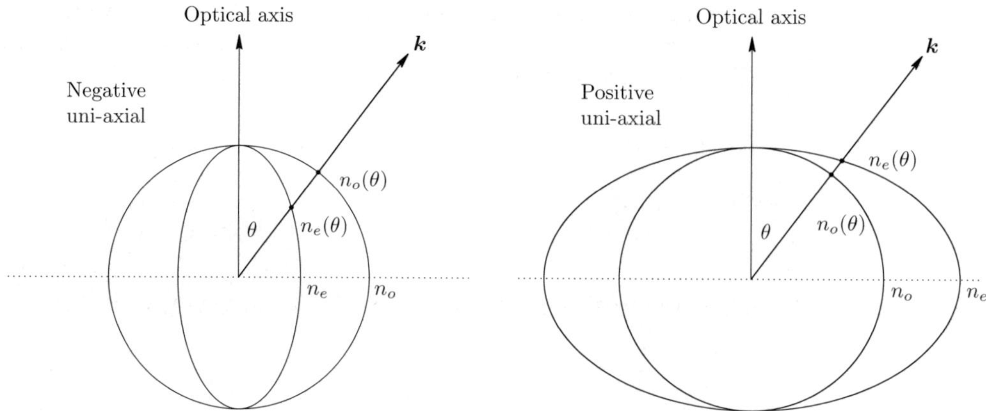

**Fig. 13.4** Normal surfaces for a positive and negative uniaxial crystal.

where $n_e = n_e(0)$ and $\theta$ is the angle that $\boldsymbol{k}$ makes with the optical axis. This equation follows immediately from the equation for an ellipse together with the relations $x = n_e(\theta)\sin\theta$ and $y = n_e(\theta)\cos\theta$, where $x$ and $y$ are the abscissa and ordinate of the two-dimensional plot.

So far, we have discussed *wave propagation*, in which the wavefront progresses along $\hat{\boldsymbol{k}}$ in a direction that is always normal to the wavefront surface. The speed of wave propagation is called the *phase velocity* ($v_\phi$). In an anisotropic medium, the direction of *energy transport* (along $\hat{\boldsymbol{S}}$) will in general be different from the direction of wave propagation. Energy is propagated along *rays*, with a speed equal to the *ray velocity*, $v_r$.

The difference between wave and ray propagation is illustrated by Fig. 13.5, which displays a construction, using Huygen's principle, of a wave travelling in an anisotropic crystal. For the ordinary wave, the Huygen's wavelets are spheres, since $v_\phi$ is indepen-

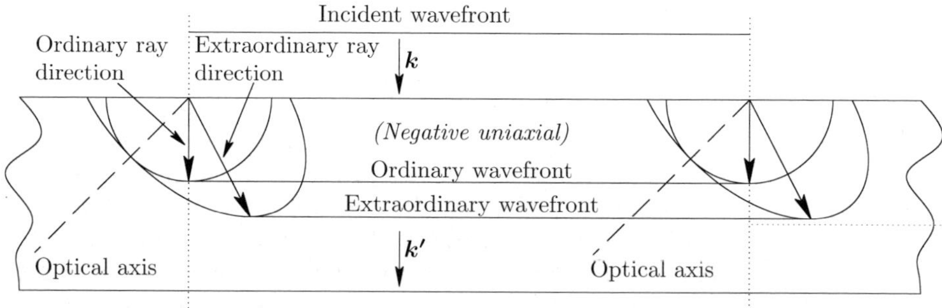

**Fig. 13.5** Huygen's construction of a wave in a birefringent crystal illustrating ray and wave propagation.

dent of $\hat{\boldsymbol{k}}$, just as in the familiar isotropic medium. One can derive Snell's law using a Huygen's construction and the ordinary wave obeys Snell's law. On the other hand, the extraordinary wavelets are ellipsoids, since different portions of the extraordinary wavefront propagate with different speeds due to the dependence of the index of refraction on $\hat{\boldsymbol{k}}$. The extraordinary wave *does not obey Snell's law*, which accounts for the name "extraordinary". If the optical axis is not perpendicular or parallel to the crystal surface, an unpolarized wave normally incident on the crystal will split into two beams: one consisting of ordinary rays and one consisting of extraordinary rays. These two beams will separate spatially as they progress through the crystal. This phenomenon is strikingly observed when a piece of birefringent calcite is placed over printed material and two separate images of the print are observed. The separation into two beams is called *double refraction*. The ray velocity, $v_r$, is related to the wave velocity, $v_\phi$, by

$$v_r = \frac{v_\phi}{\cos \phi}, \tag{13.31}$$

where $\phi$ is the angle between $\hat{\boldsymbol{k}}$ and $\hat{\boldsymbol{S}}$ (or between $\boldsymbol{D}$ and $\boldsymbol{E}$). One can actually construct an ellipsoid using $\boldsymbol{E}$ instead of $\boldsymbol{D}$ and use it to determine the *ray* behavior in much the same manner as the *wave* behavior was determined using the index ellipsoid. It can be proven that the *ray direction for any wave is normal to the index surface for that wave*. From this, it is easy to show that

$$\tan \rho = \frac{n_e^2}{n_o^2} \tan \theta, \tag{13.32}$$

where $\theta$ is the angle of $\boldsymbol{k}$ with respect to the optical axis and $\rho$ is the angle of the Poynting vector (ray direction) relative to the optical axis.

## 13.3   Second harmonic generation

We have used the linear electrical susceptibility, $\chi$, several times in this book and will remind the reader that $\epsilon_0 \chi$ is the ratio of the polarization to the electric field:

$$\text{Linear susceptibility}: \qquad \boldsymbol{P} = \epsilon_0 \chi \boldsymbol{E} \qquad \chi = \text{complex constant.} \tag{13.33}$$

Thus, a time-dependent polarization is a faithful replica (up to a multiplicative complex constant) of the time-varying electric field. If we relax the requirement that $\chi$ be constant and now let it *depend upon* the field, $\boldsymbol{E}$, the polarization is no longer a replica of the field but will contain some *distortion*, as shown in Fig. 13.6.

$$\text{Nonlinear susceptibility:} \qquad \boldsymbol{P} = \epsilon_0 \chi(\boldsymbol{E}) \boldsymbol{E}. \tag{13.34}$$

If $\chi(\boldsymbol{E})$ is *not an odd function* of $\boldsymbol{E}$, the distorted polarization will contain some second harmonic. This will only occur if the crystal is *non-centrosymmetric* (i.e., it lacks a center of symmetry). A centrosymmetric crystal will only generate odd harmonics of the incoming field, as shown in Fig. 13.7.

To treat the problem analytically, we start with the time-dependent Maxwell equations (without sources) with the *polarization* explicitly displayed:

$$\nabla \times \boldsymbol{H} = \frac{\partial \boldsymbol{D}}{\partial t} = \frac{\partial}{\partial t}(\epsilon_0 \boldsymbol{E} + \boldsymbol{P}) \tag{13.35}$$

$$\nabla \times \boldsymbol{E} = -\frac{\partial \boldsymbol{B}}{\partial t} = -\frac{\partial}{\partial t}(\mu_0 \boldsymbol{H}). \tag{13.36}$$

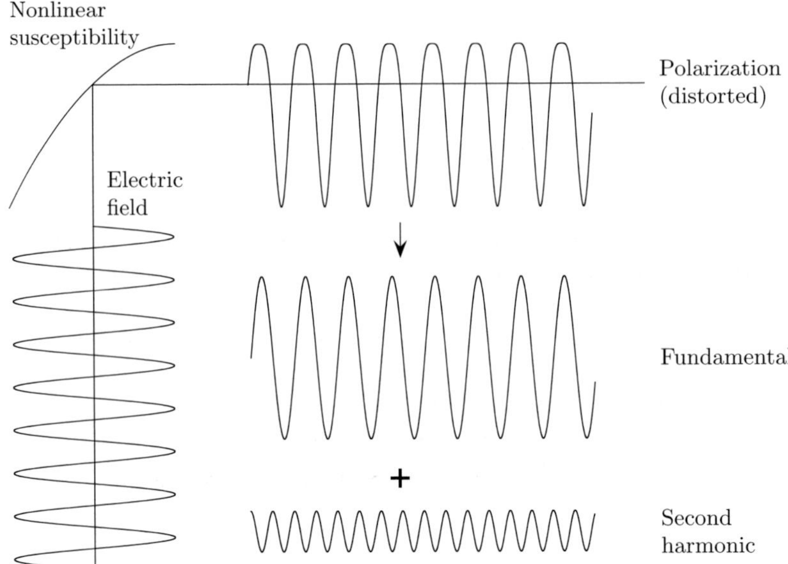



**Fig. 13.6** Illustration of distortion caused by nonlinear susceptibility. The polarization will acquire some second harmonic.

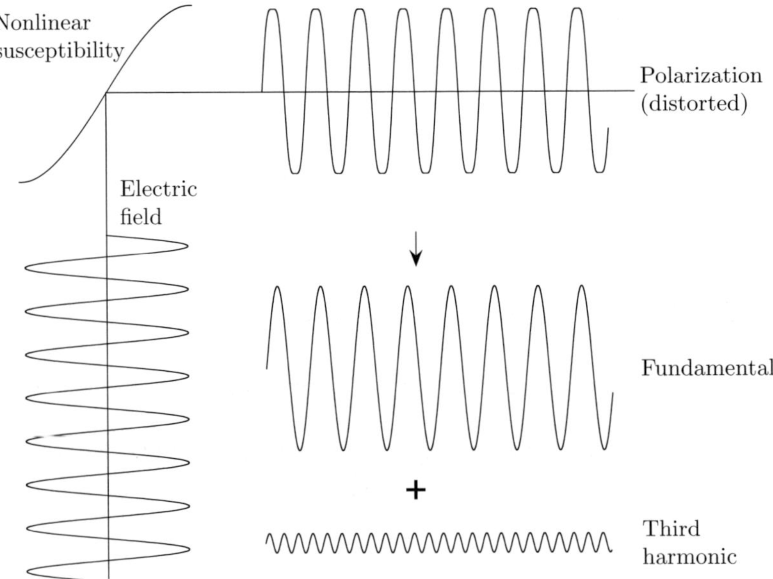



**Fig. 13.7** Illustration of distortion from a centrosymmetric crystal. The polarization will only generate odd harmonics.

The polarization has a *linear* and a *nonlinear* contribution:

$$\boldsymbol{P} = \epsilon_0 \chi_L \boldsymbol{E} + \boldsymbol{P}_{\mathrm{NL}}, \tag{13.37}$$

where $\boldsymbol{P}_{NL}$ depends *quadratically* on the field:

$$P_{NL,i} = \sum_{j,k} d_{ijk} E_j E_k \qquad i,j,k = x,y,z \tag{13.38}$$

and $d_{ijk}$ is the nonlinear optical susceptibility tensor. Including the nonlinear polarization in Ampere's law, one obtains

$$\nabla \times \boldsymbol{H} = \frac{\partial}{\partial t}\epsilon \boldsymbol{E} + \frac{\partial \boldsymbol{P}_{NL}}{\partial t}, \tag{13.39}$$

where

$$\epsilon = \epsilon_0 (1 + \chi_L). \tag{13.40}$$

Taking the curl of Faraday's law and substituting the above yields

$$\nabla^2 \boldsymbol{E} - \mu_0 \epsilon \frac{\partial^2 \boldsymbol{E}}{\partial t^2} = \mu_0 \frac{\partial^2}{\partial t^2} \boldsymbol{P}_{NL}. \tag{13.41}$$

This is a wave equation whose source is the nonlinear polarization.

This equation will be solved by assuming that $\boldsymbol{E}$ is a plane wave with frequency $2\omega$ and the source for the nonlinear polarization is a plane wave whose frequency is $\omega$:

$$\text{Output wave:} \quad E_j^{(2\omega)} = \frac{1}{2} E_{2j}(z) e^{i(2\omega t - k_2 z)} + \mathrm{cc} \tag{13.42}$$

$$\text{Input wave:} \quad E_i^{(\omega)} = \frac{1}{2} E_{1i}(z) e^{i(\omega t - k_1 z)} + \mathrm{cc} \quad (i,j = x,y), \tag{13.43}$$

where $k_1$ and $k_2$ are the input and output $k$-vectors respectively and the amplitudes $E_{1i}$ and $E_{2j}$ are *slowly varying* functions of $z$. If we assume that the $x$ and $y$ second derivatives vanish (plane wave) and $\frac{dE_{2j}}{dz} k_2 \gg \frac{d^2 E_{2j}}{dz^2}$, then we can discard the spatial second derivative and the left-hand side of the wave equation becomes

$$-\frac{1}{2}\left[ (k_2)^2 E_{2j}(z) + 2ik_2 \frac{dE_{2j}(z)}{dz} \right] e^{i(2\omega t - k_2 z)} + 4\omega^2 \mu_0 \epsilon \left[ \frac{E_{2j}}{2} e^{i(2\omega t - k_2 z)} \right] + \mathrm{cc}. \tag{13.44}$$

Using $(2\omega)^2 \mu_0 \epsilon = (k_2)^2$ and including the right-hand side (the nonlinear polarization),

$$ik_2 \frac{dE_{2j}}{dz} e^{-ik_2 z} = \mu_0 \omega^2 \sum_{ik} d_{jik} E_{1i} E_{1k} e^{-2ik_1 z}. \tag{13.45}$$

Simplifying,

$$\frac{dE_{2j}}{dz} = -\frac{i\omega}{2} \sqrt{\frac{\mu_0}{\epsilon_2}} \sum_{ik} d_{jik} E_{1i} E_{1k} e^{i\Delta k z}, \qquad \Delta k = k_2 - 2k_1, \tag{13.46}$$

where $\epsilon_2$ is the permittivity of the medium at $2\omega$.

This equation is readily solved if we use the boundary condition that $\boldsymbol{E}_2(z = 0) = 0$ and assume that there is no depletion of $\boldsymbol{E}_1$ along the crystal. Then, a simple integration yields

$$E_{2j}(L) = -\frac{\omega}{2}\sqrt{\frac{\mu_0}{\epsilon_2}}\sum_{ik} d_{jik}E_{1i}E_{1k}\frac{e^{i\Delta kL}-1}{\Delta k}, \tag{13.47}$$

where the length of the crystal is $L$. The second harmonic *intensity* (power/area) is

$$I^{(2\omega)} = \frac{\text{power}^{(2\omega)}}{\text{area}} = \frac{1}{2}\sqrt{\frac{\epsilon_2}{\mu_0}}\boldsymbol{E}_2\cdot\boldsymbol{E}_2^*, \tag{13.48}$$

which yields

$$I^{(2\omega)} = \frac{1}{8}\sqrt{\frac{\mu_0}{\epsilon_2}}\omega^2(d_{ijk})^2 E_{1i}^2 E_{1k}^2 L^2\frac{\sin^2(\Delta kL/2)}{(\Delta kL/2)^2}, \tag{13.49}$$

where we assume only one nonlinear coefficient ($d_{ijk}$) is significant and include a factor of 2 for the sum over $i$ and $k$ in the field expression. Finally, the efficiency is

$$\frac{I^{(2\omega)}}{I^{(\omega)}} = \frac{1}{2}\left(\frac{\mu_0}{\epsilon_0}\right)^{3/2}\frac{\omega^2(d_{ijk})^2 L^2}{n^3}I^{(\omega)}\frac{\sin^2(\Delta kL/2)}{(\Delta kL/2)^2}, \tag{13.50}$$

where we assume that $\epsilon \approx \epsilon_2$ and $\epsilon = n^2\epsilon_0$. This displays the characteristic dependence of the efficiency on $L^2$ and $I^{(\omega)}$ (i.e., the second harmonic intensity is proportional to the *square* of the fundamental intensity). We will show later on that the $L^2$ dependence only holds for second-harmonic generation in a guiding structure such as an optical waveguide or fiber; in free space the dependence is at best linear in $L$ when the input beam is focused optimally.

For the purpose of describing second-harmonic generation, the nonlinear optical susceptibility tensor, $d_{ijk}$, is usually written as a rectangular, two-dimensional array whose elements are defined below:

$$\begin{pmatrix} P_{NL,x} \\ P_{NL,y} \\ P_{NL,z} \end{pmatrix} = \begin{pmatrix} d_{11} & d_{12} & d_{13} & d_{14} & d_{15} & d_{16} \\ d_{21} & d_{22} & d_{23} & d_{24} & d_{25} & d_{26} \\ d_{31} & d_{32} & d_{33} & d_{34} & d_{35} & d_{36} \end{pmatrix} \begin{pmatrix} E_x^2 \\ E_y^2 \\ E_z^2 \\ 2E_zE_y \\ 2E_zE_x \\ 2E_xE_y \end{pmatrix}. \tag{13.51}$$

Despite the large number of possible elements, only a few are usually unique, depending upon the crystal geometry. While it would take us into the area of crystal symmetries to show that many of the coefficients are either zero or simply related to each other, there is a simple symmetry that is based upon a principle called *Kleinman's conjecture*. This states that, in the absence of absorption at any of the frequencies, the tensor elements, $d_{ijk}$, are invariant upon any permutation of the indices. Thus, the following index equalities for $d_{ij}$ hold:

$$
\begin{aligned}
d_{14} &= d_{25} = d_{36} \\
d_{15} &= d_{31} \\
d_{23} &= d_{34} \\
d_{12} &= d_{26} \\
d_{16} &= d_{21} \\
d_{13} &= d_{35} \\
d_{24} &= d_{32}.
\end{aligned}
\tag{13.52}
$$

A list of the non-zero $d$-values for some of the more important crystals appears below.

- **Tetragonal, class** $\bar{4}2m$ – example, ADP, AD*P, KDP, KD*P

$$
\begin{pmatrix}
0 & 0 & 0 & d_{14} & 0 & 0 \\
0 & 0 & 0 & 0 & d_{14} & 0 \\
0 & 0 & 0 & 0 & 0 & d_{36}
\end{pmatrix}
\tag{13.53}
$$

- **Trigonal, class** $3m$ – example, $\beta$-barium borate (BBO), lithium niobate, lithium tantalate

$$
\begin{pmatrix}
0 & 0 & 0 & 0 & d_{15} & -d_{22} \\
-d_{22} & d_{22} & 0 & d_{15} & 0 & 0 \\
d_{31} & d_{31} & d_{33} & 0 & 0 & 0
\end{pmatrix}
\tag{13.54}
$$

- **Hexagonal, class** $6$ – example, lithium iodate

$$
\begin{pmatrix}
0 & 0 & 0 & d_{14} & d_{15} & 0 \\
0 & 0 & 0 & d_{15} & -d_{14} & 0 \\
d_{31} & d_{31} & d_{33} & 0 & 0 & 0
\end{pmatrix}
\tag{13.55}
$$

- **Orthorhombic, class** $mm2$ – example, potassium niobate, LBO, KTP

$$\begin{pmatrix} 0 & 0 & 0 & 0 & d_{15} & 0 \\ 0 & 0 & 0 & d_{24} & 0 & 0 \\ d_{31} & d_{32} & d_{33} & 0 & 0 & 0 \end{pmatrix} \tag{13.56}$$

A tabulation of the larger nonlinear coefficients and Sellmeier coefficients (see below) for a number of popular uniaxial crystals appears in Table 13.1. A tabulation for biaxial crystals is in Table 13.2.

**Table 13.1** Nonlinear coefficients and Sellmeier coefficients for some popular uniaxial SHG crystals (Sellmeier coefficients from *Tunable Laser Handbook* (1995), nonlinear coefficients from Roberts (1992), Davis (1996) and Yariv (1989)).

| Crystal | $d_{ij}$ (pm/V) | transparency range (nm) | Sellmeier coefficients ($\lambda$ in $\mu$m) | | | | |
|---|---|---|---|---|---|---|---|
| | | | $A_o$ | $B_o$ | $C_o$ | $D_o$ | $E_o$ |
| | | | $A_e$ | $B_e$ | $C_e$ | $D_e$ | $E_e$ |
| ADP | $d_{36} = 0.41$ | 180-1500 | 1.37892 | 0.91996 | 0.01249 | .15771 | 5.76 |
| | $d_{14} = 0.63$ | | 1.35302 | 0.80752 | 0.01227 | 0.02612 | 3.3156 |
| KDP | $d_{36} = 0.39$ | 180-1550 | 1.41344 | 0.84308 | 0.01229 | 0.26923 | 10.248 |
| | $d_{14} = 0.42$ | | 1.40442 | 0.72733 | 0.01201 | 0.07974 | 12.484 |
| LiNbO$_3$ | $d_{31} = -4.3$ | 370-4500 | 2.33907 | 2.58395 | 0.04588 | 13.8169 | 519.658 |
| | $d_{33} = -27 \pm 1.0$ | | 2.35084 | 2.22518 | 0.04371 | 15.9733 | 741.146 |
| | $d_{22} = 2.1$ | | | | | | |
| BBO | $d_{22} = 2.3$ | 198-2600 | 1.71283 | 1.02790 | 0.01790 | 2.23130 | 138.65 |
| | $d_{31} = 0.1$ | | 1.50569 | 0.86544 | 0.01512 | 0.56478 | 248.36 |

The Sellmeier coefficients are used to calculate the indices of refraction using the Sellmeier equation:

$$n^2 = A + \frac{B\lambda^2}{\lambda^2 - C} + \frac{D\lambda^2}{\lambda^2 - E}. \tag{13.57}$$

There is a set of parameters for each index, where the subscript identifies the index and is $e$ and $o$ for a uniaxial crystal or $x$, $y$ and $z$ for a biaxial crystal. There are many variants of the Sellmeier equation; the one given above provides tolerable accuracy for wavelengths from the UV to the near IR.

## 13.4 Birefringent phase matching

From eqn 13.50, we see that the second harmonic power is proportional to

**Table 13.2** Nonlinear coefficients and Sellmeier coefficients for some popular biaxial SHG crystals (Sellmeier coefficients from *Tunable Laser Handbook* (1995), nonlinear coefficients from Roberts (1992)).

| Crystal | $d_{ij}$ (pm/V) | transparency range (nm) | Sellmeier coefficients ($\lambda$ in $\mu$m) | | | | |
|---|---|---|---|---|---|---|---|
| | | | $A_x$ | $B_x$ | $C_x$ | $D_x$ | $E_x$ |
| | | | $A_y$ | $B_y$ | $C_y$ | $D_y$ | $E_y$ |
| | | | $A_z$ | $B_z$ | $C_z$ | $D_z$ | $E_z$ |
| KTP | $d_{31} = 3.6$ | 350-4500 | 2.22237 | 0.78681 | 0.04746 | 0.67167 | 54.9 |
| | $d_{32} = 2.0$ | | 2.3059 | 0.72572 | 0.05387 | 1.00870 | 77.50 |
| | $d_{33} = 8.3$ | | 2.35249 | 0.96655 | 0.05812 | 1.24674 | 77.50 |
| LBO | $d_{15} = .85$ | 160-2300 | 2.07557 | 0.38193 | 0.02597 | 2.60858 | 191.04 |
| | $d_{24} = -0.67$ | | 1.61856 | 0.92347 | 0.01355 | 4.48336 | 204.16 |
| | $d_{33} = 0.04$ | | 2.00372 | 0.58147 | 0.02176 | 2.55777 | 155.84 |

$$\left(\frac{\sin(\Delta k L/2)}{(\Delta k L/2)}\right)^2, \qquad \Delta \boldsymbol{k} = \boldsymbol{k}_2 - 2\boldsymbol{k}_1. \tag{13.58}$$

This expression is unity when $\Delta k = 0$ and is zero when $\Delta k L = 2\pi$. In the interests of obtaining a useful amount of second harmonic, we require that the *phase matching condition* be satisfied:

$$\text{Phase matching:} \quad \Delta \boldsymbol{k} = 0 \implies \boldsymbol{k}_2 = 2\boldsymbol{k}_1, \tag{13.59}$$

which implies that the phase velocities and indices of refraction of the second harmonic and fundamental must be the same:

$$v_{2\omega} = v_\omega \tag{13.60}$$

$$n_{2\omega} = n_\omega. \tag{13.61}$$

If the phase matching condition is violated, the second harmonic power will have a $\sin^2(\Delta k z/2)$ dependence on $z$ as one proceeds along the doubling crystal. When $\Delta k \neq 0$, the fundamental and second harmonic will propagate at *different* speeds; after a distance of $\pi/\Delta k$ the two beams will begin to be out of phase and the newly generated second harmonic will *subtract* from the existing second harmonic. It is as though the crystal length were reduced from $L$ to $\pi/\Delta k$, which is a considerable reduction. One can consider the latter length to be a *coherence length*, $l_c$:

$$l_c = \frac{\pi}{\Delta k} = \frac{\lambda}{4(n_{2\omega} - n_\omega)}, \tag{13.62}$$

where $\lambda$ is the *fundamental* wavelength. If no phase matching techniques are employed, $l_c$ can be 100 $\mu$m, which is two orders of magnitude shorter than a typical crystal.

The phase matching condition has a finite bandwidth, which can be obtained from the $\text{sinc}^2$ factor in eqn 13.50 together with the dispersion properties of the index. A plot of the dependence of the second harmonic power on $\Delta k$ appears in Fig. 13.8. The full width at half maximum of this curve is $\delta(\Delta k L/2) = 0.886\pi$. The wavelength

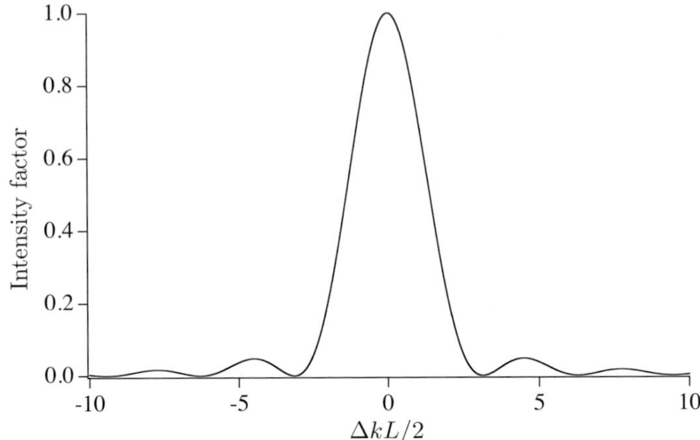

**Fig. 13.8** Plot of frequency dependence of second harmonic due to phase matching.

bandwidth is obtained by equating this to the derivative of $L\Delta k/2$ times $\Delta\lambda$:

$$\frac{L}{2}\frac{\partial \Delta k}{\partial \lambda}\Delta\lambda = 0.886\pi. \tag{13.63}$$

The result is

$$\Delta\lambda = \frac{0.443\lambda}{L}\left| \frac{1}{2}\left(\frac{\partial n}{\partial \lambda}\right)_{2\omega} - \left(\frac{\partial n}{\partial \lambda}\right)_{\omega}\right|^{-1}, \tag{13.64}$$

where the derivatives are evaluated at the fundamental and second harmonic wavelengths, and $\lambda$ is the fundamental wavelength. The derivatives can be obtained from the Sellmeier equation.

Phase matching can also be interpreted as *conservation of momentum*. Assuming that there are no phonon changes in the second-harmonic generation process, phase matching is equivalent to the requirement that the momentum of the *single outgoing photon* at $2\omega$ be equal to the sum of the momenta of the *two incoming photons* at $\omega$.

There are two commonly used methods for phase matching: exploitation of crystal *birefringence* and *quasi-phase-matching*. We will discuss the former in this section. The idea behind birefringent phase matching is to use the fact that, under *normal dispersion*, the index of refraction increases with frequency. If the fundamental polarization is aligned with the axis corresponding to the larger of the two birefringent indices (in a uniaxial crystal) and the second harmonic to the smaller, there is a wavelength where the indices of the fundamental and second harmonic are the same. We will shortly show that the optimum condition is where the fundamental beam propagates in a direction perpendicular to the optical axis of a crystal. If the crystal is *negative uniaxial*, the fundamental would be polarized perpendicular to the optical axis (the ordinary wave) and the second harmonic would be polarized along the optical axis (extraordinary wave). The scheme is illustrated in Fig. 13.9 for ADP, which is a negative uniaxial crystal. Phase matching occurs only at a fundamental wavelength of 0.526 $\mu$m.

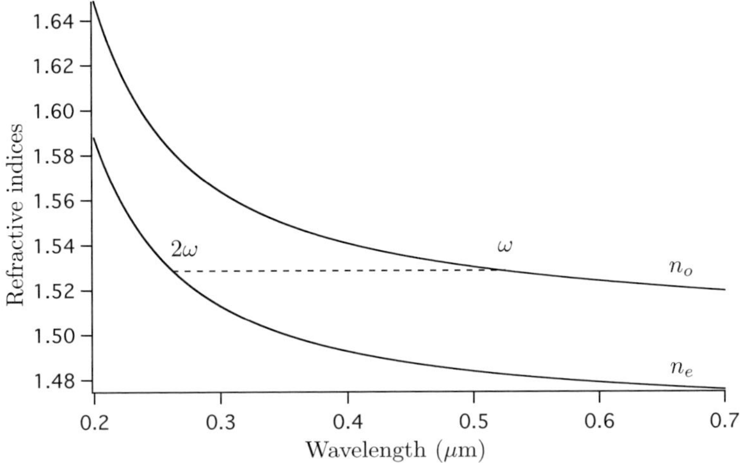


**Fig. 13.9** Index curves for ADP, illustrating phase matching for a fundamental of 0.526 $\mu$m. The indices are calculated using the Sellmeier equation and the fundamental propagates perpendicular to the optical axis.

The scheme described above has a serious drawback: it only works at a single fundamental wavelength. However, we learned that by changing the propagation direction, one can change the *extraordinary index of refraction* (but not the ordinary). Thus, one can *tune* the phase matching by changing the angle that the fundamental $k$-vector makes with the optical axis. This allows phase matching at a range of wavelengths. For a negative uniaxial crystal, the fundamental would be an ordinary wave and the second harmonic would be an extraordinary wave. The functional dependence of the extraordinary index on the angle, $\theta$, between $\boldsymbol{k}$ and the optical axis is derived from the normal surfaces:

$$\frac{1}{n_e^2(\theta)} = \frac{\cos^2\theta}{n_o^2} + \frac{\sin^2\theta}{n_e^2}. \tag{13.65}$$

Setting $n_{o,\omega} = n_{e,2\omega}$,

$$\frac{1}{n_{o,\omega}^2} = \frac{\cos^2\theta}{n_{o,2\omega}^2} + \frac{\sin^2\theta}{n_{e,2\omega}^2}. \tag{13.66}$$

This can be solved for the angle

$$\sin^2\theta = \frac{(n_{o,\omega})^{-2} - (n_{o,2\omega})^{-2}}{(n_{e,2\omega})^{-2} - (n_{o,2\omega})^{-2}}. \tag{13.67}$$

The process is illustrated in Fig. 13.10 using normal surfaces for $\omega$ and $2\omega$ plotted on the same graph. Phase matching takes place where the two solid curves intersect.

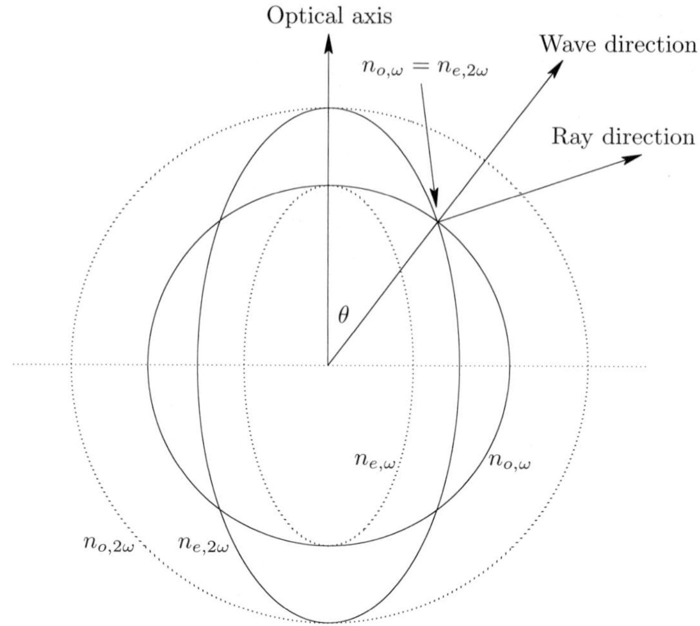



**Fig. 13.10** Illustration of birefringent phase matching in negative uniaxial crystal by plotting the normal surfaces for the fundamental and second harmonic on the same graph. Also shown is *walk-off*: the different directions between the input wave and the output ray (Poynting vector), which is perpendicular to the $2\omega$ extraordinary normal surface.

When one has a *positive uniaxial* crystal, the fundamental would be an extraordinary wave and the second harmonic would be an ordinary wave. Phase matching is obtained by varying the angle and thereby varying the index of the fundamental until it is equal to that of the second harmonic. Proceeding as before, with $n_{o,2\omega} = n_{e,\omega}$, the angle is given by

$$\sin^2 \theta = \frac{(n_{o,2\omega})^{-2} - (n_{o,\omega})^{-2}}{(n_{e,\omega})^{-2} - (n_{o,\omega})^{-2}}. \tag{13.68}$$

In our treatment so far, we have assumed that the two input fields in second harmonic generation have the same polarization. However, in eqn 13.46, the two source field components are independent of each other and second harmonic generation can occur from *orthogonal* source fields if the nonlinear coefficient has non-zero elements which couple two different polarizations. It is customary to refer to SHG from the same polarizations as *type I* and from orthogonal polarizations as *type II*. Phase matching in type II second-harmonic generation is less tunable than in type I since only one field's refractive index can be used to compensate for the index difference between $\omega$ and $2\omega$ due to normal dispersion. With a negative uniaxial crystal, the output field will be an extraordinary wave and the input field will consist of one ordinary wave and one extraordinary wave. Using a fairly obvious notation, the phase matching requirement is

$$\text{Type II:} \quad \mathbf{k}^e_{2\omega} = \mathbf{k}^o_\omega + \mathbf{k}^e_\omega \implies n_{e,2\omega} = \tfrac{1}{2}(n_{o,\omega} + n_{e,\omega}). \tag{13.69}$$

The phase matching angle can be determined in the same way as for type I, though the equations are fairly cumbersome and might need to be solved using a computer.

In order to satisfy the phase matching requirements for a given frequency, $\omega$, both the $k$-vector and the polarization are constrained to be in very specific directions. From eqn 13.46, we see that one needs to sum over all of the appropriate field components and nonlinear coefficients to obtain the second harmonic field. This sum will, of course, depend upon the orientation of the $k$-vector and polarization. It is convenient to define an , $d_{eff}$, to include these sums, and write the polarization simply as

$$P_{2\omega} = d_{eff} E_\omega E'_\omega, \tag{13.70}$$

where $E_\omega$ and $E'_\omega$ are the two input field magnitudes. One can determine the angular dependence of $d_{eff}$ using geometry. We will illustrate this with a simple example: we will determine the dependence of $d_{eff}$ on the propagation direction for a crystal of class $\bar{4}2m$ (such as ADP). It is useful to first write down *unit vectors* along the ordinary and extraordinary polarization directions when the propagation direction makes an angle $\theta$ with the $z$-axis and an angle $\phi$ with the $x$-axis. We will assume that the phase matching is type-I, so $\theta$ and $\phi$ are the same for the two waves (and $E_\omega = E'_\omega$). Using column vectors,

$$\text{Ordinary polarization unit vector:} \quad \mathbf{a} = \begin{pmatrix} \sin\phi \\ -\cos\phi \\ 0 \end{pmatrix}, \tag{13.71}$$

$$\text{Extraordinary polarization unit vector:} \quad \mathbf{b} = \begin{pmatrix} -\cos\phi\cos\theta \\ -\sin\phi\cos\theta \\ \sin\theta \end{pmatrix}. \tag{13.72}$$

From the nonlinear tensor tabulated above for this class, the polarization is

$$\begin{aligned}
P_1 &= 2d_{14}E_3 E_2 \\
P_2 &= 2d_{14}E_3 E_1 \\
P_3 &= 2d_{36}E_1 E_2.
\end{aligned} \tag{13.73}$$

We assume that the two input fields are *ordinary waves* and the output field is an extraordinary wave. The input components are $E\mathbf{a}$:

$$\begin{aligned}
E_1 &= E\sin\phi \\
E_2 &= -E\cos\phi \\
E_3 &= 0.
\end{aligned} \tag{13.74}$$

Substituting these into the polarization expressions, we obtain

$$P_1 = 0$$
$$P_2 = 0$$
$$P_3 = -2d_{36}E^2 \sin\phi\cos\phi. \tag{13.75}$$

The output extraordinary polarization, $P_e$, is the projection of the polarization along the extraordinary direction: $\boldsymbol{b}\cdot\boldsymbol{P}$

$$P_e = (-2d_{36}\sin\phi\cos\phi\sin\theta)E^2, \tag{13.76}$$

and the effective nonlinear coefficient is (simplifying $\sin\phi\cos\phi$)

$$d_{eff} = -d_{36}\sin 2\phi\sin\theta = -d_{14}\sin 2\phi\sin\theta, \tag{13.77}$$

where the last equality is due to an application of Kleinman's conjecture. A table of effective nonlinear coefficients for various crystal classes appears in Table 13.3. In

**Table 13.3** $d_{eff}$ for several crystal classes (assumes Kleinman's conjecture).

| Crystal class | $d_{eff}$ for $ooe$ | $d_{eff}$ for $eeo$ |
|---|---|---|
| Tetragonal, class $\bar{4}2m$ | $-d_{14}\sin\theta\sin 2\phi$ | $d_{14}\sin 2\theta\cos 2\phi$ |
| Trigonal, class $3m$ | $d_{15}\sin\theta - d_{22}\cos\theta\sin 3\phi$ | $d_{22}\cos^2\theta\cos 3\phi$ |
| Hexagonal, class $6$ | $d_{15}\sin\theta$ | $0$ |

the table, the expression $ooe$ refers to two input ordinary waves and a single output extraordinary wave, and $eeo$ refers to two input extraordinary waves and a single output ordinary wave.

One adverse aspect of birefringent phase matching which is illustrated in Fig. 13.10 is the fact that, in general, the directions of *energy propagation* (ray directions) are different for the fundamental and second harmonic. This will have three consequences. First, the two waves will separate in a distance called the *aperture length* and the conversion will therefore be harmed; this will be discussed in more detail in a later section and is called *walk-off*. The second effect is the greater sensitivity of the phase matching condition to the input beam propagation direction. Finally, the beam *quality* will suffer due to the walk-off. None of these effects will occur if the input beam propagates perpendicular to the optical axis: there will be no walk-off and the phase matching conditions will depend *quadratically* on the propagation direction and will therefore be insensitive to it. Phase matching of this kind is called 90° or *non-critical* phase matching and is the desirable state of affairs. Often, one *temperature tunes* the crystal in order to achieve non-critical phase matching.

## 13.5 Quasi-phase-matching

In the last section, we investigated one method of phase matching: the use of crystal birefringence together with normal dispersion to ensure that $n_\omega = n_{2\omega}$. This technique has two glaring flaws. First, there are frequency gaps in which phase matching is impossible. Second, even if phase matching can be accomplished, it often requires

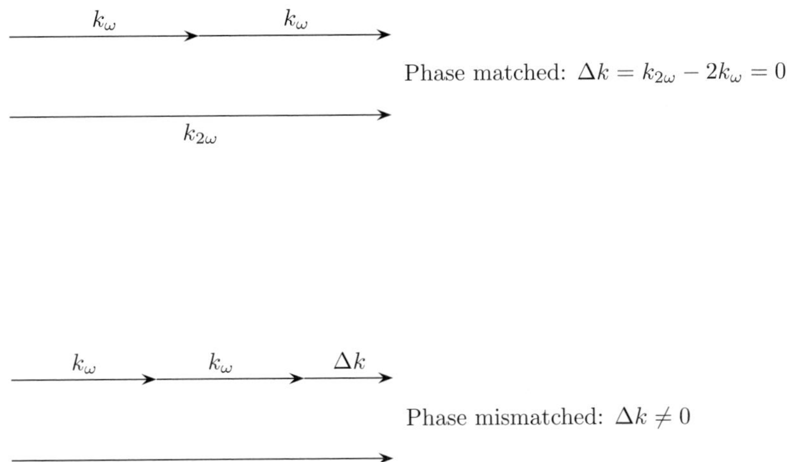


**Fig. 13.11** Vector diagrams depicting phase matching (top) and phase mismatching (bottom)

propagation angles that are far from the optimum *non-critical phase matching angle* (90° to the optical axis), with adverse effects on beam quality and conversion efficiency. The technique of *quasi-phase-matching* avoids both of these drawbacks at the expense of additional complexities in crystal fabrication.

A simple explanation of the technique can be obtained via the notion of *spatial frequency modulation*. We have already introduced this concept when we discussed the feedback mechanism in distributed feedback (DFB) semiconductor lasers. We start with a wave described by the equation

$$E(z,t) = E_0 \sin(\omega t - kz). \tag{13.78}$$

The wave vector, $k$, can considered to be the *spatial frequency* of a wave, just as $\omega$ is its *temporal frequency*. In Chapter 4, we discussed both phase and amplitude modulation by a sinusoid whose temporal frequency is $\omega_m$. For both types of modulation, the *frequency spectrum* will have sidebands on either side of the original *carrier*, displaced from the carrier by $\omega_m$. Just as temporal phase modulation can be achieved by varying the phase of the original wave by a function of time whose angular frequency is $\omega_m$, *spatial phase modulation* can be achieved by directing the wave into a periodic structure whose index of refraction varies with a period of $\Lambda$ and whose *spatial angular frequency* is therefore $2\pi/\Lambda$ (its frequency in *spatial cycles per cm* is $1/\Lambda$).

With this preliminary, we consider the case of a phase-mismatched system, shown in the bottom half of Fig. 13.11. The figure uses the momentum conservation approach to phase matching and depicts both the phase-matched condition (top) and the phase-mismatched condition (bottom). As one can see from the bottom figure, there is an excess of spatial frequency equal to $\Delta k$ at the second harmonic. The *quasi-phase-matching* scheme supplies this excess to the nonlinear polarization by generating *spatial frequency sidebands* via a phase modulation which is achieved by alternately reversing

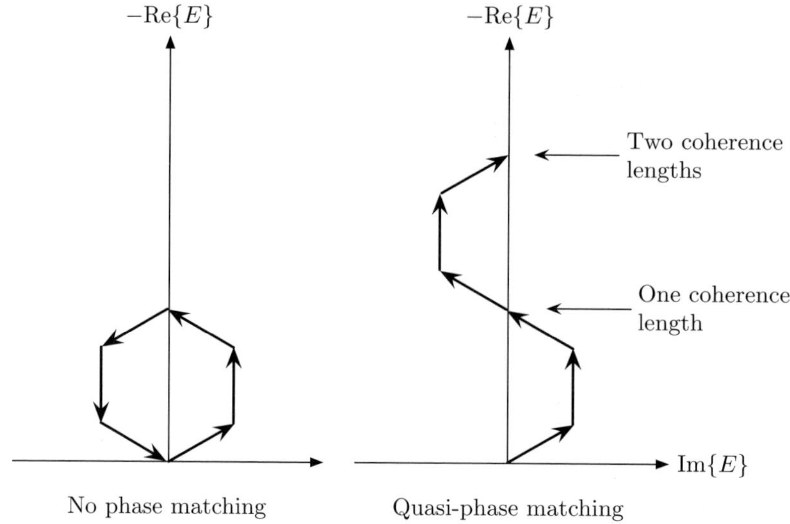



**Fig. 13.12** Phasor diagrams for electric field over two coherence lengths. On the left, complete cancellation of the field is shown when there is no phase matching. On the right, phase reversal after one coherence length in periodic structure allows the field to grow.

the sign of the nonlinear coefficient of the material with a period of $\Lambda$. Phase matching will occur when

$$\frac{2\pi}{\Lambda} + \Delta k = 0. \tag{13.79}$$

If we divide this equation by $\lambda$ and use the fact that $\Delta\lambda/\lambda = -\Delta k/k$, we obtain

$$\Lambda = \lambda\left(\frac{1}{r}\right), \qquad \text{where } r \equiv \frac{\Delta\lambda}{\lambda}, \tag{13.80}$$

and $\Delta\lambda$ is the difference between the desired wavelength and the wavelength at which phase matching is achieved. Thus the spatial period of the structure needed for quasi-phase-matching is the wavelength times the reciprocal of the fractional phase mismatch. For example, if there is a 10% phase mismatch ($\Delta\lambda/\lambda$), the *domain period, $\Lambda$,* should be 10 times the wavelength.

In Fourier analysis, there are two complementary representations for time dependent functions: the time domain and the frequency domain. Spatially dependent functions can be described in an analogous fashion. The *spatial domain* view of quasi-phase-matching which is complementary to the above spatial frequency domain view can be made with the aid of Figs. 13.12 and 13.13. In Fig. 13.12, we display on the left a phasor diagram for the electric field over two coherence lengths ($2\pi/\Delta k$). We see that the field adds to zero due to the phase mismatch: the field adds over the first coherence length and subtracts over the second. On the right we show what happens if we abruptly reverse the sign of $d_{eff}$ after *one coherence length.* The phasor diagrams

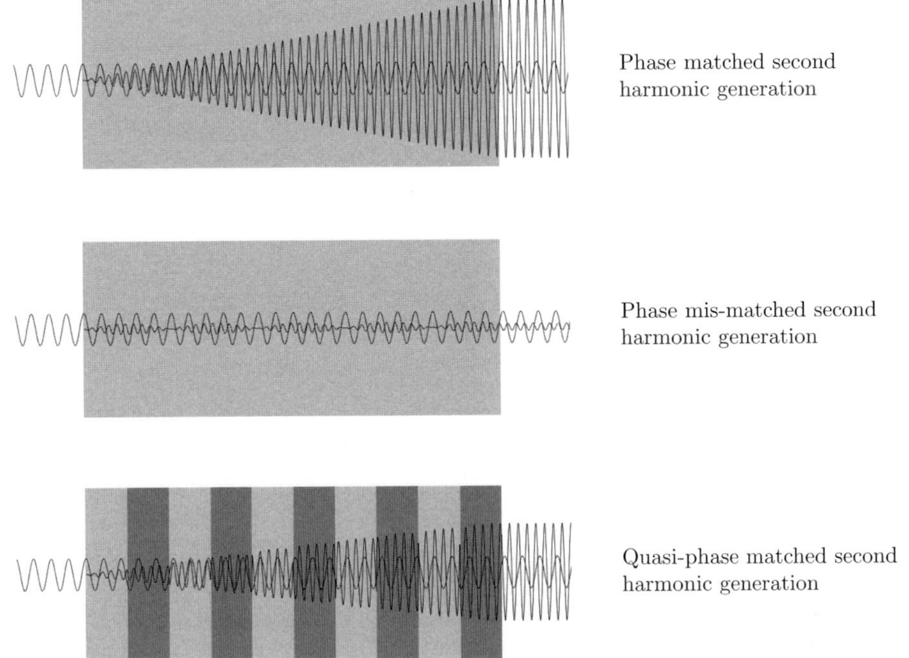


**Fig. 13.13** Snapshots of the fundamental and second harmonic *electric fields* under the conditions of phase matching (top), phase mismatching (middle) and restoration of the phase match using quasi-phase-matching (bottom).

can be more easily interpreted by considering the differential equation for the field (eqn 13.46), which, neglecting unimportant constants, is:

$$\frac{\partial E_{2j}}{\partial z} \propto i d_{eff} e^{i\Delta k z}. \tag{13.81}$$

The field begins along the imaginary axis, and successive phasors follow a circle until one coherence length is reached. At this point, $d_{eff}$ reverses sign and, for a small increment in $z$, the changes in both the real and imaginary parts of $E_{2j}$ also reverse sign; the phasors will now follow the upper semicircle in the diagram. Thus, the field will continue to grow. From the spatial frequency modulation model, a complete reversal of $d_{eff}$ at the domain boundaries is not necessary: any abrupt change in the index of refraction will cause a spatial phase modulation and will work. A structure in which $d_{eff}$ alternately reverses sign is, however, the most efficient configuration. The actual waves are depicted in Fig. 13.13 for phase matching, phase mismatching and the restoration of field growth using quasi-phase-matching. A careful inspection of the bottom graph should reveal the phase jumps in the second harmonic field at each domain boundary. One should note by comparing the middle to the bottom plot that the period, $\Lambda$, needed to restore field growth is equal to twice the coherence length.

A more rigorous analysis can be made by expanding the spatially dependent nonlinear coefficient, $d(z)$, in a Fourier series:

$$d(z) = d_{eff} \sum_{\substack{n=-\infty \\ n\ odd}}^{\infty} a_n e^{2\pi i n z/\Lambda}, \qquad a_n = \frac{2}{\pi n}, \tag{13.82}$$

which is just the series expansion for a symmetric square wave whose extreme values are $\pm d_{eff}$. Then, from the differential equation for the field growth (eqn 13.46), one has

$$\frac{dE_{2j}}{dz} \propto d(z)e^{i\Delta kz} = \frac{2d_{eff}}{\pi} \sum_{\substack{n=-\infty \\ n\ odd}}^{\infty} \frac{1}{n} e^{i(2\pi n/\Lambda + \Delta k)z}. \tag{13.83}$$

When

$$\frac{2\pi m}{\Lambda} + \Delta k = 0, \qquad m \text{ integral and odd}, \tag{13.84}$$

the $m^{th}$ term in the sum will be $1/m$ and the other terms will be periodic functions with a period of $\Lambda$. When these terms are integrated over the crystal length, $L$, they will be zero if $L$ is an exact multiple of $\Lambda$ and very small otherwise. Thus, we have

$$E_{2j} \propto \frac{2d_{eff}L}{m\pi}, \qquad \text{when } \frac{2\pi m}{\Lambda} + \Delta k = 0. \tag{13.85}$$

If the alternating domains are of equal width, $m$ must be odd; otherwise, both odd and even values will work. The second harmonic intensity, which is proportional to $|E_{2j}|^2$, will be reduced by the factor of $(2/m\pi)^2$ compared to conventional birefringent phase matching. (It is interesting to note that the field reduction factor for $m = 1$ is just equal to $2/\pi$, the ratio of the diameter to the circumference of a semicircle (Fig. 13.12)). On the other hand, the freedom one has in the direction of $\boldsymbol{k}$ will often make up for this modest factor; furthermore, the adverse problems due to double refraction are absent.

The phase matching bandwidth can be obtained in the same way as for the birefringent case. We assume that only the $m^{th}$ term contributes, and obtain, by integrating and squaring,

$$\frac{P^{(2\omega)}}{P^{\omega}} \propto \frac{\sin^2(\Delta kL/2)}{(\Delta kL/2)^2}, \tag{13.86}$$

where $\Delta k$ is the $k$-vector mismatch which includes the effect of the periodic structure:

$$\Delta k \equiv k_{2\omega} - 2k_{\omega} - K_m \qquad \text{where } K_m \equiv \frac{2\pi m}{\Lambda}. \tag{13.87}$$

From this, the bandwidth at the fundamental wavelength is

$$\Delta\lambda = \frac{0.443\lambda}{L} \left| \frac{n_{2\omega} - n_{\omega}}{\lambda} + \left(\frac{\partial n}{\partial \lambda}\right)_{\omega} - \frac{1}{2}\left(\frac{\partial n}{\partial \lambda}\right)_{2\omega} \right|^{-1} \qquad \lambda \text{ at fundamental.} \tag{13.88}$$

This reduces to eqn 13.64 for birefringent phase matching (since $n_{\omega} = n_{2\omega}$ in that case).

The phase matching condition also depends upon the temperature of the crystal through the temperature dependence of the refractive indices and the size of the period, $\Lambda$. The temperature acceptance, $\Delta T$, can be found in an analogous way as the wavelength bandwidth by solving the following equation for $\Delta T$:

$$\Delta T \frac{\partial}{\partial T}\left(\frac{\Delta k L}{2}\right) = 0.866\pi. \tag{13.89}$$

The result is

$$\Delta T = \frac{0.443\lambda}{L}\left|\frac{\partial \Delta n}{\partial T} + \alpha \Delta n\right|^{-1} \qquad \text{where } \Delta n \equiv n_{2\omega} - n_{\omega}, \tag{13.90}$$

and $\alpha$ is the thermal expansion coefficient of the material: it is defined by

$$\frac{dL}{dT} = \alpha L. \tag{13.91}$$

One can investigate other aspects of quasi-phase-matching such as the acceptance angle of the light and the effect of imperfect gratings on the doubling efficiency; such topics are well covered in the paper by Fejer et al. (1992).

The obvious disadvantage of quasi-phase-matching is its lack of tunability: each wavelength usually requires a separate grating, which is not trivial to manufacture. For a single grating at a fixed temperature, the phase-matching bandwidth is typically 1–3 nm, depending upon the crystal and its orientation relative to the polarization of the light. Figure 13.14 illustrates two geometrical approaches for addressing this limitation. The use of multiple gratings (middle figure) is a common technique employed by commercial manufacturers for extending the useful wavelength range for quasi-phase-matching. Between each grating wavelength, the crystal can be *temperature tuned*. To determine the temperature sensitivity, one solves the equation $\partial(\Delta k L/2)/\partial T = 0$ for $\partial\lambda/\partial T$:

$$\frac{\partial\lambda}{\partial T} = \alpha\lambda + \frac{\lambda}{\Delta n}\frac{\partial \Delta n}{\partial T}, \tag{13.92}$$

where we used the fact that $\partial K_m/\partial T = -\alpha K_m$, which is easily derived from the definitions of $\alpha$ and $K_m$. It is interesting that neither of the two bandwidths derived above nor the temperature sensitivity are dependent upon the grating period, $\Lambda$.

## 13.6 Second harmonic generation using a focused beam

Our treatment so far has assumed that the beams are all plane waves. A much more realistic model is to use Gaussian beams which are focused into the crystal to increase the local intensity and therefore the conversion efficiency.

We will summarize here the heuristic treatment given by Boyd and Kleinman (1968) in their classic paper. Our approach will only touch on the important points as the theory can be extremely complicated. The salient point made in this section is that the presence of *double refraction* can significantly reduce the conversion efficiency; a secondary point is that there is an optimum waist size (and consequently Rayleigh length) for the maximum conversion efficiency. The latter point is illustrated

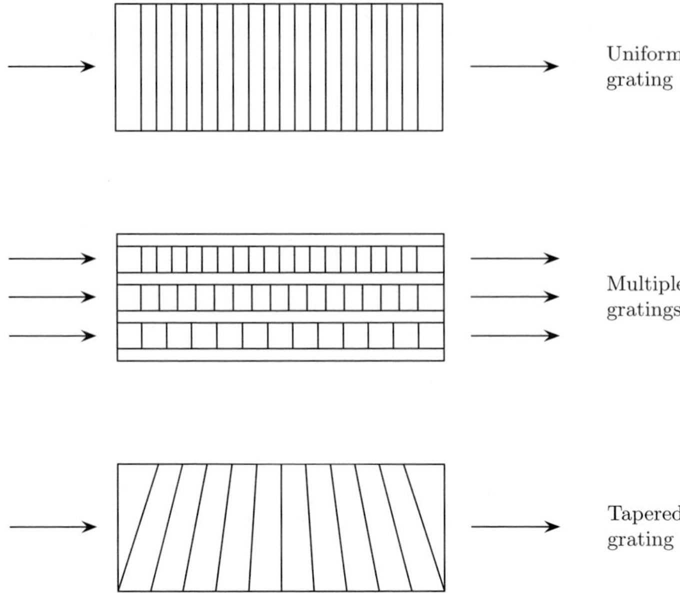



**Fig. 13.14** Two approaches to increasing the tunability of quasi-phase-matching. The top figure displays a fixed grating, useable only at a single wavelength. The middle figure shows a crystal with multiple gratings, allowing phase-matching at a number of wavelengths. The bottom figures depicts a tapered grating, allowing continuous tuning.

in Fig. 13.15, which depicts three different focusing "strengths". Clearly, one would like to focus as tightly as possible since the conversion efficiency increases with the intensity of the fundamental. However, as shown in the bottom figure, if one focuses too tightly, one will not effectively use the entire length of the crystal since the beam rapidly expands on either side of the waist and the larger beam area will not contribute much to the second harmonic generation. Focusing too weakly will use the entire crystal length (since the beam approximates a uniform cylinder) but the conversion will suffer from inadequate intensity. One can show that the optimum conversion efficiency will be obtained when the crystal length is 5.68 times the Rayleigh length for non-critical phase matching.

We will assume that we have a negative uniaxial crystal and the incident radiation is polarized in the ordinary direction and makes an angle $\theta_m$ with the optical axis. In all that follows, we will neglect any absorption of the fundamental or second harmonic. The second harmonic is an extraordinary wave which is phase matched with the fundamental. The coordinate system is shown in Fig. 13.16. In order to be consistent with the Boyd and Kleinman symbolism, we will use an expression for a Gaussian electric field which is different from eqn 1.30 but easily shown to be equivalent to it. It is given by

$$E_1 = E_0 \frac{1}{1 + i\tau} e^{ik_1 z - \frac{r^2}{\omega_0^2(1+i\tau)}}, \tag{13.93}$$

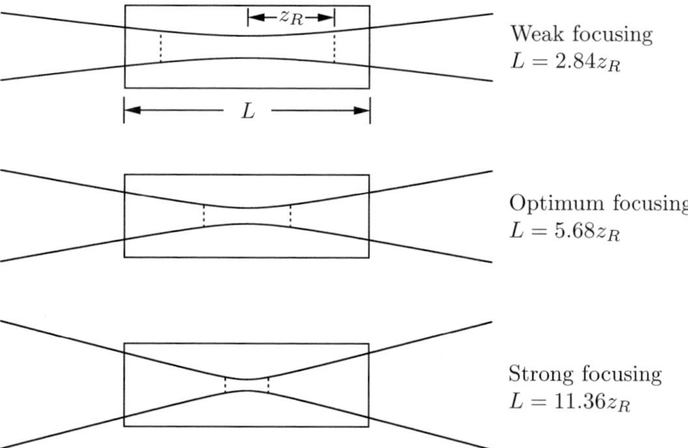



**Fig. 13.15** Three different degrees of focusing into the second harmonic generation crystal, identified by the ratio of crystal length to Rayleigh length. The vertical dashed lines show the Rayleigh distances to the waists.

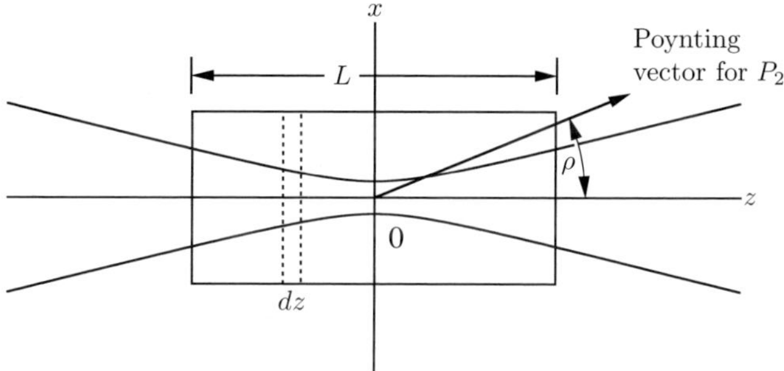



**Fig. 13.16** Geometry used in treatment of second harmonic generation with Gaussian beams. The Poynting vector of the second harmonic is shown, making an angle $\rho$ with the $z$-axis. The input $k$-vector makes an angle of $\theta_m$ with the optical axis, which is in the $xz$ plane and is not shown.

where the "1" subscript denotes the fundamental wave and $\tau$ is the $z$-coordinate expressed in units of the Rayleigh length, $z_R$:

$$\tau = \frac{z}{z_R}. \tag{13.94}$$

Figure 13.16 shows that the direction of energy propagation makes an angle $\rho$ with the $z$-axis. The phase-matching angle is at the intersection of the normal surface for the fundamental (a sphere since it is an ordinary ray) and the second harmonic (an ellipsoid). From eqn 13.32, it can be shown that the *double refraction* (or *walk-off*) angle, $\rho$, is given by

$$\tan \rho = \tfrac{1}{2} n_{o,\omega} \left\{ \frac{1}{(n_{e,2\omega})^2} - \frac{1}{(n_{o,2\omega})^2} \right\} \sin 2\theta_m. \tag{13.95}$$

Inside the crystal, the second harmonic polarization is proportional to $E_1^2$:

$$P(r,z) = P_0 \frac{1}{(1+i\tau)^2} e^{2ik_1 z - \frac{2r^2}{w_0^2(1+i\tau)}}. \tag{13.96}$$

The second harmonic field at observation point $x', y', z'$ is

$$E_2(x',y',z') = A_2(x',y',z')e^{ik_2 z'}, \tag{13.97}$$

where $A_2(x',y',z')$ is a slowly varying function of the spatial coordinates. We will neglect refraction at the faces of the crystal (or assume that it is embedded in a medium with the same refractive index). We are interested only in the direction of *energy flow* (the polarization and energy flow are along the same *ray*), so the observer coordinates are related to the source coordinates by

$$\begin{aligned}
x &= x' + \rho(z - z') & z' &< L/2 \\
x &= x' + \rho(z - L/2) & z' &> L/2 \\
y &= y' & 0 &\le z \le L/2.
\end{aligned} \tag{13.98}$$

At a source point, $x, y, z$, inside the crystal, the heuristic theory considers a thin slab, perpendicular to the $z$-axis and of width $dz$. This slab generates an infinitesimal contribution to the harmonic field given by

$$dA_2(x,y,z) = KP(x,y,z)e^{-ik_2 z}dz = KP_0 \frac{1}{(1+i\tau)^2} e^{i\Delta k z - \frac{2r^2}{w_0^2(1+i\tau)}} dz$$

$$= KP_0 \frac{e^{i\Delta k z}}{1+i\tau} \left\{ \frac{1}{1+i\tau} e^{-2\frac{x^2+y^2}{w_0^2(1+i\tau)}} \right\} dz, \tag{13.99}$$

where $\Delta k = 2k_1 - k_2$, and $K$ is a constant containing uninteresting factors. The slightly odd factorization in the second equation was done for a purpose: the bracketed factor is the *amplitude* of a Gaussian beam with the same waist size and position as the fundamental beam. This is the quantity which propagates, forming the resultant harmonic wave. We now force the latter to propagate along a ray at angle $\rho$ to the $z$-axis. Prior to summing (integrating) over all the contributions, we will substitute the ray equations, eqn 13.98, into the expression for the contribution of the incremental field to the field at the observation point, $x', y', z'$:

$$dA_2(x',y',z') = KP_0 \frac{1}{1+i\tau} \frac{e^{i\Delta k z}}{1+i\tau'} e^{-2\frac{[x'-\rho(L/2-z)]^2 + y'^2}{w_0^2(1+i\tau')}} dz \qquad z' > L/2, \tag{13.100}$$

where

$$\tau' = \frac{z'}{z_R}. \tag{13.101}$$

We finally integrate over $z$, from $-L/2$ to $L/2$:

$$E_2(x', y', z') = K P_0 \frac{e^{2ik_1 z'}}{1 + i\tau'} \int_{-L/2}^{L/2} \frac{e^{i\Delta k z}}{1 + i\tau} e^{-2\frac{[x' - \rho(L/2 - z)]^2 + y'^2}{\omega_0^2 (1 + i\tau')}} dz \qquad z' > L/2. \tag{13.102}$$

The remainder of the calculation consists of changing to a more convenient set of variables, squaring the field to obtain the second harmonic intensity and integrating over $x$ and $y$ to obtain the total power. The parameters and their significance are listed below:

$$\begin{aligned}
&\sigma = z_R \Delta k && \text{phase mismatch} \\
&\delta_0 = \omega_0/z_R = (2/z_R k_1)^{1/2} && \text{diffraction half-angle} \\
&\beta = \rho/\delta_0 && \text{normalized walk-off angle} \\
&\xi = L/2z_R && \text{crystal length in units of Rayleigh lengths.}
\end{aligned} \tag{13.103}$$

After a fair bit of algebra, the expression for the second harmonic power, $P_2$, is written in terms of the function, $h(\sigma, \beta, \kappa, \xi, \mu)$, defined by

$$h(\sigma, \beta, 0, \xi, 0) = \frac{1}{4\xi} \int_0^{2\xi} \int_0^{2\xi} \frac{e^{i\sigma(\tau - \tau') - \beta^2 (\tau - \tau')^2}}{(1 + i\tau)(1 - i\tau')} d\tau d\tau'. \tag{13.104}$$

The third and fifth arguments are the absorption and displacement of the waist from the crystal center respectively and are set equal to zero in our treatment. The second harmonic power is

$$P_2 = K P_1^2 L k_1 \cdot h(\sigma, \beta, 0, \xi, 0), \tag{13.105}$$

where $K$ is a constant which depends upon the waist sizes, indices of refraction and nonlinear coefficient. A much more useful parameter than $\beta$ (walk-off angle) is the double-refraction parameter, $B$, defined by

$$B = \beta \xi^{1/2} = \rho(L k_1)^{1/2}/2. \tag{13.106}$$

This significance of this quantity depends upon the concept of *aperture length, $L_a$*, defined by

$$L_a = \frac{\sqrt{\pi} \omega_0}{\rho}. \tag{13.107}$$

The aperture length (Fig. 13.17) is the distance over which the fundamental and second harmonic beams separate. From this, we see that the parameter $B$ is approximately the *ratio of the crystal length to the double-refraction aperture length* (assuming that $\xi \approx 2$, which we will shortly see is close to the optimum value). Putting it another way, $B$ is the inverse of the fraction of the crystal length over which the two beams overlap. Thus, a $B$ of 10 implies that the beams overlap over 10% of the crystal length. A more useful $h$-function than the one defined above is

$$h(\sigma, B, \xi) = h(\sigma, B\xi^{-1/2}, 0, \xi, 0). \tag{13.108}$$

This function depends upon three important parameters which can be optimized: the phase mismatch, the walk-off and the crystal length (normally, the only control one

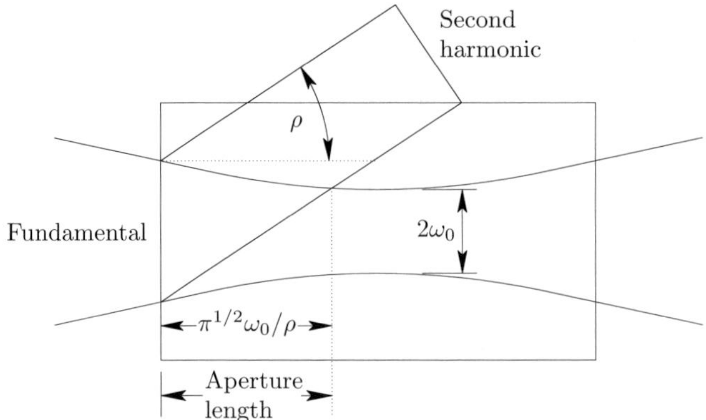



**Fig. 13.17** Illustration of *aperture length* in second harmonic generation: the axial distance over which the fundamental and second harmonic beams separate.

has over the walk-off is through the crystal length, but it is often useful to consider the former to be an *independent* parameter). We can define $h_m(B, \xi)$ to be the function $h(\ldots)$ which is optimized with respect to $\sigma$ (phase matching) at each value of $B$ and $\xi$:

$$h_m(B, \xi) = h[\sigma_m(B, \xi), B\xi^{-1/2}, 0, \xi, 0). \tag{13.109}$$

Finally, one can optimize $h_m(B, \xi)$ with respect to $\xi$ (crystal length in units of Rayleigh length) to obtain the function $h_{mm}(B)$ defined as

$$h_{mm}(B) = h_m(B, \xi(B)). \tag{13.110}$$

Curves of $h_m(B, \xi)$ have been plotted (Boyd (1968)) versus $\xi$ and display a very broad maximum. The optimum value of $\xi$ for $B = 0$ (no walkoff) is $\xi = 2.84$; this value is often mentioned in books on quantum electronics. For larger values of $B$, the optimum $\xi$ decreases, becoming about 1.4 for $B = 16$. The relative unimportance of the focusing parameter $\xi$ is suggested by the fact that $h_m(B, \xi)$ is within 10% of its maximum over the range $1.52 < \xi < 5.3$. From an experimental perspective, any value of $\xi$ in that range is probably adequate. A plot of the optimum focusing parameter, $\xi_m$, as a function of $B$ appears in Fig. 13.18.

While it is faintly useful to be aware of the optimum value of $\xi$, it is far more important to realize that walk-off can greatly reduce the second harmonic conversion efficiency. Unfortunately, one has little control over the walk-off since the parameter $\rho$ is rigidly determined by the phase matching angle which in turn depends upon the wavelength. The parameter $B$ is, however, determined also by the crystal length and an appreciation of the diminishing returns possibly obtained by increasing the crystal length can have some benefit. A plot of $h_{mm}(B)$ versus $B$ appears in Fig. 13.19. The effect of double-refraction can be seen in the following asymptotic form for $h_m(B, \xi)$ in the limit of a relatively short crystal ($\xi < 0.4$):

$$h_m(B, \xi) \approx \frac{\sqrt{\pi}}{2} \frac{\xi^{1/2}}{B} \qquad (\xi < 0.4, \ B > \sqrt{6/\xi}). \tag{13.111}$$

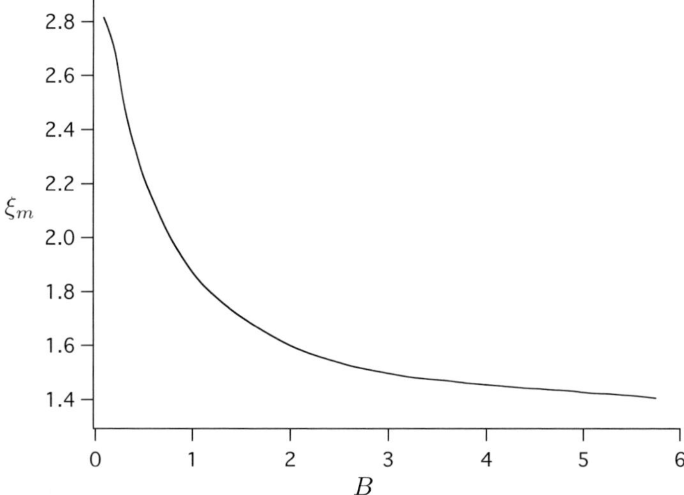

**Fig. 13.18** Plot of optimum focusing parameter, $\xi$, vs double-refraction parameter, $B$. From Boyd (1968).

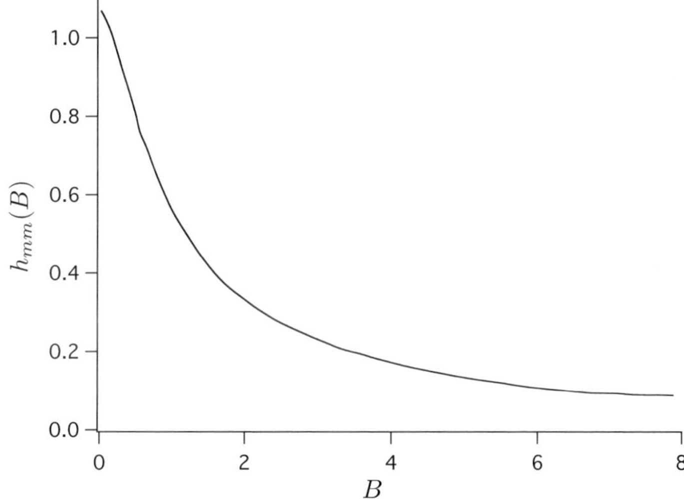

**Fig. 13.19** Plot of optimum $h_{mm}(B)$ vs double-refraction parameter, $B$. From Boyd (1968).

The inverse dependence on $B$ is suggested by the figure and suggests that there is little advantage in using long crystals when there is considerable walk-off.

We have presented a number of characteristic lengths and discussed the dependence of the doubling efficiency on the comparison between them and the crystal length. We will now summarize this data. The lengths are:

$$
\begin{aligned}
&\text{Crystal length} && L \\
&\text{Aperture length} && L_a = \pi^{1/2}\omega_0/\rho \\
&\text{Beam radius at waist} && \omega_0 \\
&\text{Effective length of focus } && L_f = \pi z_R = n\pi^2\omega_0^2/\lambda.
\end{aligned}
\tag{13.112}
$$

Boyd and Kleinman (1968) provided the following *asymptotic* behavior for the dependence of the second harmonic power on these various lengths.

$$
P_2 = \frac{KP_1^2}{\omega_0^2}
\begin{cases}
L^2 & L_a, L_f \gg L \\
LL_a & L_f \gg L \gg L_a \\
L_f L_a & L \gg L_f \gg L_a \\
4L_f^2 & L \gg L_a \gg L_f \\
4.75L_f^2 & L_a \gg L \gg L_f
\end{cases}
\tag{13.113}
$$

where the constant $K$ is different from the one used earlier. The first expression is for a very short crystal where the beam is a uniform cylinder and there is no double-refraction and shows the same $L^2$ dependence as a plane wave. The second expression is the one discussed immediately above, where double-refraction is the only limitation. Note that the remaining three expressions have no crystal length dependence, which supports the argument made above about using a short crystal when double-refraction is significant. When one uses *optimum focusing* ($L \approx L_f$), the first two asymptotic relations yield:

$$
P_2 \propto
\begin{cases}
L & L \approx L_f, L_a > L \text{ (no double refraction)} \\
L^{1/2} & L \approx L_f, L > L_a \text{ (double refraction).}
\end{cases}
\tag{13.114}
$$

The first relationship describes the usual situation: optimum focusing into a non-critically phase-matched crystal (where $\boldsymbol{k}_\omega$ is perpendicular to the optical axis). The second expression is for the case where non-critical phase-matching is not available and confirms the slower increase in efficiency with crystal length due to walk-off. In this case, a short crystal ($L \approx L_a$) is generally preferred due to the slow square root improvement in the second harmonic power when $L > L_a$. Even though we used a negative uniaxial crystal in the above, all of the essential results also hold for a positive uniaxial crystal, in which case the fundamental will experience double refraction and the second harmonic will not.

## 13.7   Second harmonic generation in a cavity

Second harmonic generation is an intrinsically inefficient process, since the laser field strength needs to be comparable to the internal atomic field for there to be a significant output. Crystals exist whose conversion efficiency (for $L = 1$ cm) is $\approx 0.01/\text{W}$, but most are in the $10^{-3}/\text{W}$ to $10^{-4}/\text{W}$ range. Fortunately, the efficiency can be greatly increased using a *resonant cavity*, which can increase the power incident on the crystal by a factor of 100 or more. Usually a ring cavity is used since it has a much smaller (destabilizing) back reflection into the laser and the traveling wave produces a *single* SHG beam instead of the two counter-propagating beams generated in a standing wave cavity. An enhancement of the fundamental by a factor of 100 will result in an enhancement of the second harmonic by a factor of 10,000, which is usually sufficient to provide a useful amount of the second harmonic. In favorable cases, a large fraction of the fundamental can be converted to the second harmonic. Large conversion efficiencies are also available using laser *intracavity* doubling, but there are many reasons for separating the laser oscillation process from the frequency doubling process, particularly when narrow linewidths are sought after.

A ring doubling cavity is shown schematically in Fig. 13.20. All of the components

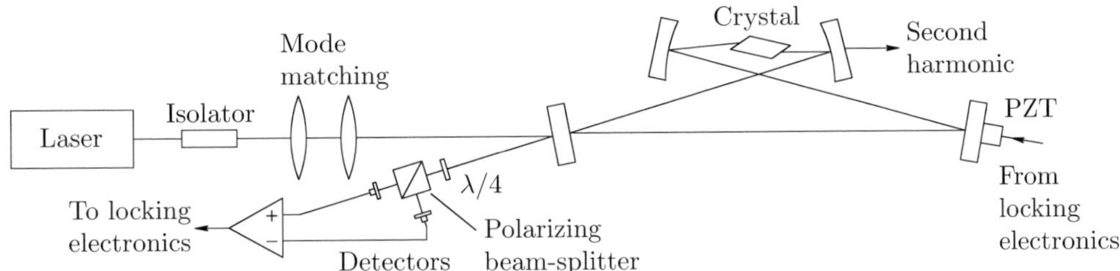


**Fig. 13.20** Schematic of ring cavity frequency doubling system.

shown in the figure (except the isolator) have already been discussed in some detail. The isolator prevents destabilization of the laser due to back reflections from the cavity or other optical components (such as the mode matching lenses). Single Faraday isolators can provide 33 dB of isolation, which is often enough to avoid feedback problems. The beam is mode matched into the lowest mode of the cavity using the two lenses shown. The cavity is designed to have a small waist size at the crystal, which allows the crystal to perform optimally, as discussed in the previous section.

There are two strategies for minimizing the loss due to the crystal. One can purchase a crystal with *anti-reflection coatings*, which can reduce the loss at each surface to $\approx 0.1\%$. Alternatively, the crystal can be cut so that the circulating beam impinges on it at Brewster's angle. The latter approach requires that the polarization be in the plane of incidence and the crystal needs to be cut so that this polarization will generate the largest amount of second harmonic. The Brewster cut approach has the advantage of allowing the complete elimination of the astigmatism in the circulating beam since the astigmatism due to the off-axis incidence on the concave mirrors can be canceled by the astigmatism due to the Brewster-cut crystal. This eases the mode matching and might improve the conversion efficiency, though in practice, a small amount of astigmatism is not a problem.

It has become standard practice to lock the cavity resonant frequency to the laser frequency using the *polarization scheme* (Chapter 4) to generate the discriminant and apply the feedback signal to a piezoelectric actuator attached to one of the mirrors. The Brewster approach facilitates the polarization scheme since the crystal is a very polarization-sensitive element. Finally, at least one mirror (the output mirror) needs a *dichroic* coating which is highly reflective at the fundamental and transmissive at the second harmonic.

With relatively efficient crystals (such as potassium niobate and KTP), it is possible to convert a large fraction (often more than 50%) of the fundamental to the second harmonic. This modifies somewhat the impedance matching strategy, since the *depletion of the fundamental* will now be a loss which must be taken into account. One can approach this problem analytically by modifying the expression for the circulating power given in eqn 3.10. Without depletion of the fundamental, the circulating intensity at resonance is

$$\text{No depletion:} \quad I_c = I_0 \frac{t_1^2}{(1 - r_1 r_m)^2}, \tag{13.115}$$

where the terms are defined in Chapter 3. If $\gamma$ is the second harmonic conversion efficiency defined by

$$I_2 = \gamma I_f^2, \tag{13.116}$$

where $I_f$ and $I_2$ are the fundamental and second harmonic intensies, then the *impedance-matched* second harmonic power will be

$$\text{No depletion, impedance matched:} \quad I_2 = \gamma \left(\frac{I_0}{T}\right)^2, \tag{13.117}$$

where $I_0$ is the input intensity to the cavity and $T = t_1^2$. The behavior when the impedance is mismatched is obtained directly from eqn 3.27:

$$\text{No depletion, impedance mismatched:} \quad I_2 = \gamma \left(\frac{I_0}{T}\right)^2 \left(\frac{4\sigma}{(\sigma + 1)^2}\right)^2, \tag{13.118}$$

where

$$\sigma \approx \frac{T}{L}, \tag{13.119}$$

and $L$ is the sum of all cavity losses exclusive of the transmission through the input coupler.

When depletion is included, the quantity $r_m$ needs to be modified. We will assume that the circulating intensity has already been established at the level, $I_c$. Each pass, of intensity, $I_1$, will generate a second harmonic *loss* of $\gamma I_1 I_c$. The *fractional loss* in intensity is $\gamma I_1 I_c / I_1 = \gamma I_c$. The *fractional remaining intensity* per pass is therefore $1 - \gamma I_c$. We will modify $r_m$ by multiplying it by the fractional *field* remaining after one round trip due to depletion (this requires a square root). Thus,

$$r_m \Longrightarrow r_m \times \sqrt{1 - \gamma I_c} \approx r_m \left(1 - \frac{\gamma I_c}{2}\right) = r_m \left(1 - \tfrac{1}{2}\sqrt{\gamma I_2}\right), \tag{13.120}$$

where we used the expression for the second harmonic intensity, $I_2 = \gamma I_c^2$, in the last equation. The circulating power with depletion is now

$$\text{Depletion:} \quad I_c = I_0 \frac{t_1^2}{[1 - r_1 r_m(1 - \tfrac{1}{2}\sqrt{\gamma I_2})]^2}, \tag{13.121}$$

and the second harmonic intensity is

$$I_2 = \frac{\gamma t_1^4 I_0^2}{[1 - r_1 r_m (1 - \frac{1}{2}\sqrt{\gamma I_2})]^4}. \tag{13.122}$$

This can be simplified somewhat if we assume that $t_1^2 \ll 1$ and $1 - r_m^2 \ll 1$. The round-trip loss (excluding depletion and transmission through the input mirror) is $L$, where

$$L \equiv 1 - r_m^2. \tag{13.123}$$

Substituting this in the equation for $I_2$ and dropping terms of second order in small quantities, the second harmonic intensity is

$$I_2 = \frac{16\gamma T^2 I_0^2}{[L + T + \sqrt{\gamma I_2}]^4} \qquad T \equiv t_1^2 = 1 - r_1^2. \tag{13.124}$$

The quantity raised to the fourth power in the denominator is just the total loss, including that due to depletion. This is a cubic equation in $\sqrt{I_2}$ and is most easily solved for $I_2$ numerically on a computer. One can, however, obtain several important insights without obtaining an explicit solution.

We recall that a cavity is *impedance matched* when the input coupling is adjusted to obtain the maximum circulating power. A more useful and equivalent definition of impedance matching is that the *reflection vanishes*. If we modify $r_m$ to include depletion, we can define $r'_m$ to be

$$r'_m = r_m \sqrt{1 - \sqrt{\gamma I_2}}. \tag{13.125}$$

When depletion is included, the resonant formulas for $I_c$ and $I_r$ are

$$I_c = I_0 \frac{T}{(1 - r_1 r'_m)^2} \tag{13.126}$$

$$I_r = \left[\frac{r'_m - r_1}{1 - r_1 r'_m}\right]^2. \tag{13.127}$$

From the second equation, we obtain the impedance matching condition (zero reflection) when $r_1 = r'_m$:

$$r_1 = r'_m \implies \sqrt{1 - T} = \sqrt{1 - L}\sqrt{1 - \frac{\gamma I_0}{T}}, \tag{13.128}$$

where, in the last equality, we use the fact that the impedance matched $I_c$ is just $I_0/T$. This is a quadratic equation in $T$ whose solution, when $L \ll 1$, is

$$T_{matched} = L/2 + \sqrt{L^2/4 + \gamma I_0}. \tag{13.129}$$

The impedance-matched second harmonic is just

$$I_2 = \gamma \left(\frac{I_0}{T_{matched}}\right)^2 = \frac{4\gamma I_0^2}{(L + \sqrt{L^2 + 4\gamma I_0})^2}. \tag{13.130}$$

Following Kimble (1993), one can define a *figure of merit*, $\epsilon_{NL} \equiv \gamma/L^2$. Using this, the *conversion efficiency* ($I_2/I_0$) is

$$\text{Conversion efficiency} = \eta = \frac{4\epsilon_{NL}I_0}{(1 + \sqrt{1 + 4\epsilon_{NL}I_0})^2} \qquad \text{(impedance matched)}. \qquad (13.131)$$

This reduces to eqn (13.117) when $4\epsilon_{NL}I_0 \ll 1$ (or $4\gamma I_0 \ll L^2$).

From the above, we can draw two conclusions. First, when the depletion loss, $\sqrt{\gamma I_2}$, is comparable to the other losses, $L + T$, the increase of second harmonic with input power (with $T$ fixed) is much slower than the squared dependence which occurs at lower powers. At high input power, the second harmonic increases as $I_0^{2/3}$. If one continually adjusts $T$ for impedance matching as the power is raised, the second harmonic is proportional to $I_0$ at high power. This can be seen from eqn 13.130, which describes the impedance matched situation.

The second conclusion is that one must change the transmission of the input coupler when using high powers to obtain optimum second harmonic (and impedance matching). As an example, if one has a potassium niobate crystal ($\gamma \approx 0.01$ /W) and a cavity with $T = 0.01$ and $L = 0.01$ (impedance matched without depletion), then depletion becomes important when $I_2 = 40$ mW, which is a modest power (one can obtain ten times this power with some effort). The input power is about 80 mW; thus the conversion efficiency is 50% from a relatively low-power laser. With this power, a 3% input coupler would be needed for impedance matching and optimum conversion. The second harmonic power using this coupler is 56 mW, a modest improvement over the impedance mismatched situation. Figure 13.21 plots the second harmonic power as a function of the input power for the above parameters. The right-hand plot is the same function on a log-log scale, showing the deviation from quadratic dependence on the input at higher powers.

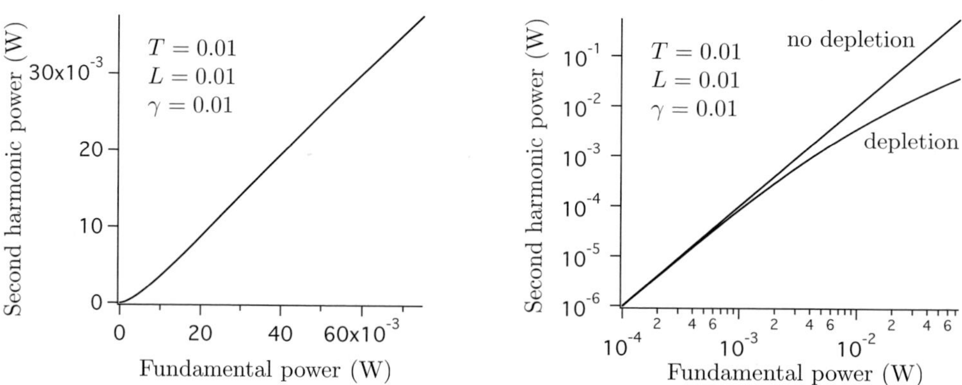


**Fig. 13.21** Second harmonic power as a function of input power on linear scale (left) and log-log scale (right).

There are a number of problems which can be troublesome when high fundamental and second harmonic power are present in a doubling crystal. For example, potassium niobate is subject to *blue induced infrared absorption*, which can greatly limit the cavity enhancement due to an increase in the absorption of the fundamental in the presence of significant *second harmonic* intensity. This phenomenon is discussed in the literature (for example, Shiv (1995)) and can be reduced by heating the crystal. With large fundamental powers, there can be significant absorption-induced heating of the crystal, which can prevent stable phase matching when the crystal possesses strongly temperature-dependent refractive indices. Active temperature stabilization of the crystal can solve this problem. The thermal effects can also produce considerable *thermal lensing* in the crystal, with consequent degradation of the beam quality. These matters are also discussed in various publications, and the reader is invited to consult the literature to learn more about them (For example, Le Targat (2005) discusses the thermal lensing problem in periodically polled KTP and is able to obtain 75% conversion efficiency by understanding and minimizing the problem.)

Before leaving this topic, we should mention a less obvious way to increase the second harmonic power using a cavity. In the foregoing, we have discussed the use of a cavity to increase the circulating power at the *fundamental*, with consequent increase in the power at the second harmonic. It turns out that one can also use a cavity to *increase the nonlinear interaction* by making the cavity *resonant at the second harmonic*. Essentially, the cavity increases the *crystal length* and thereby increases the conversion efficiency by the factor $4/T_2$, where $T_2$ is the transmission of the output coupler of the cavity at the second harmonic. This possibility was pointed out in 1966 by Ashkin (Ashkin (1966)), but was not actually demonstrated until somewhat later (e.g., by Kimble (1993)). The technique is especially applicable to low-loss crystals with small nonlinear coefficients. It can be combined with fundamental enhancement to yield good conversion efficiency with only a few milliwatts of input power (Kimble (1993)). When used alone, it is not nearly as efficient as fundamental enhancement (which provides a factor of $\approx 1/T_1^2$) and one might legitimately question whether it is worth the considerable additional equipment complexity.

## 13.8   Sum-frequency generation

The nonlinear interaction responsible for second harmonic generation can also lead to the generation of sum and difference frequencies when the two input beams are at different frequencies. The three fields are

$$\text{Output wave:} \quad E_j^{(\omega_3)} = \frac{1}{2}E_{3j}(z)e^{i(\omega_3 t - k_3 z)} + \text{cc} \tag{13.132}$$

$$\text{Input wave 1:} \quad E_i^{(\omega_1)} = \frac{1}{2}E_{1i}(z)e^{i(\omega_1 t - k_1 z)} + \text{cc} \tag{13.133}$$

$$\text{Input wave 2:} \quad E_k^{(\omega_2)} = \frac{1}{2}E_{2k}(z)e^{i(\omega_2 t - k_2 z)} + \text{cc}. \tag{13.134}$$

The equation which describes sum frequency generation is a straightforward variation of eqn 13.46,

$$\frac{dE_{3i}}{dz} = -i\frac{\omega_3}{4}\sqrt{\frac{\mu_0}{\epsilon_3}}\sum_{jk} d_{ijk}E_{1j}E_{2k}e^{i\Delta k z}, \quad \Delta k = k_3 - k_1 - k_2, \tag{13.135}$$

where the three frequencies satisfy

$$\omega_3 = \omega_1 + \omega_2. \tag{13.136}$$

The result for the *intensity* $I^{(\omega_3)}$ at the sum frequency is obtained in the same way as the second harmonic intensity:

$$I^{(\omega_3)} = \frac{1}{8}\omega_3^2 \sqrt{\frac{\mu_0^3}{\epsilon_1\epsilon_2\epsilon_3}}(d_{ijk})^2 L^2 \frac{\sin^2(\Delta k L/2)}{(\Delta k L/2)^2} I^{(\omega_1)} I^{(\omega_2)}. \tag{13.137}$$

The phase matching requirement is

$$\text{Phase matching:} \quad \boldsymbol{k}_3 = \boldsymbol{k}_1 + \boldsymbol{k}_2. \tag{13.138}$$

Phase matching can be accomplished either by using the crystal birefringence or by quasi-phase-matching, as discussed earlier in this chapter. The procedures are straightforward extensions of those used in second harmonic generation. One can use a cavity to enhance the power at *each* of the sum frequencies. A single cavity gives a sum frequency power which is *linear* in the cavity enhancement factor and one must use two overlapped cavities (e.g., Hemmati (1983)) to obtain a nonlinear enhancement which is comparable to that obtained in cavity-enhanced second harmonic generation.

## 13.9  Parametric interactions

The nonlinear susceptibility described above allows the reverse of sum frequency generation to take place: when a wave at frequency $\omega_3$ propagates in a nonlinear crystal, *amplification* of already existing waves at $\omega_1$ and $\omega_2$ can occur when momentum and energy are conserved ($\boldsymbol{k}_3 = \boldsymbol{k}_1 + \boldsymbol{k}_2$ and $\omega_3 = \omega_1 + \omega_2$). The beam at $\omega_3$ is called the *pump* and the beams at $\omega_1$ and $\omega_2$ are called the *signal* and *idler*. This phenomenon is the optical analogue of the well-known processes of radio-frequency and microwave *parametric amplification*.

Before describing optical parametric processes, we will illustrate the phenomenon with a mechanical and an electrical example. The mechanical example involves the familiar process of "pumping" a swing (it is interesting that the colloquial word *pumping* is exactly applicable to this process). The amplitude of the pendulum oscillation of a child's swing can be increased if the child lowers his center of gravity when the swing is descending and raises his center of gravity when the swing is ascending. The child is *modulating* the moment of inertia of the swing at *twice the pendulum oscillation frequency* and thereby amplifying the swing's motion. Two aspects of this process are worth noting. First, the child's behavior can only amplify an existing motion and will not start a swing which is at rest; thus, the child is *not driving the swing's oscillation*. Second, the *phase* of the child's input relative to the swing's motion is very important: if the child's timing is opposite to that described above, the swing's motion will be attenuated rather than amplified. This latter *phase sensitivity* is uniquely applicable to *degenerate* parametric amplification, where the pump is at twice the frequency of the signal being amplified. Non-degenerate parametric amplification is insensitive to the phase of the signal or idler. Finally, it should be mentioned that most *seated* swingers use a different mechanism to pump a swing: they move their bodies back and forth at the swing oscillation frequency. This is an example of a *driven oscillator* and is capable of starting a swing which is initially at rest, unlike the parametric approach.

The electrical example of parametric amplification is illustrated in Fig. 13.22, which shows a series RLC circuit whose capacitor has a variable plate separation. The res-

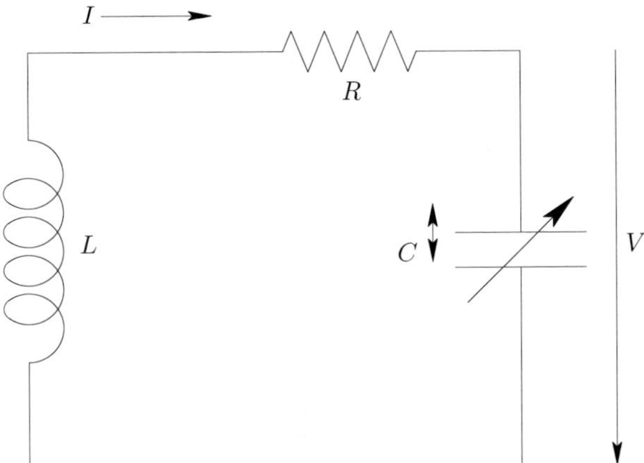


**Fig. 13.22** Circuit illustrating parametric amplification due to a modulated capacitance (via its plate separation).

onant frequency when $R \ll \omega L$ is approximately $\omega_0 = 1/\sqrt{LC}$. We assume that the capacitor is initially *charged*. If the plates are suddenly pulled apart slightly, work is done on the capacitor (due to the Coloumb attraction between the plates) and the capacitance is slightly reduced while the voltage across the capacitor is increased (since the charge does not change instantaneously and $V = Q/C$). A quarter of a cycle of $\omega_0$ later, the capacitor will be discharged and the plates are quickly restored to their original position (against zero force). After another quarter cycle, the capacitor's charge is opposite to the original charge and the plates are again pulled apart against the Coloumb attraction. If this process is repeated, we see that energy is supplied to the circuit through an external *pumping* process which occurs at twice the circuit's resonant frequency. Again, amplification will only occur for a specific phase of the pumping action. If the circuit were inert (no initial charge on the capacitor nor current through the inductor), nothing would happen when the capacitor's value is modulated. Even though our description involved *sudden* changes in the capacitor plate separation, the same phenomenon would occur if the capacitance were sinusoidally modulated. The equation satisfied by the current is

$$-L\frac{dI}{dt} = RI + \frac{1}{C}\int I \, dt. \tag{13.139}$$

Differentiating and moving the terms to the left side:

$$\frac{d^2 I}{dt^2} + \gamma \frac{dI}{dt} + \omega_0^2 I = 0 \qquad \omega_0 = \frac{1}{\sqrt{LC}}, \quad \gamma = \frac{R}{L}. \tag{13.140}$$

The capacitance modulation is

$$C = C_0 + \Delta C \sin \omega_p t. \tag{13.141}$$

The differential equation for this parametric process is thus

$$\frac{d^2 I}{dt^2} + \gamma \frac{dI}{dt} + \omega_0^2 (1 + \delta \sin \omega_p t) I = 0, \tag{13.142}$$

where, assuming $\Delta C \ll C_0$,

$$\delta = -\frac{\Delta C}{C_0}. \tag{13.143}$$

From this differential equation, the source of the term *parametric* should be clear: we are modulating a *parameter* of the circuit at frequency $\omega_p$. Actually, we are modulating a rather special kind of parameter: we are modulating a *reactance*. This modulation occurs in actual circuits by employing a *nonlinear reactance*: a reactance whose value depends upon the voltage across it. In our example circuit, we would use a *varicap* diode: a PN diode whose normal voltage-dependent capacitance, when reversed biased, is enhanced relative to ordinary diodes (varicap diodes are also called *varactor* diodes). One can also modulate the *resistance* by exploiting the nonlinear resistance of a diode. Even though *frequency mixing* can occur due to a nonlinear resistance, *parametric amplification cannot take place*. In fact, most common mixers use the nonlinear resistance of their diodes for frequency conversion. However, for an electrical realization of parametric amplification or oscillation, a modulated *nonlinear reactance* is absolutely necessary.

In anticipation of an *exponentially growing* current due to parametric resonance, we will use a trial solution for the current given by

$$I(t) = I_0 \cos(\omega t + \phi) e^{st}, \tag{13.144}$$

where $s$ is real and is the expected *gain constant*. Substituting this into the differential equation and equating terms with the same exponential time dependence, one has

$$\omega_p = 2\omega \tag{13.145}$$

$$\phi = 0 \text{ or } \pi, \tag{13.146}$$

where we have dropped the term at $3\omega$. This result confirms our claim that $\omega_p = 2\omega$ and that the amplification is phase dependent. By equating the real and imaginary parts of the coefficients of the exponential terms to zero, we obtain

$$s = \frac{\omega\delta}{4} - \frac{\gamma}{2} \tag{13.147}$$

$$\omega^2 = \omega_0^2 + s^2 + \gamma s, \tag{13.148}$$

where we assume that $\omega \approx \omega_0$. From the first equation, the condition for amplification ($s > 0$) is

$$\text{Parametric amplification:} \quad \frac{\omega_0 \delta}{2} > \gamma. \tag{13.149}$$

From the second equation, the steady-state solutions ($s = 0$) occurs when $\omega = \omega_0$.

This circuit has only a single solution at $\omega_p = 2\omega$ (degenerate) since it contains a single resonant system. One can construct a circuit with *two resonant systems*, with different resonant frequencies, which are both coupled to a nonlinear reactance. That circuit will display non-degenerate parametric amplification at *both* the signal and idler frequencies. Since our goal is to analyze *optical* parametric processes, we will now turn to the optical case which quite naturally displays the various features of non-degenerate parametric amplification.

In treating optical parametric phenomena, we start with a description of the three fields:

$$\text{Pump wave:} \quad E_j^{(\omega_3)} = \frac{1}{2} E_{3j}(z) e^{i(\omega_3 t - k_3 z)} + \text{cc} \tag{13.150}$$

$$\text{Signal wave:} \quad E_i^{(\omega_1)} = \frac{1}{2} E_{1i}(z) e^{i(\omega_1 t - k_1 z)} + \text{cc} \tag{13.151}$$

$$\text{Idler wave:} \quad E_k^{(\omega_2)} = \frac{1}{2} E_{2k}(z) e^{i(\omega_2 t - k_2 z)} + \text{cc.} \tag{13.152}$$

The pump frequency is necessarily higher than either the signal or the idler frequency for parametric amplification to take place. We will treat the pump field as a parameter and solve the equations for the growth of the signal and idler waves. These two equations are obtained in exactly the same way as the differential equation (eqn 13.135) describing sum frequency generation. The two equations are

$$\frac{dE_{1i}}{dz} = -\frac{i\omega_1}{4} \sqrt{\frac{\mu_0}{\epsilon_1}} d_{ijk} E_{2j}^* E_{3k} e^{-i\Delta k z} \tag{13.153}$$

$$\frac{dE_{2k}^*}{dz} = \frac{i\omega_2}{4} \sqrt{\frac{\mu_0}{\epsilon_2}} d_{kij} E_{1i} E_{3j}^* e^{i\Delta k z}, \tag{13.154}$$

where

$$\Delta k = k_3 - k_1 - k_2 \tag{13.155}$$

$$\omega_3 = \omega_1 + \omega_2. \tag{13.156}$$

Before solving these equations, we will simplify the notation by using coupling constants $\kappa_1$ and $\kappa_2$ and include absorption terms described by (intensity) absorption coefficients $\alpha_1$ and $\alpha_2$:

$$\begin{aligned} \text{Signal:} \quad & \frac{dE_1}{dz} = -\frac{\alpha_1}{2} E_1 - i\kappa_1 E_2^* E_3 e^{-i\Delta k z} \\ \text{Idler:} \quad & \frac{dE_2^*}{dz} = -\frac{\alpha_2}{2} E_2^* + i\kappa_2 E_1 E_3^* e^{i\Delta k z}, \end{aligned} \tag{13.157}$$

where

$$\begin{aligned} \kappa_1 &= \frac{\omega_1}{4} \sqrt{\frac{\mu_0}{\epsilon_1}} d \\ \kappa_2 &= \frac{\omega_2}{4} \sqrt{\frac{\mu_0}{\epsilon_2}} d, \end{aligned} \tag{13.158}$$

and we have dropped the coordinate subscripts to reduce clutter. In all of the above, we assume that there is little depletion of the *pump*; if this is so, we do not need an equation for $E_3$ and can treat the pump field as a constant.

To simplify matters, we will assume that $\alpha_1 = \alpha_2 = \alpha$. We will look for a solution of the form

$$E_1(z) = E_1(0)e^{Gz - i\Delta kz/2} \tag{13.159}$$

$$E_2(z) = E_2(0)e^{Gz - i\Delta kz/2}. \tag{13.160}$$

These expresions assume a *gain* equal to $G$ and include the $e^{-i\Delta kz/2}$ factor in order to remove the similar factor from the equations. Substituting and dividing out the exponentials, we obtain

$$(G - i\Delta k/2 + \alpha/2)E_1(0) + i\kappa_1 E_3 E_2^*(0) = 0 \tag{13.161}$$

$$-i\kappa_2 E_3^* E_1(0) + (G + i\Delta k/2 + \alpha/2)E_2^*(0) = 0. \tag{13.162}$$

These *homogeneous* equations will only have a non-trivial solution when the determinant of the coefficients vanishes. The solution for $G$ is

$$G = -\frac{\alpha}{2} \pm \sqrt{\kappa_1\kappa_2|E_3|^2 - \left(\frac{\Delta k}{2}\right)^2}. \tag{13.163}$$

We thus see that there is a *parametric gain*, $g$, given by

$$\text{Parametric gain:} \quad g = \sqrt{\kappa_1\kappa_2|E_3|^2 - \left(\frac{\Delta k}{2}\right)^2}, \tag{13.164}$$

where we have taken the positive square root. This gain is maximum when $\Delta k = 0$ (phase matched case). The maximum gain is proportional to the *pump field* and the square root of the product of the signal and idler frequencies. The overall gain, $G$, is reduced by absorption ($\alpha$) and phase mismatching ($\Delta k \neq 0$).

The coupled differential equations can be solved by taking linear combinations of both solutions:

$$E_1(z) = \left(E_{11}e^{gz} + E_{12}e^{-gz}\right)e^{-\alpha/2 - i\Delta kz/2} \tag{13.165}$$

$$E_2(z) = \left(E_{21}e^{gz} + E_{22}e^{-gz}\right)e^{-\alpha/2 - i\Delta kz/2}, \tag{13.166}$$

and using the boundary conditions that the $z = 0$ fields are $E_{1,2}(0)$. The result is

$$E_1(z) = E_1(0)e^{-i\Delta kz/2 - \alpha z/2}\left[\cosh gz + \frac{i\Delta k}{2g}\sinh gz\right]$$

$$- \frac{i\kappa_1}{g}E_3 E_2^*(0)e^{-i\Delta kz/2 - \alpha z/2}\sinh gz \tag{13.167}$$

$$E_2(z) = E_2(0)e^{-i\Delta kz/2 - \alpha z/2}\left[\cosh gz + \frac{i\Delta k}{2g}\sinh gz\right]$$

$$- \frac{i\kappa_2}{g}E_3 E_1^*(0)e^{-i\Delta kz/2 - \alpha z/2}\sinh gz. \tag{13.168}$$

In the following, we will simplify these somewhat cumbersome solutions.

From eqns 13.157, together with a similar one for $E_3$, one can easily show that

$$-\frac{d}{dz}|E_3|^2 = \frac{d}{dz}|E_1|^2 = \frac{d}{dz}|E_2|^2. \tag{13.169}$$

Since the squared fields are proportional to the photon flux, integrating over the length of the crystal gives the *net flux change* as one traverses the crystal; this is proportional to the power change divided by the frequency. Thus, we obtain the *Manley–Rowe relation*

$$\text{Manley–Rowe relation:} \quad -\Delta\left(\frac{P_3}{\omega_3}\right) = \Delta\left(\frac{P_2}{\omega_2}\right) = \Delta\left(\frac{P_1}{\omega_1}\right), \tag{13.170}$$

where the deltas are interpreted as the changes in the quantities between the input and output planes. The minus sign is unique to the wave at $\omega_3$, whose frequency is the sum of the other two frequencies. The Manley–Rowe relation suggests (and quantum mechanics confirms) a photon model for optical parametric interactions. Using this model, the absorption of a single photon at the highest frequency is accompanied by the *joint* creation of single photons at two other frequencies which sum to the highest frequency. This simple model can provide a great deal of insight into these processes. For example, it explains why parametric gain cannot occur at the highest of the three frequencies: if the pump is at one of the two lower frequencies, energy conservation would be violated if a photon at the highest frequency were created. Finally, it is worth mentioning that *all* of the nonlinear behaviors discussed in this chapter can be considered to be *parametric interactions* and this was recognized in the title of one of the early papers on the subject (Boyd (1968)). Thus, all of the behaviors satisfy the Manley–Rowe relationship and can be explained with a simple model involving interactions among three photons in the presence of a nonlinear medium. Of course, only *parametric amplification* can exhibit gain.

Parametric gain is a *coherent* optical gain process which, like laser gain, can lead to oscillation if a pumped medium is placed inside a resonant cavity. A *parametric oscillator* is, in principle, more widely tunable than a laser since the parametric device is merely limited by the requirement that energy and momentum be conserved. Parametric oscillation generates both a signal and an idler beam, as required by the Manley–Rowe relation. Parametric oscillators can be either *doubly resonant* or *singly resonant*, depending on whether the cavity resonates at both signal and idler frequencies or only at the signal frequency.

A doubly resonant optical parametric oscillator is shown schematically in Fig. 13.23. The oscillation conditions for a doubly resonant oscillator are obtained by requiring that the fields replicate themselves after one round trip in the cavity. The field equations for $\Delta k = 0$ are obtained from eqns 13.167 and  13.168:

$$E_1(z) = e^{-\alpha z/2}\left[E_1(0)\cosh gz - i\sqrt{\frac{\kappa_1}{\kappa_2}}E_2^*(0)\sinh gz\right] \tag{13.171}$$

$$E_2^*(z) = e^{-\alpha z/2}\left[E_2^*(0)\cosh gz + i\sqrt{\frac{\kappa_2}{\kappa_1}}E_1(0)\sinh gz\right]. \tag{13.172}$$

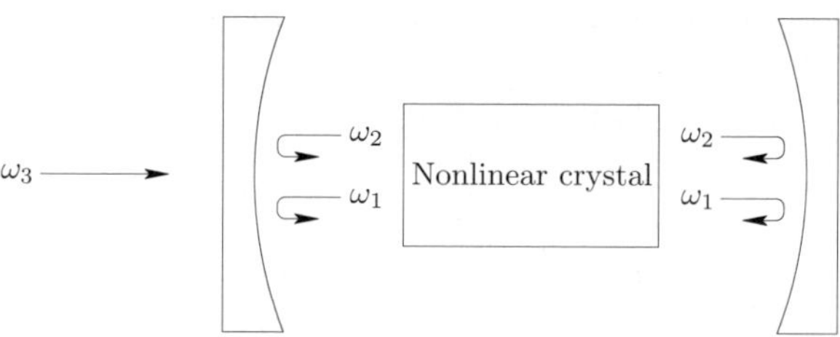


**Fig. 13.23** Schematic of doubly resonant optical parametric oscillator showing pump beam at $\omega_3$ together with signal and idler beams at $\omega_1$ and $\omega_2$.

The reflectivities are assumed to be the same for each mirror and the *squares* of the field reflectivities at the signal and idler frequencies are

$$\text{Signal:} \quad r_1^2 = R_1 e^{2i\phi_1} \tag{13.173}$$

$$\text{Idler:} \quad r_2^2 = R_2 e^{2i\phi_2}. \tag{13.174}$$

The quantities $R_1$ and $R_2$ are the *power reflectivities* at the signal and idler frequencies. One can now apply the requirement that the field replicates itself after a round trip:

$$E_1(0) = r_1^2 e^{-\alpha L - 2ik_1 L}\left[E_1(0)\cosh gL - i\sqrt{\frac{\kappa_1}{\kappa_2}}E_2^*(0)\sinh gL\right] \tag{13.175}$$

$$E_2^*(0) = (r_2^*)^2 e^{-\alpha L + 2ik_2 L}\left[E_2^*(0)\cosh gL + i\sqrt{\frac{\kappa_2}{\kappa_1}}E_1(0)\sinh gL\right], \tag{13.176}$$

where we include the $2ik_{1,2}z$ terms in the exponents since we require that the field amplitude *and* phase replicate themselves. Note that the arguments of the hyperbolic functions are $gL$ and not $2gl$ since there is *no parametric gain* in the reverse direction (there is, however, loss in both directions). This is a pair of homogeneous equations in the variables $E_1(0)$ and $E_2^*(0)$ and a non-trivial solution requires that the determinant of the coefficients vanish. From this, we obtain the oscillation condition

$$(1 - r_1^2 e^{-\alpha L - 2ik_1 L}\cosh gL)(1 - (r_2^*)^2 e^{-\alpha L + 2ik_2 L}\cosh gL)$$
$$= (r_1 r_2^*)^2 e^{-2\alpha L + 2i(k_2 - k_1)L}\sinh^2 gL. \tag{13.177}$$

We will simplify this by making the realistic assumption that the mirror transmissions are greater than the bulk losses and set $\alpha = 0$. We will also assume that the lowest threshold is obtained when *both* the signal and the idler are at axial modes of the cavity and set:

$$2k_1 L - 2\phi_1 = m \times 2\pi \tag{13.178}$$

$$2k_2 L - 2\phi_2 = n \times 2\pi \qquad m, n \ \text{integers.} \tag{13.179}$$

Then, the oscillation condition becomes

$$(1 - R_1\cosh gL)(1 - R_2\cosh gL) = R_1 R_2 \sinh^2 gL$$
$$\implies (R_1 + R_2)\cosh gL - R_1 R_2 = 1. \tag{13.180}$$

If the mirrors have $R_{1,2} \approx 1$, we can expand the hyperbolic cosine and obtain

$$\text{Doubly resonant oscillation condition:} \quad gL = \sqrt{(1 - R_1)(1 - R_2)}. \tag{13.181}$$

The threshold pumping intensity, $I_{3t}$, at $\omega_3$ can be obtained using eqns 13.158 for the coupling constants and eqn 13.48 for the intensity in terms of the field. The result is

$$\text{Pumping threshold:} \quad I_{3t} = \frac{8}{\omega_1 \omega_2 d^2 L^2} \sqrt{\frac{\epsilon_1 \epsilon_2 \epsilon_3}{\mu_0^3}} (1 - R_1)(1 - R_2). \tag{13.182}$$

An important advantage of doubly resonant parametric amplifiers is the very low threshold compared to the singly resonant oscillator. A possible disadvantage is the difficulty in obtaining two simultaneous cavity resonances at the idler and signal frequencies. If $n_1$ and $n_2$ are the indices of refraction for the signal and idler, it is easy to show that the separation between simultaneous axial resonances is about

$$\text{Separation between resonances} = \frac{\pi c}{(n_1 - n_2)L}. \tag{13.183}$$

For typical refractive indices, this separation might be 100 times the mode separation in the cavity. A small change in the cavity length can cause mode hops with large frequency excursions, which is often undesirable. This situation is not nearly as serious in a singly resonant oscillator.

A singly resonant parametric oscillator simply relaxes the requirement that the cavity be resonant at *both* signal and idler frequencies. We can derive the oscillation condition for a singly resonant oscillator from eqn 13.177 by letting $r_2 = 0$. Setting in addition $\alpha = 0$ and $\Delta k = 0$, we obtain

$$R_1 \cosh gL = 1, \tag{13.184}$$

where we assume the cavity is resonant at $\omega_1$ so that $2k_1 L - 2\phi_1 = m \times 2\pi$. When $R_1 \approx 1$, we obtain

$$\text{Singly resonant oscillation condition:} \quad gL = \sqrt{2(1 - R_1)}. \tag{13.185}$$

Thus, the ratio of thresholds for doubly to singly resonant oscillators is

$$\frac{\text{singly resonant pumping threshold}}{\text{doubly resonant pumping threshold}} = \frac{2}{1 - R_2}, \tag{13.186}$$

where the $R_2$ in the denominator is for the doubly resonant oscillator. From this, we see that the singly resonant threshold is at least 10 times greater than the doubly resonant threshold.

A singly resonant oscillator can employ a different and more flexible approach to phase matching. Since the three beams do not need to be *collinear* for momentum conservation (phase matching) to be satisfied, the pump beam is sent in at the appropriate angle to satisfy momentum conservation and ensure that the *signal* beam is along the axis of the cavity. The direction of the idler beam is unimportant, since it is discarded. Fig. 13.24 illustrates the geometry used in singly resonant oscillators. This

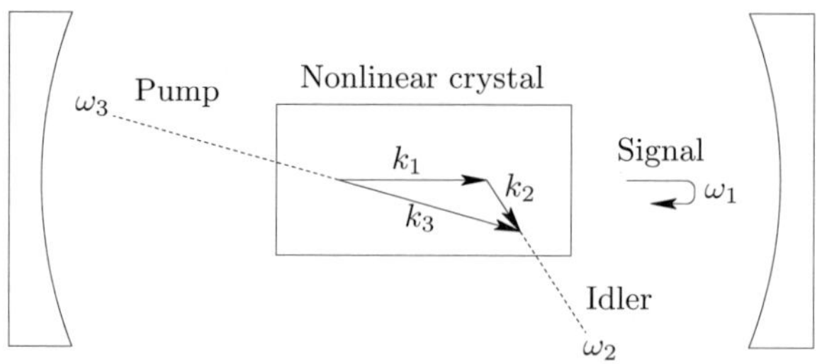

**Fig. 13.24** Schematic of singly resonant optical parametric oscillator illustrating non-collinear phase matching.

phase matching approach is called *non-collinear phase matching* and is conceptually similar to the generation of a coherent outgoing wave in Cerenkov radiation.

The *gain saturation* mechanism in optical parametric oscillators is somewhat different from saturation in a laser. The situation is actually more straightforward: when the pump intensity is sufficient to produce a threshold gain, further increases in the pumping intensity go entirely to the signal and idler beams since the gain must be *clamped* at threshold as in a laser. This has the interesting and useful result of *power limiting* the transmitted pump beam: the transmitted pump power will be independent of the incident power above the oscillation threshold. The *efficiency* can by fairly high in a parametric oscillator, particularly if the idler frequency is somewhat lower than the signal frequency. One can obtain a numerical result for the output power using the Manley–Rowe relationship. If $\Delta P = P_3 - P_{3t}$ is the pumping power excess over threshold, the *number of excess photons* is proportional to $\Delta P/\omega_3$ and these excess photons go into the signal and idler beams, whose photon numbers are proportional to $P_1/\omega_1$ and $P_2/\omega_2$ with the same proportionality constant. Equating these and solving for $P_1$,

$$\text{Power output:} \quad P_1 = \frac{\omega_1(P_3 - P_{3t})}{\omega_3}. \tag{13.187}$$

This approaches $(\omega_1/\omega_3)P_3$ when $P_3 \gg P_{3t}$.

So far, our treatment of parametric phenomena has been based entirely on *classical* principles: given the nonlinear susceptibility, everything follows without resorting to quantum mechanics. There is one parametric process which is fundamentally quantum mechanical: it is called *spontaneous parametric fluorescence*. This process is roughly analogous to *spontaneous emission* in atoms, which also must be treated quantum mechanically. Spontaneous emission can be thought of as emission which is *stimulated* by the single quantum per mode of the vacuum, and spontaneous parametric fluorescence can be considered to be parametric amplification of the vacuum field. (A similar analogy exists with spontaneous Raman scattering.)

A diagram of the apparatus demonstrating parametric fluorescence together with a vector diagram illustrating *non-collinear phase matching* appears in Fig. 13.25. Our approach will be to sum all pairs of amplified signal and idler modes assuming a single photon per signal mode (see, for example, Byer (1968) or Yariv (1989)). We begin by determining the number of signal modes whose propagation angle is between $\psi$ and

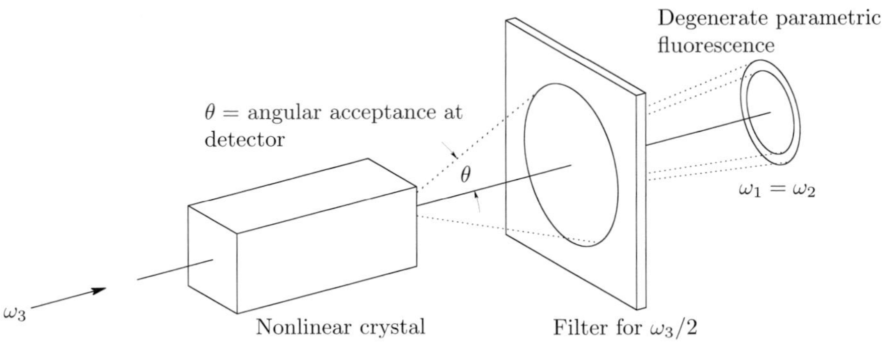



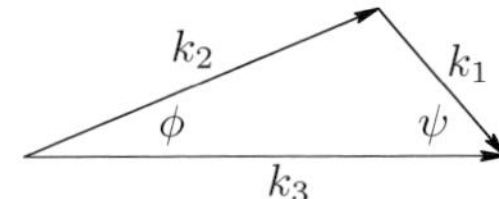



**Fig. 13.25** Schematic of the apparatus demonstrating spontaneous optical parametric fluorescence (top) and the vector diagram illustrating non-collinear phase matching (bottom). The acceptance angle of the detector is $\theta$; all of the angles are exaggerated for clarity. The *degenerate* photons are obtained using a filter whose passband is centerered on $\omega_3/2$. For type-I phase matching, photons on opposite sides of the conical surface are *entangled.*

$\psi + d\psi$ and whose wave vector magnitude is between $k_1$ and $k_1 + dk_1$. We will assume that the waves satisfy *periodic boundary conditions.* For example, the signal field is a plane wave and is given by

$$E_1(\boldsymbol{r}) = E_0 e^{i\boldsymbol{k}_1\cdot\boldsymbol{r}-i\omega_1 t} \qquad \boldsymbol{k} = \left(\frac{2\pi l}{L_x}, \frac{2\pi m}{L_y}, \frac{2\pi n}{L_z}\right), \tag{13.188}$$

where $l, m, n$ are integers, $L_{x,y,z}$ are the periods along the three directions and the volume of the crystal is $V = L_x L_y L_z$. The volume per mode in $k$-space is $(2\pi)^3/V$. The number of modes is

$$dN_1 = \frac{2\pi k_1^2 dk_1 \sin\psi d\psi}{(2\pi)^3/V}, \tag{13.189}$$

where the numerator is just the differential volume element in spherical coordinates with $\psi$ taken as the *polar angle* (with the $z$-axis along $\boldsymbol{k}_3$). Due to the cylindrical symmetry, the azimuthal integration yields $2\pi$. The intensity is obtained by multiplying by $\hbar\omega_1$ to obtain the differential energy and then by $c/n_1 V$ to obtain the differential intensity:

$$dI_1 = \frac{k_1^2 c dk_1 \hbar\omega_1 \psi d\psi}{4\pi^2 n_1}, \tag{13.190}$$

where for small $\psi$ we used $\sin\psi \approx \psi$.

The growth of the idler field due to a signal at $z = 0$ (but no idler at $z = 0$) is obtained from eqn 13.168:

$$E_2(z) = -i\kappa_2 z E_3 E_1^*(0) e^{-i\Delta k z/2} f(\Delta k, z), \qquad (13.191)$$

where we assume no bulk losses ($\alpha = 0$) and define the *phase-matching factor*, $f(\Delta k, z)$ as

$$f(\Delta k, z) \equiv \frac{\sin\sqrt{(\Delta k/2)^2 - \kappa_1\kappa_2|E_3|^2}\, z}{\sqrt{(\Delta k/2)^2 - \kappa_1\kappa_2|E_3|^2}\, z}. \qquad (13.192)$$

Note that we have converted the hyperbolic sine to a circular sine by exchanging the two terms under the square root (and we left out the "$i$" since $f(\Delta k, z)$ will be squared). The differential idler power due the differential signal intensity is obtained by squaring $E_2(L)$, converting it to a power and substituting the above expression for $dI_1$. The result is

$$dP_2 = \langle \text{Area} \rangle \times dI_2 = dI_1 \left( \kappa_2|E_3|L \right)^2 \left( \frac{\omega_2}{\omega_1} \right) f^2(\Delta k, L) A, \qquad (13.193)$$

where $A$ is the area of the idler beam and the frequency ratio is due to the Manley–Rowe relation. Using the law of sines, one can convert the $\psi$ expression in $dI_1$ to one involving $\phi$

$$\psi d\psi = \left( \frac{k_2}{k_1} \right)^2 \phi d\phi \qquad (\psi, \phi \ll 1). \qquad (13.194)$$

The total idler power is obtained by integrating $dP_2$ over $\phi$ from 0 to $\theta$ and over all $\omega_2$ (using $d\omega_2 = -d\omega_1$):

$$P_2 = \int_{-\infty}^{\infty} \int_0^{\theta} d\omega_2 d\phi (dP_2). \qquad (13.195)$$

Since the integral is over $\omega_2$ and $\phi$, we need to convert $\Delta k$ and the $k$ values to functions of these variables. The quantity $\Delta k$ can be obtained using the law of sines when $\phi, \psi \ll 1$ as

$$\Delta k = k_3 - k_2 \cos\phi - k_1 \cos\psi \approx \frac{k_2 k_3}{k_1} \frac{\phi^2}{2} + k_3 - k_2 - k_1. \qquad (13.196)$$

The $k$ values can be expanded around their *colinearly phase matched* values (where $k_{30} = k_{10} + k_{20}$) as

$$\begin{aligned}
k_2 &= k_{20} + \frac{\partial k_2}{\partial \omega_2}(\omega_2 - \omega_{20}) \\
k_1 &= k_{10} + \frac{\partial k_1}{\partial \omega_1}(\omega_1 - \omega_{10}),
\end{aligned} \qquad (13.197)$$

where the $\omega_{10}, \omega_{20}$ frequencies correspond to the colinearly phase matched $k$ values. The integral can be evaluated with these substitutions and with the assumption that

$\kappa_1\kappa_2|E_3|^2$ can be ignored in comparison to $\Delta k$ in the $f^2(\Delta l, L)$ expression. The result is

$$P_2 = \pi\frac{\beta L P_3}{|b|}\theta^2,\tag{13.198}$$

where

$$b \equiv \left(\frac{\partial k_2}{\partial\omega_2}\right)_{\omega_{20}} - \left(\frac{\partial k_1}{\partial\omega_1}\right)_{\omega_{10}}\tag{13.199}$$

$$\beta \equiv \frac{\hbar\omega_1\omega_2^4 n_2 d^2}{\pi^2 c^5 n_1 n_3 \epsilon_0^3}.\tag{13.200}$$

The reason for the asymmetry in $\omega_1$ and $\omega_2$ is that we are starting with $\omega_1$ and are summing over all contributions of $\omega_2$ that satisfy the conservation laws and have $k_2$ values which make an angle less than $\theta$ with the $z$-axis. One of the references mentioned above began with $\omega_2$ and obtained the same result with the identity of the idler and signal reversed from ours. Of course, $\omega_1$ and $\omega_2$ are not independent of each other, since they must satisfy $\omega_1 + \omega_2 = \omega_3$.

The *bandwidth* (range of $\omega_2$ values) can be obtained from our above expression for $\Delta k$:

$$\Delta k = \frac{k_2 k_3}{k_1}\frac{\phi^2}{2} + k_3 - k_2 - k_1 = -b(\omega_2 - \omega_{20}) + a\phi^2,\tag{13.201}$$

where

$$a = \frac{k_2 k_3}{2k_1},\tag{13.202}$$

and $b$ is defined in eqn 13.199 and we used the relation $\omega_2 - \omega_{20} = -(\omega_1 - \omega_{10})$. We will define the bandwidth by the separation between the first nodes of the sinc function $(\sin(\Delta k L/2)/(\Delta k L/2))$. Thus

$$\Delta k L = \left(a\phi^2 - b(\omega_2 - \omega_{20})\right)L = 2\pi \implies \Delta\omega = \omega_2 - \omega_{20} = \frac{a\phi^2 L - 2\pi}{bL}.\tag{13.203}$$

Therefore, the bandwidth is

$$\Delta\omega = \frac{2\pi}{|b|L} \qquad \text{when } \theta = \phi_{max} < \sqrt{\frac{2\pi}{aL}}\tag{13.204}$$

$$\Delta\omega = \frac{a\theta^2}{|b|} \qquad \text{when } \theta > \sqrt{\frac{2\pi}{aL}}.\tag{13.205}$$

In typical cases, the bandwidth is many tens of nm.

The principal use of parametric fluorescence in atomic physics is as a source of *entangled* photons. These can be obtained when the signal and idler frequencies are the same; this situation is called *degenerate spontaneous parametric fluorescence* (or, more commonly, *degenerate parametric down-conversion*). In the degenerate case, the fluorescent beam (at $\omega_1 = \omega_2$) becomes a conical surface shown in Fig. 13.25 for the case when the signal and idler have the same polarization (type-I, non-collinear phase matching). If the polarizations of the two beams are orthogonal (type-II phase matching), there will be two cones which are displaced from each other and which contain photons of one or the other polarization. The geometries for both type-I and type-II phase matching are shown in Fig. 13.26.

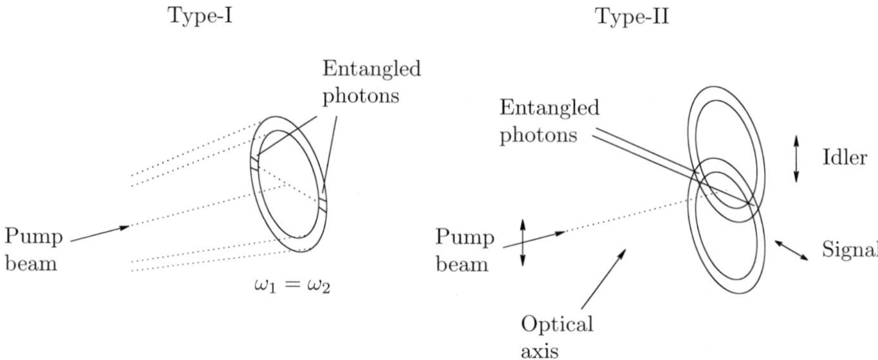


**Fig. 13.26** Detail of beams from degenerate parametric fluorescence for type-I and type-II phase matching. A negative uniaxial crystal is assumed, so the pump is polarized in the plane of the optical axis (extraordinary wave). The polarizations for the type-II case are indicated by the arrows.

Even though non-collinear phase matching offers great tunability, one must still consider the crystal birefringence when designing an experiment using degenerate parametric fluorescence. Since $k_1 = k_2$ and $\omega_3 = 2\omega_1$, the phase matching condition is

$$\boldsymbol{k}_3 = 2\boldsymbol{k}_1 \implies \frac{n_3\omega_3}{c} = 2\frac{n_1\omega_1}{c}\cos\phi \implies \cos\phi = \frac{n_3}{n_1}. \tag{13.206}$$

This cannot be satisfied unless $n_3 \leq n_1$. Therefore, when using a negative uniaxial crystal (such as BBO), the pump polarization must be along the extraordinary direction (in the plane of the wave vector and the optical axis). For type-I phase matching, the signal and idler are ordinary waves. With type-II phase matching, both polarizations are generated and each wave of a given polarization will have its own conical surface. The two conical surfaces are displaced from each other by an angle which depends upon the angle that the pump beam makes with the optical axis.

The photons generated by degenerate parametric fluorescence can be *entangled*. The idea of entanglement is best understood using the photon picture of the parametric process. A pump photon spontaneously splits into signal and idler photons of the same frequency. Each individual process must conserve both momentum and energy, so the fluorescent photons have half the energy of the pump photons, and momentum conservation requires that the two outgoing photons be symmetrically displaced about the $k$-vector of the pump beam. The entanglement of the two emitted photons lies in the fact that all of the properties of one photon of the emitted pair can be determined from the measured properties of the other photon. Entangled photons are mutually coherent and therefore can generate interference fringes. The location of the entangled photons on opposite sides of the cone for type-I or at the intersections of the cones for type-II is a result of the requirement that momentum be conserved. The photons at the intersections of the cones in type-II phase matching are, in addition, *polarization-entangled.*

## 13.10  Further reading

The books by Yariv (1989) and Davis (1996) have fairly complete sections on nonlinear optics while the books by Ghatak and Thyagarajan (1989) and by Milonni and Eberly (1988) have shorter but quite satisfactory discussions of the subject. In addition, the aforementioned books by Yariv and Davis have excellent sections on anisotropic crystals, although the best treatment in the author's opinion is in the classic book by Born and Wolf (1980). A very complete treatment of frequency tuning and frequency and temperature tolerances in quasi-phase-matching appears in a paper by Fejer, et al. (1992) while an early paper on quasi-phase-matching itself is by Somekh and Yariv (1972). The classic paper on harmonic generation using focused beams is by Boyd and Kleinman (1968) and the classic paper on the use of a (standing wave) cavity to enhance second harmonic generation is by Ashkin, et al. (1966). The latter paper also discusses the use of a cavity resonant at the *second harmonic* for nonlinear enhancement. The paper by Ou and Kimble (1993) also discusses the benefits of resonance on the second harmonic and the related paper by Polzik and Kimble (1991) discusses the effects of depletion and thermal lensing in an enhancement cavity. A good table of Sellmeier coefficients for a number of nonlinear materials appears in the *Tunable Laser Handbook* (Duarte (1995)) and a tabulation of nonlinear coefficients is in the paper by Roberts (1992). A classic treatment of spontaneous parametric fluorescence is in the paper by Byer and Harris (1968) and a discussion of the applications of entangled photons is in the paper by Zeilinger (1999).

## 13.11  Problems

(13.1) Using the Sellmeier coefficients given in the text, determine the phase matching angle for generating second harmonic radiation at 231 nm from BBO. (BBO is a negative uniaxial crystal.)

(13.2) For a 1 cm length of BBO used in problem 1, what is the aperture length and the *double refraction parameter, B*?

(13.3) In a negative uniaxial crystal, assume that there exists a phase matching angle, $\theta_m$, at which the ordinary index at $\omega$ is equal to the extraordinary index at $2\omega$. For small deviations of the propagation angle about $\theta_m$, derive an expression for the corresponding change, $\Delta k(\theta)L$, in the total phase difference between the fundamental and the second harmonic. (This allows one to determine the maximum *angular acceptance* of the fundamental beam.)

(13.4) A bow-tie cavity is used to enhance the second harmonic generation in a potassium niobate crystal. Assume that the input coupler power transmission is 0.02 and the remaining losses are also 0.02. If the crystal conversion efficiency is 0.01/W and the input power is 100 mW, how much second harmonic power will be generated? What is the optimum input coupling transmission and what will be the second harmonic power if the optimum input coupling is used?

# 14
# Frequency and amplitude modulation

## 14.1   Introduction

Many of the applications described in this book require the ability to modulate the frequency, phase or amplitude of laser radiation. The devices used for these purposes fall into two broad categories: those that exploit the dependence of the indices of refraction of certain crystals on an applied electric field (the *electro-optic effect*) and those that use an acoustic wave to modulate the index of refraction (via the *acousto-optic* effect). In this chapter, we will describe both categories of modulators and the underlying processes that enable them to function.

## 14.2   The linear electro-optic effect

The *linear electro-optic effect* refers to the linear dependence of the refractive indices on an applied electric field in certain crystals. In the absence of an electric field, the crystal is described in some coordinate system by the equation for the *index ellipsoid*,

$$\frac{x^2}{n_x^2} + \frac{y^2}{n_y^2} + \frac{z^2}{n_z^2} = 1, \tag{14.1}$$

where $x, y$ and $z$ are the axes (*principal axes*) along which $\boldsymbol{E}$ is parallel to $\boldsymbol{D}$. (These matters are discussed in some detail in the previous chapter.) When an electric field is present, the index ellipsoid will change to

$$\left(\frac{1}{n^2}\right)_1 x^2 + \left(\frac{1}{n^2}\right)_2 y^2 + \left(\frac{1}{n^2}\right)_3 z^2 + 2\left(\frac{1}{n^2}\right)_4 yz + 2\left(\frac{1}{n^2}\right)_5 xz + 2\left(\frac{1}{n^2}\right)_6 xy = 1. \tag{14.2}$$

The six indexed quantities (of the form $1/n^2$) are parameters which describe the effect of the electric field and reduce to $1/n_{x,y,z}^2$ or zero when the field is not present:

$$\begin{aligned}
\left(\frac{1}{n^2}\right)_1 &= \frac{1}{n_x^2} \qquad\qquad \text{(when } E = 0) \\[2mm]
\left(\frac{1}{n^2}\right)_2 &= \frac{1}{n_y^2} \\[2mm]
\left(\frac{1}{n^2}\right)_3 &= \frac{1}{n_z^2} \\[2mm]
\left(\frac{1}{n^2}\right)_4 &= \left(\frac{1}{n^2}\right)_5 = \left(\frac{1}{n^2}\right)_6 = 0.
\end{aligned} \tag{14.3}$$

Recall that the $x, y$ and $z$ in the above two equations refer to possible *polarization directions* of the optical wave. The electro-optic effect is described by the *changes* in the $(1/n^2)_i$ due to the electric field:

$$\Delta\left(\frac{1}{n^2}\right)_i = \sum_{j=1}^{3} r_{ij} E_j \qquad i = 1 \ldots 6. \tag{14.4}$$

Written out, the equation is

$$\begin{pmatrix} \Delta\left(\frac{1}{n^2}\right)_1 \\ \Delta\left(\frac{1}{n^2}\right)_2 \\ \Delta\left(\frac{1}{n^2}\right)_3 \\ \Delta\left(\frac{1}{n^2}\right)_4 \\ \Delta\left(\frac{1}{n^2}\right)_5 \\ \Delta\left(\frac{1}{n^2}\right)_6 \end{pmatrix} = \begin{pmatrix} r_{11} & r_{12} & r_{13} \\ r_{21} & r_{22} & r_{23} \\ r_{31} & r_{32} & r_{33} \\ r_{41} & r_{42} & r_{43} \\ r_{51} & r_{52} & r_{53} \\ r_{61} & r_{62} & r_{63} \end{pmatrix} \begin{pmatrix} E_x \\ E_y \\ E_z \end{pmatrix}. \tag{14.5}$$

The $6 \times 3$ matrix $r_{ij}$ is called the *electro-optic tensor* and is zero if the crystal is symmetric under a coordinate system inversion. To see why this is so, when $\boldsymbol{r} \to -\boldsymbol{r}$, $\boldsymbol{E} \to -\boldsymbol{E}$, but in a symmetric crystal $r_{ij}$ and $\Delta(1/n^2)_i$ are unchanged. Thus we have $\Delta(1/n^2)_i \to -\Delta(1/n^2)_i$ which can only be true if $r_{ij} = 0$ for all $i, j$.

Only a few elements of the electro-optic tensor are independent. Using the crystal symmetry properties, one can list the non-zero values of $r_{ij}$ for the more important crystal classes (as was done for the nonlinear coefficients, $d_{ij}$, in the last chapter):

- **Tetragonal, class** $\bar{4}2m$ – example, ADP, AD*P, KDP, KD*P

$$\begin{pmatrix} 0 & 0 & 0 \\ 0 & 0 & 0 \\ 0 & 0 & 0 \\ r_{41} & 0 & 0 \\ 0 & r_{41} & 0 \\ 0 & 0 & r_{63} \end{pmatrix} \tag{14.6}$$

- **Trigonal, class** $3m$ – example lithium niobate, lithium tantalate

$$\begin{pmatrix} r_{11} & 0 & r_{13} \\ -r_{11} & 0 & r_{13} \\ 0 & 0 & r_{33} \\ 0 & r_{42} & 0 \\ r_{42} & 0 & 0 \\ 0 & -r_{11} & 0 \end{pmatrix} \tag{14.7}$$

- **Hexagonal, class** 6 – example, lithium iodate

$$
\begin{pmatrix}
0 & 0 & r_{13} \\
0 & 0 & r_{13} \\
0 & 0 & r_{33} \\
r_{41} & r_{42} & 0 \\
r_{42} & -r_{41} & 0 \\
0 & 0 & 0
\end{pmatrix}
\tag{14.8}
$$

- **Orthorhombic, class** $mm2$ – example, potassium niobate, LBO

$$
\begin{pmatrix}
0 & 0 & r_{13} \\
0 & 0 & r_{23} \\
0 & 0 & r_{33} \\
0 & r_{42} & 0 \\
r_{51} & 0 & 0 \\
0 & 0 & 0
\end{pmatrix}
\tag{14.9}
$$

Table 14.1 lists the electro-optic tensor elements for the more important modulator materials.

**Table 14.1** Electro-optic tensor elements for a few important crystals (from Yariv (1989))

| Crystal | Point group symmetry | $r_{ij}$ (in $10^{-12}$ m/V) |
| --- | --- | --- |
| ADP | $\bar{4}2m$ | $r_{41} = 28$ |
| | | $r_{63} = 8.5$ |
| KDP | $\bar{4}2m$ | $r_{41} = 8.6$ |
| | | $r_{63} = 10.6$ |
| LiNbO$_3$ | $3m$ | $r_{13} = 8.6$ |
| | | $r_{22} = 3.4$ |
| | | $r_{33} = 30.8$ |
| | | $r_{42} = 28$ |
| LiTaO$_3$ | $3m$ | $r_{13} = 5.7$ |
| | | $r_{33} = 30.3$ |

## 14.3 Bulk electro-optic modulators

The *formal* approach to determine the behavior of an electro-optic crystal when an electric field is applied is to first *diagonalize* the quadratic form appearing in the equation for the index ellipsoid (eqn 14.2). This is simply a transformation to a coordinate system where the $yz$, $xz$ and $xy$ terms are zero. Then one can obtain the change in the relevant refractive index due to the field (using eqn 14.4) and the consequent change in the propagation behavior of a beam traversing the crystal. Due to the sparse nature of the electro-optic tensor, one can often determine the new axes by inspection. A simple example should illustrate the approach.

The material KDP is a tetragonal, class $\bar{4}2m$ crystal with only two independent electro-optic tensor elements. It is also a uniaxial crystal, whose ellipsoid equation in the presence of an electric field is

$$\frac{x^2 + y^2}{n_o^2} + \frac{z^2}{n_e^2} + 2r_{41}E_x yz + 2r_{41}E_y xz + 2r_{63}E_z xy = 1, \tag{14.10}$$

where the factors of 2 are due to their presence in the terms $(1/n^2)_i$ in the index ellipsoid equation for $i = 4, 5, 6$. We will first analyze a *longitudinal* electro-optic modulator, where both the field and wave propagation are along the $z$-axis. The equation for the index ellipsoid is

$$\frac{x^2 + y^2}{n_o^2} + \frac{z^2}{n_e^2} + 2r_{63}E_z xy = 1. \tag{14.11}$$

From the symmetry of this equation about both the $z$-axis and a plane through the $z$-axis that is tilted at $45°$ to the $x$-axis, it should be evident that the $xy$ term will be eliminated by a $45°$ rotation about the $z$-axis. The new coordinates (primed) are related to the original ones by

$$\begin{pmatrix} x' \\ y' \end{pmatrix} = \frac{1}{\sqrt{2}} \begin{pmatrix} 1 & 1 \\ -1 & 1 \end{pmatrix} \begin{pmatrix} x \\ y \end{pmatrix}. \tag{14.12}$$

Performing the substitution yields

$$\left( \frac{1}{n_0^2} + r_{63}E_z \right) x'^2 + \left( \frac{1}{n_o^2} - r_{63}E_z \right) y'^2 + \frac{z^2}{n_e^2} = 1. \tag{14.13}$$

The electric field breaks the degeneracy in the $xy$-plane, producing two new and *different* indices whose principal axes are at $45°$ to the original $x$ and $y$ axes. Assuming that the index change is small, and using

$$\Delta n = -\frac{n^3}{2} \Delta \left( \frac{1}{n^2} \right), \tag{14.14}$$

one obtains,

$$n_{x'} = n_o - \frac{n_o^3}{2} r_{63}E_z \tag{14.15}$$

$$n_{y'} = n_o + \frac{n_o^3}{2} r_{63}E_z, \tag{14.16}$$

while $n_z$ is unchanged. If we launch a wave along the $z$-axis that is polarized along the original $x$ direction, the waves whose polarizations are projected along the $x'$- and

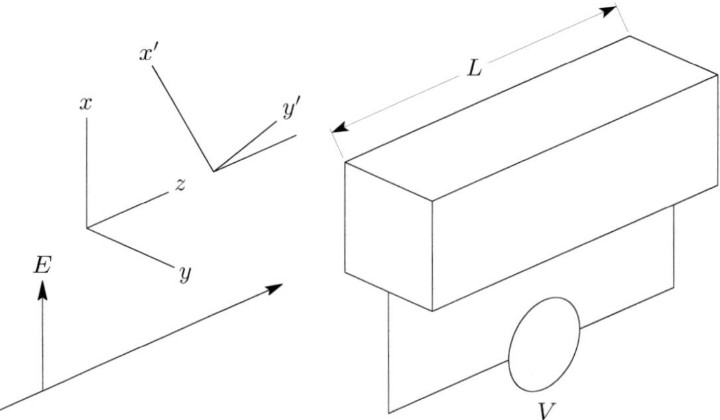

**Fig. 14.1** Figure illustrating geometry of longitudinal electro-optic modulator.

$y'$-axes will propagate at *different* speeds, generating an *elliptically* polarized wave. The two components are

$$E_{x'} = \frac{E_0}{\sqrt{2}} e^{i\omega t - i(\omega/c)n_{x'} z} \tag{14.17}$$

$$E_{y'} = \frac{E_0}{\sqrt{2}} e^{i\omega t - i(\omega/c)n_{y'} z}. \tag{14.18}$$

The *phase difference* between these waves is

$$\Delta\phi = \frac{\omega}{c}(n_{y'} - n_{x'})z = \frac{\omega n_o^3 r_{63} E_z}{c} z, \tag{14.19}$$

where the wave enters the crystal at $z = 0$. If the crystal length is $L$, the longitudinal field is approximately $V/L$ for an applied voltage $V$ and the phase difference is $\pi$ radians when

$$\text{Longitudinal:} \quad V_\pi = \frac{\lambda}{2n_o^3 r_{63}}. \tag{14.20}$$

This quantity is independent of the crystal length and is called the *half-wave voltage* of the crystal for operation as a longitudinal modulator. For KDP at the HeNe emission wavelength (633 nm), $V_\pi \approx 9$ kV, a very large value. The geometry is illustrated in Fig. 14.1. If a polarizer is placed after the crystal and aligned along the $y$-axis, the transmission of a vertically polarized beam will be zero when $V = 0$ and at its maximum value when $V = V_\pi$ since at the half wave voltage the modulator behaves like a half-wave-plate and rotates the polarization by twice the angle (90°) the fast axis (the $x'$-axis) makes with the electric field vector. Thus, the modulator/polarizer combination provides *amplitude modulation*. The modulation will not be a linear function of the voltage for large voltage excursions. If the polarization of the laser beam is along the $x'$- or $y'$-axis, the modulator will produce pure phase modulation with a phase deviation of

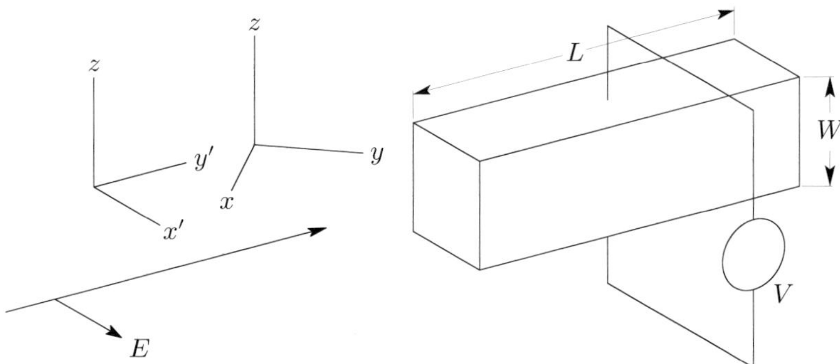

**Fig. 14.2** Figure illustrating geometry of transverse electro-optic modulator set up for pure phase modulation.

$$\text{Longitudinal:} \quad \phi_{deviation} = \frac{\pi n_o^3 r_{63} L}{\lambda} E_z = \frac{\pi n_o^3 r_{63}}{\lambda} V. \tag{14.21}$$

Longitudinal modulators are rarely used because of the large and impractical half-wave voltage. A significant reduction in $V_\pi$ can be obtained using the *transverse* configuration, where the electric field is applied perpendicular to the wave propagation direction. If this is done, the phase shift can be increased by increasing the *aspect ratio* (length/width) of the modulator until the aperture is too small to allow easy passage of the laser beam. A possible geometry in KDP is for the field to be applied along the $z$-axis and for the beam to propagate along either the $x'$- or $y'$-axis. The crystal will need to be *cut* so that the faces which are normal to the beam propagation direction are at 45° to the original $x$- or $y$-axis. The geometry is shown in Fig. 14.2. When used as a pure phase modulator (polarization perpendicular to applied field), the change in index is

$$|\Delta n'| = \frac{n_o^3}{2} r_{63} E_z, \tag{14.22}$$

and the phase deviation is

$$\text{Transverse:} \quad \phi_{deviation} = \frac{\pi n_o^3 r_{63}}{\lambda} \left( \frac{L}{W} \right) V, \tag{14.23}$$

where $W$ is the width of the crystal along the $z$-axis. Unlike for a longitudinal modulator, the phase shift depends upon the aspect ratio. The half-wave voltage is

$$\text{Transverse:} \quad V_\pi = \frac{\lambda}{n_o^3 r_{63}} \frac{W}{L}. \tag{14.24}$$

The half-wave voltage can be reduced to a practical value by using a sufficiently large aspect ratio.

Bulk modulators (those whose dimensions are much larger than the wavelength of light) typically do not have aspect ratios of greater than 20 and therefore have half-wave voltages of 100 volts or greater. For fixed frequency applications, a simple

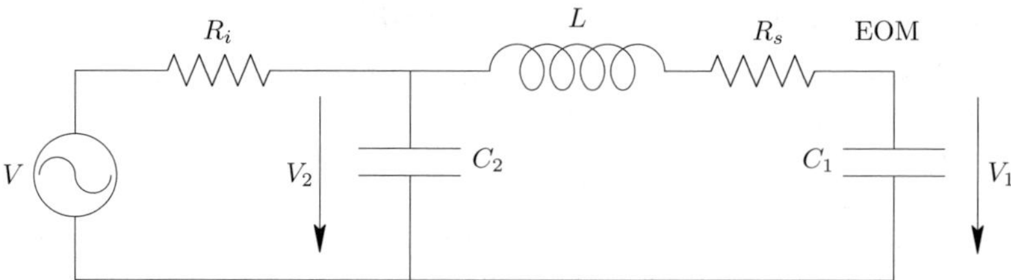

**Fig. 14.3** A simple circuit for driving an electro-optic modulator at a fixed frequency.

resonant step-up transformer is usually sufficient to drive such a modulator. A variant of the transformer which might be simpler to construct is a coil with a capacitive divider. It is actually exactly analogous to a step-up transformer, where the turns ratio is equal to the ratio of the capacitor values. Geometrically, it is a capacitively loaded pi-network and is shown in Fig. 14.3. The electro-optic modulator capacitance is $C_1$. The coil is chosen to resonate with $C_1$ at approximately the operating frequency. The input capacitor ($C_2$) is quite a bit larger than $C_1$ and is chosen to impedance match the network to the generator whose internal impedance is $R_i$. Matching will take place when

$$R_i = Q^2 \left(\frac{C_1}{C_2}\right)^2 R_s = R_p \left(\frac{C_1}{C_2}\right)^2 , \qquad (14.25)$$

where $R_p$ is the parallel resistance of the coil, which is $Q^2$ times the series resistance, $R_s$. The $Q$ of the coil is equal to its reactance divided by its series resistance. When the network is matched, the step-up ratio ($V_1/V_2$) is simply $C_2/C_1$. The criterion for the validity of this approach is

$$Q \gg \frac{C_2}{C_1}. \qquad (14.26)$$

This relation also establishes the maximum step-up ratio ($\approx Q$). These relations are easily derived from the rules governing the combination of impedances. However, there is a simple explanation which avoids algebra if we accept the analogy to a step-up transformer with turns ratio $C_2/C_1$. The usual transformation law from series to parallel resistance of a circuit element is $R_p = Q^2 R_s$ (when $Q \gg 1$). We assume that most of the voltage is dropped across $C_1$ (compared to that across $C_2$) so the parallel resistance can be considered to appear across $C_1$. The rule for transforming impedances across a transformer is to multiply by the square of the turns ratio, yielding the above expression for the matching to the generator. If we assume that the circulating current in the series resonant circuit consisting or $L$, $C_1$ and $C_2$ is much greater than the current injected by the generator (this is equivalent to the condition $Q \gg C_2/C_1$), then the two capacitors behave like a voltage divider giving us the step-up ratio of $C_2/C_1$.

This circuit is perfectly adequate for driving an electro-optic modulator at a fixed frequency whose associated wavelength is much greater than the length of the modulator. However, for very high frequencies or a large bandwidth, one must use a *traveling wave* modulator, described in the next section.

## 14.4 Traveling wave electro-optic modulators

At very high frequencies, the applied field can change significantly during the transit of the modulator by the laser beam. This *transit-time* effect will reduce the size of the phase deviation and occurs when the length of the modulator approaches the wavelength of the applied radio-frequency field. The solution is a *traveling wave modulator*: the modulating signal is launched onto a transmission line which is in close contact with the electro-optic material and travels in the same direction as the optical beam. Clearly, if the speed of the radio-frequency wave is the same as the optical wave, there will be no transit time effects. This is closely analogous to *phase matching* in sum frequency generation.

In analyzing a traveling wave modulator, we assume that the *optical phase change*, $d\phi$, due to a *radio-frequency* field, $E_{rf}(z)$, over a distance $dz$ is

$$d\phi = \alpha E_{rf}(z)dz, \tag{14.27}$$

where $\alpha$ is a constant. The *total* phase change, $\Delta\phi$, is obtained by integrating

$$\Delta\phi = \alpha \int_0^L E_{rf}(z)dz, \tag{14.28}$$

where $L$ is the crystal length and the entrance to the crystal is at $z = 0$. The radio-frequency traveling wave is a function of both $t$ and $z$ and is given by

$$E_{rf}(z,t) = E_{rf,0}e^{i(\omega_m n_m/c)z - i\omega_m t}, \tag{14.29}$$

where $n_m$ and $\omega_m$ are the refractive index and frequency of the traveling radio-frequency wave. One must eliminate the time dependence in performing the integral. This is done by recognizing that the *optical* wave has progressed a distance $z$ in time $t$ given by

$$z = \frac{c}{n}t \implies t = \frac{n}{c}z, \tag{14.30}$$

where we assume that it enters the crystal at $t = 0$ and where $n$ is the index for the *optical* wave. If we substitute the $t$ expression into the field equation, we are assuming that we are traveling at the same speed as an optical wavefront. The result of the substitution is

$$E_{rf}(z) = E_{rf,0}e^{i\omega_m z(n_m/c - n/c)} = E_{rf,0}e^{i\omega_m z\Delta n/c}, \tag{14.31}$$

where $\Delta n = n_m - n$. Integrating, we obtain

$$\begin{aligned}
\Delta\phi &= \frac{\alpha E_{rf,0}c}{i\omega_m \Delta n}(e^{i\omega_m \Delta n L/c} - 1) \\
&= \frac{2\alpha E_{rf,0}c}{\omega_m \Delta n}e^{i\omega_m \Delta n L/2c}\sin(\omega_m \Delta n L/2c) \\
&= \alpha E_{rf,0}Le^{i\omega_m \Delta n L/2c}\text{sinc}(\omega_m \Delta n L/2c).
\end{aligned} \tag{14.32}$$

Note that $\Delta\phi = \alpha E_{rf,0}L$ when either $\omega_m = 0$ (DC case) or $\Delta n = 0$ (perfectly phase-matched case). One can assume that the maximum *useful* bandwidth is from $\omega_m = 0$ to the frequency, $\omega_c$, where the argument of the sinc is $\pi/2$ ($\text{sinc}(\pi/2) = 0.64$).

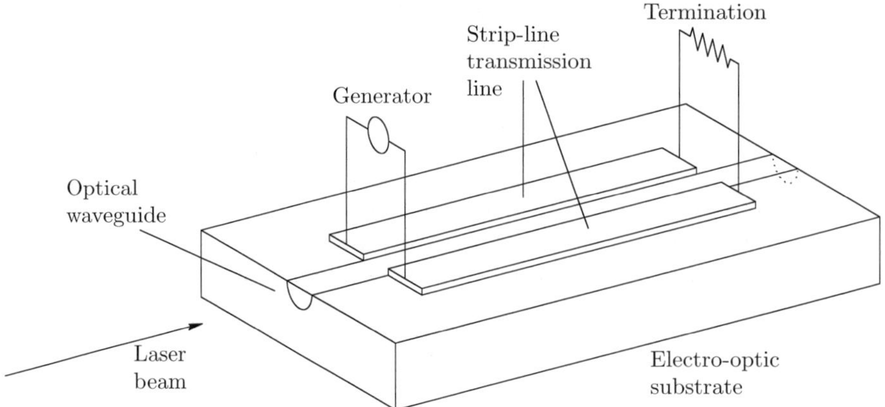



**Fig. 14.4** A traveling wave electro-optic modulator using an optical waveguide. It combines a bandwidth greater than 1 GHz with a half-wave voltage of about 10 V.

$$\text{Bandwidth:} \quad \omega_c = \frac{\pi c}{\Delta n L}. \tag{14.33}$$

The transit time limitation without a traveling wave is obtained by setting $n_m = 0$ (infinite RF speed; i.e., no traveling wave). Then the maximum frequency is

$$\text{Transit time limitation:} \quad \omega_c = \frac{\pi c}{n L}. \tag{14.34}$$

Observe that at the latter value of $\omega_c$, $L = \lambda/2$.

A typical traveling wave phase modulator is shown in Fig. 14.4. It is manufactured with a *strip line* transmission line for the radio-frequency wave and an optical waveguide for the optical field. The substrate is often lithium niobate, which works well at about 1.5 $\mu$m. In addition to the large bandwidth (at least several GHz), the small width of the strip line produces a very large electric field and allows a half-wave voltage as low as 10 V. One application of this device is in sweeping a frequency-locked laser over a wide frequency range. The laser can be locked by making one of the phase modulation sidebands resonant with one of the reference cavity's modes; changing the radio-frequency will sweep the laser without harming its frequency lock.

## 14.5 Acousto-optic modulators

In the previous section, we described phase modulation due to an electric field which causes a first-order change in the indices of refraction of a non-centrosymmetric crystal. A high frequency sound wave can also modulate the indices of refraction of a medium via the *acousto-optic effect* and a number of devices, called *acousto-optic modulators*, exploit this phenomenon to modulate the frequency, amplitude or propagation direction of a laser beam.

The amplitude of a sound wave in an elastic medium is described by the *strain tensor*, $S_{ij}$, a second-order tensor defined by

$$S_{ij} = \begin{cases} \dfrac{\partial \xi_i}{\partial x_j} + \dfrac{\partial \xi_j}{\partial x_i} & i \neq j \\[2ex] \dfrac{\partial \xi_i}{\partial x_i} & i = j, \end{cases} \tag{14.35}$$

where $\xi_i$ is the displacement of the medium in the direction of $x_i$ due to the sound wave and $x_i$ is a Cartesian coordinate. This is a *symmetric tensor* and therefore has only six independent elements. When describing the acousto-optic effect, it is the usual practice to redefine the strains in analogy with the six coefficients of the general equation for an index ellipsoid:

$$\begin{aligned} S_1 &= S_{11} \\ S_2 &= S_{22} \\ S_3 &= S_{33} \\ S_4 &= S_{23} \\ S_5 &= S_{31} \\ S_6 &= S_{12}. \end{aligned} \tag{14.36}$$

Hopefully, the distinction between the singly subscripted quantities and doubly subscripted quantities will not be confusing. Off-diagonal strains ($S_{ij}, i \neq j$) describe a *shear wave*, and diagonal strains describe a *compression wave*. The type of wave propagating in the medium will depend upon the characteristics of the *transducer* which launches the wave. We describe the acousto-optic effect using an equation similar to that for the electro-optic effect:

$$\Delta \left( \frac{1}{n^2} \right)_i = \sum_j p_{ij} S_j, \tag{14.37}$$

where the material constants, $p_{ij}$, are called the *strain-optic coefficients* and the $1/n^2$ quantities are defined in the same way as in eqn 14.2. In an isotropic solid, most of these constants vanish and one can readily show that the non-vanishing coefficients satisfy

$$p_{11} = p_{22} = p_{33} \tag{14.38}$$

$$p_{12} = p_{21} = p_{13} = p_{31} = p_{23} = p_{32} \tag{14.39}$$

$$p_{44} = p_{55} = p_{66} = (p_{11} - p_{12})/2. \tag{14.40}$$

All of the remaining coefficients are zero. Thus, there are only two independent coefficients ($p_{11}$ and $p_{12}$) in this case.

The geometry of a simple acousto-optic device is shown in Fig. 14.5. The sound wave, whose frequency, wave vector and wavelength are $\Omega, K$ and $\Lambda$, is launched by a transducer in the positive $x$-direction. An absorber at the top of the modulator ensures that the wave is a *traveling wave*. The laser beam is normally incident from the left along the $z$-direction. We will consider light polarizations along the $y$-axis (perpendicular to the plane of the figure) or along the $x$-axis. The spatial refractive index

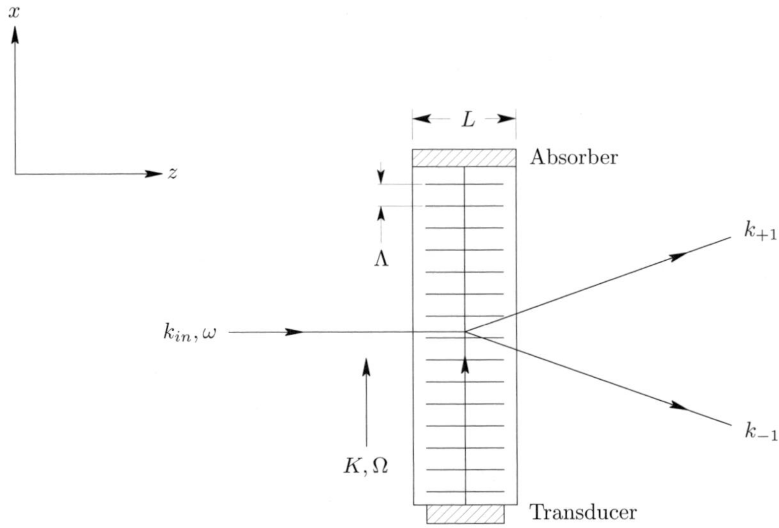



**Fig. 14.5** Geometry of acousto-optic modulator in *Raman–Nath* regime.

modulation described by eqn 14.37 will form a moving, *phase grating* and initially the grating will be assumed to be very thin, which is the requirement for so-called *Raman–Nath* (also called Debye–Sears) diffraction. We will shortly quantify the meaning of a "thin" grating.

We will assume that the sound field is a *compressive* wave traveling in an isotropic medium, so that the only non-zero strain component is $S_1$. From eqn 14.37, the changes to the refractive index for the two possible light polarizations are

$$x\text{-polarization:} \quad \Delta n_1 = -\tfrac{1}{2}n_1^3 p_{11} S_1 \tag{14.41}$$

$$y\text{-polarization:} \quad \Delta n_2 = -\tfrac{1}{2}n_2^3 p_{12} S_1, \tag{14.42}$$

where we have used $\Delta(1/n^2) = -2\Delta n/n^3$. To allow both cases, we will lump the constants into a constant $C$:

$$\Delta n(x, z, t) = CS(x, z, t), \tag{14.43}$$

where $S$ is the sound wave amplitude and we include explicit time and space dependencies in $\Delta n$ and $S$. For a sound wave traveling in the $x$-direction, $S$ is

$$S(x, z, t) = S_0 e^{iKx - i\Omega t}. \tag{14.44}$$

The electric field, $E(x, z, t)$, of a light wave propagating in the $z$-direction in the absence of a sound wave is

$$\text{No sound wave:} \quad E(x, z, t) = E_0 e^{ink_0 z - i\omega t}, \tag{14.45}$$

where $n$ is the refractive index of the medium and $k_0$ is the wave vector $(\omega/c)$ of the light. In the presence of a sound wave,

$$\text{With sound wave:} \quad E(x, z, t) = E_0 e^{i(n+\Delta n)k_0 z - i\omega t}. \tag{14.46}$$

The explicit time and spatial dependence of $\Delta n$ is

$$\Delta n = CS_0 e^{iKx - i\Omega t} = \Delta n_0 e^{iKx - i\Omega t}. \tag{14.47}$$

If we let $\phi(x, L, t)$ be the total phase shift experienced by the light as it traverses the crystal,

$$\phi(x, L, t) = nk_0 L + k_0 \int_0^L \Delta n(x, z, t)dz = nk_0 L + k_0 L \Delta n_0 e^{iKx - i\Omega t}, \tag{14.48}$$

where we assume that $L \ll 2\pi/K$ in evaluating the integral. By taking the real part of this expression, we see that there is a sinusoidal "corrugation" of the phase shift in the $x$-direction; this structure acts like a *phase grating*. The electric field at the exit plane is

$$E(x, L, t) = E_0 e^{i\phi(x,L,t) - i\omega t} = E_0 e^{ink_0 L - i\omega t} e^{ik_0 L \Delta n_0 \cos(Kx - \Omega t)}, \tag{14.49}$$

where we have replaced the exponential expression for the phase shift by its real part. The second exponential can be expanded in Bessel functions using the well-known relation,

$$e^{i\beta \cos\theta} = \sum_{m=-\infty}^{\infty} i^m J_m(\beta) e^{im\theta}. \tag{14.50}$$

The result is

$$E(x, L, t) = E_0 e^{-i\omega t} \sum_{m=-\infty}^{\infty} i^m J_m(k_0 L \Delta n_0) e^{i(nk_0 L + mKx - m\Omega t)}. \tag{14.51}$$

The result of the modulation will be the generation of a number of beams at different frequencies and angles relative to the input beam. At the exit face of the medium, the $k$-vector of the $m^{th}$ beam has an $x$-component given by $mK$ and a $z$-component given by $nk_0$. Converting these to angles relative to the input beam, we have the following association between the beam frequencies and exit angles, $\theta_m$:

$$\omega + m\Omega \Longleftrightarrow \theta_m = \tan^{-1} \frac{mK}{nk_0} \approx \frac{m\lambda}{\Lambda}, \tag{14.52}$$

where $\lambda = 2\pi/nk_0$ and we assume that $\theta_m \ll 1$. The intensities of the various orders are determined by the argument to the Bessel functions; this argument is called the *Raman–Nath parameter*, $\nu$:

$$\text{Raman–Nath parameter:} \quad \nu = k_0 L \Delta n_0. \tag{14.53}$$

The electric field, $E_m$, of the $m^{th}$ order at $z = L$ is

$$E_m = E_0 (i)^m J_m(\nu) e^{ink_0 L + imKx - i(\omega + m\Omega)t}. \tag{14.54}$$

Two assumptions were made in the above calculation. First, it was assumed that the width of the sound field ($L$ in the figure) is small enough to ignore optical diffraction.

We assume that the optical wave is a parallel plane wave and does not suffer significant diffraction. Quantitatively, a "piece" of the wave of width $\delta x$ will suffer an angular spread of $\lambda/\delta x$ and will have a width of $\delta x + L\lambda/\delta x$ at the exit face. We choose $\delta x$ to minimize the total width at $z = L$ and find that $\delta x = \sqrt{L\lambda}$ and the total width at the exit face is $2\sqrt{L\lambda}$. We can ignore optical diffraction when the spread is less than the wavelength $\Lambda$ of the sound wave:

$$\text{Criterion for Raman–Nath diffraction:} \quad L \ll \frac{\Lambda^2}{4\lambda}. \tag{14.55}$$

It is customary to use the *Klein–Cook parameter*, $Q$, to distinguish Raman–Nath diffraction from the alternative (called *Bragg* diffraction):

$$\text{Klein–Cook parameter:} \quad Q \equiv \frac{2\pi L\lambda}{\Lambda^2} = L\frac{K^2}{k}. \tag{14.56}$$

Using $Q$, Raman–Nath diffraction occurs when

$$\text{Raman–Nath diffraction:} \quad Q \ll 1. \tag{14.57}$$

The second assumption is that $\Delta n_0$ is small enough that there is no *additional bending* of the optical wave due to the transverse refractive index gradient. We will not belabor this point since our focus is on Bragg diffraction where this phenomenon is largely absent. In the Raman–Nath regime, we can ignore index gradients when

$$\text{Criterion for ignoring ray bending:} \quad \nu \ll 2/Q. \tag{14.58}$$

Since $Q \ll 1$ in the Raman–Nath regime, the Raman–Nath parameter can be somewhat greater than unity and many diffraction orders can be present in the output beam (only two are shown in the figure). As an example of Raman–Nath diffraction, consider $TeO_2$, a common material in acousto-optic modulators. The sound velocity, $v_m$, in this material is $4.2 \times 10^3$ m/s and the refractive index is about 2.26. The maximum modulation frequency, $f_m$, at 633 nm for $L = 1$ cm is

$$\text{Raman–Nath diffraction in 1 cm } TeO_2: \quad f_m < v_m\sqrt{\frac{n}{2\pi L\lambda_0}} = 31.7 \text{ MHz}, \tag{14.59}$$

where $v_m$ is the speed of the sound wave in the medium. We thus see that Raman–Nath diffraction is a relatively low-frequency phenomenon compared to Bragg diffraction, which occurs at frequencies greater than $\approx 100$ MHz.

Before leaving the topic of Raman–Nath diffraction, we will discuss a *figure of merit* which will help in comparing the diffraction efficiencies of various modulator materials. From eqn 14.42, we see that $\Delta n \propto n^3 pS$. For *weak* diffraction ($\nu < 1$), only first order diffraction will occur and $J_1(\nu) \propto \nu$. Thus, the *power*, $P$, in the diffractive wave is proportional to

$$P \propto (\Delta n)^2 \propto n^6 p^2 S^2, \tag{14.60}$$

where the first proportionality is due to $E \propto J_1(\nu) \propto \nu \propto \Delta n$ when $\nu < 1$. From acoustics, the intensity, $I_s$, of the acoustical wave is given by

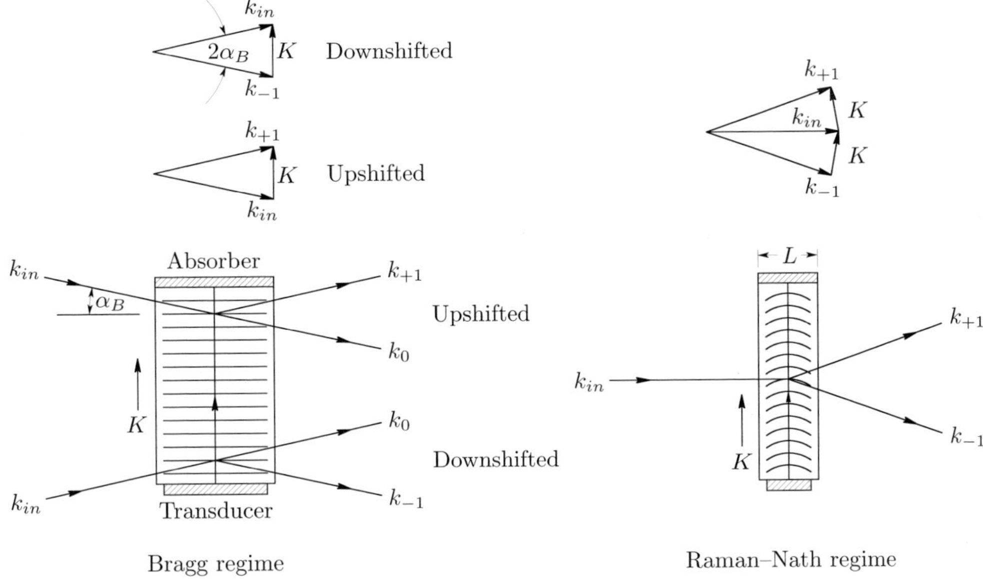

Bragg regime         Raman–Nath regime

**Fig. 14.6** Geometry of acousto-optic modulator in Bragg regime (left) and Raman–Nath regime (right).

$$I_s = \tfrac{1}{2}\rho v_m^3 S^2, \tag{14.61}$$

where $\rho$ is the density of the material. The ratio of the diffracted beam power to the acoustical intensity is proportional to the figure of merit, $M_2$:

$$\text{Figure of merit:} \quad M_2 = \frac{n^6 p^2}{\rho v_m^3}. \tag{14.62}$$

The strong dependence on the refractive index favors optically dense materials for acousto-optic devices.

At higher acoustical frequencies, the above approximations break down and we are in the so-called Bragg regime. Before addressing the theory of Bragg diffraction analytically, we will obtain the basic results using the model of a *photon–phonon collision*. This model is illustrated in Fig. 14.6 for both the Raman–Nath and Bragg regimes. We require that momentum be conserved in a collision between the incoming photon and the phonon due to the acoustical wave:

$$\hbar \boldsymbol{k}'_{photon} = \hbar \boldsymbol{k}_{photon} + \hbar \boldsymbol{K}_{phonon}, \tag{14.63}$$

where the prime refers to the photon after the collision. Since the photon momenta ($\hbar k$, $\hbar k'$) are much greater than the phonon momentum ($\hbar K$), the *magnitude* of the photon momentum is essentially unchanged in the collision while the angular deflection is $2\alpha_B$. From this, we obtain the *Bragg condition*

$$\text{Bragg condition:} \quad \sin \alpha_B = \frac{K}{2k} = \frac{\lambda}{2\Lambda}, \tag{14.64}$$

where we have dropped the subscripts and distinguish the photon quantities from the phonon quantities by case. This is the familiar expression for Bragg diffraction of X-rays in crystals where the $\Lambda$ is one of the lattice plane spacings. One difference from the X-ray case is that acousto-optic Bragg scattering only generates a *single* diffracted order while there are many higher orders in X-ray diffraction. This is due to the *sinusoidal* modulation of the refractive index in acousto-optic diffraction; in X-ray diffraction, the scattering from each diffraction plane occurs over a very small distance which, from Fourier theory, results in additional scattering from higher order harmonics of the inverse of the lattice plane spacing (a *spatial frequency*). Of course, in the Raman–Nath regime, multiple orders are possible since the medium behaves like a thin phase grating and the resulting *phase modulation* can generate multiple orders when the modulation index ($\nu$) is large enough. Unlike Raman–Nath diffraction, where a normally incident beam will *always* generate a set of diffracted beams, in the Bragg regime the input beam direction must be close to the Bragg angle ($\alpha_B$) for a significant diffracted beam to emerge (the allowed angular error is a user exercise). Note that we are ignoring the additional angular changes that occur when the beam exits the modulator medium due to refraction. These angles are easily determined using Snell's law.

The *frequency shift* of the diffracted beam can be explained by the *Doppler shift* due to the moving sound wave (it can also be explained using *energy conservation* in the scattering process). An observer facing the diffracted wave will see a component of the sound velocity given by $v_m \sin \alpha_B$. The Doppler shift of the light due to this is

$$\text{Doppler shift} = \frac{c}{\lambda} \frac{v_m \sin \alpha_B}{c} = \frac{c}{\lambda} v_m \frac{\lambda}{2\Lambda c} = \frac{\Omega}{2}. \tag{14.65}$$

The incoming wave will see the *negative* of this same Doppler shift with the result that the net Doppler shift will be $\Omega$ for the upshifted beam (and $-\Omega$ for the downshifted one).

The right side of Fig. 14.6 illustrates Raman–Nath diffraction using the same collision model. It is interesting to note that an alternative (but equivalent) explanation for the criterion for Raman–Nath diffraction can be obtained from this model. Instead of considering diffraction of the *light beam*, we examine the diffraction of the *sound wave*. For small $L$ or low acoustical frequencies, the sound wavefronts will no longer be plane but will have some curvature, as shown in the figure. From the vector diagram, one can see that we will obtain two simultaneous diffraction orders when the angle between $k_{+1}$ and $k_{-1}$ is equal to the angle between the two $K$ vectors which are normal to the sound wavefronts on each side of the curved wavefronts. From simple diffraction theory, the latter is about $\Lambda/L = 2\pi/KL$ and, from geometry, the former is about $2K/k$. Equating these, we obtain

$$\text{Raman–Nath:} \quad \frac{2\pi}{LK} \approx \frac{2K}{k} \implies L \approx \frac{\pi k}{K^2}, \tag{14.66}$$

which is essentially the same upper limit on the length as was obtained earlier. Thus, Bragg diffraction will occur when the sound wavefronts are planes and Raman–Nath diffraction will occur when there is enough wavefront curvature to allow two or more simultaneous diffracted orders. The criterion for Bragg diffraction can be expressed using the Klein–Cook parameter:

$$\text{Bragg regime:} \quad Q \gg 1. \tag{14.67}$$

A theoretical analysis of Bragg diffraction is somewhat more complicated than that of Raman–Nath diffraction. We will use the *coupled wave* approach, which is often used to describe the propagation of two coupled modes in an optical waveguide. We used a very similar approach in analyzing a distributed feedback semiconductor laser. We will make the *small angle approximation,* which assumes that the incoming wave makes an angle of much less than 1 radian with the $z$-axis. We start with the scalar wave equation for each electric field component and substitute into it the sum of the three possible wave solutions (incoming wave and two possible diffracted waves) together with a spatially modulated dielectric permittivity. The latter is given by

$$\epsilon = \epsilon_u + \Delta\epsilon \sin(\Omega t - Kx), \tag{14.68}$$

where $\epsilon_u$ is the permittivity in the absence of the sound wave. In the absence of sources, each electric field component satisfies the *scalar wave equation*

$$\nabla^2 E - \frac{1}{v_\phi^2}\frac{\partial^2 E}{\partial t^2} = \nabla^2 E - \mu_0\epsilon\frac{\partial^2 E}{\partial t^2} = 0, \tag{14.69}$$

where $E$ is any field component and $v_\phi$ is the phase velocity of the wave in the medium. The use of the scalar wave equation obscures polarization issues but it is simple and will provide the desired result. If we substitute the spatially modulated permittivity, we obtain

$$\nabla^2 E - \mu_0\epsilon_u\frac{\partial^2 E}{\partial t^2} \approx \mu_0\Delta\epsilon \sin(\Omega t - Kx)\frac{\partial^2 E}{\partial t^2}, \tag{14.70}$$

where we made the simplifying assumption that $\Omega \ll \omega$ and consider the right-hand side to be a *source* term (it is actually a spatially modulated polarization).

Using our knowledge of what to expect from Bragg diffraction, we assume there are only two possible outgoing waves: one at frequency $\omega + \Omega$ and a second at frequency $\omega - \Omega$. The total electric field will therefore be of the form

$$E = E_0 + E_+ + E_-, \tag{14.71}$$

where

$$E_0 = A_0(x, z)e^{i\omega t - i\boldsymbol{k}\cdot\boldsymbol{r}} \tag{14.72}$$

$$E_+ = A_+(x, z)e^{i(\omega+\Omega)t - i\boldsymbol{k}^+\cdot\boldsymbol{r}} \tag{14.73}$$

$$E_- = A_-(x, z)e^{i(\omega-\Omega)t - i\boldsymbol{k}^-\cdot\boldsymbol{r}}, \tag{14.74}$$

where $A_0$ and $A_\pm$ are slowly varying functions of the coordinates. We will leave out some of the details in the following, but will undertake the following operations:

- We will eliminate the $y$-dependence in the differential equation.

- We will substitute $E$ into the differential equation and neglect terms proportional to $\partial^2 A/\partial x^2$ and $\partial^2 A/\partial z^2$ since the amplitudes are slowly varying over the distance of an optical wavelength.
- We will move everything to the left-hand side and equate the coefficients of $e^{i\omega t}$ and $e^{i(\omega \pm \Omega)t}$ to zero since these exponentials are linearly independent.
- Finally, we will assume that the amplitudes depend only on $z$ and therefore $\partial A/\partial x \approx 0$

It turns out (as expected) that it is not possible to obtain a solution which includes *both* $E_+$ and $E_-$. In the following we will treat the case of a non-zero $E_+$ $(E_- = 0)$.

After a bit of algebra, two coupled differential equations emerge:

$$\frac{d\tilde{A}_0}{dz} = \kappa \tilde{A}_+ e^{i\Delta k_z z} \tag{14.75}$$

$$\frac{d\tilde{A}_+}{dz} = -\kappa \tilde{A}_0 e^{-i\Delta k_z z}, \tag{14.76}$$

where

$$\kappa = \frac{\omega^2 \mu_0 \Delta\epsilon}{4\sqrt{k_z k_z^+}} \tag{14.77}$$

$$\Delta k_z = k_z - k_z^+ \tag{14.78}$$

$$\tilde{A}_0 = \sqrt{\frac{k_z}{2\omega\mu_0}} A_0 \tag{14.79}$$

$$\tilde{A}_+ = \sqrt{\frac{k_z^+}{2\omega\mu_0}} A_+. \tag{14.80}$$

The quantity $\kappa$ is the *coupling constant* between the incident and diffracted waves due to the sound wave and the square modulus of the tilde amplitudes are equal to the *powers* in the incident and diffracted waves.

The two coupled equations are solved by differentiating one and substituting the other to obtain a second-order differential equation. The latter is solved in the usual way subject to the initial conditions:

$$\tilde{A}_0(z = 0) = 1 \tag{14.81}$$
$$\tilde{A}_+(z = 0) = 0, \tag{14.82}$$

where we assume that the incident power is unity. The solutions are

$$P_0(z) = |\tilde{A}_0|^2 = \cos^2 \delta z + \left(\frac{\Delta k_z}{2\delta}\right)^2 \sin^2 \delta z \tag{14.83}$$

$$P_+(z) = |\tilde{A}_+|^2 = \left(\frac{\kappa}{\delta}\right)^2 \sin^2 \delta z, \tag{14.84}$$

where

$$\delta = \sqrt{\kappa^2 + (\Delta k_z)^2/4}. \tag{14.85}$$

When the diffractive wave makes the same angle with the $z$-axis as the incident wave (i.e., satisfies the Bragg condition), $\Delta k_z = 0$ and the powers are

$$P_0(z) = \cos^2 \kappa z \tag{14.86}$$

$$P_+(z) = \sin^2 \kappa z. \tag{14.87}$$

From eqn 14.42 and the fact that $\epsilon = \epsilon_0 n^2$:

$$|\Delta \epsilon| = \epsilon_0 n^4 p S. \tag{14.88}$$

Substituting this into eqn 14.77, we obtain

$$\kappa = \frac{\omega n^3 p S}{4c \cos \alpha_B}, \tag{14.89}$$

where $k_z = k_z^+$ and we are assuming incidence at the Bragg angle, so $k_z = k_z^+ = (\omega/c) n \cos \alpha_B$. Using the expressions for the sound intensity (eqn 14.61) and the figure of merit (eqn 14.62), we obtain

$$\kappa = \frac{\pi}{\sqrt{2}\lambda_0 \cos \alpha_B} \sqrt{M_2 I_s}. \tag{14.90}$$

One will obtain unity diffraction efficiency (theoretically) when $\kappa L = \pi/2$. The corresponding sound intensity is

$$\text{Maximum Bragg diffraction:} \quad I_s = \frac{\lambda_0^2 \cos^2 \alpha_B}{2 M_2 L^2}. \tag{14.91}$$

Finally, the acoustical power is obtained by multiplying this by $LH$ where $H$ is the width of the modulator (its extent in the $y$-direction):

$$\text{Maximum Bragg diffraction:} \quad P_s = \frac{\lambda_0^2 \cos^2 \alpha_B}{2 M_2} \frac{H}{L}. \tag{14.92}$$

The acoustical power requirements are proportional to $\lambda_0^2$; hence, it is more difficult to modulate infrared beams than visible beams. Since $P_s$ is inversely proportional to $M_2$, one would like a medium with a large index of refraction ($M_2 \propto n^6$) and a low sound speed (though this will harm the modulation bandwidth).

Acousto-optic modulators have two *bandwidths*. One is the *modulation bandwidth*, which is the maximum modulation frequency of an amplitude modulated drive signal for which the modulation will be transferred to the diffracted beam. The second bandwidth is the *range of drive frequencies* that will produce a diffracted beam. The first is approximately the inverse of the transit time of the sound wave across the (focused) laser beam. If one focuses to a waist $\omega_0$, the modulation bandwidth is approximately

$$\text{Modulation bandwidth} \approx \frac{v_m}{2\omega_0}. \tag{14.93}$$

Increasing the modulation bandwidth by tightening the focus can have the undesired side effect of reducing the diffraction efficiency since the angular spread of the input

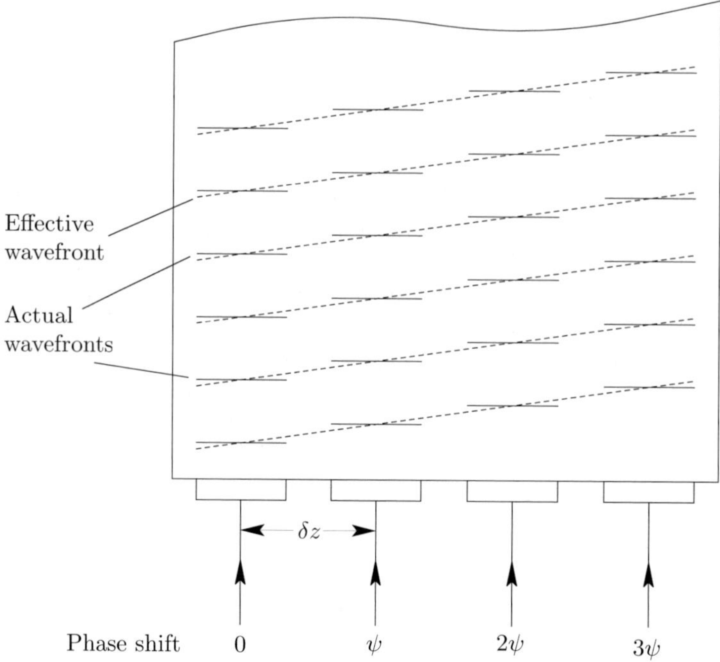


**Fig. 14.7** Phased array of transducers used to increase frequency range of Bragg regime acousto-optic modulator. The dashed line is the *effective* tilted wavefront generated by the several phase-shifted transducers.

beam will be large enough that a portion of the input beam will not satisfy the Bragg condition. The other bandwidth (range of drive frequencies) is due to the violation of the Bragg condition when the drive frequency is changed. Although the spread of the sound wave is not as extreme as in a Raman–Nath modulator, there is some angular spread, given by $\approx \Lambda/L$. The change of the Bragg angle due to a change in drive frequency of $\Delta f_m$ is $(\lambda/2v_m)\Delta f_m$ where $f_m$ is the average drive frequency. Equating these and solving for $\Delta f_m$

$$\text{Tuning bandwidth:} \quad \Delta f_m \approx \frac{2\Lambda v_m}{\lambda L} = \frac{2\Lambda^2}{\lambda L} f_m. \tag{14.94}$$

One unfortunate consequence of this relationship is that $L$ must be decreased to obtain a large $\Delta f_m$ and this reduces the diffraction efficiency. A solution is to use a *phased array transducer*, shown schematically in Fig. 14.7. Four transducers are shown in the figure; a larger or smaller number can be used. Phase shifters are provided for each transducer so that adjacent transducers differ in phase by $\psi$ radians. A phase shift of $\psi$ corresponds to a *wavefront* displacement of $\Lambda(\psi/2\pi)$ and an angular *tilt* of the composite wavefront by angle $(\Lambda/\delta z)(\psi/2\pi) = \psi v_m/(2\pi\Omega\delta z)$, where $\delta z$ is the separation between the transducers. This angle is inversely proportional to the sound frequency, $\Omega$. Since the Bragg angle is an arcsine of the frequency, the tracking will not be exact. A tolerable strategy is to match the angles at the end of the range, $\Delta f_m$. The transducer separation needed for this is

$$\text{Tracking at endpoints of range:} \quad \delta z = \frac{n v_m^2}{(f_0^2 - \Delta f_m^2/4)\lambda}, \tag{14.95}$$

where $f_0 = \Omega_0/2\pi$ is the center frequency in Hz.

We will end this section with a discussion of the applications of acousto-optic modulators. The most obvious application is as an *amplitude modulator*. The radio-frequency drive signal is modulated (perhaps using an inexpensive double-balanced mixer) and the power in the diffracted beam will be a monotonic function of the signal sent to the radio-frequency modulator. Unfortunately, the modulation is not *linear* since the power in the diffracted beam is proportional to $\cos^2 \kappa L$ and $\kappa$ is proportional to the *amplitude* of the transducer drive. For small signal modulation, one should operate the modulator at the most linear point in its diffraction efficiency function: this occurs when $\kappa L \approx \pi/4$. Acousto-optic modulators have been very successfully used in this way to stabilize the intensity of a laser beam.

Bragg modulators are also used to deflect laser beams by a controlled angle. The input (and diffracted) angle is set to the middle of the acoustical frequency range and the input beam angle is held constant as the sound frequency is swept. The *change* in the diffracted beam angle, $\delta\alpha$, for a change in the sound frequency, $\delta\Omega$, is simply

$$\delta\alpha = \frac{\lambda}{v_m}\delta\Omega. \tag{14.96}$$

The maximum angular sweep range and the modulation bandwidth are determined by the bandwidth considerations discussed above.

Perhaps the most common application of an acousto-optic modulator in an atomic physics laboratory is as a *frequency shifter*. The diffracted beam will be frequency shifted by exactly $\pm\Omega$ as discussed above. One of the undesirable side effects of sweeping the frequency of a laser beam in this way is the accompanying *angular change* in the diffracted beam. The usual solution is to use the modulator in the *double-pass configuration* shown in Fig. 14.8. The basic idea is to reflect the diffracted beam back along its path and into the modulator for a second pass. Since the second pass makes exactly the same angle with the acoustical wave as the first pass, the beam will receive the same magnitude and sign of frequency shift as it did in the first pass. The beam, as it emerges the second time, will be collinear with incident beam for all frequency shifts. Thus, the net frequency shift will be twice the acoustical frequency and the angular shift will be zero. The beam can be separated from the incident beam using the combination of polarizing beam splitter and quarter-wave plate shown in the figure; its operation is explained in Chapter 4.

The key to the operation of the device is a means of ensuring that the beam be exactly retro-reflected back into the modulator for any diffraction angle. This is accomplished using a modified *cat's-eye retro-reflector* consisting of the lens and mirror shown in the figure. Usually, cat's-eye retro-reflectors position the mirror one focal length from the lens so that a collimated beam incident on the lens will yield a reflected collimated beam at the same angle as the incident beam but displaced from it by a

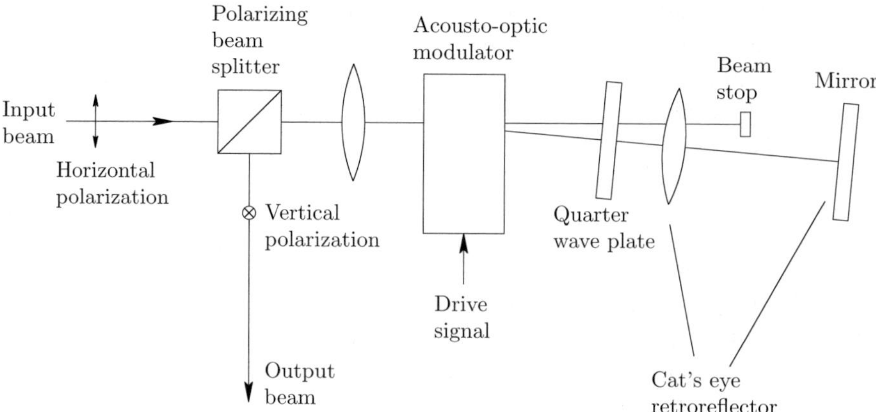


**Fig. 14.8** Double-pass scheme for doubling frequency shift and eliminating angular shift in acousto-optic modulator used as a frequency shifter.

small distance. A better arrangement is to place the lens *one focal length from the center of the modulator*. Since the incident beam is focused into the modulator, the beam will emerge collimated from the second lens and it can be sent normal to the mirror. The reflected beam will then enter the modulator at the ideal angle for the second pass. One can replace the mirror/lens with a concave mirror whose center of curvature is at the center of the modulator.

## 14.6 Further reading

The electro-optic effect and associated devices are well covered in the books by Yariv (1989) and Davis (1996). In particular, the latter has a good, though short, discussion of traveling wave electro-optic modulators. Acousto-optic devices are adequately described in the two above texts. An excellent discussion of the distinction between the Raman–Nath and Bragg regimes together with a treatment of the Bragg regime using a coupled wave analysis appear in the book by Ghatak and Thyagarajan (Ghatak (1989)). A useful book devoted entirely to acousto-optics is by Korpel (1997). There are numerous papers on acousto-optics; two which were useful are Korpel (1981) and Young and Yao (1981). A description of a double-pass acousto-optic modulator appears in the paper by Donley, et al. (2005).

## 14.7 Problems

(14.1) Lithium niobate is a negative uniaxial crystal which is often used as an electro-optic phase modulator. Derive an expression for the phase shift in a crystal of length $l$ and width $d$ in terms of the electro-optic coefficients $r_{13}$ and $r_{33}$ and the indices of refraction ($n_e$ and $n_o$) for a wave propagating in the $y$-direction and whose polarization makes a $45°$ angle with the $z$-axis in the $xz$-plane. The electric field is applied in the $z$-direction, which is also the optical axis. Using this expression, find the half-wave voltage for a HeNe laser beam ($\lambda = 633$ nm) in a crystal where $l = 10$ mm and $d = 0.5$ mm (use the Sellmeier equation to determine the refractive indices).

(14.2) For the modulator in problem 1, the light is now polarized along the $z$-axis, which will result in pure phase modulation (unlike in the previous problem). What is the half-wave voltage for this case?

(14.3) Assume that we have a 6 MHz acoustical wave propagating in water at a speed of 1500 m/s. What is the maximum width of the modulator "cell" for Raman–Nath diffraction? If the cell is 1 cm wide, show that the diffraction is Raman–Nath and calculate the angle between the direct beam (perpendicular to the sound wave) and the first diffracted wave.

(14.4) An acousto-optic modulator is constructed from dense flint glass, which has the following parameters: $n = 1.92$, $p = 0.25$, $v_m = 3.1 \times 10^3$ m/s and $\rho = 6.3$ kg/m$^3$. Assume that the modulator is 50 mm long and 2 mm wide and the acoustic frequency is 40 MHz. Calculate the figure of merit, $M_2$, and the Bragg diffraction angle for a HeNe beam (633 nm). What is the acoustical power for maximum diffraction?

(14.5) Determine the allowed *angular* deviation from the Bragg condition of an acousto-optic modulator where the diffraction efficiency is reduced by a factor of two.

# References


Ashcroft, N. W. and Mermin, N. D., *Solid State Physics* (Brooks Cole, 1976).

Ashkin, A., Boyd, G. D. and Dziedzic, J. M., "Resonant Optical Second Harmonic Generation and Mixing", IEEE Journ, Quant. Elect. **2**, 109 (1966).

Born, M. and Wolf, E., *Principles of optics* (6th Ed, Pergamon Press, 1980).

Boyd, G. D. and Kleinman, D. A., "Parametric Interaction of Focused Gaussian Light Beams", Journ. Appl. Phys. **39**, 3597 (1968).

Byer, R. L. and Harris, S. E., "Power and Bandwidth of Spontaneous Parametric Emission", Phys. Rev. **168**, 1064 (1968).

Byer, R. L., "Diode-pumped Solid-State Lasers", Science **239**, 742 (1988).

Davis, C. C., *Lasers and Electro-Optics* (Cambridge University Press, 1996).

Day, T., Gustafson, E. K., and Byer, R. L., "Sub-Hertz Relative Frequency Stabilization of Two-Diode Laser-Pumped Nd:YAG Lasers Locked to a Fabry-Perot Interferometer", IEEE Journ. Quant. Elect. **28**, 1106 (1992).

Donley, E. A., Heavner, F. L., Tataw, M. O. and Jefferts, S. R., "Double-pass acousto-optic modulator system", Rev. Sci. Instrum. **76** 063112 (2005).

Dorf, R. C., *Modern Control Systems*, (Addison-Wesley, 1967).

Drever, R. W. P., Hall, J. L., Kowalski, F. V., Hough, J., Ford, G. M. and Ward, H., "Laser Phase and Frequency Stabilization Using an Optical Resonator", Appl. Phys. B **31**, 97 (1983).

Duarte, F. J. ed., *Tunable Lasers Handbook*, (Academic Press, 1995).

Elliot, D. S., Roy, R. and Smith, S. J., "Extracavity laser band-shape and bandwidth modification", Phys. Rev. A **26**, 12 (1982).

Fejer, M. M., Magel, G. A., Jundt, D. H. and Byer, R. L., "Quasi-Phase-Matched Second Harmonic Generation: Tuning and Tolerances", IEEE Journ. Quant. Elect., **28**, 2631 (1992).

Foot, C. J., *Atomic Physics* (Oxford University Press, 2005).

Fork, R. L., Martinez, O. E. and Gordon, J. P., "Negative dispersion using pairs of prisms", Opt. Lett. **9** 150 (1984).

Freed, C. and Haus, H. A., "Photoelectron Statistics Produced by a Laser Operating Below and Above the Threshold of Oscillation", IEEE Journ. Quant. Elect. **2**, 190 (1966).

Gardner, F. M., *Phaselock Techniques* (3rd ed., Wiley-Interscience, 2004).

Ghatak, A. K. and Thyagarajan, K., *Optical electronics*, (Cambridge University Press, 1989).

Hall, J. L., Hollbery, L., Baer, T. and Robinson, H. G., "Optical heterodyne saturation spectroscopy", Appl. Phys. Lett. **39**, 680 (1981).

Hall, J. L, Taubman, M. S. and Ye, J., "Laser Stabilization", in Optical Society of America, *Handbook of Optics, Volume IV* (McGraw Hill, 2000).

Hansch, T. W. and Couillaud, B., "Laser Frequency Stabilization by Polarization Spectroscopy of a Reflecting Reference Cavity", Optics. Comm. **35**, 441 (1980).

Harris, M. L., Cornish, S. L., Tripathi, A. and Hughes, I. G., "Optimization of sub-Doppler DAVLL on the rubidium D2 line", J. Phys. B At. Mol. Phys. **41**, 085401 (2008).

Hecht, E., *Optics*, (4th Ed, Addison Wesley, 2002).

Helmcke, J., Lee, S. A. and Hall, J. L., "Dye laser spectrometer for ultrahigh spectral resolution: design and performance" Applied Optics **21**, 1686 (1982).

Hemmati, H., Bergquist, J. C., Itano, W. M., "Generation of continuous-wave 194-nm radiation by sum-frequency mixing in an external ring cavity", Opt. Lett. **8**, 73 (1983).

Henry, Charles H., "Theory of the Lindwidth of Semiconductor Lasers", IEEE Journ. Quant. Elect. **18**, 259 (1982).

Henry, C. H. and Kazarinov, R. F., "Instability of Semiconductor Lasers Due to Optical Feedback from Distant reflectors", IEEE Journ. Quant. Elect. **22**, 294 (1986).

Jenkins, F. A. and White, H. E., *Fundamentals of Optics*, (3rd Ed, McGraw-Hill, 1957).

Joyce, W. B and Dixon, R. W., "Analytic approximation for the Fermi energy of an ideal Fermi gas", Appl. Phys. Lett. **31**, 354 (1977).

Kapon, E. ed., *Semiconductor Lasers I* (Academic Press, 1999).

Ou, Z. Y. and Kimble, H. J., "Enhanced conversion efficiency for harmonic generation with double resonance", Opt. Lett. **18**, 1053 (1993).

Kittel, C., *Introduction to Solid State Physics*, (8th Ed, Wiley, 2004).

Kogelnik, H. and Li, T., "Laser Beams and Resonators", Appl. Opt. **5**, 1550 (1966).

Kogelnik, H. and Shank, C. V., "Coupled-Wave Theory of Distributed Feedback Lasers", J. Appl. Phys. **43**, 2327 (1972).

Korpel, A., "Acousto-Optics – A Review of Fundamentals", Proc. of IEEE **69**, 48 (1981).

Korpel, A., *Acousto-Optics* (Marcel Dekker, Inc., 1997).

Labachelerie, M. de and Passedat, G., "Mode-hop suppression of Littrow grating-tuned lasers", Appl. Optics **32**, 269 (1993).

Lang, R. L., Park, R., Mehuys, D., O'Brien, S., Major, J. and Welch, D., "Numerical Analysis of Flared Semiconductor Laser Amplifiers", IEEE Journ. Quant. Elect. **29** 2044 (1993).

Lathi, B. P., *Signals, Systems and Communication*, (John Wiley and Sons, 1965).

Liu, Y., Lin, J., Huang, G., Guo, Y. and Duan, C., "Simple empirical analytic approximation to the Voigt profile", J. Opt. Soc. Am. B **18**, 666 (2001).

Loudon, R., *The Quantum Theory of Light* (2nd Ed, Clarendon Press, Oxford, 1983).

Ma, K.-S., Jungner, P., Ye, J. and Hall, J. L., "Delivering the same optical frequency at two places: accurate cancellation of phase noise indroduced by an optical fiber or other time-varying path", Opt. Lett. **19**, 1777 (1994).

Maguire, L. P., van Bijnen, R. M. W., Mese, E. and Scholten, R. E., "Theoretical calculation of saturated absorption for multi-level atoms", J. Phys. B **39**, 2709 (2006).

Milonni, P. W. and Eberly, J. H., *Lasers*, (John Wiley and Sons, 1988).

Mogensen, F., Olesen, H. and Jacobsen, G., "Locking Conditions and Stability Properties for a Semiconductor Laser with External Light Injection", IEEE Journ. Quant. Elect. **21**, 184 (1985).

Mollenauer, L. F. and Stolen, R. H., "The soliton laser", Opt. Lett. **9**, 13 (1984).

Mor, O. and Arie, A., "Performance Analysis of Drever-Hall Laser Frequency Stabilization Using a Proportional + Integral Servo", IEEE Journ. Quant. Elect. **33**, 532 (1997).

Nilsson, A. C., Gustafson, E. K. and Byer, R. L., "Eigenpolarization Theory of Monolithic Nonplanar Ring Oscillators", IEEE Journ. Quant. Elect. **25**, 767 (1989).

Paschotta, R., Spühler, G. J., Sutter, D. H., Matuschek, N. and Keller, U., "Double-chirped semiconductor mirror for dispersion compensation in femtosecond lasers", Appl. Phys. Lett. **75** 2166 (1999).

Pearman, C. P., Adams, C. S., Cox, S. G., Griffin, P. F., Smith, D. A. and Hughes, I. G., "Polarization spectroscopy of a closed atomic transition: applications to laser frequency locking", **35**, 5141 (2002).

Petermann, K., *Laser Diode Modulation and Noise*, (Springer, 1991).

Petitbon, I., Gallion, P., Debarge, G. and Chabran, C., "Locking Bandwidth and Relaxation Oscillations of on Injection-Locked Semiconductor Laser" IEEE Journ. Quant. Elect. **24**, 148 (1988).

Polzik, E. S. and Kimble, H. J., "Frequency doubling with $KNbO_3$ in an external cavity", Optics Lett. **16**, 1400 (1991).

Richter, L. E., Mandelberg, H. I., Kruger, M. S. and McGrath, P. A., "Linewidth Determination from Self-Heterodyne Measurements with Subcoherence Delay times", IEEE Journ. Quant. Elect. **22**, 2070 (1986).

Roberts, D. A., "Simplified Characterization of Uniaxial and Biaxial Nonlinear Optical Crystals: A Plea for Standardization of Nomenclature and Conventions", IEEE Journ. Quant. Elect. **28**, 2057 (1992).

Siegman, A. E., *Lasers* (University Science Books, 1986).

Shiv, L., Sorensen, J. L., Polzik, E. S. and Mizell, G., 'Inhibited light-induced absorption in $KNbO_3$", Optics Letters **20**, 2270 (1995).

Somekh, S and Yariv, A., "Phase Matching by Periodic Modulation of the Nonlinear Optical Properties", Opt. Comm. **6**, 301 (1972).

Suhara, T., *Semiconductor Laser Fundamentals* (Marcel Dekker, Inc., 2004).

Svelto, O., *Principles of Lasers*, (4th ed, Springer, 2004).

Le Targat, R., Zondy, J. J., Lemonde, P., "75%-Efficiency blue generation from an intracavity PPKTP frequency doubler", Optics Comm. **247**, 471 (2005).

Tkach, R. W. and Chraplyvy, "Regimes of Feedback Effects in 1.5-$\mu$m Distributed Feedback Lasers", IEEE Journ. Lightwave Tech. **LT-4**, 1675 (1986).

Torgerson, J., Nagourney, W., "Tunable cavity coupling scheme using a wedged plate", Optics Comm. **161**, 264 (1999).

Verdeyen, Joseph T., *Laser Electronics* (3rd Ed, Prentice Hall, 1995).

Yariv, A., *Quantum Electronics* (3rd ed., John Wiley and Sons, 1989).

Ye, J. and Cundiff, S. T (ed.), *Femtosecond Optical Frequency Comb Technology* (Springer, 2005).

Young, E. H. Jr. and Yao, S.-K., "Design Considerations for Acousto-Optic Devices", Proc. of IEEE **69**, 54 (1981).

Yue, C, Pend, J. D., Liao, Y. B. and Zhou, B. K., "Fiber ring resonator with finesse of 1260", Electron. Lett. **24**, 622 (1988).

Zeilinger, A., "Experiment on the foundations of quantum physics", Rev. of Modern Phys. **71**, S288 (1999).

Zimmermann, M., Gohle, C., Holzwarth, R., Udem, T. and Hänsch, T. W., "Optical clockwork with an offset-free difference-frequency comb: accuracy of sum- and difference-frequency generation", Opt. Lett. **29**, 310 (2004).

$M^2$ parameter, 32

ABCD
  matrix, 8
    free-space, 8
    slab, 9
    spherical mirror, 9
    thin lens, 8
    unimodular property, 9
  rule, 10
Absorption, 88
Acousto-optic modulation
  acousto-optic effect, 360
  beam deflection application, 371
  Bragg regime, 365
    coupled wave analysis, 367
  Doppler shift interpretation, 366
  double pass using cat's-eye retroreflector, 371
  figure of merit, $M_2$, 365
  frequency shifting application, 371
  Klein-Cook parameter, 364
  modulation bandwidth, 369
  modulator geometry, 361
  phased array transducer, 370
  Raman-Nath (Debye-Sears) regime, 362
  Raman-Nath parameter, 363
  ray bending due to index gradients, 364
  strain tensor, 360
  strain-optic coefficients, 361
    for isotropic solid, 361
  tuning bandwidth, 370
Amplitude modulation, 62
Anisotropic crystal, 301
  biaxial, 306
  dielectric tensor, 302
    symmetry of, 302
  double refraction, 309
  extraordinary polarization, 306
  Fresnel equation, 304
  index ellipsoid, 305
  normal surfaces, 306
  optical axes, 306
  ordinary polarization, 306
  ray and wave propagation, 308
  uniaxial, 306
Astigmatism
  defined, 27
  orthogonal, 27
  tangential and sagittal planes, 27

Band theory, 142
Birefringent filter, 129
Bloch equation
  Bloch vector, 76
  classical, 78
  damping
    elastic collisions, 81
    inelastic collisions, 81
    radiative, 80
    relaxation times, $T_1$ and $T_2$, 83
  effective field, 76, 78
  interpretation, 78
  optical, 76
  resonant behavior, 78
  steady-state solution, 81, 83
Bloch function, 139
Bloch theorem, 139
Bloch-Siegert shift (AC Stark shift), 79

Cavities, optical, 13
  alternative representation of loss, 48
  Fresnel-Kirchoff analysis, 13
  Hermite-Gaussian modes, 20
    angular spread of, 21
    half-width of, 20
  self-consistent approach, 14
  stability criterion, 15
  standing wave, 13
    arbitrary two-mirror, 17
    circulating field, 38, 44
    confocal, 18
    equivalent lens sequence, 15
    free spectral range (FSR), 22
    intensities, 39
    reflected field, 37, 44
    resonant frequencies, 21
    stability criterion, 16
    stability diagram, 16
    symmetric two-mirror, 18
  traveling wave (ring), 23
    astigmatism in crystal at Brewster's angle, 28
    astigmatism in curved mirror, 27
    astigmatism, effect on stability, 30
    free spectral range (FSR), 25
    reflective coupling, 45
    resonant frequencies, 25
    stability criterion, 24
    stability range, 24
    waist radii ratio, 25

Collisions, *see* Bloch equation, damping
Control systems, 235
  closed-loop error, 237, 247
  closed-loop gain, 237
  controller and plant, 256
  differentiator, 259
  integrator, 259
  laser stabilization, 235, 258, 266
    linewidth of stabilized laser, 240
    noisy source, 283
    of DFB laser, 272
    of NPRO, 272
    shot noise limitation, 238
    simple considerations, 235
    using piezo-electric frequency control,
      272
  lead-lag, 260
  linear systems, 241
    impulse response, 246
    Laplace transform, 242
    natural modes, 246
    negative feedback, 247
    partial fraction expansion, 244
    transfer function, 241
    transient response, 250
  open-loop gain, 237, 248
  phase-locked loop, 257
  PID, 261
    Ziegler-Nichols procedure, 264
  return difference, 248
  stability, 245
    Bode plot, 252
    effect of time delays, 256
    phase margin, 253
    unity gain instability in op-amp, 254
  temperature control, 258, 262
    resistive heater nonlinearity, 264
    thermistor bridge, 265
  tracking, 248
Coulomb gauge, 73

Density matrix, 75
Density of electron energies, 139
Density of states (electrons), 139, 170
Dipole approximation, 74
Dipole matrix element, 74
Discriminator, atomic and molecular, 287
  polarization spectroscopy, 295
  saturation spectroscopy, 287
  side of line, 298
  sub-Doppler dichroic atomic vapor locking
    (DAVLL), 294
Discriminator, optical, 53
  cavity pole location, 70
  frequency response, 67
  polarization technique, 58
  Pound-Drever-Hall (PDH) method, 63
  side of line, 55
  simple example, 53
Doppler broadening, 93

effective nonlinear coefficient, 319
Einstein $A$ and $B$ coefficients, 88, 151
Electro-optic modulation
  electro-optic tensor, 353
    symmetry properties, 353
    values for important crystals, 353
  linear electro-optic effect, 352
  longitudinal modulator, 355
  step-up transformer to drive, 358
  transverse modulator, 357
  traveling wave modulator, 359
    transit time limitation, 360
Enhancement, measurement of, 49
Equivalent mass, 142

Fermi energy, 137
Fermi-Dirac distribution, 137
  Pauli exclusion principle, 138
Finesse, 40
  measurement of, 48
Franck-Condon principle, 127
Frequency metrology using mode-locked
    lasers, 228
  carrier envelope offset, 230
    controlling, 232
  microstructure fiber, 230
Frequency modulation, 61
  Bessel function expansion, 63
  deviation, 63
  modulation index, $\beta$, 63
  phasor representation, 61

Gaussian beam, 1
  beam radius, 2
  complex beam parameter, $q$
    definition, 2
    free-space transformation, 2, 7
    interpretation of $R$ and $\omega$, 2
    transformation across slab, 7
    transformation across thin lens, 7
    transformation across thin mirror, 7
  confocal parameter, 4
  dependence of $\omega$ on $z$, 4
  dependence of $R$ on $z$, 4
  function, 2
  Gouy phase, 5, 19, 22
  Rayleigh length, 4
  waist, 3, 5
  wavefront radius, 3
Gouy phase, *see* Gaussian beam, Gouy phase
Group velocity, 222
Gyromagnetic ratio, 77

Hamiltonian
  of two-level system, 72
  interaction term, 74
  with electromagnetic field, 73
Hole burning, 94
Holes, *see* Semiconductor, positive charge
    carriers (holes)

Homogeneous broadening, *see* Spectral
    broadening, homogeneous

Impedance matching (into cavity), 41
Impedance mismatching (into cavity), 42
Inhomogeneous broadening, *see* Spectral
    broadening, inhomogeneous
Inversion, population, 79

Jones calculus, 56
  effect of rotations, 58

Kronig-Penney model, 140

Lamor frequency, 77
Laser, 72
  amplifier, 203
    tapered, 205
  argon ion, 123
    two-step pumping, 124
  diode-pumped solid-state YAG (DPSS),
    135
  external cavity diode, 192
    correcting pointing changes in Littrow,
      200
    frequency modulation, 197
    instabilities, 197
    linewidth, 196
    Littman configuration, 200
    Littrow configuration, 198
  feedback sensitivity, 196
  four-level, 105
  frequency pulling, 107
  gain, 87, 92
  gain clamping, 101
  helium neon (He-Ne), 121
    population saturation, 123
    pumping mechanism, 122
  injection locked semiconductor, 205
  mode-locked, 212
    AM approach, 217
    chirped mirror, 224
    dispersion compensating prisms, 224
    FM approach, 217
    group delay dispersion (GDD), 223
    group velocity dispersion (GVD), 223
    in homogeneously broadened medium,
      217
    self phase modulation, 226
    theory of, 212
    Ti-sapphire laser, 225
    using optical Kerr effect, 221
    using saturable absorber, 219
    using saturating semiconductor
      (SESAM), 220
  multimode, 108
  optimum output coupling, 103
  organic dye, 126
    jet stream, 127
    ring configuration, 129

    triplet quenching, 127
  oscillation condition, 101
  oscillation frequencies, 106
  power output, 103
  Schawlow-Townes expression for linewidth,
    118
  semiconductor, 159
    circularization of mode, 208
    confinement factor, 166
    distributed Bragg reflection, 178
    distributed feedback, 174
    double heterostructure, 162
    frequency modulation, 189
    gain-guided and index-guided, 166
    homojunction, 159
    linewidth, 189
    quantum well, 167
    relaxation oscillations, 186
    response to amplitude modulation, 187
    temperature control of frequency, 188
    threshold current density, 166
    wavelength versus material, 208
  spatial hole burning, 110
  three and four level comparison, 106
  three-level, 104
  threshold gain, 101
  titanium-sapphire, 130
  YAG non-planar ring oscillator (NPRO),
    134
  yttrium-aluminum-garnet (YAG), 132
Laser frequency stability, 277
  Allan variance, 278
  delayed heterodyne measurement, 282
  delayed homodyne measurement, 282
  fractional frequency and phase
    fluctuations, 278
  frequency spectral density to Allan
    variance, 280
  measurement of using two sources, 281
Law of mass action, 144

Mode matching, 30
  simplified two lens approach, 31
  with single lens, 30
  with two lenses, 31

Optical fiber: noise cancellation, 275

Parametric processes, 338
  examples
    pumping child's swing, 338
    RLC circuit, 339
  optical
    doubly resonant parametric oscillator,
      343
    Manley-Rowe relation, 343
    non-collinear phase matching, 346
    pump, signal, idler definitions, 341
    singly resonant parametric oscillator, 345

spontaneous parametric fluorescence, 346
phase sensitivity of degenerate amplification, 338, 340
Paraxial
approximation, 1, 3
ray optics, 8
ray vector, 8
wave equation, 2
Pauli exclusion principle, 138
Periodic boundary conditions, 89
Phase velocity, 223
Photon lifetime (in cavity), 46
Photon occupation number above threshold, 111
Photon statistics, 114
Bose-Einstein distribution, 116
chaotic light, 114
coherent light, 114
Poisson distribution, 115
variances, 117
Planck formula (blackbody radiation), 88
Polarization, atomic, 85
Pound-Drever-Hall (PDH), *see* Discriminator, optical

Quality factor, $Q$, 47
relation to finesse, 47

Rabi frequency, 74
generalized, 78
Radiative damping, *see* Bloch equation, damping
Ray optics, *see* Paraxial, ray optics
Ring (traveling wave) cavity, *see* Cavity, optical
Ring-down technique, 50
Rotating frame, 77
Rotating wave approximation, 74, 79

Saturation, 81, 94
intensity of, 95
Schrödinger equation, 72
coupled equations (two-level system), 73, 75
Second harmonic generation, 309
$d$-values for important crystals, 313
aperture length, 329
birefringent phase matching, 315
bandwidth, 316
cavity enhancement of, 332
effect of depletion, 334
resonance at second harmonic, 337
coherence length, 315
double-refraction parameter, 329
effective nonlinear coefficient, 320
field and intensity of second harmonic, 312
focused beam, 325

effect of walk-off, 330
Kleinman's conjecture, 313
nonlinear susceptibility, 311
quasi-phase-matching, 321
bandwidth, 324
temperature dependence, 325
Sellmeier equation, 314
two-dimensional susceptibility tensor, 312
type II, 318
walk-off, 320
Semiconductor, 142
Bohr electron model, 145
depletion region, 147
direct, 149
doping, 144
equality of Fermi levels, 147
extrinsic, 146
galium arsenide (GaAs), 159
intrinsic, 143
joint density of states, 152, 171
lattice matching, 163
law of mass action, 144
linewidth enhancement factor, $\alpha$, 183
optical gain, 148, 172
pn junction, 147
rectification in, 148
positive charge carriers (holes), 144
quasi-Fermi levels, 150
approximate determination, 154
rate equation for carriers, 184
rate equation for electric field, 182
rate equation for phase, 183
selection rules for optical transition, 150, 171
spontaneous decay, 156
ternary and quarternary, 163
transparency (optical) condition, 155
Sommerfeld model, 139
Spectral broadening
homogeneous, 92
inhomogeneous, 92
Spontaneous emission, 88
rate of, 87
Stimulated emission, 88
Stokes, treatment of reflection, 36
Sum-frequency generation, 337
Susceptibility, atomic, 84

Temporal coherence, 112
coherence length, 114
coherence time, 114
measured using Michelson interferometer, 113

Voigt profile, 94

Wave equation, 1
Wiener-Khintchine theorem, 82, 240, 279